T0314214

Methods of Soil Enzymology

Methods of Soil Enzymology

Richard P. Dick, editor

Number 9 in the Soil Science Society of America Book Series
Published by Soil Science Society of America, Inc., Madison, Wisconsin, USA

Soil Science
Society of America

Soil Science Society of America, Inc.
5585 Guilford Road, Madison, WI 53711-5801 USA

ISBN: 978-0-89118-854-4

Library of Congress Control Number: 2011934696

Cover: PDB ID-3R31, courtesy of Agarwal, Almo, and Swaminathan. (Betaine aldehyde dehydrogenase from *Agrobacterium tumefaciens* catalyzes the NAD dependent synthesis of betaine.)

Printed in the United States of America.

Contents

Foreword

Soil enzymes are critically important to the functioning of the soil. Soil enzymes are the product of soil microorganisms and are the drivers for biogeochemical cycling in soils. Soil enzymes are also involved in remediation of contaminated soils. As a result, soil enzymes have been considered indicators for soil ecosystem health. While the Soil Science Society of America has previously published a methods book on soil microbiology and biochemistry, this book is devoted to soil enzymes. Given the critical nature of carbon, nitrogen, and other nutrient cycles in soil as it relates to plant growth and environmental issues, this book provides the standardization and appropriate application of soil enzymes. The authors have provided a comprehensive review of soil enzymes. This SSSA book will provide guidance for those scientists interested in soil ecosystems. Soil scientists will find this book to be a key reference. Ecologists and microbiologists will find this book to be a useful tool for application of soil enzymes to their work. Students interested in soils will find this book to be a useful reference for their education and research. The Soil Science Society of America is pleased to provide this work for the science community and hope that you will find this book useful.

Charles W. Rice, President of the Soil Science Society of America

Preface

This book follows a series of methods books related to soil microbiology published under the auspices of the Soil Science Society of America (SSSA). A number of soil enzyme activity assays were published in Part 2 of the Agronomy Monograph 9, *Methods of Soil Analysis* that had two editions. Subsequently, another methods book devoted solely to soil microbiology and biochemistry was published in the SSSA Book Series (1994) entitled *Methods of Soil Analysis. Part 2. Microbiological and Biochemical Properties*, SSSA Book Series 5 (Weaver et al., editors). In this book, M.A. Tabatabai authored a chapter entitled "Soil Enzymes" (p. 801–834) (Tabatabai, 1994), which provided an important foundation for this new book, for which he co-authored three chapters.

The impetus to develop a book solely devoted to soil enzymology came about for several reasons. One is that a comprehensive set of soil enzyme activity assays (Chapters 6 to 12), representing the spectrum of published methods, does not exist in a single publication. Rather, methods are scattered throughout journal articles or are embedded in soil methods books as chapters alongside other soil biological or chemical methods. Additionally for activity measures more methods have been developed since the "Soil Enzyme" chapter was published in the SSSA Book Series 5.

The contents and organization of Chapters 6 to 12 for the enzyme activity assays were based on ecological functions rather than a strict Enzyme Commission, numerical classification scheme (EC number). This was done purposefully to guide operators to suites of enzymes that are related to a given biogeochemical process such as nutrient cycling or other soil functions. However, it should be noted that many enzymes may not have a single process they participate in (e.g. hydrolytic enzymes from the C and N cycles may be involved in both cycles). Therefore it important to be aware of all the potential roles that a given enzyme may play in biochemical processes. Nonetheless, all of these chapters include the EC classification number for each enzyme the first time they are specifically discussed or introduced where the EC number is based on the chemical reactions they catalyze.

Another component of this book is the inclusion of enzymological methods that are not classical activity assays. In this regard, there are chapters on extraction of enzymes from soils (Chapter 16) and methods for stabilizing enzymes on solid supports (Chapter 15) for potential applications in remediation of polluted soils or other environmental applications. Chapter 13, which can be characterized as semi-quantitative, provides important methods for studying fine-scale (mm) spatial distribution of enzyme activities in soil and rhizospheres in undisturbed soil profiles. To accomplish this goal, Chapter 13 presents methods that make in situ imprints of soil enzyme activity profiles and for assaying of individual plant roots tips.

Beyond presenting enzymological methods, this book provides a context for scientists and practitioners to use soil enzymological methods effectively (Chapters 1 to 5). The book reports on the rich history of soil enzymology that stretches back to the very late 19th century (Chapter 1). Building on this, and to put the methods within a research and applications context, Chapter 2 gives important information on the microbial ecology of extracellular enzyme in soils. Kinetics of soil enzymes is reported in Chapter 3, which provides guidance on proper methods and interpretation of the Michaelis–Menten equation. Chapter 4 presents the basics for developing new enzyme activity assays. Chapter 5 gives guidance on sampling and storing soils with specific recommendations for soil sample pretreatment on enzyme assays.

An important criterion for including a given method in this book was whether the method had been rigorously tested and provides quantitative measures of activities under optimal reaction conditions (substrate saturation, pH, required cofactors if needed, and so forth). This was done to allow development of data that can be compared across studies, independent of the operator. Some have proposed that pH should not be buffered to allow results to better reflect activities under in situ conditions. This would be a viable option if the goal is to study pH effects or functional capacity of the enzymes in the natural environment. However, as a standardized method to allow comparison among various soil samples within and among studies, the enzyme assay should be conducted under buffered conditions at optimal pH.

With these criteria in mind some assays that may be found in the literature were not included in the book. One example is that we did not include bench-scale methods that use the methylumbelliferyl (MUF) labeled substrate. This is because there is severe quenching of fluorescence by humic substances. Fluorescence is also sensitive to temperature, pH, and many other environmental factors that impact its detection. For quantitative detection of fluorescence, it is essential to develop a standard curve for every sample and every batch of analyses—making it impractical from a labor and cost perspective. Nonetheless, MUF-based microplate assays are generating increasing interest due, in part, to its high sensitivity and the capacity to simultaneously assay multiple enzymes using a small amount of samples. We included a microplate MUF-based method (Chapter 14) because the use of multichannel pipettes enable a standard curve be conducted for each sample and each batch analysis with adequate replication to reduce variability. There is evidence that the proposed microplate method can pick up treatment effects if run under a very strict and consistent set of protocols. However, it is less clear if data can be compared from one study or another due. The method does not necessarily represent an optimized method, and thus, should be used and interpreted with caution.

This book will be useful to microbiologists, biochemists, and ecologists involved in studying microbial communities and their ability to deliver ecological services, soil biochemical processes (e.g. nutrient transformations and soil organic matter formation), and potential functional capacity of the microbial community and soil ecosystems. These methods can complement or be correlated with other measures of microbial biomass and/or diversity to provide insights into biochemical functions of soils. The overall objective of the book is to provide a set of standardized enzymological methods, enabling meta analysis of data from the literature and enhancing global collaboration and interactions among scientists. The activity assays could be used by commercial labs with potential applications to assess soil quality or for environmental soil remediation.

I am honored that the leading scientists who have extensive, in-depth experience and expertise in soil enzymology took the time and effort to develop these excellent chapters. This select group of scientists is uniquely suited to write these chapters and have first-hand knowledge of the enzymology methods they have presented. This ensures that the methods presented are current, relevant, and readily applicable. I want to thank all contributing authors for their diligence and patience in bringing this book to fruition with such collegiality.

Richard P. Dick, Ohio State University

REFERENCE

Tabatabai, M.A. 1994. Soil enzymes. p. 801–834. *In* R.W. Weaver et al. (ed.) Methods of soil analysis. Part 2. Microbiological and biochemical properties. SSSA Book Ser. 5. SSSA, Madison, WI.

Contributors

Acosta-Martinez, V.   USDA-ARS, Cropping Systems Research Lab., 3810 4th St., Lubbock, TX 79415 (veronica.acosta-martinez@ars.usda.gov)

Bilen, Serdar   Atatürk Univ., Fac. of Agriculture, Dep. of Soil Science, Erzurum, Turkey (sbilen@atauni.edu.tr)

Brooks, Denise D.   Forest Sciences Dep., Univ. of British Columbia Vancouver Campus, 2424 Main Mall, Vancouver, British Columbia, Canada; current address: Okinawa Inst. of Science and Technology, 1919-1 Tancha, Onna-son, Kunigami, Okinawa, Japan 904-0412 (brooks.denise@oist.jp)

Burns, Richard G.   Professor of Environmental Microbiology, School of Agriculture and Food Sciences, The Univ. of Queensland, Brisbane, Queensland 4072, Australia (r.burns@uq.edu.au)

Courty, Pierre-Emmanuel   INRA-Nancy, 54280 Champenoux, France; current address: Univ. of Basel-Botanical Inst., Hebelstrasse 1, CH-4056 Basel, Switzerland (pierre.courty@unibas.ch)

Deng, Shiping   Oklahoma State Univ., Dep. of Plant and Soil Sciences, 368 Agricultural Hall, Stillwater, OK 74078 (shiping.deng@okstate.edu)

Dick, Richard P.   Ohio Eminent Scholar, Professor of Soil Microbial Ecology, The School of Environment & Natural Resources, The Ohio State Univ., Columbus OH 43210 (Richard.Dick@snr.osu.edu)

Dick, Warren A.   Professor, Soil Science, School of Environment and Natural Resources, The Ohio State Univ., 101A Hayden Hall, 1680 Madison Ave., Wooster, OH 44691 (dick.5@osu.edu)

Dudal, Yves   ENVOLURE SAS, Campus La Gaillarde, Bât 12, 2 place Pierre Viala, 34060 Montpellier cedex 2, France (yves.dudal@envolure.com)

Fornasier, Flavio   C.R.A.–R.P.S. Consiglio per la Ricerca e la Sperimentazione in Agricoltura–Centro di Ricerca per lo Studio delle Relazioni tra Pianta e Suolo, 23, Via Trieste, 34170 Gorizia, Italy (flavio.fornasier@entecra.it)

Frankenberger, William T.   2326 Geology, Dep. of Environmental Sciences, Univ. of California, Riverside, CA 92521 (william.frankenberger@ucr.edu)

Freeman, Chris   Bangor University, School of Biological Sciences, Bangor University Deiniol Road, Bangor, Gwynedd LL57 2DG, UK (c.freeman@bangor.ac.uk)

Garbaye, Jean   INRA-Nancy, 54280 Champenoux, France (garbaye@nancy.inra.fr)

Giagnoni, Laura   Dep. of Plant, Soil and Environmental Science, Piazzale Cascine 15, 50144, Univ. of Firenze, Firenze, Italy (laura.giagnoni@unifi.it)

Gianfreda, Liliana   Univ. di Napoli Federico II, Via Università 100, Portici, Italy (liliana.gianfreda@unina.it)

Grierson, Pauline F.   Ecosystems Research Group, School of Plant Biology M090, Univ. of Western Australia, 35 Stirling Highway, Crawley WA 6009 Australia (pauline.grierson@uwa.edu.au)

Jones, Melanie D.   Biology and Physical Geography Unit, Science Building, Univ. of British Columbia, Okanagan Campus, 3333 University Way, Kelowna, BC V1V 1V7 Canada (Melanie.Jones@ubc.ca)

Kandeler, Ellen   Inst. of Soil Science and Land Evaluation, Soil Biology Section (310b), Univ. of Hohenheim, Emil-Wolff-Str. 27, D-70593 Stuttgart, Germany (Ellen. Kandeler@uni-hohenheim.de)

Kang, Hojeong   School of Civil and Environmental Engineering, Yonsei Univ., Seoul, Korea

Kim, Jeong Jin   Dep. of Earth and Environmental Sciences, Andong National Univ., Andong, Korea (jjkim@andong.ac.kr)

Klose, Susanne   Chiquita Brands North America, James Lugg Research Ctr., 607 Brunken Ave., Salinas, CA 93901 (sklose@chiquita.com)

Landi, Loretta   Dep. of Plant, Soil, and Environmental Science, Piazzale Cascine 15, 50144, Univ. of Firenze, Firenze, Italy (loretta.landi@unifi.it)

Lorenz, Nicola   The School of Environment & Natural Resources, The Ohio State Univ., Columbus, OH 43210 (nicola.lorenz.64@googlemail.com)

Nanniperi, Paolo   Dep. of Plant, Soil, and Environmental Science, Piazzale Cascine 15, 50144, Univ. of Firenze, Firenze, Italy (paolo.nannipieri@unifi.it)

Poll, Christian   Inst. of Soil Science and Land Evaluation, Soil Biology Section (310b), Univ. of Hohenheim, Emil-Wolff-Str. 27, D-70593 Stuttgart, Germany (Christian. Poll@uni-hohenheim.de)

Popova, Inna   Dep. of Plant and Soil Sciences, Oklahoma State Univ., 368 Agricultural Hall, Stillwater, OK 74078 (inna.popova@okstate.edu)

Pritsch, Karin   Inst. of Soil Ecology, Helmholtzzentrum München, German Research Ctr. for Environmental Health, Ingolstaedter Landstraße 1, 85764 Neuherberg, Germany (pritsch@helmholtz-muenchen.de)

Prosser, Jennifer A.   ESR Ltd., 34 Kenepuru Dr., Porirua 5240, New Zealand (jennifer.prosser@ esr.cri.nz)

Quiquampoix, Hervé   INRA, UMR Eco&Sols, 2 place Pierre Viala, 34060 Montpellier, cedex 2, France (Francequiquampoix@inra.montpellier.fr)

Rao, Maria A.   Univ. di Napoli Federico II, Via Università 100, Portici, Italy (mariarao@ unina.it)

Renella, Giancarlo   Dep. of Plant, Soil, and Environmental Science, Piazzale Cascine 15, 50144, Univ. of Firenze, Firenze, Italy (giancarlo.renella@unifi.it)

Speir, Tom W.   ESR Ltd., 34 Kenepuru Dr., Porirua 5240, New Zealand (tom.speir@esr.cri. nz)

Stott, Diane E.   USDA-ARS, National Soil Erosion Research Lab., 275 S. Russell St., West Lafayette, IN 47907 (diane.stott@ars.usda.gov).

Tabatabai, M. Ali   Dep. of Agronomy, Iowa State Univ., 2403 Agronomy Hall, Ames, IA 50011 (malit@iastate.edu)

Wallenstein, Matthew D.   Natural Resource Ecology Lab., Colorado State Univ., Fort Collins, CO 80523 (matthew.wallenstein@colostate.edu)

Conversion Factors
for SI and Non-SI Units

To convert Column 1 into Column 2 multiply by	Column 1 SI unit	Column 2 non-SI unit	To convert Column 2 into Column 1 multiply by
		Length	
0.621	kilometer, km (10^3 m)	mile, mi	1.609
1.094	meter, m	yard, yd	0.914
3.28	meter, m	foot, ft	0.304
1.0	micrometer, μm (10^{-6} m)	micron, μ	1.0
3.94×10^{-2}	millimeter, mm (10^{-3} m)	inch, in	25.4
10	nanometer, nm (10^{-9} m)	Angstrom, Å	0.1
		Area	
2.47	hectare, ha	acre	0.405
247	square kilometer, km^2 (10^3 m)^2	acre	4.05×10^{-3}
0.386	square kilometer, km^2 (10^3 m)^2	square mile, mi^2	2.590
2.47×10^{-4}	square meter, m^2	acre	4.05×10^3
10.76	square meter, m^2	square foot, ft^2	9.29×10^{-2}
1.55×10^{-3}	square millimeter, mm^2 (10^{-3} m)^2	square inch, in^2	645
		Volume	
9.73×10^{-3}	cubic meter, m^3	acre-inch	102.8
35.3	cubic meter, m^3	cubic foot, ft^3	2.83×10^{-2}
6.10×10^4	cubic meter, m^3	cubic inch, in^3	1.64×10^{-5}
2.84×10^{-2}	liter, L (10^{-3} m^3)	bushel, bu	35.24
1.057	liter, L (10^{-3} m^3)	quart (liquid), qt	0.946
3.53×10^{-2}	liter, L (10^{-3} m^3)	cubic foot, ft^3	28.3
0.265	liter, L (10^{-3} m^3)	gallon	3.78
33.78	liter, L (10^{-3} m^3)	ounce (fluid), oz	2.96×10^{-2}
2.11	liter, L (10^{-3} m^3)	pint (fluid), pt	0.473
		Mass	
2.20×10^{-3}	gram, g (10^{-3} kg)	pound, lb	454
3.52×10^{-2}	gram, g (10^{-3} kg)	ounce (avdp), oz	28.4
2.205	kilogram, kg	pound, lb	0.454
0.01	kilogram, kg	quintal (metric), q	100
1.10×10^{-3}	kilogram, kg	ton (2000 lb), ton	907
1.102	megagram, Mg (tonne)	ton (U.S.), ton	0.907
1.102	tonne, t	ton (U.S.), ton	0.907

continued

To convert Column 1 into Column 2 multiply by	Column 1 SI unit	Column 2 non-SI unit	To convert Column 2 into Column 1 multiply by

Yield and Rate

To convert Column 1 into Column 2 multiply by	Column 1 SI unit	Column 2 non-SI unit	To convert Column 2 into Column 1 multiply by
0.893	kilogram per hectare, kg ha^{-1}	pound per acre, lb acre^{-1}	1.12
7.77×10^{-2}	kilogram per cubic meter, kg m^{-3}	pound per bushel, lb bu^{-1}	12.87
1.49×10^{-2}	kilogram per hectare, kg ha^{-1}	bushel per acre, 60 lb	67.19
1.59×10^{-2}	kilogram per hectare, kg ha^{-1}	bushel per acre, 56 lb	62.71
1.86×10^{-2}	kilogram per hectare, kg ha^{-1}	bushel per acre, 48 lb	53.75
0.107	liter per hectare, L ha^{-1}	gallon per acre	9.35
893	tonne per hectare, t ha^{-1}	pound per acre, lb acre^{-1}	1.12×10^{-3}
893	megagram per hectare, Mg ha^{-1}	pound per acre, lb acre^{-1}	1.12×10^{-3}
0.446	megagram per hectare, Mg ha^{-1}	ton (2000 lb) per acre, ton acre^{-1}	2.24
2.24	meter per second, m s^{-1}	mile per hour	0.447

Specific Surface

10	square meter per kilogram, m^2 kg^{-1}	square centimeter per gram, cm^2 g^{-1}	0.1
1000	square meter per kilogram, m^2 kg^{-1}	square millimeter per gram, mm^2 g^{-1}	0.001

Density

1.00	megagram per cubic meter, Mg m^{-3}	gram per cubic centimeter, g cm^{-3}	1.00

Pressure

9.90	megapascal, MPa (10^6 Pa)	atmosphere	0.101
10	megapascal, MPa (10^6 Pa)	bar	0.1
2.09×10^{-2}	pascal, Pa	pound per square foot, lb ft^{-2}	47.9
1.45×10^{-4}	pascal, Pa	pound per square inch, lb in^{-2}	6.90×10^3

Temperature

1.00 (K − 273)	kelvin, K	Celsius, °C	1.00 (°C + 273)
(9/5 °C) + 32	Celsius, °C	Fahrenheit, °F	5/9 (°F − 32)

Energy, Work, Quantity of Heat

9.52×10^{-4}	joule, J	British thermal unit, Btu	1.05×10^3
0.239	joule, J	calorie, cal	4.19
10^7	joule, J	erg	10^{-7}
0.735	joule, J	foot-pound	1.36
2.387×10^{-5}	joule per square meter, J m^{-2}	calorie per square centimeter (langley)	4.19×10^4
10^5	newton, N	dyne	10^{-5}
1.43×10^{-3}	watt per square meter, W m^{-2}	calorie per square centimeter minute (irradiance), cal cm^{-2} min^{-1}	698

continued

To convert Column 1 into Column 2 multiply by	Column 1 SI unit	Column 2 non-SI unit	To convert Column 2 into Column 1 multiply by
Transpiration and Photosynthesis			
3.60×10^{-2}	milligram per square meter second, mg m^{-2} s^{-1}	gram per square decimeter hour, g dm^{-2} h^{-1}	27.8
5.56×10^{-3}	milligram (H$_2$O) per square meter second, mg m^{-2} s^{-1}	micromole (H$_2$O) per square centimeter second, μmol cm^{-2} s^{-1}	180
10^{-4}	milligram per square meter second, mg m^{-2} s^{-1}	milligram per square centimeter second, mg cm^{-2} s^{-1}	10^4
35.97	milligram per square meter second, mg m^{-2} s^{-1}	milligram per square decimeter hour, mg dm^{-2} h^{-1}	2.78×10^{-2}
Plane Angle			
57.3	radian, rad	degrees (angle), °	1.75×10^{-2}
Electrical Conductivity, Electricity, and Magnetism			
10	siemen per meter, S m^{-1}	millimho per centimeter, mmho cm^{-1}	0.1
10^4	tesla, T	gauss, G	10^{-4}
Water Measurement			
9.73×10^{-3}	cubic meter, m^3	acre-inch, acre-in	102.8
9.81×10^{-3}	cubic meter per hour, m^3 h^{-1}	cubic foot per second, ft^3 s^{-1}	101.9
4.40	cubic meter per hour, m^3 h^{-}1	U.S. gallon per minute, gal min^{-1}	0.227
8.11	hectare meter, ha m	acre-foot, acre-ft	0.123
97.28	hectare meter, ha m	acre-inch, acre-in	1.03×10^{-2}
8.1×10^{-2}	hectare centimeter, ha cm	acre-foot, acre-ft	12.33
Concentration			
1	centimole per kilogram, cmol kg^{-1}	milliequivalent per 100 grams, meq 100 g^{-1}	1
0.1	gram per kilogram, g kg^{-1}	percent, %	10
1	milligram per kilogram, mg kg^{-1}	parts per million, ppm	1
Radioactivity			
2.7×10^{-11}	becquerel, Bq	curie, Ci	3.7×10^{10}
2.7×10^{-2}	becquerel per kilogram, Bq kg^{-1}	picocurie per gram, pCi g^{-1}	37
100	gray, Gy (absorbed dose)	rad, rd	0.01
100	sievert, Sv (equivalent dose)	rem (roentgen equivalent man)	0.01
Plant Nutrient Conversion			
	Elemental	Oxide	
2.29	P	P$_2$O$_5$	0.437
1.20	K	K$_2$O	0.830
1.39	Ca	CaO	0.715
1.66	Mg	MgO	0.602

Chapter 1

A Brief History of Soil Enzymology Research

Richard P. Dick* and Richard G. Burns

1–1 INTRODUCTION

The subject of soil enzymology has intrigued and excited scientists for more than one hundred years. Today, its relevance to climate change assessment and prediction, understanding soil fertility and plant growth, and the protection and rehabilitation of polluted and degraded land spurs interest in extracellular enzymes. Coupling this with the application of emerging novel techniques enables the opportunity for breakthroughs on the many functions of extracellular enzymes in microbial ecology. In the last decade, over 15,000 papers and patents have been published that link soil with enzymes in one way or another.

In this brief history, we concentrate on the early years because the more recent research is contained within other chapters of this book and points to a future in which our understanding of substrate catalysis and nutrient turnover in soils (and water columns and sediments) will serve as a blueprint for the rational management of our most precious natural resource.

1–2 THE EARLY YEARS

At the dawn of the 20th century, awareness and understanding of much of microbiology was driven by human health needs. Nonetheless, there were parallel efforts to apply classical laboratory-based microbiological techniques to the study of soils, not only as sources of infectious diseases but also as the location for transformations, critical to the nitrogen cycle and the nutrition of plants.

The earliest known report of soil enzymes was by Woods (1899) who recorded the activity of oxidizing enzymes, such as peroxidases, and concluded that these were persistent extracellular enzymes that arose through the decaying of roots and other plant parts. This was communicated that year at the annual meeting of the American Association for the Advancement of Science in Columbus, Ohio and led rapidly to the widespread acceptance that soils exhibit enzymatic activity;

Richard P. Dick, Ohio Eminent Scholar, Professor of Soil Microbial Ecology, The School of Environment & Natural Resources, Ohio State University, Columbus OH 43210 (Richard.Dick@snr.osu.edu), *corresponding author;
Richard G. Burns, Professor of Environmental Microbiology, School of Agriculture and Food Sciences, The University of Queensland, Brisbane, Queensland 4072, Australia (r.burns@uq.edu.au).

doi:10.2136/sssabookser9.c1

questions on the locations, functions, and significance of this activity quickly followed. Two years later, Herbert Conn (1901), one of soil biology's pioneers, expanded on the observations of Woods and concluded:

"Metabolism of bacteria or yeasts, or by enzymes secreted by these organisms or by higher plants, are of vital importance in agricultural processes. Without their agency in breaking up organic compounds the soil would rapidly become unfit for supporting life."

Soil scientists in the first decade of the 20th century tended to view the soil solution as akin to animal blood or a living tissue because of its ability to carry out catalytic reactions (Quastel, 1946). At that time, catalase was the most widely studied soil enzyme, but this was almost by default because of the ease of its detection using manometric procedures and the absence of methods for assaying other enzymes. Soon, however, soil peroxidase was found by Cameron and Bell (1905) and later investigated by Schreiner and Sullivan (1910) using improved methods. König et al. (1906, 1907) and May and Gile (1909), in forward-looking research, used biological inhibitors (e.g., cyanide) to distinguish between catalase activity associated with the soil organic fractions and that due to active microbial cells. Soon proteases, and then amidases, were measured; the first report on proteolytic activity in soil extracts was made by Fermi (1910).

Around this time more papers on catalase activity in soils appeared (Wachtel, 1912; Penkava, 1913a,b; Kappen, 1913) and gave rise to questions on the role of enzymes in soil functions and their possible relevance to agriculture. Gerretsen (1915) used an iodometric method to show that soils can perform oxidizing reactions although many believed that these transformations were mediated by inorganic, rather than biological, components. These assumptions were, in part, due to the poor correlations between biological measurements and enzyme activity, and unfortunately led to the idea that inorganic "catalytic fertilizers" could be used to increase soil fertility (Lyon et al., 1917; Sullivan and Reid, 1912). The significance of biologically based enzymatic activity was later shown by Chouchack (1924) who reported that hydrogen peroxide degradation was severely reduced by soil sterilization.

1–3 SOIL ENZYME METHODOLOGY

As stated previously, for the first part of the 19th century catalase was the dominant enzyme studied. Early mechanistic studies suggested that the reaction was inorganic (Osugi, 1922) and that it contributed to humus turnover (Balks,1925). Stoklasa (1924) described the determination of catalase activity in Abderhalden's *Handbuch der Biologischen Arbeitsmethoden*. Waksman and Dubos (1926) even proposed catalase activity as a soil fertility index but quickly discarded the idea as too simplistic to reflect the overall catalytic complexity of soils (Waksman, 1927). Nonetheless, in the 1930s, soil enzyme research continued to focus on catalase as a biological indicator of soil fertility, notably Kurtyakov (1931) in the Soviet Union; Radu (1931) in Romania; Rotini (1931) and Galetti (1932) in Italy; Matsuno and Ichikava (1934) in Japan; and Scharrer in Germany (Scharrer, 1927, 1928a,b, 1936). Scharrer was, in effect, the first to publish a paper on the kinetics of soil catalase (Scharrer, 1933) and, by the end of the 1930s, there was general agreement that catalase activity was driven by microorganisms.

This conclusion was underpinned by the research of Baeyens and Livens (1936) in Belgium and Vály (1937) in Hungary who found that catalase activity in high organic matter soils was reduced by antiseptics and that soils to which manure had been added had elevated activity, presumed due to increased microbial activities (Kobayashi, 1940). By the early 1960s, a large number of papers provided overwhelming evidence that most soil enzyme activity was due to microorganisms and that inorganic catalysis was minimal (Hofmann, 1963; Hofmann and Hoffmann, 1955a).

From the 1930s to the 1950s, the number of enzyme assays increased greatly and many others were refined and optimized. These included pyrophosphatase (Rotini 1933), urease (Rotini 1935a), phosphatases (Rogers et al., 1941, 1942; Rogers 1942a,b), and tyrosinase, lipase, asparaginase, and amylase (see Kuprevich and Shcherbakova, 1956). The presence of extracellular asparaginase was first demonstrated by Drobník (1957) in toluene-treated soils. Further improvements for protease and amidase assays were made by Ladd and Butler (1975) and Ladd and Paul (1973) who used substrates that rapidly degrade, such as benzoylarginineamide, dipeptides, casein, and hemoglobin. This permitted short incubation times and reduced the complications due to microbial growth. In other words, the intention was to measure the preexisting activity and not to confuse it with that due to new microbial proliferation and the resulting enzyme synthesis following substrate addition.

Parallel (and often complementary) developments were taking place in aquatic biochemistry and protease and lipase activities were measured in freshwater sediments by Messineva (1941) who also demonstrated a differential depth distribution of protease and catalase activities. The former was not recorded below 2-m depth compared with the latter that was found as far down as 7.5 m.

In the 1950s and 1960s, soil enzymology research was dominated by Western and Eastern European groups, resulting in a plethora of new assays that are only summarized here. For a full review of this era the reader is referred to the excellent history of soil enzyme research authored by John Skujiņš (1978). The most productive laboratories were those of Kuprevich in Russia (Kuprevich and Shcherbakova, 1971), E. Hofmann and G. Hoffmann in Germany (Hofmann and Hoffmann, 1966), Galstyan in Armenia (Galstyan, 1959), and Kiss and coworkers in Romania (Kiss et al., 1975a,b). The work in this period was directed to developing methodologies for assaying the increasingly large number of extracellular enzymes that were being identified.

Most attention was focused on essential carbon, nitrogen, and phosphorus cycle enzymes such as carbohydrases, proteases, urease, and phosphatases. In addition, the well-established catalase assay was further optimized and a dehydrogenase assay for use in soils was developed by Lenhard (1956). Invertase was widely measured because of its ease of detection using the method first published by Hofmann and Seegerer (1950) and subsequently modified by Kuprevich (1951). These researchers also showed that sucrose hydrolysis takes place in the presence of the microbial inhibitor, toluene. Later, Gettkandt (1956) developed a photometric method to detect the product glucose.

During this period there was a strong emphasis on carbohydrase assays with G. Hofmann and E. Hoffmann providing leadership. They developed methods for the measurement of α-glucosidase (maltase) using α-phenylglucoside as the substrate (Hofmann and Hoffmann, 1953, 1954) and β-glucosidase (a.k.a. emulsin, cellobiase, gentiobiase) in toluene-treated soils using salicin, arbutin, β-phenylglucoside (Hofmann and Hoffmann, 1953), or cellobiase (Hoffmann, 1959b) as substrates. Amylase activity in soil was first detected by Hofmann and Seegerer (1951), an assay that was improved subsequently by Hofmann and Hoffmann (1955b) and Drobník (1955).

These papers showed that β-amylase activity was considerably greater than that of α-amylase (Hofmann and Hoffmann, 1955a,b). Much later, Galstyan (1965) concluded that the activities of both amylases were due to what he termed "accumulated" extracellular enzymes—presumably those that had built up over a period of time and were not necessarily related to the extant microbial biomass. Other carbohydrases detected in soils around this time included xylanase (Sorensen, 1955), lichenase (Kiss et al., 1962b), and inulase (Hoffmann, 1959b).

Although it was assumed, a priori, that cellulase existed extracellularly in soil, it was not until Markus (1955) showed the hydrolysis of cellophane in toluene-treated soil that its functional location was definitively demonstrated. Later, in the 1960s and 1970s, a number of investigators used various substrates, including cellophane, cellulose powder, avicel, filter papers, and carboxymethylcellulose, to detect what became an increasing number of extracellular cellulases (Kong et al., 1971; Kong and Dommergues, 1972). However, many of these methods were either insensitive or inappropriate as evidenced by the work of Kiss et al. (1962b) who used many of these substrates and showed that cellulase activity was only detected in 1 out of 10 soils! More accurate methods developed later showed (not surprisingly) that every soil has the capability for the breakdown of cellulose (Kiss et al., 1975a; Deng and Tabatabai, 1994) and, indeed, most other polysaccharides (Burns, 1982a).

Nilsson (1957) was the first to report nucleotide hydrolysis (i.e., nucleases), while extracellular lipase activity in peat and aquatic sediments was measured by Pokorna (1964). Extracellular arylsulphatase was reported by Tabatabai and Bremner (1970a,b) using p-nitrophenyl sulfate as the substrate. Subsequent research showed that for a range of soils, 39.6 to 73.1% of the arylsulfatase activity was bound to the cell constituents of viable organisms.

In the USA there was limited soil enzymology research up until the late 1960s. Thus, there are no references to soil enzymes in such classics as Waksman's *Soil Microbiology* (1952) and Alexander's *Introduction to Soil Microbiology* (1961). In 1965, however, a method for the enzymatic (urease) examination of soils was included in the *Methods of Soil Analysis* published by the American Society of Agronomy (Porter, 1965). Nonetheless, there was important research taking place at the University of California at Berkeley led by a protein photochemist, Douglas McLaren. He reported a trypsin-like activity in soil by using a benzoylargininea-mide as the substrate (McLaren et al., 1957).

McLaren was the first to use irradiation techniques to distinguish what was later termed "abiontic" activity by John Skujiņš (see below) from activity directly associated with the cell surface or intracellular enzymes of viable and proliferating microorganisms (McLaren et al., 1957; Skujiņš et al., 1962). McLaren also was conducting pioneering studies of enzyme activities before and after sorption to clays and other soil-like supports (McLaren, 1954; McLaren and Estermann, 1956) and it was when working with McLaren that Richard Burns first became intrigued with soil enzymes (Burns et al., 1972a,b).

Esterases were always of interest to soil biologists and crop scientists because many esterases are involved in the release or mineralization of essential plant nutrients. Rogers and coworkers were the first to make major progress in detecting and developing methods for phosphatase activity (Rogers, 1942a,b; Rogers et al., 1941). They showed that dephosphorylation of calcium glycerophosphate occurred rapidly and suggested that the source of phosphatase was plant roots.

These and other early assays (e.g., Jackman, 1951; Jackman and Black, 1951, 1952a,b; Mortland and Gieseking, 1952) measured soluble phosphate as the end product. Subsequently, this was recognized as an unsatisfactory approach because the amount of phosphate fixation could vary from soil to soil. This required careful measurement and the application of a correction factor when estimating the amount of product released. In response to this, synthetic substrates, such as n-phenyl phosphate, were evaluated because they enabled the rapid detection of a nonadsorbed (and, therefore, easily and reliably extracted) colored product, phenol (Kroll et al., 1955; Kramer and Erdei, 1958, 1959). However, Tabatabai and Bremner (1969) reported that phenol was not quantitatively extracted from all soils and that the color formed was not stable. In response to this, an optimized method for measuring phosphatase, using p-nitrophenyl phosphate, was developed by Bertrand and de Wolf (1968), and independently by Tabatabai and Bremner (1969), with the latter method being widely used to this day. A large number of assays now have been described for esterases (i.e., the EC 3 group) that utilize substrates to produce the product p-nitrophenol (pNP). The advantages of pNP are that it is not naturally occurring in soils (hence very low background readings) and is not sorbed (except to a minor extent on high organic soils such as Histosols; Klose et al., 2003).

However, the relevance of activities using synthetic substrates to those using natural substrates continues to be debated (Gianfreda and Ruggiero, 2006) as do the more general concerns about the relationships of enzyme kinetics measured under optimal conditions of temperature, pH, substrate concentration, etc. (sometimes referred to as maximum potential activities) to those actually expressed in the soil (Chapter 2, Wallenstein and Burns, 2011, this volume).

Historically, and certainly since the 1950s, the majority of soil enzymology research has focused on hydrolytic rather than oxidative enzymes (e.g., phenol oxidases and peroxidases). This is due, in part, to the limitations of the available methods. In the early to mid-20th century, oxidative activity was measured by oxygen consumption and later by using L-3,4-dihydroxyphenylalanine (DOPA) as a substrate for phenol oxidases (Sinsabaugh and Linkins, 1988; Pind et al., 1994; Sinsabaugh et al., 2008). The phenol oxidation assay using this substrate apparently does not provide consistent results among soils and is especially variable due to length of incubation and substrate concentration (Sinsabaugh, 2010). In part, this is due to interference during the assay from chemical oxidation by Mn and Fe species (Rabinovich et al., 2004) and consumption of oxidation products. Similarly, information about soil peroxidase has limitations because its activity is most commonly measured by adding DOPA and H_2O_2 to soil suspensions, which is then compared with DOPA alone. Thus, this combines the problem of DOPA oxidation with reactions competing for H_2O_2.

In response to the issues with DOPA, other substrates have been proposed that include guaiacol (Nannipieri et al., 1991), catechol (Perucci et al., 2000; Benitez et al., 2006), ABTS (laccase substrate) (Luis et al., 2005; Floch et al., 2007), pyrogallol (Allison and Vitousek, 2004), or o-toluidine (Fioretto et al., 2000; Di Nardo et al., 2004). However, Sinsabaugh et al. (2008) in their review of the literature, found that ranges for mean and extreme values of soil phenol oxidase activities using these substrates were similar to when DOPA was used as the substrate.

Oxidative enzymes perform important ecological functions by degrading lignin, humifying and mineralizing organic C inputs to soils, and releasing dissolved organic C. These processes are particularly important in understanding C transformation and storage in soils, which is critical for managing the global C cycle and ultimately for predicting climate change relative to the activity of these enzymes and CO_2 emissions from soils. However, phenol oxidases and peroxidases are typically uncorrelated with soil organic matter or hydrolase activities and have a high spatial and temporal variation that obscures their relationships with environmental variables and ecological processes (Sinsabaugh, 2010). This may be due in part to the methodological issues raised above and the fact that these enzymes are less stable than hydrolases in the environment (Huang, 1990; Allison, 2006; Laiho, 2006). Nonetheless, Sinsabaugh (2010) gives insights into the role of phenol oxidases and peroxidases in affecting ecosystem functions as follows: the enzymes are often correlated with decomposition rates and lignin content of substrates added to soils; nitrogen amendment appears to affect the activity of phenol oxidases and peroxidases that corresponds to decomposition rates; nitrogen amendments seem to reduce oxidative rates in basidiomycete-dominated soils of temperate and boreal forests whereas activities in grassland soils are dominated by glomeromycota and ascomycetes show little net response; and land use that leads to loss of soil organic matter tends to increase oxidative activities.

In the early 1970s, a sensitive assay for measuring lipases in soil was developed using the butyrate ester of 4-methyl umbelliferone (MUF) as a substrate (Pancholy and Lynd, 1972). The product, 4-methyl umbelliferone, could be measured easily using fluorometric techniques. Improvements to this method followed (Cooper and Morgan, 1981; Darrah and Harris,1986) and methods using these substrates and multiwell microplates have been developed (Marx et al., 2001). However, analyses using MUF methods suffer from serious quenching, which requires running a standard with each sample (see Chapter 14, Deng et al., 2011, this volume), and the methods have not been rigorously optimized.

Because of the agricultural importance of urea as a fertilizer, urease has been one of the most widely studied of all soil enzymes and a number of methods for its determination in soil have been developed. Soil urease was first examined by Rotini (1935a) and subsequent early research was dominated by Conrad and coworkers (Conrad and Adams, 1940; Conrad 1940a,b; Conrad 1942a,b). Tabatabai and Bremner (1972) thoroughly evaluated the urease assay (including buffers used to optimize the reaction) and showed it could be used even in soils that fix the measured product, ammonium.

In the modern era (i.e., since the late 1960s), M.A. Tabatabai, with his research group at Iowa State University, is an acknowledged leader for many major contributions in the development of enzyme assays and other biochemical methods applicable to soils. Methods developed by this laboratory were thoroughly vetted, reliable, clearly described (including helpful comments and caveats), and quantitative. A wide range of assays have been developed (indeed, many are found in this book) involved in the C, N, P, and S cycles in soils and other biogeochemical processes. This laboratory has been a "breeding ground" for many graduate students and postdocs who have further advanced the studies of soil enzymology (e.g., Warren Dick, William Frankenberger, Jr., Richard Dick, Dianne Stott, Susanne Klose, Veronica Acosta-Martínez, and Shiping Deng).

1–4 ORIGINS AND DISTRIBUTION OF ENZYMES IN SOIL

From the early years it has been of interest to determine the plant and microbial origins of "free" enzymes in soils. As stated previously, the first report on soluble enzymes in soil (Woods, 1899) pointed to roots as the source and Knudson and Smith (1919) recorded the excretion of amylase by plant roots. Soon others indicated the plant origins of invertase (Knudson, 1920), phosphatase (Rogers et al., 1942), and any number of enzymes including catalase, tyrosinase, asparaginase, urease, amylase, invertase, protease, and lipase (Kuprevich, 1949). Very soon the release of enzymes by plant roots (Knudson and Smith, 1919) became well established (see Dick and Tabatabai, 1986), but this led to an erroneous conclusion that this was the *major* source of extracellular enzymes (Knudson and Smith, 1919; Rogers, 1942b; Rogers et al., 1942; Kuprevich, 1949). As mentioned in the previous section, it is now well known that microorganisms are the principal source of most soil enzymes either through direct synthesis and secretion or as generators of signals that induce enzyme production by other microorganisms and plants. Furthermore, indirect evidence presented by Ladd (1978), Dick et al. (1983), and Nannipieri et al. (1983) suggests that enzymes from plant debris are rapidly decomposed and used for microbial growth in incubated soils.

Of course, the synergistic impact of the rhizosphere on enzyme secretion by both components is likely to be very important. In recent years, the extraction of DNA from soil and the application of molecular techniques has enabled us to start targeting functional genes and is giving us a better idea of the origins of many soil enzymes (see Krsek et al., 2006; and Chapter 2, Wallenstein and Burns, 2011, this volume).

1–4.1 The Quest to Discover the Locations of Soil Enzymes

As detailed above, there was early recognition of the existence of microbial-free extracellular enzymes in soils and some good indications from assays that inhibited microbial activity and proliferation. Much later, Skujiņš (1976) coined the word "abiontic" to represent soil extracellular enzymes. Abiontic is derived from the Greek a-, the α privative meaning "removal or absence of a quality," and the Greek suffix "biontic," meaning "having a form of life."

Abiontic enzymes then, are those exclusive to, and unregulated by, viable cells. This includes enzymes: excreted by living cells into soil solution during cell growth and division; attached to cell debris and dead cells; and released into the soil solution from extant cells or lysed cells but whose original functional location was on, or within, the cell. Additionally (and importantly), abiontic enzymes can exist in stabilized forms primarily in two locations: adsorbed to internal or external clay surfaces and complexed with humic colloids through adsorption, entrapment, or copolymerization during humic matter genesis (McLaren, 1975; Boyd and Mortland, 1990; Quiquampoix et al., 2002).

Clearly, abiontic enzymes are no longer subject to cellular regulation and control processes; they have become independent of living cells and the external influences on them. However, it is clear to us today that these various sites of enzymes are far from fixed: extracellular enzymes are undergoing frequent changes to location, which inevitably causes changes in both activity and resistance to denaturation and degradation.

The initial approach to distinguishing between the various origins of these extracellular enzymes was to use antiseptics and antibiotics to either block microbial activity or inhibit microbial growth during the assay. The ideal inhibitory agent should inhibit microbial activity (but not damage preexisting cells) and not affect those enzyme activities that are no longer associated with viable cells. Earlier research commonly used the inhibitory agents of phenol, acetone, thymol, chloroform, ether, and toluene (Rotini, 1935a). The use of toluene in soil enzyme studies was controversial during the 1950s and 1960s with considerable debate as to whether (and in what ways) toluene affected microorganisms and enzymatic activity (Claus and Mechsner, 1960; Hofmann and Hoffmann, 1961; Kiss et al., 1962a). Further research by Beck and Poschenrieder (1963) and Kiss and Boaru (1965) demonstrated that toluene is a highly effective agent for inhibition of microbial proliferation without lysing cells. It is useful for enzymatic determinations because it prevents the synthesis of new enzymes. Toluene is still one of the most popular choices because of its ability to inhibit growth during the assay incubation (Foght, 2008)

Another approach was to use antibiotics to inhibit microorganisms and thus protein production, which was first reported by Kuprevich (1951) (the exact antibiotic was not revealed as the paper only mentioned "BIN No.7"; Kiss et al., 1975a). Later investigations (Kiss 1958a,b,c; Kiss, 1961; Kiss and Drøgan-Bularda, 1968; Kunze, 1970; Benefield, 1971; Kiss et al., 1972) using a variety of antibiotics showed that, although activities may be reduced, activity was maintained at some level. The conclusions, although derived indirectly, were that soils contained extracellular enzyme activity that was independent of viable cells. Such experiments, however, do not resolve the problem of enzymatic activity arising from lysed cells. Furthermore, these compounds may be substrates for certain subpopulations that remain viable or flourish and alter enzyme activity, making it difficult to conclusively isolate abiontic activity.

Until the early 1950s there seemed to be a limited understanding or consideration that enzymes could be stabilized in the soil matrix; rather, the emphasis was on differentiating between those enzymes secreted into the soil solution and those directly associated with viable cells (i.e., intracellular or located at the cell surface). However, in the late 1950s and into the1960s, studies showed that enzymes can be strongly sorbed by clays which will affect their activity, kinetics, and stability (see review by McLaren and Packer, 1970). Several investigators were prompted to fractionate soil to obtain information on the localization of enzyme activities according to texture. A study of esterase activity showed that the clay fraction had the highest activity, with intermediate activity in the silt fraction but very little in the sand (Haig, 1955). Fractionation experiments were also performed by Hoffmann (1959a) who demonstrated that carbohydrolase activity peaked in the silt fraction whereas urease activity peaked in the clay fraction. Because the clay fraction contained a minimal number of microorganisms, it was concluded that extracellular urease had originated elsewhere in the soil but had been adsorbed and remained active on the clay fraction.

An interesting study by Ramirez-Martinez and McLaren (1966) empirically demonstrated there must be large amounts of catalytic enzymes outside of viable cells. They measured phosphatase activity on bacterial and fungal cultures and compared it to the activity in soils. They found that 1 g of fungal biomass or 1×10^9 or 1×10^{12} phosphatase-producing bacterial cells would be required to equal phosphatase activity in a typical soil. Since these microbial numbers far exceed what is possible in soil, the only conclusion can be that there are large amounts of accumulated phosphatases in the soil matrix other than those originating from viable microorganisms.

During the 1970s, research was initiated to extract enzymes from soil to better understand their origins and functional locations. For example, Chalvignac and Mayaudon (1971) found tryptophan decarboxylase activity associated with the extracted and dialyzed humic acid fractions. In this case, the humic fraction (NaOH extraction followed by precipitation in HCl) retained half of its activity. Ladd (1972) demonstrated that proteases may be extracted from soils with a simple (Tris) buffer. Proteases were studied also by Mayaudon et al. (1975) who concluded that the polysaccharides, also present in the extracts, contributed to the stabilization of the humic–enzyme systems in soils.

Two methods for extraction of urease were developed by McLaren and his colleagues. One method (Burns et al., 1972a) involved dispersion of soil organic matter with urea, whereas the other (Burns et al., 1972b) employed sonication. In the first method, the extracted urea was later decomposed by the urease isolated from soil. The authors concluded that soil urease is present in the organic colloidal particles with pores large enough for water, urea, and ammonia to pass freely, but sufficiently small to exclude protease enzymes. Subsequently, McLaren et al. (1975) concluded that both methods showed urease activity in soil that may be regarded as a part of the native humus soil fraction.

In Australia, electron microscopy to visualize the location of enzymes was used by Ralph Foster and another soil enzyme pioneer, Jeff Ladd (Foster and Martin, 1981; Ladd et al., 1996). This gave clues as to the distribution of acid phosphatase and catalase between the soil organic and inorganic fractions

Mayaudon and Sarkar (1974a,b) extracted diphenol oxidases, which were separated on a DEAE cellulose column in three fractions. The conclusion from this study was that diphenol oxidases were associated with a nucleoprotein and also that they were in a form of humic acid–enzyme complexes (Mayaudon and Sarkar, 1975).

Later, Tabatabai and Fu, (1992) reported that alkaline solutions yield high amounts of humic substances that contain humic–enzyme complexes (Nannipieri et al., 1996). Fornasier and Margon (2007) developed a new detergent-based approach that obtained extracts that were rich in enzyme activities but were colorless to pale yellow, suggesting negligible co-extraction of humic substances (Chapter 16, Fornasier et al., 2011, this volume).

An alternative approach for isolating abiontic activity is to use irradiation to sterilize soils and denature all enzymes of viable microbial cells but allow free enzymes to be stabilized (and thus resistant to denaturation) to remain active. Interestingly, as early as the late 1920s Scharrer (1928a) attempted to sterilize soils with ultraviolet radiation to measure residual (abiontic) activities. The irradiation had minimal effects on catalase activity. Much later, Dommergues (1960) exposed soil to continuous infrared irradiation for 7 d and found that the microbial count was reduced by 50%, while invertase activity remained unchanged in one soil. In several other soils, however, invertase activity declined by up to 71%.

Effective radiation methods for sterilizing soils did not occur until the introduction of high-energy electron beams or γ radiation in the 1950s. Dunn et al. (1948) were the first to use ionizing radiation for sterilization of small quantities of soil. Later it was shown that a 5 to 10 MeV electron beam and hard x-rays and γ-rays from a ^{60}Co source (a 2 Mrep dose) could sterilize soil (McLaren et al., 1957; 1962). Microorganisms could not be cultured from these soils and yet urease (McLaren et al., 1957; Skujiņš and McLaren, 1969), phosphatase (Skujiņš et al., 1962), and other enzymes (McLaren, 1969) remained active. The effects of 40 and 90 MI-Iz microwave irradiation on soils was tested by Voets and Dedeken (1965). The irradiation resulted in a considerable decrease in microbial counts but also in invertase and protease activities.

In the context of these studies, it is apparent that enzymes differ widely in their sensitivity to irradiation. The dosage necessary for microbial inactivation also reduces or eliminates some enzyme activities—for example, urease (McLaren et al., 1957; Skujiņš and McLaren, 1969; Vela and Wyss, 1962). Irradiation will also depolymerize humates and polysaccharides (He et al., 2010) in which enzymes are trapped and protected. Nonetheless, the important conclusion of these radiation studies (from most of the enzymes studied) was that a substantial amount of activity came from enzymes stabilized by soil colloids and humic substances.

An alternative sterilization approach uses microwave irradiation. Speir et al. (1986) used this method and showed that soil microbial biomass was much more susceptible to microwave irradiation than was phosphatase activity. Knight and Dick (2004) applied microwave irradiation at a range of known energy levels to identify the lowest amount of energy needed to sterilize soils. This allowed them to isolate the activity of abiontic and viable cell enzyme fractions for β-glucosidase. It was inferred that sterilization of soil with this treatment denatured most of the enzymes associated with viable cells. They showed that differences in β-glucosidase activity due to soil management within the same soil type are due to the stabilized fraction, not the activity associated with viable cells. This suggested that the abiontic fraction of β-glucosidase is not fixed but is a dynamic property that can be reduced by intensive land management. This work further supported the potential for β-glucosidase activity to be a soil quality indicator as the abiontic fraction is a major source of its activity. Its activity would, therefore, change slowly over time due to management with less chance for wide variability due to seasonal or environmental factors.

Beginning in the 1970s, but particularly in the 1990s, other approaches than high-energy electron beams or γ-radiation methods were used and further confirmed the presence and catalytic capabilities of abiontic enzymes in soils. Busto and Perez-Mateos (1995) extracted humic compounds from the soil and showed them to contain as much as 50% of the total β-glucosidase activity.

Another approach to understanding the formation of stabilized enzymes in soil has involved synthesis of humic–enzyme (Sarkar and Burns, 1984; Ruggiero and Radogna, 1988) and clay–enzyme (Boyd and Mortland, 1990; Dick and Tabatabai, 1987; Ruggiero et al., 1989) complexes. The former were created by copolymerizing enzymes with likely humic constituents such as phenols, tannins, and resorcinol; the latter involved adsorption to various homo- and hetero-ionic kaolinites and montmorillonites These studies showed that model immobilized enzymes certainly had some of the properties of the natural soil–enzyme complexes and were examined for their survival and activities when introduced to soil.

Others showed a variety of enzymes including acid phosphatase (Rao et al., 1996; Nannipieri et al., 1988; Rao et al., 2000), β-glucosidase (Hayano and Katami, 1977; Busto and Perez-Mateos, 2000), urease (Gianfreda et al., 1995), and others (Sarkar, 1986; Ruggiero et al., 1989; Grego et al., 1990; Lähdesmäki and Piispanen, 1992) that can be stabilized on humic and/or clay colloids and retain much of their activity after extraction (see Chapter 16, Fornasier et al., 2011, this volume).

Of course, this is all part of a much bigger story concerning humic matter formation, the interactions of proteins and amino acids with soil components (Quiquampoix, 2000), and soil as a catalytic system (Ruggiero et al., 1996).

These immobilized enzymes appear to be protected against denaturation by both proteolytic enzymes and heat (Hayano and Katami, 1977; Hope and Burns, 1987; Lähdesmäki and Piispanen, 1992; Deng and Tabatabai, 1994; Nannipieri, 2006; Rao et al., 2000). Nannipieri et al. (1996) suggested this may be due to steric interference in the case of proteolysis and an increase in the rigidity of the protein structure when complexed with humic colloids. Ladd and Butler (1975) indicated that enzymes could be bound to humus by ionic, hydrogen or covalent binding and that most of the abiontic enzymes in soil are stabilized as humic–protein complexes.

More recent developments in the use of monoclonal and polyclonal antibodies, immunogold and immunofluorescent labeling, and reporter genes are providing sensitive ways to locate enzymes in the soil matrix and are providing new information on the ecology of extracellular enzymes (See Chapter 2, Wallenstein and Burns, 2011, this volume). Atomic force microscopy also has the potential to help as well as giving information about the strength of association, configuration, mode of action, and reactivity of the enzymes with regard to sorbed and insoluble substrates (Lee et al., 2007). Amino acid–, protein–, and enzyme–clay complexes have been much studied (e.g., Quiquampoix, 2000; Quiquampoix and Burns, 2007) and, in recent times, the extraction of proteins from soil and its components and their subsequent separation and purification with gel chromatography, has given us greater insight into the locations of large numbers of enzymes in soil (Nannipieri, 2006).

1–4.2 Potential Ecological Importance of Stabilized Soil Enzymes

From the early irradiation research, and experiments on humic and/or clay–enzyme complexes (Paul and McLaren, 1975; Boyd and Mortland, 1990), a conceptual model emerged in the 1980s that abiontic enzymes may have important relationships with soil organisms. Burns (1982, 1983) described the various possible locations of extracellular enzymes and hypothesized that stable humic–enzyme complexes are important for substrate catalysis. He suggested that for some microorganisms, substrates may not be directly available as energy sources because they are too large or insoluble for cell uptake. Therefore, a microorganism may have no means to detect the presence of the potential substrate and thus will not be induced to synthesize and secrete the enzymes necessary to hydrolyze the compound. However, the enzymes associated with a humic fraction would not be under the same regulatory control as they would be if they were intercellular. In turn, the product released by the complexed soil enzyme could be further broken down by other soil enzymes stabilized in the soil matrix or be taken up by microorganisms.

Stabilized enzymes exclusive of viable cells could be very important for the ecological functioning of soil microbial communities. It may be an advantageous situation for a microbial cell to be located on the surface of a humic colloid containing a number of enzyme molecules for which they would not have to expend energy to produce the necessary hydrolytic enzymes. Indeed, for some species their success in a soil environment, where a large proportion of the potential carbon and energy source is in a high molecular mass polymeric form, may depend on a close association with humic–enzyme complexes (see Chapter 2, Wallenstein and Burns, 2011, this volume).

1–4.3 Spatial and Temporal Distribution of Soil Enzymes

Kandeler and Dick (2007) reviewed the literature on spatial distribution of enzyme activities. At the microscale (μm to mm), where enzymes are expressed at the molecular level and interact with the microbial habitat, is a fundamental process scale that ultimately controls biogeochemical cycles and microbial ecology. The location of enzymes in intracellular or extracellular locations and their interactions and spatial variation with the physical surroundings are the basis for the distribution of enzyme activities at higher spatial scales.

Studying enzyme activities across particle size fractionation of soil (e.g., Poll et al., 2003) has shown that there is a distribution of enzymes among various inorganic and organic soil components. The spatial and physical size distribution patterns of C sources control the location of microbial community members and associated enzymes among particle size fractions. These individual properties of enzymes cause a preferential binding either to clay–humic–enzyme complexes or to particulate organic matter. For example, xylanase activity and fungal-derived phospholipid fatty acids showed a very close relationship to the particulate organic matter of the coarse sand fraction (Kandeler et al., 1999a,b,c; 2000; 2001; Stemmer et al., 1998; Poll et al., 2003).

Limited insights on enzyme location at these microscales also has been made by studying model humic–enzyme copolymers (Ruggiero et al., 1996), chemical extraction of enzymes from the soil matrix (Tabatabai and Fu, 1992), and by modeling the immobilization of enzymes on various matrices (Quiquampoix et al., 2002). Generally, pico- and nanoscale studies have been performed using artificial systems (e.g., studies of adsorption of specific proteins onto clay particles) (Quiquampoix et al., 2002). Recently, techniques using polyclonal antibodies, immunofluorescent staining and immunogold labeling (Daniel et al., 2007), coupled with a range of advanced microscopic techniques, is allowing the identification of enzyme locations and functions at the soil and substrate microsite level.

Mesoscale is the intermediate spatial pattern that ranges on the order of meters which includes the distribution of soil enzyme activities both vertically (soil profile) and horizontally at the field scale under either agricultural or forest plots that receive relatively uniform long-term management. The vertical distribution of enzyme activities are highest at the surface and decrease with depth, which has been shown for many enzymes and on every soil studied so far (Juma and Tabatabai, 1978; Dick, 1984; Speir and Ross, 1984; Taylor et al., 2002).

Field-scale investigations have shown that organic matter turnover rates, microbial biomass, and enzyme activities of soil samples vary with vegetation and soil types (Bonmati et al., 1991; Vaughan et al., 1994; Bahri and Berndtsson 1996; Stork and Dilly, 1998; Schutter et al., 2001). Relatively little is known about topographic, pedogenic, soil mineralogy, and other properties that control microbial and enzyme distribution at landscape levels (Bergstrom and Monreal., 1998; Stork and Dilly, 1998; Wirth, 1999) and the characterization of these interactions is essential to achieve a better understanding of complex ecosystem processes (Goovaerts, 1998).

The aggregation of micro- and mesospatial soil biogeochemical processes to large- or macroscales of terrestrial ecosystems has broad implications for global ecology and the environment. Tscherko and Kandeler (1999) investigated a subset of microbial factors at the mesoscale and found these were correlated and sensitive to site factors (land use, soil type, and pollution). These researchers went on to determine the importance of factors that influence microbial biomass, N-mineralization and enzyme activities (xylanase, urease, phosphatase, arylsulfatase) on a large continental scale of Central Europe. Soil type turned out to be the most important site factor as it integrates climatic, topographical and geological conditions, soil acidification, and vegetation factors (Tscherko, 1999) that, collectively, control enzyme activities at the macroscale. At this landscape scale, significant correlations have been found between organic C and enzyme activities (Dutzler-Franz, 1977; Bergstrom and Monreal., 1998; Wirth, 1999). Decker et al. (1999) evaluated the patterns of variation of β-glucosidase, chitinase, phenol oxidase, and acid phosphatase in oak forest soils at spatial scales from 10s of km to <3 m. Spatial scale differences were detected at the regional, topographic, and single tree scale that were mainly due to strong nutrient availability gradients.

Seasonal variations in enzyme activities do occur, and have been related to temperature and moisture availability patterns (Bandick and Dick, 1999; Ndiaye et al., 2000). Some have described the distribution of enzyme activities (and in particular phosphatases) between bulk and rhizosphere soils (Naseby and Lynch, 2002) using fractionation or thin sectioning (Ladd et al., 1996).

1–5 RELATIONSHIP OF ENZYME ACTIVITY TO BIOLOGICAL PROPERTIES IN SOILS

From the early days of soil enzyme research, and particularly through the 1950s, there were numerous studies that attempted to relate enzyme assays to other soil biological properties; many of them gave confusing and often contradictory results (see Skujiņš, 1967). As an example, Turková and Srogl (1960) showed a low correlation between carbon dioxide production (soil respiration) and amylase or invertase activities. In this case different plant communities induced varying rates of activity on the identical soil types.

One line of evidence for this counterintuitive relationship between enzyme activities and microbial properties was produced by measuring activities in the presence or absence of toluene. A number of early studies focused on invertase activity (Peterson and Astafyeva, 1962; Daragan-Sushcheva et al., 1963; and Galstyan, 1965) and the hydrolysis of sucrose (Kiss, 1958a) and showed that toluene made no difference to enzyme activities. This suggested that the inhibition of microorganisms was not affecting the enzymes and that, therefore, most of the activity must be due to preexisting extracellular enzymes. Later work by Kiss et al. (1971, 1972), Voets and Dedeken (1965), and Voets et al. (1965) demonstrated that invertase activity accumulated in soils as an extracellular enzyme and that even if proliferation of microorganisms was allowed during the short assay, it made very little contribution to the total activity measured.

Skujiņš (1967) reviewed much of this research and reported very few direct correlations of activities of various soil enzymes with CO_2 evolution or direct counts of microbes. This would be expected for several reasons including: (i) methodological inadequacies in counting microorganisms and measuring carbon dioxide evolution (e.g., most early studies were correlating activities with respiration that we now know has high spatial variability, or direct counts that are highly selective); (ii) specificity of many enzyme assays (i.e., only a small fraction of the population may possess and express the measured enzyme at any given time); and (iii) the activity of many enzymes in soil is a composite of activities from many sources and locations with a proportion having no direct relationship to extant viable organisms.

However, from the mid 1970s the relationships between microbial properties and enzymes became better established, particularly that involving dehydrogenase, protease, cellulase, phosphatase, and urease (Laugesen, 1972; Laugesen and Mikkelsen, 1973; Nannipieri et al., 1978; Ross and Cairns, 1982; Tiwari et al., 1989). Alef and Kleiner (1986) proposed arginine ammonification as a simple assay and biological index. On 22 soils they found high correlations of activity with biomass C, heat output, soil ATP content, or soil protease activity but not with viable microbial counts. Frankenberger and Dick (1983) evaluated 11 enzymes in 10 diverse soils for their potential relationships with microbial respiration, biomass, viable plate counts, and other soil properties. They found that alkaline phosphatase, amidase, and catalase activities were highly correlated with both microbial respiration and total biomass in glucose-amended soils. The lack of correlation with microbial plate counts possibly indicates that enzyme activities were associated with microorganisms that were active in soil but whose numbers were not accurately reflected by the viable plate counts.

Nonetheless, there were many exceptions where various microbial measures were not closely related to enzyme activities. To be fair, a lack of correlation should not be surprising because many, if not the majority, of the total measured activities in soils contain a significant contribution from extracellular enzymes. Furthermore, most enzymes perform a very specific reaction, and therefore only a small fraction of the microbial population may possess that enzyme at any given time.

If a measure of the currently active microbial population is the objective, then a more appropriate enzyme is one that is directly associated with viable cells and microbial activity and that is rapidly inactivated following cell death. In this context, the most widely studied enzyme indicator of soil biological activity is dehydrogenase. However, even then, dehydrogenase has not consistently correlated with microbial activity (Skujiņš, 1978; Frankenberger and Dick, 1983). Howard (1972) reviewed the literature and presented data that provided a potential explanation. Using oxygen uptake to calculate theoretical dehydrogenase activity, he showed that the observed activity was substantially more than the theoretical activity. He hypothesized that extracellular phenol oxidases, which are common in soil, also may carry out the dehydrogenase reaction; this would explain the overestimation. Another problem with the dehydrogenase assay is that Cu can interfere with the analytical procedure so that soils high in solution Cu (or that have received Cu-contaminated amendments) will show artificially low dehydrogenase activity levels (Chander and Brookes, 1991).

Protease often is correlated with microbial biomass (Nannipieri et al., 1978; Asmar et al., 1992) because it apparently exists only in microbial cells or, if it is excreted into the soil solution, may not remain for more than a few days (Nannipieri et al., 1978; Asmar et al., 1992).

1–6 APPLICATIONS AND ECOSYSTEMS SERVICES

Soil enzymology, as a subdiscipline within soil science, is not only important for understanding the functioning of soils but also has made contributions toward practical applications. This has included attempts to assess soil quality that can be utilized to assist land managers in: (i) determining the degree of soil degradation, (ii) guidance on sustainable practices, (iii) optimizing agricultural productivity, and (iv) use in environmental applications.

1–6.1 Enzyme Activity as an Index of Soil Fertility or Quality

In the 1950s, there was increasing recognition that enzyme activity varied with soil type and land management as well as displaying a vertical and horizontal variability. This quickly led to the proposal that enzyme activity could be used as a measure of soil functionality. Specifically, it was hoped that enzyme assays could be practical tools to guide crop production. There were two possible options for using enzyme activity as a soil indicator—one as an index of soil fertility relative to crop production and the other as an index of the microbial community.

Rippel (1931) proposed that enzyme activity (catalase activity in particular) could be used to indicate soil fertility. In the 1950s, considerable effort was focused on the possible connection between invertase activity and soil fertility, particularly related to crop productivity (e.g., Kroll, 1953; Hofmann and Bräunlich, 1955; Drobník and Seifert, 1955) although Koepf (1954a,b,c) had reported there was no close association between nutrient levels and invertase activity. Similarly, Drobník (1957) and Galstyan (1959) found poor relationships between soil enzyme activities and crop production. Conversely, Verstraete and Voets (1977) showed that some soil enzymes (phosphatase, invertase, β-glucosidase, and urease) were correlated with crop yields and that these were a superior measure to that achieved with microbial enumeration.

There are a number of factors contributing to these somewhat contradictory results. First, the early studies were seeking a direct relationship between enzyme activity and plant nutrient availability. This is an unrealistic expectation as enzyme assays are often conducted under "optimal" conditions that cannot reflect seasonal and year-to-year climatic variations that strongly affect nutrient mineralization rates under field conditions. Furthermore, enzyme assays are specific and, as mentioned before, most measure a significant amount of accumulated extracellular activity that may not reflect the current microbial generation of plant nutrients. All of these factors will limit the value of using enzyme activities to predict nutrient release and plant responses in agroecosystems. There are countless other factors that may confound a possible relationship including external nutrient inputs and water availability. It is likely that a combination of soil, plant, microbial community, and climate override any singular relationship between enzyme activities and soil fertility or site productivity.

A further example of why this is such a difficult relationship to establish, even at a case-by-case level, is given by Yarochevich (1966) who showed that yields were the same for manure-amended and inorganic fertilizer-amended soils because nutrients were not limiting crop yield. Yet, at the same time, enzyme activities were significantly greater in manured soils than in inorganic nutrient-amended soils.

On the other hand, the relationship between enzymatic activity and plant growth in unmanaged soils with native plants either under natural conditions (Pancholy and Rice, 1973a,b; Pancholy et al., 1975) or in highly disturbed land-scapes (Ross et al., 1992; Kiss et al., 1993) is more predictable. For example, Ross et al. (1992) showed a strong relationship of arylsulfatase and invertase with grass–clover pasture production during remediation of a simulated lignite mining soil over a 5-yr period in New Zealand. In contrast, total C had a much smaller or limited response to remediation, suggesting enzyme activities can be early detectors of positive impacts on soil quality properties that are relevant to plant growth.

Throughout most of the 1970s and 1980s, research on soil enzymology was focused on method development or on assays as corollary measures within a battery of microbial analyses. The goal was to develop insights into biogeochemical or biological responses among soil types, rhizospheres, and soils from manipulated laboratory and field experiments (see reviews Dick, 1992; Dick, 1994; Dick, 1997; Kandeler and Dick, 2007). However, with the strong interest in ways to assess soil quality in the 1990s, enzyme assays again have become candidates for assessing soil fertility and health.

A conceptual model for enzyme assays as indicators of soil quality was articulated by Dick (1994). Enzyme activities hold potential as soil quality indicators because the assays often are simple and rapid, are integrative in nature, and are sensitive to land management (Ross and Cairns, 1982; Dick, 1994, 1997; Bandick and Dick, 1999; Ndiaye et al., 2000). Detailed discussions of the combinations of enzyme activities (and other soil properties) that are most likely to reflect soil fertility and the weighting of the various activities to generate soil fertility indices have been developed and discussed by many (e.g., Trasar-Cepeda et al., 2000; Bastida et al., 2008; Gil-Sotres et al., 2005; Shaw and Burns, 2006).

The accumulated extracellular enzymes, in particular, may provide a useful indicator of long-term or even permanent changes in soil quality because they are likely to be complexed and protected (and therefore, long-lived) over a period of time due to their association with the humic and clay complexes. The interpretation from a soil quality perspective is that soil management that promotes the formation and stabilization of organic matter and associated structural properties (e.g., aggregation and porosity) would also promote stabilization of enzymes and that many of these would have an important role to play in maintaining soil quality. But since enzyme activities are much more sensitive to changes than are most measures of soil organic matter, they can be viewed as early and sensitive indicators of the decline or improvement of soil quality (Bandick and Dick, 1999; Ndiaye et al., 2000).

Soils that have been managed to promote soil quality (e.g., minimum tillage, organic amendments, crop rotations, etc.) would be expected to have higher microbiological activity. This would be reflected in greater enzyme production and, in time, a buildup of stabilized and protected enzymes complexed in the soil matrix (Bandick and Dick, 1999).

In this context, the emphasis on using soil enzymes (and other soil properties) was to detect soil management effects rather than to reflect or predict crop productivity. Indeed, this relationship has been found to hold true for agricultural practices (tillage, organic inputs) and the negative effects of some heavy metal and organic pollutants (see reviews by Dick 1994; Dick 1997; Nannipieri et al., 2002).

Enzymes vary in their degree of inhibition by trace elements. The nature and degree of this inhibition by heavy metals are strongly related to soil type (Speir et al., 1992). Greater inhibition of arylsulfatase and phosphatase has been shown on soil with low surface area, CEC, and organic matter content, which are likely to reduce the potential of the soil to inactivate metals via complexation or sorption reactions and increases the availability of metals to affect enzyme processes (Speir et al., 1995).

Of the soil enzymes tested so far, arylsulfatase appears to be the most sensitive to trace elements whereas acid phosphatase, urease, and invertase are less affected by metals (Al-Khafaji and Tabatabai, 1979; Bardgett et al., 1994; Yeates et al., 1994). Thus, arylsulfatase activity could be regarded as a sensitive indicator of soil pollution by heavy metals and other trace elements.

In many cases, enzyme activities can be early predictors of the effects of soil management on soil quality and indicators of how rapid these changes would be expected to occur. For example, the response of xylanase activities in the top soil of a chernozem was detectable within the first year after a reduced tillage system was put in place, whereas 4 yr were required before significant changes were measured in microbial biomass and N-mineralization (Kandeler et al. (1999a). Ndiaye et al. (2000) showed that β-glucosidase and arylsulfatase activities can respond within 2 to 3 yr after initiation of winter cover-cropping compared with other physical and chemical (e.g., total organic C) properties, which showed minimal measurable effects (Buller, 1999).

These observations show the potential of soil enzyme activities to reflect soil management or environmental impacts but a remaining and major challenge is to calibrate key enzymes assays. This is because the soil type can have much greater effects on soil enzyme levels than the more subtle, but significant, effects due to management within a soil type. Research is needed to enable interpretation of enzyme assays that are independent of soil type without the need for comparison to an unmanaged control.

1–6.2 Detoxification of Polluted Soils

Many approaches to soil bioremediation concentrate on using either microbial inocula or stimulating the indigenous, but underperforming, microbial community (Adriano et al., 1999; Jain et al., 2005; Whiteley and Lee, 2006). Of course, the hope is that this results in increased enzyme production by the flush of existing microbes or as a result of the added microbes. These enzymes may then initiate, or accelerate, the degradation of the target pollutant. This is a valuable approach in some instances (Ford et al., 2007).

However, the survival in soil of added nonindigenous microbes is often poor as is the expression of their desired catabolic activities. As many of the target pollutants will require (at least initially) extracellular enzymes to degrade them, we must consider their efficacy even if the additional microbial cells survive. This is again problematic for all the reasons outlined previously; enzymes are rapidly degraded by microorganisms and inactivated by adsorption on mineral surfaces and complexation within organic matter.

These problems have prompted many to consider alternatives that involve stabilizing the catabolic enzymes and, if this could be achieved, entirely bypassing the use of microbial inoculants. We know that microbial extracellular enzymes have the potential to transform or inactivate a wide range of organic contaminants (Gramss et al., 1999; Dec and Bollag, 2000; Ruggaber and Talley, 2006). Prominent among these enzymes are laccases, peroxidases, hydroxylases, dioxygenases, lipases, nitrilases, and esterases as well as hydroxyl radical production mechanisms. Both bacterial and fungal enzymes are involved and the genus *Burkholderia* (Vial et al., 2007) and the ligninolytic white rot fungi, such as *Phanerochaete chrysosporium* Burds. 1974, *Trametes versicolor* (L. : Fr.) Pilát 1920 [1921], *Irpex lacteus* (Fr. : Fr.) Fr. 1828, and *Pleurotus ostreatus* (Jacq. : Fr.) P. Kumm. 1871, are valuable sources of potential remediating catalysts (Novotny et al., 2004; Tortella et al., 2005; Asgher et al., 2008; Baldrian, 2008; Novotny et al., 2009).

The pollutants that are targets for remediation include: pesticides (e.g., 2,4-D, DDT, atrazine, parathion); BTEX hydrocarbons (benzene, toluene, ethylbenzene, xylene) from gasoline and many industrial solvents; polycyclic aromatic hydrocarbons (e.g., naphthalene, anthracene, phenanthrene) from fossil fuel wastes; polychlorinated biphenyls from electrical insulators, chloroderivatives, chlorophenols, and chlorobenzene from paper mill effluents (Gianfreda and Rao, 2008). Jean-Marc Bollag in the USA and Liliana Gianfreda in Italy have done extensive research toward the development of in situ remediation with enzyme preparations often stabilized on solid phases (Gianfreda and Bollag, 2002).

The preferred method has been to trap the enzyme on a solid support using various chemical and physical immobilization techniques such as entrapment, encapsulation, covalent bonding, adsorption, and cross-linking (Gianfreda and Bollag, 2002). Both new and established methods of enzyme immobilization and entrapment that offer protection against proteolysis and denaturation have been evaluated for their capacity to remediate polluted soils (Husain et al., 2009). Enzymes entrapped in a support will depend on the diffusion of an organic substrate into the matrix before degradation can occur and thus could be used only if the pollutant is soluble. On the other hand, slow release or decaying formulations would release the enzymes into the soil and allow them to interact with both soluble and insoluble compounds. But, as before, their survival is a problem and, added to this, is the well-discussed topic of pollutant bioavailability (Semple et al., 2007), which will have a big influence on rates and total decay of the target compounds.

The economics and efficacy of this type of bioremediation when applied on a field scale are problematic. Inevitably, production costs are high although selection and transgenics can achieve the overexpression of enzymes and thus the yields before immobilization and application can be much improved (Ahuja et al., 2004; Wackett, 2004; Copley, 2009). Another positive aspect is that the immobilized enzyme is stable, has a long shelf life, and may even be re-used (especially if applied to ex situ clean up).

Enzyme reactions can have a major impact in remediating toxic levels of metals and metalloids, including selenium, arsenic, chromium, and mercury. Frankenberger and his coworkers (Thompson-Eagle and Frankenberger, 1992; Frankenberger and Karlson, 1994a,b; Frankenberger et al., 2004) have studied volatilization, methylation, oxidation, and reduction of these elements in terms of their toxicity, mobility, and transformations into different species. Bioremediation of selenium was demonstrated in which selenate and selenite reductases and a methyltransferase converted the aqueous selenium spp. into a methylated nontoxic product, dimethylselenide, which has a high vapor pressure. This transformation was demonstrated in the field and optimized to remove inorganic selenium from seleniferous dewatered sediments. Other studies have characterized the enzymes involved in methylated arsenic (Frankenberger, 2001) and mercury (Dungan and Frankenberger, 2002) and the biotransformation of hexavalent chromium to the trivalent state (Losi et al., 1994).

1–6.3 Soil Enzymes and Plant Growth Responses

A more recent development uses soil enzymes to produce biologically active compounds in soil for agricultural applications. One approach is to add substrates that, on hydrolysis, produce plant hormones to stimulate plant growth or allelochemicals that could benefit crops by suppressing disease organisms or weeds.

Plant hormones such as ethylene, auxins, cytokinins, and gibberellins, which influence plant growth and development, can be produced by microorganisms in soil, especially within the plant root rhizosphere (Frankenberger and Arshad, 1995). Frankenberger and his coworkers (Frankenberger and Arshad, 1995; Arshad and Frankenberger, 1997, 2002) reported the success of adding precursors of phytohormones catalyzed by specific enzymes of an inoculum to the rhizosphere. With autoradiography, Martens and Frankenberger (1994) demonstrated that plants do take up exogenous sources of phytohormones within the root zone. This novel approach and its economic and agricultural advances have been demonstrated with auxins, gibberellins, and ethylene. By using specific precursors, the enzymatic synthesis of a particular phytohormone can be promoted and controlled within the rhizosphere.

Phytotoxins or allelochemicals may be of either microbial or plant origin which can be produced with the addition of precursors that are transformed to form these biologically active compounds by soil enzyme activity. One such example is myrosinase (thioglucoside glucohydrolase, EC 3.2.3.1) that breaks the β-thioglucoside bond of glucosinolate molecules, producing glucose, sulfate, and a diverse group of products. Further nonenzymatic, intramolecular rearrangement of these diverse products releases isothiocyanates, thiocyanates or nitriles, which are allelochemicals. Glucosinolates have been found to inhibit weed-seed germination and some pathogens in soil (Angus et al., 1994; Brabban and Edwards, 1995; Brown and Morra, 1997).

1–7 SUMMARY

Soil enzymes are central to ecosystem processes because they mediate innumerable reactions that have biogeochemical significance in soils. Catalase was the first enzyme detected in soils in 1896 and for the next 50 years or so, most soil enzymology research was focused on developing assays and increasing the number of enzymes measured—especially carbohydrolases, proteases, phosphatases, and urease. From the 1970s a large number of assays were developed, most notably on N cycling enzymes and, in comparatively recent times, more attention has been drawn toward the all-important laccases and phenol oxidases (Sinsabaugh, 2010).

There are now over 100 enzymes that have been assayed, and at least partially characterized, in soils. A relatively small amount of any given enzyme can be extracted directly from soil and subsequently purified, therefore individual enzymes are studied mainly by measuring activity in situ and selectively inhibiting or destroying enzymes associated with specific fractions within the matrix.

The early soil enzymology research achieved this by using antiseptics to separate activity arising from extant viable cells from that of previously produced extracellular enzymes. This work had limited success because of inconsistent results and the difficulty in interpreting outcomes. A major breakthrough came in the late 1950s and early 1960s when ionizing radiation was used to sterilize soil. This showed that for many enzymes, a significant amount of activity was due to the stabilized extracellular fraction. Even after killing all viable cells, soil enzymes remained active at high levels for many enzymes. From this work it was established that, with the exception of dehydrogenase and possibly a few other enzymes, nearly all soil extracellular enzymes exist in association with both viable cells as well as away from the cell, complexed in various ways within the soil matrix.

The multiple locations of extracellular enzymes and how these might impact the ecology of microbial communities were not well understood (or even much thought about) until the 1970s (Burns, 1978) but have attracted a great deal of attention in recent times (see Chapter 2, Wallenstein and Burns, 2011, this volume; Allison, 2005, 2006). The idea that enzyme activity could be an index (or at least a rough guide) of soil quality as well as resilience was an exciting prospect and has been actively pursued. However, there were inconsistent results and typically weak or no correlations with crop yields. Similarly, there has been long-standing interest in using soil enzyme activities as indexes of microbial community process capability and diversity, but again with mixed results. Although the explanations for these outcomes are complex, a major reason has to be that the "background" stabilized extracellular enzyme fractions make a significant contribution to the total activity measurements and respond to external stresses in different ways than the extant microbial population.

In recent years, the emphasis has been toward using enzyme activities as indicators not of crop productivity but rather for detecting the longer-term negative or positive effects of land management. Soil enzyme assays have emerged as technological tools for various applications in environmental and ecosystems management. Several enzymes have shown sensitivity in reflecting early changes in soil quality due to soil management long before there are measurable changes in total organic C levels. This holds potential to assist land managers in managing ecosystems for long-term sustainability. Enzyme assays can detect the level of degradation and recovery of soils in highly disturbed landscapes such as reclaimed strip mine landscapes. The bioavailability of certain heavy metals in soils can be gauged by enzyme activity measurements. In contrast, stabilized enzymes incorporated into soils have been shown to degrade certain contaminants and thus provide an avenue to remediate contaminated soils in situ.

In conclusion, in the last decade, advances in molecular biology, microscopy, and analytical techniques have begun to provide new insights into the functions and locations of extracellular enzymes. Another research driver is the need to understand the details of how extracellular enzymes function in a large number of industrial, medical, and environmental processes. For example, those concerned with composting, wastewater and sludge treatment, and the conversion of plant materials, including wood and straw residues, to fermentable sugars for bioethanol production need to define the functions of extracellular enzymes and seek ways to improve their activities. The paper and pulp industry invests resources into the study of extracellular-enzyme-producing microorganisms as does the food industry, which must control post-harvest spoilage and manage the polysaccharidic wastes arising from many processes. The invasive, destructive, and economically damaging activities of phytopathogens and the complex enzymology of ruminant digestion all require a detailed knowledge of organic polymer solubilization and mineralization. The list is endless.

In recent times, concerns regarding the consequences of climate change on soil processes have stimulated experimental research, models, and theories on soil organic matter formation and its decomposition (Kirschbaum, 2004; Eliasson et al., 2005; Fang et al., 2005; Jones et al.,2005; Knorr et al., 2005; Bradford et al., 2008). The recalcitrant polyphenolic and polysaccharidic soil organic matter fraction of soil is a long-term repository for sequestered carbon and is essential for aggregate structure, soil stability, plant nutrient and water retention, microbial diversity and activities, and a host of properties that contribute to soil fertility and plant productivity.

Increasing soil (and water) temperatures, elevated atmospheric carbon dioxide and more frequent wetting and drying cycles will change microbial numbers and community composition and accelerate growth and enzyme activities either directly or following their impact on plants and increases in rhizodeposits. Therefore, those attempting to assess the outcomes of climate change by generating predictive carbon cycle models (Luo, 2007) must take into account any predicted increases in soil enzyme activities and the associated decline in the hitherto recalcitrant soil organic matter (Davidson and Janssens, 2006; Kuzyakov, 2010).

The importance of soil enzyme research in helping to understand the consequences of climate change on soil enzyme activities may be the most important challenge that soil biologists ever have faced.

REFERENCES

Adriano, D.C., J.-M. Bollag, W.T. Frankenberger, and R.C. Sims (ed.). 1999. Bioremediation of contaminated soils. Agron. Monogr. 37. ASA, CSSA, and SSSA, Madison WI.

Ahuja, S.K., G.M. Ferreira, and A.R.I. Moreira. 2004. Utilization of enzymes for environmental applications. Crit. Rev. Biotechnol. 24:125–154. doi:10.1080/07388550490493726

Alef, K., and D. Kleiner. 1986. Arginine ammonification, a simple method to estimate the microbial activity potential in soils. Soil Biol. Biochem. 18:233–235. doi:10.1016/0038-0717(86)90033-7

Alexander, M. 1961 Introduction to soil microbiology. John Wiley & Sons, New York.

Al-Khafaji, A.A., and M.A. Tabatabai. 1979. Effects of trace elements on arylsulfatase activity in soils. Soil Sci. 127:129–133. doi:10.1097/00010694-197903000-00001

Allison, S.D. 2005. Cheaters, diffusion and nutrients constrain decomposition by microbial enzymes in spatially structured environments. Ecol. Lett. 8:626–635. doi:10.1111/j.1461-0248.2005.00756.x

Allison, S.D. 2006. Soil minerals and humic acids alter enzyme stability: Implications for ecosystem processes. Biogeochemistry 81:361–373. doi:10.1007/s10533-006-9046-2

Allison, S.D., and P.M. Vitousek. 2004. Extracellular enzyme activities and carbon chemistry as drivers of tropical plant litter decomposition. Biotropica 36:285–296.

Angus, J.F., P.A. Grander, J.A. Kirkegaard, and J.M. Desmarchelier. 1994. Biofumigation: Isothiocyanates released from Brassica roots inhibit growth of the take-all fungus. Plant Soil 162:107–112. doi:10.1007/BF01416095

Arshad, M., and W.T. Frankenberger, Jr. 1997. Plant growth regulating substances in the rhizosphere: Microbial production and function. Adv. Agron. 62:45–151. doi:10.1016/S0065-2113(08)60567-2

Arshad, M., and W.T. Frankenberger, Jr. 2002. Ethylene: Agricultural sources and applications. p. 342. Kluwer Academic/Plenum Publishers, New York.

Asgher, M., H.N. Bhatti, M. Ashraf, and R.L. Legge. 2008. Recent developments in biodegradation of industrial pollutants by white rot fungi and their enzyme system. Biodegradation 19:771–783. doi:10.1007/s10532-008-9185-3

Asmar, F., F. Eiland, and N.E. Nielsen. 1992. Interrelationship between extracellular enzyme activity, ATP content, total counts of bacteria and CO_2 evolution. Soil Biol. Fertil. 14:288–292. doi:10.1007/BF00395465

Baeyens, J., and J. Livens. 1936. Catalytic power of a soil and fertility. Agricoltura 30:145–155.

Bahri, A., and R. Berndtsson. 1996. Nitrogen source impact on the spatial variability of organic carbon and nitrogen in soil. Soil Sci. 161:288–297. doi:10.1097/00010694-199605000-00004

Baldrian, P. 2008. Wood-inhabiting ligninolytic basidiomycetes in soils: Ecology and constraints for applicability in bioremediation. Fungal Ecol. 1:4–12. doi:10.1016/j.funeco.2008.02.001

Balks, R. 1925. Research on the formation and degradation of humus in soil. Landw. Versuchs-Stationen 103:221–258.

Bandick, A.K., and R.P. Dick. 1999. Field management effects on soil enzyme activities. Soil Biol. Biochem. 31:1471–1479. doi:10.1016/S0038-0717(99)00051-6

Bardgett, R.D., T.W. Speir, D.J. Ross, G.W. Yeates, and H.A. Kettles. 1994. Impact of pasture contamination by copper, chromium, and arsenic timber preservative on soil microbial properties and nematodes. Biol. Fertil. Soils 18:71–79. doi:10.1007/BF00336448

Bastida, F., A. Zsolnay, T. Hernandez, and C. Garcia. 2008. Past, present and future of soil quality indices: A biological perspective. Geoderma 147:159–171. doi:10.1016/j.geoderma.2008.08.007

Beck, T., and H. Poschenrieder. 1963. Experiments on the effect of toluene on the soil microflora. Plant Soil 18:346–357. doi:10.1007/BF01347234

Benefield, C.B. 1971. A rapid method for measuring cellulose activity in soils. Soil Biol. Biochem. 3:325–329. doi:10.1016/0038-0717(71)90042-3

Benitez, E., R. Nogales, M. Campos, and F. Ruano. 2006. Biochemical variability of olive-orchard soils under different management systems. Appl. Soil Ecol. 32:221–231. doi:10.1016/j.apsoil.2005.06.002

Bergstrom, D.W., and C.M. Monreal. 1998. Increased soil enzyme activities under two row crops. Soil Sci Soc. Am. J. 62:1295–1301.

Bertrand, D., and A. de Wolf. 1968. Effect of microelements applied as complementary fertilizers on the soil microflora. C. R. Acad. Agric. Fr. 54:1130–1133.

Bonmati M.C, B. Ceccanti, and P. Nannipieri. 1991. Spatial variability of phosphatase, urease, protease, organic carbon and total nitrogen in soil. Soil Biol. Biochem. 23:391–396. doi:10.1016/0038-0717(91)90196-Q

Boyd, S.A., and M.M. Mortland. 1990. Enzymes interactions with clays and clay-organic matter complexes. p. 1–28. In J.M. Bollag and G. Stotzky (ed.) Soil biochemistry. Vol. 6. Marcel Dekker, New York.

Brabban, A.D., and C. Edwards. 1995. The effects of glucosinolates and their hydrolysis products on microbial growth. J. Appl. Bacteriol. 79:171–177.

Bradford, M.A., C.A. Davies, S.D. Frey, T.R. Maddox, J.M. Melillo, J.E. Mohan, J.F. Reynolds, K.K. Treseder, and M.D. Wallenstein. 2008. Thermal adaptation of soil microbial respiration to elevated temperature. Ecol. Lett. 11:1316–1327. doi:10.1111/j.1461-0248.2008.01251.x

Brown, P.D., and M.J. Morra. 1997. Control of soil-borne plant pests using glucosinolate-containing plants. Adv. Agron. 61:167–231. doi:10.1016/S0065-2113(08)60664-1

Buller, G. 1999. Effects of cover crops on soil physical properties. MS thesis. Oregon State Univ., Corvallis, OR.

Burns, R.G. 1978. Enzymes in soil: Some theoretical and practical considerations. p. 295–339. In R.G. Burns (ed.) Soil enzymes. Academic Press, New York.

Burns, R.G. 1982a. Carbon mineralization by mixed cultures. p. 475–543. In A.T. Bull and J.H. Slater (ed.) Microbial interactions and communities. Academic Press, London.

Burns, R.G. 1982b. Enzyme activity in soil: Location and a possible role in microbial activity. Soil Biol. Biochem. 14:423–427. doi:10.1016/0038-0717(82)90099-2

Burns, R.G. 1983. Extracellular enzyme-substrate interactions in soil. p. 249–298. *In* J.H. Slater, R. Whittenbury, and J.W.T. Wimpenny (ed.) Microbes in Their Natural Environment. 34th Symposium of the Society for General Microbiology. Cambridge Univ. Press.

Burns, R.G., A.H. Puķīte, A.D. Mc, and A.D. Laren. 1972b. Concerning the location and persistence of soil urease. Soil Sci. Soc. Am. Proc. 36:308–311. doi:10.2136/sssaj1972.03615995003600020030x

Burns, R.G., M.H. El-Sayed, and A.D. McLaren. 1972a. Extraction of an urease-active organocomplex from soil. Soil Biol. Biochem. 4:107–108. doi:10.1016/0038-0717(72)90048-X

Busto, M.D., and M. Perez-Mateos. 1995. Extraction of humic-β-glucosidase fractions from soil. Biol. Fertil. Soils 20:77–82. doi:10.1007/BF00307845

Busto, M.D., and M. Perez-Mateos. 2000. Characterization of β-D-glucosidase extracted from soil fractions. Eur. J. Soil Sci. 51:193–200. doi:10.1046/j.1365-2389.2000.00309.x

Cameron, F.K., and J.M. Bell. 1905. The mineral constituents of the soil solution. Bull. No. 30. USDA Bur. of Soils, Washington, DC.

Chalvignac, M.A., and J. Mayaudon. 1971. Extraction and study of soil enzymes metabolizing tryptophan. Plant Soil 34:25–31. doi:10.1007/BF01372757

Chander, K., and P.C. Brookes. 1991. Is the dehydrogenase assay invalid as a method to estimate microbial activity in copper-contaminated soils. Soil Biol. Biochem. 23:909–915. doi:10.1016/0038-0717(91)90170-O

Chouchack, D. 1924. L'analyse du sol par les bactéries. C.R. Acad. Sci. 178:2001–2002.

Claus, D., and K. Mechsner. 1960. The usefulness of E. Hofmann's methods for the determination of enzymes in soil. Plant Soil 12:195–198. doi:10.1007/BF01377370

Conn, H.W. 1901. Agricultural bacteriology; A study of the relation of bacteria to agriculture, with special reference to the bacteria in the soil, in water, in the dairy, in miscellaneous farm products, and in plants and domestic animals. Blakiston's Son and Co. Philadelphia, PA.

Conrad, J.P. 1940a. Hydrolysis of urea in soils by thermolabile catalysis. Soil Sci. 49:253–263. doi:10.1097/00010694-194004000-00002

Conrad, J.P. 1940b. The nature of the catalyst causing the hydrolysis of urea in soils. Soil Sci. 50:119–134. doi:10.1097/00010694-194008000-00005

Conrad, J.P. 1942a. The occurrence and origin of urease-like activities in soils. Soil Sci. 54:367–380. doi:10.1097/00010694-194211000-00012

Conrad, J.P. 1942b. Enzymatic versus microbial concepts of urea hydrolysis in soils. J. Am. Soc. Agron. 34:1102–1113.

Conrad, J.P., and C.N. Adams. 1940. Retention by soils of the nitrogen of urea and some related phenomena. J. Am. Soc. Agron. 32:48–54.

Cooper, A.B., and H.W. Morgan. 1981. Improved fluorometric method to assay for soil lipase activity. Soil Biol. Biochem. 13:307–312. doi:10.1016/0038-0717(81)90067-5

Copley, S.D. 2009. Evolution of efficient pathways for degradation of anthropogenic chemicals. Nat. Chem. Biol. 5:559–567. doi:10.1038/nchembio.197

Daniel, G., J. Volc, L. Filonova, O. Plihal, E. Kubatova, and P. Halada. 2007. Characteristics of *Gloeophyllum trabeum* alcohol oxidase, an extracellular source of H_2O_2 in brown rot decay of wood. Appl. Environ. Microbiol. 73:6241–6253. doi:10.1128/AEM.00977-07

Daragan-Sushcheva, A.Yu., and R.S. Katsnelson. 1963. Effect of meadow grasses on enzyme activity of soils. Trudy Bot. Inst. Akad. Nauk SSSR 14:160–171.

Darrah, P.R., and P.J. Harris. 1986. A fluorometric method for measuring the activity of soil enzymes. Plant Soil 92:81–88. doi:10.1007/BF02372269

Davidson, E.A., and I.A. Janssens. 2006. Temperature sensitivity of soil carbon decomposition and feedbacks to climate change. Nature 440:165–173. doi:10.1038/nature04514

Dec, J., and J.M. Bollag. 2000. Phenoloxidase-mediated interactions of phenols and anilines with humic materials. J. Environ. Qual. 29:665–676. doi:10.2134/jeq2000.00472425002900030001x

Decker, K.L.M., R.E.J. Boerner, and S.J. Morris. 1999. Scale-dependent patterns of soil enzyme activity in a forest landscape. Can. J. For. Res. 29:232–241. doi:10.1139/cjfr-29-2-232

Deng, S., H. Kang, and C. Freeman. 2011. Microplate fluorimetric assay of soil enzymes. p. 311–318. *In* R.P. Dick (ed.) Methods of soil enzymology. SSSA Book Ser. 9. SSSA, Madison, WI. (This volume.)

Deng, S.P., and M.A. Tabatabai. 1994. Cellulase activity of soils. Soil Biol. Biochem. 26:1347–1354. doi:10.1016/0038-0717(94)90216-X

Dick, R.P. 1992. A review: Long-term effects of agricultural systems on soil biochemical and microbial parameters. Agric. Ecosyst. Environ. 40:25–36. doi:10.1016/0167-8809(92)90081-L

Dick, R.P. 1994. Soil enzyme activities as indicators of soil quality. p. 107–124. In J.W. Doran, D.C. Coleman, D.F. Bezdicek, and B.A. Stewart (ed.) Defining soil quality for a sustainable environment. SSSA Spec. Pub. SSSA, Madison, WI.

Dick, R.P. 1997. Enzyme activities as integrative indicators of soil health. p. 121–156. In C.E. Pankhurst, B. Doube and V. Gupta (ed.) Bioindicators of soil health. CAB International, Oxon, UK.

Dick, R.P., and M.A. Tabatabai. 1986. Hydrolysis of polyphosphates by corn roots. Plant Soil 94:247–256. doi:10.1007/BF02374348

Dick, R.P., and M.A. Tabatabai. 1987. Polyphosphates as sources of phosphorus for plants. Fert. Res. 12:107–118. doi:10.1007/BF01048912

Dick, W.A. 1984. Influence of long-term tillage and crop rotation combinations on soil enzyme activities. Soil Sci. Soc. Am. J. 48:569–574. doi:10.2136/sssaj1984.03615995004800030020x

Dick, W.A., N.G. Juma, and M.A. Tabatabai. 1983. Effects of soils on acid phosphatase and inorganic pyrophosphatase of corn roots. Soil Sci. 136:19–25. doi:10.1097/00010694-198307000-00003

Di Nardo, C., A. Cinquegrana, S. Papa, A. Fuggi, and A. Fioretto. 2004. Laccase and peroxidase isoenzymes during leaf litter decomposition of Quercus ilex in a Mediterranean ecosystem. Soil Biol. Biochem. 36:1539–1544. doi:10.1016/j.soilbio.2004.07.013

Dommergues, Y. 1960. Effect of infra-red and solar radiations on the inorganic nitrogen content and on some biological characteristics of soils. Agron. Trop. (Nogent-sur-Marne) 15:381–389.

Drobník, J. 1955. The hydrolysis of starch by the enzymatic complex of soils. Folia Biol. 1:29–40.

Drobník, J. 1957. Biological transformations of organic substances in the soil. Pochvovedenie 12:62–71.

Drobník, J., and J. Seifert. 1955. The relationship of enzymatic inversion in soil to some soil-microbiological tests. Folia Biol. 1:41–47.

Dungan, R.S., and W.T. Frankenberger, Jr. 2002. Enzyme-mediated transformations of heavy metals/metalloids. p. 539–565. In R.G. Burns and R.P. Dick (ed.) Enzymes in the environment. Marcel Dekker, New York.

Dunn, C.G., W.L. Campbell, H. Fram, and A. Hutchins. 1948. Biological and photochemical effects of high energy, electrostatically produced Roentgen rays and cathode rays. J. Appl. Phys. 19:605–616. doi:10.1063/1.1698179

Dutzler-Franz, G. 1977. Enzyme activities of different soil types under the influence of some chemical and physical soil properties. J. Plant Nutr. Soil Sci. 140:329–350. doi:10.1002/jpln.19771400307

Eliasson, P.E., R.E. McMurtrie, D.A. Pepper, M. Stromgren, S. Linder, and G.I. Agren. 2005. The response of heterotrophic CO_2 flux to soil warming. Glob. Change Biol. 11:167–181. doi:10.1111/j.1365-2486.2004.00878.x

Fang, C.M., P. Smith, J.B. Moncrieff, and J.U. Smith. 2005. Similar response of labile and resistant soil organic matter pools to changes in temperature. Nature 436:881. doi:10.1038/nature04044

Fermi, C. 1910. Sur la presence des enzymes dans le sol, dans le eaux et dans les poussieres. Zentralbl. Bakteriol. Parasitenkd. Abt. II. 26:330–335.

Fioretto, A., S. Papa, E. Curcio, G. Sorrentino, and A. Fuggi. 2000. Enzyme dynamics on decomposing leaf litter of Cistus icanus and Myrtus communis in a Mediterranean ecosystem. Soil Biol. Biochem. 32:1847–1855. doi:10.1016/S0038-0717(00)00158-9

Floch, C., E. Alarcon-Gutiérrez, and S. Criquet. 2007. ABTS assay of phenol oxidase activity in soil. J. Microbiol. Methods 71:319–324. doi:10.1016/j.mimet.2007.09.020

Foght, J. 2008. Anaerobic biodegradation of aromatic hydrocarbons: Pathways and prospects. J. Mol. Microbiol. Biotechnol. 15:93–120. doi:10.1159/000121324

Ford, C.I., M. Walter, G.L. Northott, H.J. Di, K.C. Cameron, and T. Trower. 2007. Fungal inoculum properties: Extracellular enzyme expression and pentachlorophenol removal in highly contaminated field soils. J. Environ. Qual. 36:1599–1608. doi:10.2134/jeq2007.0149

Fornasier, F., Y. Dudal, and H. Quiquampoix. 2011. Enzyme extraction from soil. p. 371–384. In R.P. Dick (ed.) Methods of soil enzymology. SSSA Book Ser. 9. SSSA, Madison, WI. (This volume.)

Fornasier, F., and A. Margon. 2007. Bovine serum albumin and Triton X-100 greatly increase phosphomonoesterases and arylsulphatase extraction yield from soil. Soil Biol. Biochem. 39:2682–2684. doi:10.1016/j.soilbio.2007.04.024

Foster, R.C., and J.K. Martin. 1981. In situ analysis of soil components of biological origin. Soil Biochem. 5:75–110.

Frankenberger, W.T., Jr. 2001. Environmental chemistry of arsenic. p. 391. Marcel Dekker, New York.

Frankenberger, W.T., Jr., C. Amrhein, T.W.M. Fan, D. Flaschi, J. Glater, E. Kartinen, Jr., K. Kovac, E. Lee, H.M. Ohlendorg, L. Owens, N. Terry, and A. Toto. 2004. Advanced treatment technologies in the remediation of seleniferous drainage waters and sediments. Irrig. Drain. Syst. 18:19–41. doi:10.1023/B:IRRI.0000019422.68706.59

Frankenberger, W.T., Jr., and M. Arshad. 1995. Phytohormones in soils: Microbial production and function. p. 503. Marcel Dekker, New York.

Frankenberger, W.T., Jr., and W.A. Dick. 1983. Relationships between enzyme activities and microbial growth and activity indices in soil. Soil Sci. Soc. Am. J. 47:945–951. doi:10.2136/sssaj1983.03615995004700050021x

Frankenberger, W.T., Jr., and U. Karlson. 1994a. Microbial volatilization of selenium from soils and sediments. p. 369–387. In W.T. Frankenberger, Jr. and S. Benson (ed.) Selenium in the environment. Marcel Dekker, New York.

Frankenberger, W.T., Jr., and U. Karlson. 1994b. Campaigning for bioremediation. CHEMTECH 24:45–51.

Galetti, A.C. 1932. A rapid and practical test for the oxidizing power of soil. Ann. Chim. Appl. 22:81–83.

Galstyan, A.Sh. 1959. Some questions of the study of soil enzymes. Soobshch. Lab. Agrokhim. Akad. Nauk Arm. SSR 2:19–25.

Galstyan, A.Sh. 1965. Method of determining activity of hydrolytic enzymes of soil. Pochvovedenie 2:68–74.

Gerretsen, F.C. 1915. Het oxydeerend vermogen van dem bodem in verband met het uitzuren. Meddl. Proefsta. Java Suikerind. 5:317–331.

Gettkandt, G. 1956. Colorimetric determination of glucose in soil solutions and its application to Hofmann's enzyme method. Landw. Forsch. 9:155–158.

Gianfreda, L., and J.M. Bollag. 2002. Isolated enzymes for the transformation and detoxification of organic pollutants. p. 495–538. In R.G. Burns and R.P. Dick (ed.) Enzymes in the environment: Activity, ecology, and applications. Marcel Dekker, New York.

Gianfreda, L., A. DeCristofaro, M.A. Rao, and A. Violante. 1995. Kinetic behavior of synthetic organo- and organo-mineral-urease complexes. Soil Sci. Soc. Am. J. 59:811–815. doi:10.2136/sssaj1995.03615995005900030025x

Gianfreda, L., and M.A. Rao. 2008. Interactions between xenobiotics and microbial and enzymatic soil activity. Crit. Rev. Environ. Sci. Technol. 38:269–310. doi:10.1080/10643380701413526

Gianfreda, L., and P. Ruggiero. 2006. Enzyme activities in soil. p. 257–311. In P. Nannipieri and K. Small (ed.) Nucleic acids and proteins in soil. Springer, Berlin.

Gil-Sotres, F., C. Trasar-Cepeda, M.C. Leiros, and S. Seoane. 2005. Different approaches to evaluating soil quality using biochemical properties. Soil Biol. Biochem. 37:877–887. doi:10.1016/j.soilbio.2004.10.003

Goovaerts, P. 1998. Geostatistical tools for characterizing the spatial variability of microbiological and physico-chemical soil properties. Biol. Fertil. Soils 27:315–334. doi:10.1007/s003740050439

Gramss, G., B. Kirsche, K.D. Voigt, T. Gunther, and W. Fritsche. 1999. Conversion rates of five polycyclic aromatic hydrocarbons in liquid cultures of fifty-eight fungi and the concomitant production of oxidative enzymes. Mycol. Res. 103:1009–1018. doi:10.1017/S0953756298008144

Grego, S., A. D'Annibale, M. Luna, L. Badalucco, and P. Nannipieri. 1990. Multiple forms of synthetic pronase-phenolic copolymers. Soil Biol. Biochem. 22:721–724. doi:10.1016/0038-0717(90)90021-Q

Haig, A.D. 1955. Some characteristics of esterase- and urease-like activity in the soil. Ph.D. diss. Univ. of California, Davis.

Hayano, K., and A. Katami. 1977. Extraction of β-glucosidase activity from pea field soil. Soil Biol. Biochem. 9:349–351. doi:10.1016/0038-0717(77)90008-6

He, Z., O. Tsutom, D.C. Olk, and W. Fengchang. 2010. Geoderma 156:143–151. doi:10.1016/j.geoderma.2010.02.011

Hofmann, E. 1963. The origin and importance of enzymes in soil. Rec. Progr. Microbiol. 8:216–220.

Hofmann, E., and K. Bräunlich. 1955. The saccharase content of soils as affected by various factors of soil fertility. Z. Pflanzenernaehr. Dueng. Bodenkd. 70:114–123.

Hofmann, E., and G. Hoffmann. 1953. α- and β-glucosidases in the soil. Naturwissenschaften 40:511. doi:10.1007/BF00629068

Hofmann, E., and G. Hoffmann. 1954. Enzyme system of cultivated soil. V. α- and β-galactosidase and α-glucosidase. Biochem. Z. 325:329–332.

Hofmann, E., and G. Hoffmann. 1955a. The origin, determination and significance of enzymes in soil. Z. Pflanzenernaehr. Dueng. Bodenkd. 70:9–16.

Hofmann, E., and G. Hoffmann. 1955b. The enzyme system of cultivated soil. VI. Amylase. Z. Pflanzenernaehr. Dueng. Bodenkd. 70:97–104.

Hofmann, E., and G. Hoffmann. 1961. The reliability of E. Hofmann's methods for determining enzyme activity in soil. Plant Soil 14:96–99. doi:10.1007/BF01343774

Hofmann, E., and G. Hoffmann. 1966. Determination of, soil biological activity with enzymatic methods. Adv. Enzymol. 28:365–390.

Hofmann, E., and A. Seegerer. 1950. Soil enzymes as measure of biological activity. Biochem. Z. 321:97–99.

Hofmann, E., and A. Seegerer. 1951. Enzyme systems of our cultivated soil. I. Saccharase. Biochem. Z. 322:174–179.

Hoffmann, G. 1959a. Distribution and origin of some enzymes in soil. Z. Pflanzenernaehr. Dueng. Bodenkd. 85:97–104.

Hoffmann, G. 1959b. Investigations on the synthetic effects of enzymes in soil. Z. Pflanzenernaehr. Dueng. Bodenkd. 85:193–201.

Hope, C.F.A., and R.G. Burns. 1987. Activity, origins and location of cellulases in a silt loam soil. Biol. Fertil. Soils 5:164–170. doi:10.1007/BF00257653

Howard, P.J.A. 1972. Problems in the estimation of biological activity in soil. Oikos 23:235–240. doi:10.2307/3543411

Huang, P.M. 1990. Role of soil minerals in transformations of natural organics and xenobiotics in soil. p. 29–116. In J.M. Bollag and G. Stotsky (ed.) Soil biochemistry. Vol. 6. Marcel Dekker, New York.

Husain, Q., M. Husain, and Y. Kulshrestha. 2009. Remediation and treatment of organopollutants mediated by peroxidases: A review. Crit. Rev. Biotechnol. 29:94–119. doi:10.1080/07388550802685306

Jackman, R.H. 1951. Phosphorus status of some pumice soils. p. 176–183. In Proc. New Zealand Grassland Assoc. 13th Ann. Conf.

Jackman, R.H., and C.A. Black. 1951. Hydrolysis of iron, aluminium, calcium, and magnesium inositol phosphates by phytase at different pH values. Soil Sci. 72:261–266. doi:10.1097/00010694-195110000-00002

Jackman, R.H., and C.A. Black. 1952a. Phytase activity in soils. Soil Sci. 73:117–125. doi:10.1097/00010694-195202000-00004

Jackman, R.H., and C.A. Black. 1952b. Hydrolysis of phytate phosphorus in soils. Soil Sci. 73:167–171. doi:10.1097/00010694-195203000-00001

Jain, R.K., M. Kapur, S. Labana, B. Lal, P.M. Sarma, D. Bhattacharya, and I.S. Thakur. 2005. Microbial diversity: Application of microorganisms for the biodegradation of xenobiotics. Curr. Sci. 89:101–112.

Jones, C., C. McConnell, K. Coleman, P. Cox, P. Falloon, D. Jenkinson, and D. Powlson. 2005. Global climate change and soil carbon stocks; predictions from two contrasting models for the turnover of organic carbon in soil. Glob. Change Biol. 11:154–166. doi:10.1111/j.1365-2486.2004.00885.x

Juma, N.G., and M.A. Tabatabai. 1978. Distribution of phosphomonoesterases in soils. Soil Sci. 126:101–108. doi:10.1097/00010694-197808000-00006

Kandeler, E., and R.P. Dick. 2007. Soil enzymes: Spatial distribution and function in agroecosystems. p. 263–287. In G. Benckiser and S. Schnell (ed.) Biodiversity in agricultural production systems. Taylor and Francis, New York.

Kandeler, E., J. Luxh, D. Tscherko, and J. Magid. 1999a. Xylanase, invertase and protease activities at the soil-litter interface of a sandy loam. Soil Biol. Biochem. 31:1171–1179. doi:10.1016/S0038-0717(99)00035-8

Kandeler, E., S. Palli, M. Stemmer, and M.H. Gerzabek. 1999b. Tillage changes microbial biomass and enzyme activities in particle-size fractions of a Haplic Chernozem. Soil Biol. Biochem. 31:1253–1264. doi:10.1016/S0038-0717(99)00041-3

Kandeler, E., M. Stemmer, and E.M. Klimanek. 1999c. Response of soil microbial biomass, urease and xylanase within particle size fractions to long-term soil management. Soil Biol. Biochem. 31:261–273. doi:10.1016/S0038-0717(98)00115-1

Kandeler, E., D. Tscherko, K.D. Bruce, M. Stemmer, P.J. Hobbs, R.D. Bardgett, and W. Amelung. 2000. The structure and function of the soil microbial community in microhabitats of a heavy metal polluted soil. Biol. Fertil. Soils 32:390–400. doi:10.1007/s003740000268

Kandeler, E., D. Tscherko, M. Stemmer, S. Schwarz, and M.H. Gerzabek. 2001. Organic matter and soil microorganisms—Investigations from the micro- to the macro-scale. Bodenkultur 52:117–131.

Kappen, H. 1913. Die katalytische Kraft des Ackerbodens. Fühlings Landw. Ztg. 62:377–392.

Kirschbaum, M.U.F. 2004. Soil respiration under prolonged soil warming: Are rate reductions caused by acclimation or substrate loss? Glob. Change Biol. 10:1870–1877. doi:10.1111/j.1365-2486.2004.00852.x

Kiss, S. 1958a. Soil enzymes. p. 495–622. In M.J. Csapo (ed.) Talajtan. Agro-Silvica. Bucharest, Romania.

Kiss, S. 1958b. New data regarding the identity of soil saccharase and soil α-glucosidase (maltase). Stud. Univ. Bahej-Bolyai Sir. Biol. 3:51–55.

Kiss, S. 1958c. Experiments on the production of saccharase in soil. Z. Pflanzenernaehr. Dueng. Bodenkd. 81:117–125. doi:10.1002/jpln.19580810204

Kiss, S. 1961. Presence of levan sucrase in soils. Naturwissenschaften 48:700. doi:10.1007/BF00595949

Kiss, S., and M. Boaru. 1965. Some methodological problems of soil enzymology. p. 115–127. In Symp. On Methods in Soil Biology. Bucharest, Romania.

Kiss, S., I. Bosica, and M. Pop. 1962a. Effectiveness of toluene as an antiseptic agent in the determination of the activity of enzymes in the soil. Stud. Univ. Babel-Bolyai, Sir. Biol. 2:65–70.

Kiss, S., I. Bosica, and M. Pop. 1962b. Enzymic degradation of lichenin in soil. Contrib. Bot. Cluj. 335–340.

Kiss, S., and M. Drøgan-Bularda. 1968. Levan sucrase activity in soil under conditions unfavorable for the growth of microorganisms. Rev. Roum. Biol. Sér. Bot. 13:435–438.

Kiss, S., M. Drøgan-Bularda, and F.H. Khaziev. 1972. The influence of chloromycetin on the activity of some oligases in soil. p. 451–462. In Proc. Lucr. Conf. Nat. Ştiinţa Solului.

Kiss, S., M. Drøgan-Bularda, and D. Pasca. 1993. Enzymology of technogenic soils. Casa Cartii de Stiinta, Cluj, Romania.

Kiss, S., M. Drøgan-Bularda, and D. Radulescu. 1971. Biological significance of enzymes accumulated in soil. Contrib. Bot. Cluj. 377–397.

Kiss, S., M. Drøgan-Bularda, and D. Rødulescu. 1975a. Biological significance of enzymes in soil. Adv. Agron. 27:25–87. doi:10.1016/S0065-2113(08)70007-5

Kiss, S., G. Ştefanic, and M. Drøgan-Bularda. 1975b. Soil enzymology in Romania II. Contrib. Bot. Cluj. 197–207.

Klose, S., K.D. Wernecke, and F. Makeschin. 2003. Microbial biomass and enzyme activities in coniferous forest soils under long-term fly ash pollution. Biol. Fertil. Soils 38:32–44. doi:10.1007/s00374-003-0615-4

Knight, T.R., and R.P. Dick. 2004. Differentiating microbial and stabilized β-glucosidase activity relative to soil quality. Soil Biol. Biochem. 36:2089–2096. doi:10.1016/j.soilbio.2004.06.007

Knorr, W., I.C. Prentice, J.I. House, and E.A. Holland. 2005. Long-term sensitivity of soil carbon turnover to warming. Nature 433:298–301. doi:10.1038/nature03226

Knudson, L. 1920. The secretion of invertase by plant roots. Am. J. Bot. 7:371–379. doi:10.2307/2435227

Knudson, L., and R.S. Smith. 1919. Secretion of amylase by plant roots. Bot. Gaz. 68:460–466. doi:10.1086/332584

Kobayashi, M. 1940. Effects exerted by the continuous use of manures upon the buffer capacity and catalytic action of soil. J. Sci. Soil, Japan 14:789–796.

Koepf, H. 1954a. Investigations on the biological activity in soil. I. Respiration curves of the soil and enzyme activity under the influence of fertilizing and plant growth. Z. Ackerund Pflanzenbau 98:289–312.

Koepf, H. 1954b. Experimental study of soil evaluation by biochemical reactions. I. Enzyme reactions and CO_2 evolution in different soils. Z. Pflanzenernaehr. Dueng. Bodenkd. 67:262–270.

Koepf, H. 1954c. Experimental study of soil evaluation by biochemical reactions. II. Enzyme reactions and CO_2 evolution in a static fertilizer trial and with three principal cultural systems. Z. Pflanzenernaehr. Dueng. Bodenkd. 67:271–277.

Kong, K.T., J. Balandreau, and Y. Dommergues. 1971. Measurement of the activity of cellulases in organic soils. Biol. Sol. 13:26–27.

Kong, K.T., and Y. Dommergues. 1972. Limitation of the cellulolysis in organic soils. II. Study of the enzymes in soil. Rev. Ecol. Bio. Sol 9:629–640.

König, J., J. Hasenbäumer, and E. Coppenrath. 1906. Several new properties of cultivated soils. Landw. Versuchs-Stationen 63:471–478.

König, J., J. Hasenbäumer, and E. Coppenrath. 1907. Relationships between the properties of soil and nutrient uptake by plants. Landw. Versuchs-Stationen 66:401–461.

Kramer, M., and S. Erdei. 1958. Investigation of the phosphatase activity of soils by means of disodium monophenyl phosphate. I. Method. (In Hungarian, with English abstract.) Agrokem. Talajt. 7:361–366.

Kramer and Erdei. 1959. The application of the method of phosphatase activity determination in agricultural chemistry. Poehvovedenie 9:99–102.

Kroll, L. 1953. A biochemical method for determining the biological activity of soils. Agrokém. Talajt. 2:301–306.

Kroll, L., M. Kramer, and E. Lorince. 1955. The application of enzyme analysis with phenylphosphate to soils and fertilizers. Agrokbn. Talajt. 4:173–182.

Krsek, M., W.H. Gaze, N.Z. Morris, and E.M.H. Wellington. 2006. Gene detection, expression and related enzyme activity in soil. p. 217–255. In P. Nannipieri and K. Smalla (ed.) Nucleic Acids and Proteins in Soil. Springer-Verlag, Berlin.

Kunze, C. 1970. The effect of streptomycin and aromatic carboxylic acids on the catalase activity in soil samples. Zentralbl. Bakteriol. Parasitenkd. Abt. II 124:658–661.

Kuprevich, V.F. 1949. Extracellular enzymes of roots of autotrophic higher plants. Dokl. Akad. Nauk SSSR 78:953–956.

Kuprevich, V.F. 1951. The biological activity of soil and methods for its determination. Dokl. Akad. Nauk SSSR 79:863–866.

Kuprevich, V.F., and T.A. Shcherbakova. 1956. Determination of the invertase and catalase activity of soils. Vestsi Akad. Navuk Belarusk SSR. Ser. Biyal. 2:115–116.

Kuprevich, V.F., and T.A. Shcherbakova. 1971. Comparative enzymatic activity in diverse types of soil. p. 167–201. In A.D. McLaren and J. Skujinš (ed.) Soil biochemistry. Vol. 2. Marcel Dekker, New York.

Kurtyakov, N.I. 1931. Characterization of the catalytic power of soil. Pochvovedenie 3:34–48.

Kuzyakov, Y. 2010. Priming effects: Interactions between living and dead organic matter. Soil Biol. Biochem. 42:1363–1371. doi:10.1016/j.soilbio.2010.04.003

Ladd, J.N. 1972. Properties of proteolytic enzymes extracted from soil. Soil Biol. Biochem. 4:227–237. doi:10.1016/0038-0717(72)90015-6

Ladd, J.N. 1978. Origin and range of enzymes in soil. p. 51–96. In R.G. Burns (ed.) Soil enzymes. Academic Press, New York.

Ladd, J.N., and J.H.A. Butler. 1975. Humus-enzyme systems and synthetic organic polymer-enzyme analogs. p. 143–194. In A.D. McLaren and E.A. Paul (ed.) Soil biochemistry. Vol. 5. Marcel Dekker, New York.

Ladd, J.N., R.C. Foster, and P. Nannipieri. 1996. Soil structure and biological activity. p. 23–78. In G. Stotzky and J.M. Bollag (ed.) Soil biochemistry. Vol. 9. Marcel Dekker, New York.

Ladd, J.N., and E.A. Paul. 1973. Changes in enzyme activity and distribution of acid soluble, amino acid nitrogen in soil during nitrogen immobilization and mineralization. Soil Biol. Biochem. 5:825–840. doi:10.1016/0038-0717(73)90028-X

Lähdesmäki, P., and R. Piispanen. 1992. Soil enzymology: Role of protective colloid systems in the preservation of exoenzyme activities in soil. Soil Biol. Biochem. 24:1173–1177. doi:10.1016/0038-0717(92)90068-9

Laiho, R. 2006. Decomposition in peatlands: Reconciling seemingly contrasting results on the impacts of lowered water levels. Soil Biol. Biochem. 38:2011–2024. doi:10.1016/j.soilbio.2006.02.017

Laugesen, K. 1972. Urease activity in Danish soils. Danish J. Plant Soil Sci. 76:221–229.

Laugesen, K., and J.P. Mikkelsen. 1973. Phosphatase activity in Danish soils. Danish J. Plant Soil Sci. 77:252–257.

Lee, J.M., J.A. Heitmann, and J.J. Pawlak. 2007. Local morphological and dimensional changes of enzyme-degraded cellulose materials measured by atomic force microscopy. Cellulose 14:643–653. doi:10.1007/s10570-007-9172-6

Lenhard, G. 1956. Die Dehydrogenaseaktivität des Bodens als Maß für die Mikroorganismentätigkeit im Boden. Z. Pflanzenernaehr. Dueng. Bodenkd. 73:1–11. doi:10.1002/jpln.19560730102

Losi, M.E., C. Amrhein, and W.T. Frankenberger, Jr. 1994. Environmental biochemistry of chromium. Rev. Environ. Contam. Toxicol. 136:91–121.

Luis, P., H. Kellner, B. Zimdars, U. Langer, F. Martin, and F. Buscot. 2005. Patchiness and spatial distribution of laccase genes of ectomycorrhizal, saprotrophic and unknown basidiomycetes in the upper horizons of a mixed forest Cambisol. Microb. Ecol. 50:570–579. doi:10.1007/s00248-005-5047-2

Luo, Y.Q. 2007. Terrestrial carbon-cycle feedback to climate warming. Annu. Rev. Ecol. Evol. Syst. 38:683–712. doi:10.1146/annurev.ecolsys.38.091206.095808

Lyon, T.L., E.O. Fippin, and H.O. Buckman. 1917. Soils: Their properties and management. Macmillan Co., New York.

Markus, L. 1955. Determination of carbohydrates from plant materials with anthrone reagent. II. Assay of activity in soil and farmyard manure. Agrokem. Talajt. 4:207–216.

Martens, D.A., and W.T. Frankenberger, Jr. 1994. Assimilation of exogenous 2'-^{14}C-indoleacetic acid and 3'-^{14}C-tryptophan exposed to the roots of three wheat varieties. Plant Soil 166:281–290. doi:10.1007/BF00008341

Marx, M.C., M. Wood, and S.C. Jarvis. 2001. A microplate fluorimetric assay for the study of enzyme diversity in soils. Soil Biol. Biochem. 33:1633–1640. doi:10.1016/S0038-0717(01)00079-7

Matsuno, T., and C. Ichikava. 1934. Catalytic action of Japanese soils. Res. Bull. Gifu Imp. Coll. Agr. 37:22.

May, D.W. and P.L. Gile. 1909. The catalase of soils. Puerto Rico Agr. Exp. Sta. Circular No. 9:3–13.

Mayaudon, J., L. Batistic, and J.M. Sarkar. 1975. Properties of proteolytically active extracts from fresh soils. Soil Biol. Biochem. 7:281–286. doi:10.1016/0038-0717(75)90067-X

Mayaudon, J., and J.M. Sarkar. 1974a. Study of diphenol oxidases extracted from a forest litter. Soil Biol. Biochem. 6:269–274. doi:10.1016/0038-0717(74)90030-3

Mayaudon, J., and J.M. Sarkar. 1974b. Chromatography and purification of the diphenol oxidases of soil. Soil Biol. Biochem. 6:275–285. doi:10.1016/0038-0717(74)90031-5

Mayaudon, J., and J.M. Sarkar. 1975. Polyporus versicolor laccases in the soil and the litter. Soil Biol. Biochem. 7:31–34. doi:10.1016/0038-0717(75)90027-9

McLaren, A.D. 1954. Adsorption and reactions of enzymes and proteins on kaolinite—II. The action of chymotrypsin on lysozyme. Soil Sci. Soc. Am. Proc. 18:170–174. doi:10.2136/sssaj1954.03615995001800020014x

McLaren, A.D. 1969. Radiation as a technique in soil biology and biochemistry. Soil Biol. Biochem. 1:63–73. doi:10.1016/0038-0717(69)90035-2

McLaren, A.D. 1975. Soil as a system of humus and clay immobilized enzymes. Chem. Scr. 8:97–99.

McLaren, A.D., and E.F. Estermann. 1956. The adsorption and reactions of enzymes and proteins on kaolinite. III. The isolation of enzyme-substrate complexes. Arch. Biochem. Biophys. 61:158–173. doi:10.1016/0003-9861(56)90328-9

McLaren, A.D., R.A. Luse, and J.J. Skujiņš. 1962. Sterilization of soil by irradiation and some further observations on soil enzyme activity. Soil Sci. Soc. Am. Proc. 26:371–377. doi:10.2136/sssaj1962.03615995002600040019x

McLaren, A.D., and L. Packer. 1970. Some aspects of enzyme reactions in heterogeneous systems. Adv. Enzymol. 33:245–308.

McLaren, A.D., A.H. Puķīte, and I. Barshad. 1975. Isolation of humus with enzymatic activity from soil. Soil Sci. 119:178–180. doi:10.1097/00010694-197502000-00011

McLaren, A.D., L. Reshetko, and W. Huber. 1957. Sterilization of soil by irradiation with an electron beam, and some observations on soil enzyme activity. Soil Sci. 83:497–502. doi:10.1097/00010694-195706000-00011

Messineva, M.A. 1941. Enzymatic activity in the deposits of Lake Zaluchye. Tr. Labor. Genesisa Sapropelya 2:61–71.

Mortland, M.M., and J.E. Gieseking. 1952. The influence of clay minerals on the enzymatic hydrolysis of organic phosphorus compounds. Soil Sci. Soc. Am. Proc. 16:10–13. doi:10.2136/sssaj1952.03615995001600010004x

Nannipieri, P. 2006. Role of stabilised enzymes in microbial ecology and enzyme extraction from soil with potential applications in soil proteomics. Soc. Biol. 8:75–94. doi:10.1007/3-540-29449-X_4

Nannipieri, P., B. Ceccanti, and D. Bianchi. 1988. Characterization of humus-phosphatase complexes extracted from soil. Soil Biol. Biochem. 20:683–691. doi:10.1016/0038-0717(88)90153-8

Nannipieri, P., A. Gelsomino, and M. Felici. 1991. Method to determine guaiacol oxidase activity in soil. Soil Sci. Soc. Am. J. 55:1347–1352. doi:10.2136/sssaj1991.03615995005500050025x

Nannipieri, P., R.L. Johnson, and E.A. Paul. 1978. Criteria for measurement of microbial-growth and activity in soil. Soil Biol. Biochem. 10:223–227. doi:10.1016/0038-0717(78)90100-1

Nannipieri, P., E. Kandeler, and P. Ruggiero. 2002. Enzyme activities as a research tool for microbiological and biochemical processes in soil. p. 1–33. In R.G. Burns and R.P. Dick (ed.) Enzymes in the environment: Activity, ecology and applications. Marcel Dekker, New York.

Nannipieri, P., L. Muccini, and C. Ciardi. 1983. Microbial biomass and enzyme activities: Production and persistence. Soil Biol. Biochem. 15:679–685. doi:10.1016/0038-0717(83)90032-9

Nannipieri, P., P. Sequi, and P. Fusi. 1996. Humus and enzyme activity. p. 293–328. In A. Piccolo (ed.) Humic substances in terrestrial ecosystems. Elsevier, New York.

Naseby, D.C., and J.M. Lynch. 2002 p. 109–123. In R.G. Burns and R.P. Dick (ed.) Enzymes and microorganisms in the rhizosphere. Enzymes in the environment: Activity, ecology, and applications.

Ndiaye, E.L., J.M. Sandeno, D. McGrath, and R.P. Dick. 2000. Integrative biological indicators for detecting change in soil quality. Am. J. Altern. Agric. 15:26–36. doi:10.1017/S0889189300008432

Nilsson, P.E. 1957. Influence of crop on biological activities in soil. K. Lantbrukshoegsk. Ann. L. 23:175–218.

Novotny, C., T. Cajthaml, K. Svobodova, M. Susla, and V. Sasek. 2009. Irpex lacteus, a white-rot fungus with biotechnological potential—review. Folia Microbiol. (Praha) 54:375–390. doi:10.1007/s12223-009-0053-2

Novotny, C., K. Svobodova, P. Erbanova, T. Cajthaml, A. Kasinath, E. Lang, and V. Sasek. 2004. Ligninolytic fungi in bioremediation: Extracellular enzyme production and degradation rate. Soil Biol. Biochem. 36:1545–1551. doi:10.1016/j.soilbio.2004.07.019

Osugi, S. 1922. The catalytic action of soil. Ber. Ohara Inst. Landw. Forsch. Japan 2:197–210.

Pancholy, S.K., and J.Q. Lynd. 1972. Quantitative fluorescence analysis of soil lipase activity. Soil Biol. Biochem. 4:257–259. doi:10.1016/0038-0717(72)90018-1

Pancholy, S.K., and E.L. Rice. 1973a. Soil enzymes in relation to old field succession: Amylase, cellulase, invertase, dehydrogenase, and urease. Soil Sci. Soc. Am. Proc. 37:47–50. doi:10.2136/sssaj1973.03615995003700010018x

Pancholy, S.K., and E.L. Rice. 1973b. Carbohydrases in soils as affected by successional stages of revegetation. Soil Sci. Soc. Am. Proc. 37:227–229. doi:10.2136/sssaj1973.03615995003700020021x

Pancholy, S.K., E.L. Rice, and J.A. Turner. 1975. Soil factors preventing revegetation of a denuded area near an abandoned zinc smelter in Oklahoma. J. Appl. Ecol. 12:337–342. doi:10.2307/2401736

Paul, E.A., and A.D. McLaren. 1975. Biochemistry of the soil subsystem. p. 1–36. In E.A. Paul and A.D. McLaren (ed.) Soil biochemistry. Marcel Dekker, New York.

Penkava J. 1913a. Neuere Ansichten über die Bedeutung des Eisens und des Kalkes im Boden. Zemed. Arch. 4:1–12. Cited in Kiss, 1958a.

Penkava J. 1913b. Neuere Ansichten über die Bedeutung des Eisens und des Kalkes im Boden. Zemed. Arch. 4:99–106. Cited in Kiss, 1958a.

Perucci, P., C. Casucci, and S. Dumontet. 2000. An improved method to evaluate the o-diphenol oxidase activity of soil. Soil Biol. Biochem. 32:1927–1933. doi:10.1016/S0038-0717(00)00168-1

Peterson, N.V., and E.V. Astafyeva. 1962. Determination of saccharase activity in soil. Mikrobiol. 1:918–922.

Pind, A., C. Freeman, and M.A. Lock. 1994. Enzymatic degradation of phenolic materials in peatlands—measurement of phenol oxidase activity. Plant Soil 159:227–231. doi:10.1007/BF00009285

Pokorna, V. 1964. Method of determining the lipolytic activity of upland and lowland peats and muds. Sov. Soil Sci. 1:85–87.

Poll, C., A. Thiede, N. Wermbter, A. Sessitsch, and E. Kandeler. 2003. Micro-scale distribution of microorganisms and microbial enzyme activities in a soil with long-term organic amendment. Eur. J. Soil Sci. 54:715–724. doi:10.1046/j.1351-0754.2003.0569.x

Porter, L.K. 1965. Enzymes. p. 1536–1549. In C.A. Black, D.D. Evans, J.L. White, L.E. Ensminger, and F.E. Clark (ed.) Methods of soil analysis. Part 2. ASA, Madison, WI.

Quastel, J.H. 1946. Soil metabolism. The Royal Institute of Chemistry of Great Britain and Ireland, London.

Quiquampoix, H. 2000. Mechanisms of protein adsorption on surfaces and consequences for exacellular enzyme activity in soil. p. 171–206. In J.M. Bollag and G. Stotzky (ed.) Soil biochemistry. Vol. 10. Marcel Dekker, New York.

Quiquampoix, H., and R.G. Burns. 2007. Environmental and health consequences of protein interactions with soil mineral surfaces. Elements 3:401–406.

Quiquampoix, H., S. Servagent-Noinville, and M.H. Baron. 2002. Enzyme adsorption on soil mineral surfaces and consequences for the catalytic activity. p. 285–306. In R.G. Burns and R.P. Dick (ed.) Enzymes in the environment: Activity, ecology and applications. Marcel Dekker, New York.

Rabinovich, M.L., A.V. Bolobova, and L.G. Vasilchenko. 2004. Fungal decomposition of natural aromatic structures and xenobiotics: A review. Appl. Biochem. Microbiol. 40:1–17. doi:10.1023/B:ABIM.0000010343.73266.08

Radu, I.F. 1931. The catalytic power of soils. Landw. Versuchs-Stationen 112:45–54.

Ramirez-Martinez, J.R., and A.D. McLaren. 1966. Determination of soil phosphatase activity by a fluorimetric technique. Enzymologia 30:243–253.

Rao, M.A., L. Gianfreda, F. Palmiero, and A. Violante. 1996. Interactions of acid phosphatase with clays, organic molecules and organo-mineral complexes. Soil Sci. 161:751–760. doi:10.1097/00010694-199611000-00004

Rao, M.A., A. Violante, and L. Gianfreda. 2000. Interaction of acid phosphatase with clays, organic molecules and organo-mineral complexes: Kinetics and stability. Soil Biol. Biochem. 32:1007–1014. doi:10.1016/S0038-0717(00)00010-9

Rippel, A. 1931. Bakteriologisch-chemische Methoden der Fruchtbarkeitsbestimmung. p. 670–671 In E. Blank (ed.) Handbuch der Bodenlehre. Verlag Springer, Berlin.

Rogers, H.T. 1942a. Dephosphorylation of organic phosphorus compounds by soil catalysts. Soil Sci. 54:439–446. doi:10.1097/00010694-194212000-00005

Rogers, H.T. 1942b. The availability of certain forms of organic phosphorus to plants and their dephosphorylation by exoenzyme systems of growing roots and by soil catalysts. Iowa State Coll. J. Sci. 17:108–110.

Rogers, H.T., R.W. Pearson, and W.H. Pierre. 1941. Absorption of organic phosphorus by corn and tomato plants and the mineralizing action of exoenzyme systems of growing roots. Soil Sci. Soc. Am. Proc. 5:285–291. doi:10.2136/sssaj1941.036159950005000C0053x

Rogers, H.T., R.W. Pearson, and W.H. Pierre. 1942. The source and phosphatase activity of exoenzyme systems of corn and tomato roots. Soil Sci. 54:353–366. doi:10.1097/00010694-194211000-00011

Ross, D.J., and A. Cairns. 1982. Effect of earthworms and ryegrass on respiratory and enzyme activities of soil. Soil Biol. Biochem. 14:583–587. doi:10.1016/0038-0717(82)90091-8

Ross, D.J., T.W. Speir, J.C. Cowling, and C.W. Feltham. 1992. Soil restoration under pasture after lignite mining: Management effects on soil biochemical properties and their relationships with herbage yields. Plant Soil 140:85–97. doi:10.1007/BF00012810

Rotini, O.T. 1931. Sopra il potere catalasico del terreno. Ann. Labor. Ric. Ferm. Spallanzani 2:333–351.

Rotini, O.T. 1933. La presenza e l'attività delle pirofosfatasi in alcuni substrati organici e nel terreno. Atti Soc. Ital. Progr. Sci., 21. Riun. 2:1–11.

Rotini, O.T. 1935a. Enzymatic transformation of urea in soil. Ann. Labor. Ferment. L. Spallanzani 3:173–184.

Ruggaber, T.P., and J.W. Talley. 2006. Enhancing bioremediation with enzymatic processes: A review. Pract. Period. Hazard. Toxic Radioact. Waste Manage. 10:73–85. doi:10.1061/(ASCE)1090-025X(2006)10:2(73)

Ruggiero, P., J. Dec, and J.M. Bollag. 1996. Soil as a catalytic system. In G. Stotzky and J.M. Bollag (ed.) Soil biochemistry. Vol. 9. Marcel Dekker, New York.

Ruggiero, P., and V.M. Radogna. 1988. Humic acids tyrosinase interactions as a model of soil humic enzyme complexes. Soil Biol. Biochem. 20:353–359. doi:10.1016/0038-0717(88)90016-8

Ruggiero, P., J.M. Sarkar, and J.M. Bollag. 1989. Detoxification of 2,4-dichlorophenol by a laccase immobilized on soil or clay. Soil Sci. 147:361–370. doi:10.1097/00010694-198905000-00007

Sarkar, J.M. 1986. Formation of [^{14}C]cellulase-humic complexes and their stability in soil. Soil Biol. Biochem. 18:251–254. doi:10.1016/0038-0717(86)90057-X

Sarkar, J.M., and R.G. Burns. 1984. Synthesis and properties of beta-D-glucosidase phenolic copolymers as analogs of soil humic-enzyme complexes. Soil Biol. Biochem. 16:619–625. doi:10.1016/0038-0717(84)90082-8

Scharrer, K. 1927. Zur Kenntniss der Hydroperoxyd spaltenden Eigenschaft der Böden. Biochem. Z. 189:125–149.

Scharrer, K. 1928a. Beiträge zur Kenntnis der Wasserstoffperoxyd zersetzenden Eigenschaft des Bodens. Landw. Versuchs-Stationen 107:143–187.

Scharrer, K. 1928b. Katalytische Eigenschaften der Böden. Z. Pflanzenernaehr. Dueng. Bodenkd. 12:323–329.

Scharrer, K. 1933. Reaction kinetics of the hydrogen peroxide decomposing properties of soils. Z. Pflanzenernaehr. Dueng. Bodenkd. Teil A. 31:27–36.

Scharrer, K. 1936. Catalytic characteristics of the soil. Forschungsdient 1:824–831.

Schreiner, O., and M.X. Sullivan. 1910. Studies in soil oxidation. Bull. 73. USDA, Bur. of Soils, Washington, DC.

Schutter, M., J. Sandeno, and R.P. Dick. 2001. Seasonal, soil type, and alternative management influences on microbial communities of vegetable cropping systems. Biol. Fertil. Soils 34:397–410. doi:10.1007/s00374-001-0423-7

Semple, K.T., K.J. Doick, L.Y. Wick, and H. Harms. 2007. Microbial interactions with organic contaminants in soil: Definitions, processes and measurement. Environ. Pollut. 150:166–176. doi:10.1016/j.envpol.2007.07.023

Shaw, L.J., and R.G. Burns. 2006. Enzyme activity profiles and soil health: Methodology, assay interpretation and future directions. p. 158–182. In J. Bloem, D.W. Hopkins, and A. Benedetti (ed.) Microbiological methods for assessing soil quality. CABI, Oxfordshire, UK.

Sinsabaugh, R.L. 2010. Phenol oxidase, peroxidase and organic matter dynamics of soil. Soil Biol. Biochem. 42:391–404. doi:10.1016/j.soilbio.2009.10.014

Sinsabaugh, R.L., C.L. Lauber, M.N. Weintraub, B. Ahmed, S.D. Allison, C.L. Crenshaw, A.R. Contosta, D. Cusack, S. Frey, M.E. Gallo, T.B. Gartner, S.E. Hobbie, K. Holland, B.L. Keeler, J.S. Powers, M. Stursova, C. Takacs-Vesbach, M. Waldrop, M. Wallenstein, D.R. Zak, and L.H. Zeglin. 2008. Stoichiometry of soil enzyme activity at global scale. Ecol. Lett. 11:1252–1264.

Sinsabaugh, R.L., and A.E. Linkins. 1988. Exoenzyme activity associated with lotic epilithon. Freshwater Biol. 20:249–261. doi:10.1111/j.1365-2427.1988.tb00449.x

Skujiņš, J. 1967. Enzymes in soil. p. 371–414. In A.D. McLaren and G.H. Peterson (ed.) Soil biochemistry. Marcel Dekker, New York.

Skujiņš, J. 1976. Extracellular enzymes in soil. CRC Crit. Rev. Microbiol. 4:383–421. doi:10.3109/10408417609102304

Skujiņš, J. 1978. History of abiontic soil enzyme research. p. 1–49. In R.G. Burns (ed.) Soil enzymes. Academic Press, New York.

Skujiņš, J.J., L. Braal, and A.D. McLaren. 1962. Characterization of phosphatase in a terrestrial soil sterilized with an electron beam. Enzymologia 25:125–133.

Skujiņš, J.J., and A.D. McLaren. 1969. Assay of urease activity using ^{14}C-urea in stored, geologically preserved soils. Soil Biol. Biochem. 1:89–99. doi:10.1016/0038-0717(69)90038-8

Sorensen, H. 1955. Xylanase in the soil and the rumen. Nature 176:74. doi:10.1038/176074a0

Speir, T.W., J.C. Cowling, G.P. Sparling, A.W. West, and D.M. Corderoy. 1986. The effects of microwave radiation on the microbial biomass, phosphatase activity and the levels of extractable N and P in a low fertility soil under pasture. Soil Biol. Biochem. 18:377–382. doi:10.1016/0038-0717(86)90041-6

Speir, T.W., H.A. Kettles, A. Parshotam, P.L. Searle, and L.N.C. Vlaar. 1995. A simple kinetic approach to derive the ecological dose value, ED_{50}, for the assessment of Cr(VI) toxicity to soil biological properties. Soil Biol. Biochem. 27:801–810. doi:10.1016/0038-0717(94)00231-O

Speir, T.W., and D.J. Ross. 1984. Spatial variability of biochemical properties in a taxonomically-uniform soil under grazed pasture. Soil Biol. Biochem. 16:153–160. doi:10.1016/0038-0717(84)90106-8

Speir, T.W., D.J. Ross, C.W. Feltham, V.A. Orchard, and G. Yeates. 1992. Assessment of the feasibility of using CCA (cooper, chromium and arsenic)-treated and boric acid-treated sawdust as soil amendments. II. Soil biochemical and biological properties. Plant Soil 142:249–258. doi:10.1007/BF00010970

Stemmer, S., M. Gerzabek, and E. Kandeler. 1998. Invertase and xylanase activity of bulk soil and particle-size fractions during maize straw decomposition. Soil Biol. Biochem. 31:9–18. doi:10.1016/S0038-0717(98)00083-2

Stoklasa, J. 1924. Methoden zur biochemischen Untersuchung des Bodens. p. 1–262. In E. Abderhalden (ed.) Handbuch der biologischen Arbeitsmethoden. Abt. XI. Teil 3. Heft 1. Urban und Schwarzenberg, Berlin and Vienna.

Stork, R., and O. Dilly. 1998. Maßstabsabhängige räumliche Variabilität mikrobieller Bodenkenngrößen in einem Buchenwald. Z. Pflanzenernaehr. Bodenkd. 161:235–242.

Sullivan, M.X., and F.R. Reid. 1912. Studies in soil catalysis. Bull. 86. USDA, Bur. of Soils, Washington, DC.

Tabatabai, M.A., and J.M. Bremner. 1969. Use of p-nitrophenyl phosphate for assay of soil phosphatase activity. Soil Biol. Biochem. 1:301–307. doi:10.1016/0038-0717(69)90012-1

Tabatabai, M.A., and J.M. Bremner. 1970a. Factors affecting soil arylsulfatase activity. Soil Sci. Soc. Am. Proc. 34:427–429. doi:10.2136/sssaj1970.03615995003400030023x

Tabatabai, M.A., and J.M. Bremner. 1970b. Arylsulfatase activity of soils. Soil Sci. Soc. Am. Proc. 34:225–229. doi:10.2136/sssaj1970.03615995003400020016x

Tabatabai, M.A., and J.M. Bremner. 1972. Assay of urease activity in soils. Soil Biol. Biochem. 4:479–487. doi:10.1016/0038-0717(72)90064-8

Tabatabai, M.A., and M.H. Fu. 1992. Extraction of enzymes from soils. p. 197–227. In G. Stotzky and J.M. Bollag (ed.) Soil biochemistry. Vol. 7. Marcel Dekker, New York.

Taylor, J.P., B. Wilson, M.S. Mills, and R.G. Burns. 2002. Comparison of microbial numbers and enzymatic activities in surface soils and subsoils using various techniques. Soil Biol. Biochem. 34:387–401. doi:10.1016/S0038-0717(01)00199-7

Thompson-Eagle, E.T., and W.T. Frankenberger, Jr. 1992. Bioremediation of soils contaminated with selenium. p. 261–310. *In* R. Lal and B.A. Stewart (ed.) Advances in soil science. Springer-Verlag, New York.

Tiwari, M.B., B.K. Tiwari, and R.R. Mishra. 1989. Enzyme activity and carbon dioxide evolution from upland and wetland rice soils under three agricultural practices in hilly regions. Biol. Fertil. Soils 7:359–364. doi:10.1007/BF00257833

Tortella, G.R., M.C. Diez, and N. Duran. 2005. Fungal diversity and use in decomposition of environmental pollutants. Crit. Rev. Microbiol. 31:197–212. doi:10.1080/10408410500304066

Trasar-Cepeda, C., M.C. Leiros, and F. Gil-Sotres. 2000. Biochemical properties of acid soils under climax vegetation (Atlantic oakwood) in an area of the European temperate-humid zone (Galicia, NW Spain): Specific parameters. Soil Biol. Biochem. 32:747–755. doi:10.1016/S0038-0717(99)00196-0

Tscherko, D. 1999. The response of soil microorganisms to environmental change. p. 69–87. Ph.D. diss., Universität für Bodenkultur, Vienna.

Tscherko, D., and E. Kandeler. 1999. Biomonitoring of soils—Microbial biomass and enzymatic processes as indicators for environmental change. Bodenkultur 50:215–226.

Turková, V., and M. Srogl. 1960. The relationship between enzymatic and other soil-biological tests in the same habitat. Rostl. Výroba. 33:1431–1438.

Vály, F. 1937. Catalase activity in soils. Mezograzd. Kutatasok 10:195–203.

Vaughan, D., B.G. Ord, S.T. Buckland, E.I. Duff, and C.D. Campbell. 1994. Distribution of soil invertase in relation to the root systems of *Picea sitchensis* (Bong.) Carr. and *Acer pseudoplatanus* L. during development of young plants. Plant Soil 167:73–77. doi:10.1007/BF01587601

Vela, G. R., and O. Wyss. 1962. The effect of gamma radiation on nitrogen transformation in soil. Bacteriol. Proc. 62:24

Verstraete, W., and J.P. Voets. 1977. Soil microbial and biochemical characteristics in relation to soil management and fertility. Soil Biol. Biochem. 9:253–258. doi:10.1016/0038-0717(77)90031-1

Vial, L., M.C. Groleau, V. Dekimpe, and E. Deziel. 2007. Burkholderia diversity and versatility: An inventory of the extracellular products. J. Microbiol. Biotechnol. 17:1407–1429.

Voets, J.P., and M. Dedeken. 1965. Influence of higher-frequency and gamma irradiation on the soil microflora and the soil enzymes. Meded. Landbouwhogesch. Opozoekingsta. Staat Gent 30:2037–2049.

Voets, J.P., M. Dedeken, and E. Bessems. 1965. The behaviour of some amino acids in gamma irradiated soils. Naturwissenschaften 52:476. doi:10.1007/BF00626239

Wachtel, P. 1912. Die Wasserstoffperoxidkatalyse durch Boden. Ph.D. diss., University of Jena, Germany. Cited in Kiss, 1958a.

Wackett, L.P. 2004. Evolution of new enzymes and pathways: Soil microbes adapt to s-triazine herbicides. Pesticide Decontam. Detox. 863:37–48. doi:10.1021/bk-2004-0863.ch004

Waksman, S.A. 1927. Principles of soil microbiology. Williams and Wilkins Co., Baltimore, MD.

Waksman, S.A. 1952. Soil microbiology. John Wiley & Sons, New York.

Waksman, S.A., and R.J. Dubos. 1926. Microbiological analysis as an index of soil fertility. X. The catalytic power of the soil. Soil Sci. 22:407–420. doi:10.1097/00010694-192612000-00001

Wallenstein, M.D., and R.G. Burns. 2011. Ecology of extracellular enzyme activities and organic matter degradation in soil: A complex community-driven process. p. 35–56. *In* R.P. Dick (ed.) Methods of soil enzymology. SSSA Book Ser. 9. SSSA, Madison, WI. (This volume.)

Whiteley, C.G., and D.J. Lee. 2006. Enzyme technology and biological remediation. Enzyme Microb. Technol. 38:291–316. doi:10.1016/j.enzmictec.2005.10.010

Wirth, S.J. 1999. Soil microbial properties across an encatchment in the Moraine, agricultural landscape of Northeast Germany. Geomicrobiol. J. 16:207–219. doi:10.1080/014904599270596

Woods, A.F. 1899. The destruction of chlorophyll by oxidizing enzymes. Zentralbl. Bakteriol. Parasitenkd. Abt. 2(5):745–754.

Yaroshevich, I.V. 1966. Effect of fifty years' application of fertilizers in a rotation on the biological activity of a chernozem. Agrokhimiya 6:14–19.

Yeates, G.W., V.A. Orchard, T.W. Speir, J.L. Hunt, and M.C.C. Hermans. 1994. Reduction in soil biological activity following pasture contamination by copper, chromium, arsenic timber preservative. Biol. Fertil. Soils 18:200–208. doi:10.1007/BF00647667

Ecology of Extracellular Enzyme Activities and Organic Matter Degradation in Soil: A Complex Community-Driven Process

Matthew D. Wallenstein* and Richard G. Burns

2–1 INTRODUCTION

Microorganisms produce enzymes to perform many biogeochemical processes that include various inorganic and redox reactions. Another very important function is the production of hydrolytic enzymes for decomposition and mineralization of nutrients. This process is critical for the functioning of ecosystems and for industrial and environmental applications. A majority of enzyme assays in this book measure the rate of hydrolytic reactions. Therefore, the objective of this chapter is to present the functions and ecology of enzymes involved in decomposition to provide context and assistance for interpreting enzyme assay outcomes involved in decomposition.

In fertile soils, heterotrophic microorganisms are supplied with detritus from plant and other biomass that is rich in carbon and the nutrients that are required for cell maintenance and growth. However, microorganisms cannot directly transport these large macromolecules into the cytoplasm. Rather, they rely on the activities of the myriad of enzymes that they synthesize and release into their immediate environment. These extracellular enzymes depolymerize organic compounds and generate soluble, low-number oligomers and monomers that are then recognized by cell-wall receptors and transported across the outer membrane and into the cell.

Protein synthesis and enzyme production and secretion is energetically expensive and requires nitrogen (Schimel and Weintraub, 2003) and is ultimately debilitating unless there are equivalent nutritional rewards. Thus, the allocation of cell resources to enzyme synthesis and secretion must involve a dynamic balance between the investment of the precious resources allocated to the production of enzymes with the energy and nutrients gained as a result of their activity. However, soil is an inherently hostile environment for extracellular enzymes because once they leave the cell they are subject to denaturation, degradation, and inactivation

Matthew D. Wallenstein, Natural Resource Ecology Laboratory, Colorado State University, Fort Collins, CO 80523 (matthew.wallenstein@colostate.edu), *corresponding author;
Richard G. Burns, School of Agriculture and Food Sciences, University of Queensland, Brisbane, Australia (r.burns@uq.edu.au).

doi:10.2136/sssabookser9.c2

through both biotic and abiotic mechanisms. At first glance, the microbial break-down of organic macromolecules in soil looks to be an impossible task!

The locations and functions of enzymes in soil have been researched and discussed for decades (Burns 1978, 1982; Burns and Dick, 2002; Nannipieri et al., 2002; Caldwell, 2005; Wallenstein and Weintraub, 2008), have been the subject of recent international conferences (Grenada, Spain; Prague, The Czech Republic; Viterbo, Italy; and most recently Bad Nauheim, Germany), and are now the focus of the Enzymes in the Environment Research Coordination Network (http://enzymes. nrel.colostate.edu [verified 23 Feb. 2011]). In the last decade, advances in molecular biology, microscopy, and analytical techniques plus some imaginative rethinking (Allison, 2005; Bouws et al., 2008; Wallenstein and Weintraub, 2008) have begun to provide new insights into the ecology of extracellular enzymes. Another motivation for the many scientific advances in this subject is the need to understand the detail of how enzymes function in a large number of industrial, medical, and environmental processes. For example, those concerned with composting (Crecchio et al., 2004; Raut et al., 2008), waste water (Shackle et al., 2006) and sludge treatment (Alam et al., 2009), and the conversion of plant materials, including wood and straw residues, to fermentable sugars for bioethanol production (Wackett, 2008) are trying hard to understand the functions and how to improve the efficiency of the many enzymes involved. The paper and pulp industry invests resources into the study of extracellular enzyme-producing microorganisms (Witayakran and Ragauskas, 2009) as does the food industry, which must control post-harvest spoilage and manage the polysaccharidic wastes arising from many processes (Bayer et al., 2007). The invasive, destructive, and economically disastrous activities of phytopathogens (Kikot et al., 2009) and the complex enzymology of ruminant digestion (Morrison et al., 2009) also require a detailed knowledge of organic polymer solubilization and mineralization. Many potential organic pollutants are chemically complex and/or poorly soluble and these require extracellular catalysis (often by "ligninases") before uptake, catabolism, and detoxification (Nannipieri and Bollag, 1991). The rational and successful bioremediation of contaminated soil will depend on a thorough understanding of these enzymatic processes (Asgher et al., 2008; Wackett, 2009). Microbial enzymes to degrade adhesins, disrupt biofilms, and repel colonizers are of interest to the antifouling industry (Kristensen et al., 2008) and are the subject of countless patents. The range of interests in extracellular enzymes seems endless.

Our well-founded concerns regarding the consequences of climate change on soil processes have stimulated a great deal of experimental research, modeling, and theorizing on soil organic matter formation and decomposition (Davidson et al., 2000; Kirschbaum 2004; Eliasson et al., 2005; Fang et al., 2005; Jones et al., 2005; Knorr et al., 2005; Bradford et al., 2008). The recalcitrant polyphenolic and polysaccharidic fraction of soil is a long-term repository for sequestered carbon as well as being essential for crumb structure, soil stability, plant nutrient and water retention, microbial diversity and activities, and a host of properties that contribute to soil fertility and plant productivity. Increasing soil (and water) temperatures, elevated atmospheric carbon dioxide (Finzi et al., 2006), and more frequent wetting and drying cycles will change microbial community composition and accelerate growth and enzyme activities (Henry et al., 2005; Chung et al., 2007; Allison and Martiny, 2008) either directly or following their impact on plants. The consequences of these changes may include a decline in the humic component: the carbon sink becomes a flux (Melillo et al., 2002; Jones et al., 2003). Those attempting to assess the outcomes of global warming by generating predictive carbon cycle models (Luo, 2007) must take into account any predicted increases in soil enzyme activities and the associated decline in the hitherto recalcitrant humic matter (Davidson and Janssens, 2006).

The objective of this chapter is to describe the complexity and diversity of soil enzymes and the macromolecules that they degrade. We will also discuss some aspects of the regulation of extracellular enzyme synthesis and secretion and the many locations and multiple fates of these enzymes after they are released from the cytoplasm. The chemical, physical, and biological properties of soil all affect enzyme diffusion, survival, and substrate turnover, as well as the proportion of the product that is available to and assimilated by the producer cells. The ways in which microorganisms and their extracellular enzymes attempt to overcome the generally destructive or inhibitory properties of the soil matrix and the various strategies they adopt for effective substrate detection and utilization will be described.

2–2 SUBSTRATE AND ENZYME DIVERSITY IN SOILS

Extracellular enzymes in soils catalyze the degradation of organic matter primarily through hydrolytic and oxidative reactions. Hydrolytic enzymes are substrate-specific, in that their conformation enables them to catalyze reactions that cleave specific bonds (e.g., C–O and C–N bonds) that link monomers. On the other hand, oxidative enzymes that act on broader classes of substrates that share similar bonds (e.g., C–C and C–O–C) use either oxygen (oxygenases) or hydrogen peroxide (peroxidases) as electron acceptors.

Although soil organic matter is traditionally classified into several broad groups based on solubility and molecular mass, its chemical structure varies widely at the molecular level (Piccolo, 2001; Kelleher and Simpson, 2006). Thus, an equally diverse suite of enzymes has evolved to access the carbon and nutrients contained, but not immediately bioavailable, in plant, animal, and microbial detritus (Caldwell, 2005). Another factor contributing to the need for an enormous variety of soil extracellular enzymes is their interactions with the abiotic environment. For example, temperature is a strong controller of the three-dimensional physical structure of enzymes (conformation) as well as their configuration at surfaces such as clays (Daniel et al., 2008). The active site on an enzyme (both before and after complexing and sorption) may only be exposed and accessible to substrates under a narrow temperature range, resulting in thermal optima for the activity of each enzyme (Daniel et al., 2001) that is different from that when the enzyme is in the solution phase. Similarly, pH can affect the conformation of enzymes and thus their activity (Niemi and Vepsalainen, 2005), and again, especially at surfaces where hydrogen ion concentrations may be different from those in the bulk phase. Soil ion exchange capacity and charge density and distribution also will influence substrate–enzyme interactions and kinetics. Thus, the same or different microorganisms may, by necessity, produce not only an arsenal of enzymes that target the same substrate but also multiple variants of individual enzymes.

For complex organic material, such as plant debris (composed of cellulose, hemicelluloses, pectins, starch, and lignins), extracellular degradation requires the simultaneous and/or sequential activities of a large number of hydrolytic and oxidoreductive enzymes produced by a diverse community of bacteria and fungi. In the case of a plant leaf or a dead microbial cell (composed of peptidoglycans, lipoteichoic acids, lipopolysaccharides, glycoproteins, chitin, glucans, and mannans), it is likely that more than 50 different extracellular enzymes are involved even before all the organic matter is transformed into the low-molecular-mass carbon and energy sources that can enter the cell. Intuitively, the multienzyme processes necessary for successful biopolymer degradation in soils are improbable: cascades of enzymes secreted in an organized sequence and surviving long enough to function in concert to produce pentoses, hexoses, phenols, amino acids, amino sugars, etc. Despite this unlikely series of events, and the necessary contribution of diverse members of the decomposer community, the turnover of organic macromolecules is a constant and usually effective process. The complexity of soil organic matter degradation and the multiple roles of extracellular enzymes are well illustrated by the microbial ecology and enzymology of cellulose and lignin decay.

Cellulose, the most abundant form of fixed carbon, is a chemically simple yet structurally complex and insoluble polymer that is composed of linear chains of 5000 or more glucose units held together with H-bonds to form rigid microfibrils. The microfibrils are then linked to hemicelluloses, pectins, glycoproteins, and lignins. Different microorganisms have developed different and sometimes complex strategies to deal with cellulose in this natural state. But the rewards are great: an abundance of glucose. Basidiomycete and ascomycete fungi are major degraders of cellulose, typically employing a battery of extracellular hydrolytic enzymes (Baldrian and Valáková, 2008) including endo-1.4-β- glucanases or endocellulases (EC 3.2.1.4), cellobiohydrolases or exocellulases (EC 3.2.1.91) and β-glucosidases (EC 3.2.1.21). Many fungi secrete all three of these enzymes although some bacteria retain β-glucosidases within the cytoplasm because they are able to transport cellobiose through the cell wall. In addition to the three principal cellulases, the overall oxidative decomposition of cellulose is a function of cellobiose dehydrogenase (EC 1.1.99.18) and enzymically generated hydrogen peroxide and the resulting hydroxyl radicals. Other ill-defined extracellular enzymes (e.g., "expansins," "swollenins"), which help to loosen the structure of cell walls before cellulase penetration, have been recorded (Tsumuraya, 1996; Kim et al., 2009). Some anaerobic bacteria use cellobiose phosphorylase (EC 2.4.1.20) to convert the dimer to glucose and glucose-1-phosphate.

The best-known (and now sequenced) cellulose degrader, *Trichoderma reesei* E.G. Simmons (*Hypocrea jecorina* Berk. & Broome) has 30 or more glycosyl hydrolases, including seven endoglucanases, and a secretome containing greater than 100 proteins (Martinez et al., 2008). Its close relative, *Trichoderma harzianum* Rifai, produces more than 250 extracellular proteins when grown on chitin or fungal cell walls (Suarez et al., 2005). Cellulolysis is less common in bacteria but there are many important strains, especially ruminant anaerobes and those found in high-temperature environments. However, as with all microorganisms, growth conditions and substrates will strongly impact the synthesis and secretion of enzymes. The enzymology of cellulolysis and some of the strategies for degrading plant cell walls have been reviewed (Aro et al., 2005; Wilson, 2008) and a large number of cellulolytic fungi and bacteria studied in detail.

Once in contact with their substrate, cellulases have a variety of ways in which they not only maintain their stability but also increase their activity. One mechanism relates to the all-important cellulose binding moieties (CBM) and their binding affinities (Hilden and Johansson, 2004)). Fungal and bacterial CBMs belong to many different families (Wilson, 2008) and serve to anchor the enzyme to its substrate at appropriate sites for the catalytic domain (CD) to cleave the β-1,4-linkages (Boraston et al., 2003). The CBMs may also detach and slide across the fibrillar surface and thus move the associated CD along the cellulose chain. This relocates the CD and processively hydrolyzes the substrate (Jervis et al., 1997; Bu et al., 2009).

Not all cellulolytic microorganisms secrete the full complement of endo- and ectocellulases and they must, therefore, rely on other microorganisms to successfully degrade cellulose. This observation reinforces the notion of a community-driven process and it has been reported that endo- and ectocellulases from unrelated micro-organisms can act synergistically sometimes with specific activities up to 15× those shown by the individual enzymes (Irwin et al., 1993). Logically, endocellulases should precede ectocellulases because they will expose more sites for attack.

The CAZy web site (http://www.cazy.org [verified 24 Feb. 2011]) currently lists 14 families containing cellulases. Why do fungi and bacteria express so many cellulases? There are many possible answers to this, as suggested previously, but certainly the physical diversity of the plant cell wall with its amorphous and crystalline cellulose regions, the many differing contributions of and associations with other structural polymers, and the variable chemical and physical properties of soil all play a part. The microbial community needs cellulases for every occasion and in every situation to exploit the huge carbon and energy resource offered by plant residues.

Lignin, with which cellulose is usually associated, is a complex phenylpropanoid and one of the major structural components of plant litter. In soil, the recalcitrant acid-insoluble humic fraction is composed of lignin degradation products (vanillin, ferulic acid, guaiacol, etc.) and many condensed polymers of these aromatics. Not surprisingly, lignin and its chemical cousin humic matter are among the most refractory components of soil organic matter.

Lignin degradation is performed mainly by basidiomycetes (Hatakka, 1994; Osono, 2007), some ascomycetes, such as *Xylaria* spp. (Kellner et al., 2007) and certain actinobacteria (Kirby, 2006). Both saprotrophic (Valaskova et al., 2007) and ectomycorrhizal fungi (Chen et al., 2001) can produce lignin-degrading enzymes. Lignin is broken down by a suite of oxidative enzymes including laccases (EC 1.10.3.2), manganese peroxidases (EC 1.11.1.13), and lignin peroxidases (EC 1.11.1.14). The laccase gene, in particular, is widespread among bacteria (Alexandre and Zhulin, 2000; Claus, 2003), and the importance of bacteria in lignin degradation may be underestimated (Kellner et al., 2008). Lignin peroxidase (EC 1.11.1.14), cellobiose dehydrogenases (EC 1.1.99.18), and pyranose-2-oxidases (EC 1.1.3.10) are also involved in lignin degradation (Baldrian et al., 2006; Nyanhongo et al.,2007). The well-known contribution of Fenton chemistry to the overall sequence demands the input of enzymes generating hydrogen peroxide (e.g., glucose oxidase and glyoxal oxidase) as well as Fe^{2+} and Mn^{2+}. Clearly, as is the situation with cellulose, lignin degradation is a complex process involving a large number of enzymes (Wong, 2009).

There are three classically described lignin degradation sequences involving white-rot, soft-rot, and brown-rot fungi (Osono, 2007). White-rot fungi are the only degraders able to completely mineralize lignin. They accomplish this through a combination of hydroxylation and demethylation, followed by oxidative degradation of the remaining aromatic rings by Mn-peroxidase. In contrast, brown-rot fungi modify lignin by removing methoxyl groups, but do not oxidase the aromatic rings. Soft-rot fungi, such as ascomycetes and deuteromycetes, break down the middle lamella of the cell wall, which acts to soften the lignin structure. Soft-rot fungi appear to produce peroxidase enzymes that are specific to lignin in hardwoods.

The white-rot fungus *Phanerochaete chrysosporium* Burds. has more than 100 glycosyl hydrolases, in excess of 20 "ligninases" and a secretome of nearly 800 proteins (Vanden Wymelenberg et al., 2006) including lignin peroxidases (8), manganese peroxidases (8), other oxidoreductases (87), and glycoside hydrolases (90). The hydrogen peroxide required for peroxidase activity is produced by a suite of enzymes including glyoxal oxidases (EC 1.1.3. -), superoxide dismutases (EC 1.15.1.1), and aryl alcohol oxidases (1.1.3.7). The lignin degrader *Coprinopsis cinerea* (Schaeff.) Redhead, Vilgalys & Moncalvo is predicted to secrete 1769 proteins and, to date, 76 of these have been identified as enzymes (Bouws et al., 2008).

The physical structure of detritus is an important determinant of which compounds are enzymatically degraded first. Carbohydrates are often located on the outer fiber structures of litter, making them susceptible to the early colonizers of litter. On the other hand, some constituents that are otherwise relatively easily degraded can persist in litter until later stages because they are physically protected. Lignin acts as a barrier around cellulose and needs to be broken down to increase the access of cellulases to their substrate. Furthermore, lignin degradation per se is thought to be energetically unfavorable; thus, it is possible that brown-rot and soft-rot fungi degrade lignin only to access cellulose, other polymers, or sources of nitrogen. Nonetheless, the relative amounts of acid-insoluble substances tend to increase as lignocellulose decomposition occurs (Osono, 2007). It is probable that lignin degradation has to be coupled to cellulolysis to generate an adequate carbon and energy supply and that fungi and bacteria act in concert to achieve this end.

2–3 LOCATION AND STABILITY OF EXTRACELLULAR SOIL ENZYMES

Microorganisms and their extracellular enzymes must be capable of detecting, moving toward, and transforming organic debris to soluble monomers (or short oligomers) that are subsequently transported into the cytoplasm. As stated previously, the macromolecular components of living and dead plant, animal, and microbial tissues are often physically and chemically associated with clays, minerals, or other organic compounds. This presents a barrier that restricts microbial and extracellular enzyme access even to the otherwise vulnerable soluble constituents of organic matter. The substrates themselves may also be sequestered within soil components and this will reduce their accessibility to enzymes (Jastrow et al., 2007). Thus, degradation of macromolecules and more easily metabolized organics in soils requires not only enzyme production, but also physical contact of enzymes with their target substrates and the sites of catalysis. Furthermore, abiotic conditions must be within a range where enzymes can survive and activity can occur.

Extracellular enzymes may be associated with the microbial cell's plasma membrane, contained within and attached to the walls of the periplasmic space, cell wall, and glycocalyx, or released into the soil aqueous phase (Fig. 2–1; Sinsabaugh, 1994). The periplasm may provide Gram-negative bacteria with a reservoir of activity that is retained until an external signal for secretion is received—perhaps

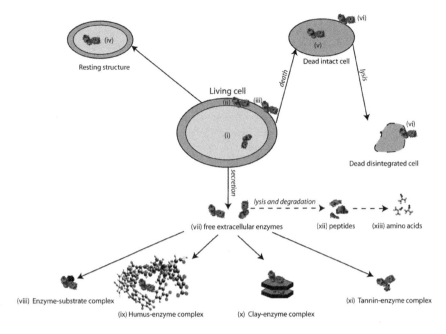

Fig. 2–1. Locations of enzymes in soil: (i) enzymes functioning within the cytoplasm of proliferating microbial, animal, and plant cells; (ii) enzymes restricted to the periplasmatic space of proliferating Gram-negative bacteria; (iii) enzymes attached to the outer surface of viable cells which active sites extend into the soil environment; (iv) enzymes within nonproliferating cells such as fungal spores, protozoa cysts, plant seeds, and bacteria endospores; (v) enzymes attached to an entire dead cell and cell debris; (vi) enzymes leaking from intact cells or released from lysed cells, originally located on the cell membrane or within the cell, which may survive for a short period in the soil solution; (vii) free extracellular enzymes in soil pore water; (viii) enzymes associated temporarily in soluble or insoluble enzyme–substrate complexes; and (ix) enzymes complexed with humic colloids by absorption, entrapment, or copolymerization during humification; (x) enzymes absorbed to the external or internal (i.e., within the lattices of 2:1 layer silicates) surfaces of clay minerals; (xi) enzymes bound to condensed tannins leached from plant leaves or roots. Extracellular enzymes can be converted abiotically or through the activity of proteolytic enzymes into (xii) peptides and then degraded by peptidases into their constituent amino acids (xiii) that are assimilated by microbes and plants.

an efficient and rapid method for responding to the appearance of a potential substrate. Periplasmic enzymes may also process their proteins and enzymes en route to the extracellular environment to prepare them for the hostile "real world." For example, glycosylation may take place in the periplasm (Feldman et al., 2005). With regard to catalysis within the periplasm, the targets must be limited to low-molecular-mass structures that can be transported through the outer membrane. Thus, enzymes that degrade sugar dimers, such as cellobiose and mannose, and some phosphatases and nucleases are often (but not always) restricted to the periplasmic space and further transform their substrates immediately before transport into the cytoplasm.

Some extracellular enzymes are retained on the outer surface of the cell wall (i.e., mural or peripheral enzymes) and are likely to be configured in such a way that their active sites are exposed and the zones that are most subject to attack by proteases are protected. Two other types of nondiffusing extracellular enzymes exist: those within the polysaccharidic coat that covers the cell or is part of a multicellular biofilm (Flemming and Wingender, 2002; Romani et al., 2008) and those organized within microscopically visible structures attached to but protruding from the cell wall. The latter, called cellulosomes, were first described for the anaerobic thermophile *Clostridium thermocellum* Viljoen et al., and have been much researched since (Bayer et al., 2004; Gold and Martin, 2007; Bayer et al., 2008; Peer et al., 2009). *Clostridium thermocellum* produces dozens of extracellular enzymes, including endoglucanases, exoglucanases, β-glucosidases, xylanases, lichenases, laminarinases, xylosidases, galactosidases, mannosidases, pectin lyases, pectin methylesterases, polygalacturonate hydrolases, cellobiose phosphorylases, and cellodextrin phosphorylases (Lamed et al., 1983; Demain et al., 2005) and many of them may be contained within cellulosomes.

We now know that many anaerobic ruminant and some soil bacteria (including species belonging to the genera *Bacillus, Clostridium,* and *Bacteroides*), archaea, and fungi package their hydrolases at the cell surface. The enzymes are housed within somewhat rigid structures that vary from around 20 nm to greater than 200 nm in size and have a molecular mass in excess of one million. Cellulases (as well as hemicellulases, pectinases and xylanases) are arranged on a protein scaffold that provides an architecture that facilitates endo- and ectocleavage of polysaccharides. Cellulosomes may also contain proteases and secrete antibiotics (Schwarz and Zverlov, 2006) to give the microorganisms some protection against competitors (see below). Based on our rapidly advancing knowledge of the bacterial proteome, the possibility of constructing cellulosome-like chimeras is under investigation (Mingardon et al., 2007). Some pathogenic bacteria actually bypass the risky secretion phase by translocating the extracellular enzymes directly into the target cell using a secretory needle (Pastor et al., 2005). It is not known if such a mechanism functions in soil organic matter degradation.

Some enzymes that dissociate from the cell and are released into the aqueous phase will survive for a time. In fact, many extracellular enzymes are inherently more stable than their intracellular counterparts because they are glycosylated or have disulfide bonds. These modifications provide thermostability, a broad pH range for activity, and also some resistance to proteolysis. Other enzymes become stabilized through interactions with clay minerals and soil organic matter (Fig. 2–1) (Burns, 1982; Nannipieri et al., 2002) or tannins (Joanisse et al., 2007). In fact, much of the activity of some enzymes is associated with organic and inorganic colloids rather than being free in solution (Kandeler 1990). Many of these bound enzymes are not extracellular sensu stricto but rather they have become externalized as a result of the death and lysis of their parent cell. Indeed, the same enzyme may be found in different locations. Regardless of their origins, stabilized enzymes often have reduced in situ activity, as complexation can restrict substrate accessibility, occlude active sites, and cause conformational changes (Allison and Jastrow, 2006; Nannipieri, 2006; Quiquampoix and Burns, 2007).

Enzymes stabilized as components of humic matter and associated with organo–mineral complexes may be catalytic (Tate, 2002). Evidence for this source of catalytic enzymes was first discovered using ionizing radiation to sterilize small quantities of soil. Later it was shown that a 5 to 10 MeV electron beam and hard x-rays and γ-rays from a 60co source (a 2 Mrep dose) could sterilize soil (Dunn et al., 1948; McLaren et al., 1957, 1962).

The catalytic ability of enzymes varies between specific types. For example, Gianfreda et al. (1995) found that when urease complexes with tannic acid are formed in the presence of ferric ions and aluminum hydroxide species, the enzyme retains its conformation and function. Pflug (1982) reported that clay-bound cellulase enzymes maintained activity, whereas starch-degrading α-amylase and amyloglucosidase activities were completely inhibited following adsorption to clays. The same enzyme may even express different activities depending on its distribution between the various adsorptive soil fractions (Marx et al., 2005).

Stabilization also may protect enzymes against proteases and other denaturing agents (Nannipieri et al., 1978; Nannipieri et al., 1988). It is possible that enzymatic function may be restored if enzymes are detached from organo–mineral complexes. However, irreversible deactivation is common with enzymes adsorbed to surfaces as a result of changes in protein conformation. If the enzyme is unfolded and the number of points of contact with the surface increase, more energy will be required to reverse the unfolding of the adsorbed enzyme, and may exceed the thermal energy available (Quiquampoix et al., 2002).

For the reasons mentioned above, the turnover time of extracellular enzymes complexed with humic molecules or adsorbed by clay minerals (or a combination of both) is likely to be longer than those free in the aqueous phase although their capacity to diffuse to distantly located substrates will be much reduced (Hope and Burns, 1985). However, turnover rates of enzymes in soils have not yet been measured—an important gap in our understanding of soil enzymology.

The active proportion of stabilized soil enzymes represents a reservoir of potential enzyme activity that may be important under some conditions. Indeed, it may represent the first catalytic response to changes in substrate availability in soils, and may serve as the originator of signaling molecules for the microbial community. Soil-bound enzymes also may be a reservoir of potential activity in soils during periods when microbial biomass is low or shut down due to stressed conditions (Stursova and Sinsabaugh, 2008). Furthermore, it may enable some microorganisms to take advantage of stabilized enzymes to produce products without their having to expend energy to produce these enzymes.

Identifying the functional locations of extracellular enzymes in soils and quantifying their individual contribution to the catalysis of a particular substrate are challenges yet to be overcome by soil biochemists (Wallenstein and Weintraub, 2008). The majority of research on soil enzymes has focused on quantifying potential enzyme activity rates in soil slurries where substrate supply is nonlimiting. In undisturbed soils, enzyme and substrate diffusion are restricted by the microscale spatial structure of the soil which reduces water movement and isolates microsites.

Several studies have examined the distribution of enzymes relative to aggregate size (Henry et al., 2005; Dorodnikov et al., 2009) and Dong et al. (2007) used a novel approach to demonstrate that β-glucosidase activity was concentrated around plant roots. There has been some progress in imaging substrates and enzymes in soils since the early influential studies using scanning electron microscopy (Foster and Martin, 1981; Foster, 1985) and confocal laser scanning microscopy (Alvarez et al., 2006) and atomic force microscopy show potential (Nigmatullin et al., 2004; Kaiser and Guggenberger, 2007). The use of autofluorescent proteins and enzymes warrants further investigation (Larrainzar et al., 2005). However, electron-dense soil minerals and humic substances and sequestration within clays and humates interfere with the visualization of enzymes. A potential advance that could overcome many of these limitations involves the use of Quantum Dots (Qdots), which are nanoscale crystals that emit in the near-infrared wavelengths and are more photostable than current fluorophores (Michalet et al., 2005). Qdots that are quenched until they bind to their target enzyme and then continue to emit (Blum et al., 2005; Blum et al., 2007) already are used widely in biomedical research to detect protease activity, and this technology could be adapted to soils (Whiteside et al., 2009). Direct measurement of in situ enzyme activity would be the ultimate tool to fully understand the complex interactions between microorganisms, enzymes, and soil organic matter (SOM).

2–4 Thermal Controls on Enzyme Activities

The conditions chosen to conduct enzyme assays in the laboratory have long been debated because the outcome will be determined by factors such as pH, substrate concentration, water content, and agitation. In addition to these methodological uncertainties, there are temporal and spatial variations ranging from the soil microenvironmental to the geographic that render the interpretation of the data very difficult. A good example of the problem is the influence of temperature on enzyme activities (Wallenstein and Weintraub, 2008). In most ecosystems, soil temperatures vary on diel to seasonal time scales, and change in response to long-term climate trends. If we assume that over a certain range, enzyme activity roughly doubles for every 10°C increase in temperature (i.e., $Q_{10} = 2$), then the effect of temperature may have a greater impact on in situ activity rates than seasonal fluctuations in enzyme potential at most sites.

The assumption that all enzymes are equally sensitive to temperature, or even that the same class of enzyme exhibits a consistent temperature sensitivity within a single site, has not been borne out in the literature. In fact, several studies have demonstrated that the temperature sensitivity of extracellular enzymes changes seasonally (Fenner et al., 2005; Koch et al., 2007; Trasar-Cepeda et al., 2007; Wallenstein et al., 2009). The most likely explanation is that the measured enzyme pool consists of contributions from different enzymes and isoenzymes and these change with time, as do their microbial source(s) (Loveland et al., 1994; Sanchez-Perez et al., 2008). Consistent with this hypothesis, Di Nardo et al. (2004) found temporal changes in laccase and peroxidase isoenzymes during leaf litter decomposition.

There is also some evidence for biogeographical patterns in enzyme temperature sensitivity. For example, many studies have observed that enzymes from microorganisms inhabiting cold environments have unusually low temperature optima (Huston et al., 2000; Coker et al., 2003; Feller and Gerday, 2003). Nonetheless, these observations suggest that microorganisms producing enzymes that maintain optimal activity under native soil conditions are favored. Thus, soil microbial community composition is likely controlled to some extent through feedbacks with enzyme efficacy.

The accumulated evidence of numerous studies suggests a wide range in temperature sensitivities for different enzymes, and measured Q_{10} values are often much less than two (McClaugherty and Linkins,1990; Frankenberger and Tabatabai, 1991a, b; Wirth and Wolf, 1992; Criquet et al., 1999; Parham and Deng, 2000; Elsgaard and Vinther, 2004). For example, Trasar-Cepeda et al. (2007) measured the Q_{10} of nine different enzymes in three different soils and found that the Q_{10} at 20°C exceeded 2.0 for only a single enzyme (β-glucosidase) in one of the soils. Most of the enzymes had a Q_{10} closer to 1.5. The apparent temperature sensitivity of enzymes in laboratory assays with unlimited amounts of substrate and without constraints to diffusion may differ markedly from in situ temperature sensitivities. Even despite methodological concerns, there is insufficient data to assess the degree to which enzyme temperature sensitivity varies across spatial gradients or in response to other environmental factors. However, it is clear that temperature sensitivities differ within a single environment. For example, Koch et al. (2007) found that at low temperatures, the relative temperature sensitivity of C-degrading enzymes was greater than aminopeptidases (which degrade N-rich proteins), suggesting that relative N availability could be decreased directly by temperature. Similarly, in the study by Wallenstein et al. (2009), N-degrading enzymes tended to have a lower Q_{10} (overall mean of 1.59) than did C-degrading enzymes (overall mean of 2.07). Thus, without any changes in enzyme pools, the relative in situ activity of these enzymes would change along with temperature, resulting in higher rates of C-mineralization relative to N-mineralization. Because different enzymes have different temperature sensitivities, changes in soil temperature also may alter the relative rates of decomposition for different components of SOM. Therefore, seasonal changes in temperature can alter the balance of SOM components contributing to soil respiration *without any changes in soil enzyme pools* (or measured enzyme potentials). Natural or human-driven changes in climate also could alter the relative rate of decomposition of SOM components and ultimately the quantity and composition of SOM.

Well-established biochemical principles predict that the temperature sensitivity of enzyme-mediated reactions is directly related to the activation energy required to initiate that reaction. In other words, we would expect enzymatic reactions with higher activation energies to exhibit a greater degree of temperature sensitivity, all other factors being equal. If we extend this principle to organic matter decomposition, we would predict that the decomposition of low quality litter (i.e., litter with high molecular complexity), or specific steps in the decomposition sequence that have high activation energies, to be more temperature-sensitive than the decomposition of high quality (labile) litter or that involving enzymatic steps with low activation energies.

This has been demonstrated under laboratory conditions by Fierer et al. (2005) who found that the decomposition of less-labile litter is more temperature-sensitive than is recalcitrant litter and that the temperature sensitivity of decomposition increases as decomposition progresses. Recent work suggests that this principle may also explain some of the observed variability in the temperature sensitivity of soil organic matter decomposition (Conant et al., 2008).

It is important to note that organic carbon quality is likely to be only one factor regulating the apparent temperature sensitivity of decomposition in the field as litter carbon availability can be controlled by abiotic processes (soil aggregation, organo–mineral interactions) that may obscure the apparent carbon quality–temperature sensitivity interactions. As temperatures increase, the proportion of assimilated substrate that is allocated to new biomass (substrate use efficiency) decreases (Steinweg et al., 2008). In other words, more of the substrate C is lost through respiration at higher temperatures. This has important implications for the long-term fate of detritus and could affect the proportion of detritus that is humified and stored in stable soil C pools versus that proportion returned to the atmosphere as CO_2, thus affecting the global C budget.

2–5 REGULATION OF EXTRACELLULAR ENZYME PRODUCTION

Our understanding of the microbial regulation of extracellular enzyme production in soil comes from either pure-culture studies, from patterns of enzyme potential across C and nutrient gradients, or in response to experimental additions (Carreiro et al., 2000; Saiya-Cork et al., 2002; Michel and Matzner, 2003; Gallo et al., 2004; Allison and Vitousek, 2005). In the case of pure-culture studies, microbial physiology in artificial media does not reflect in situ behavior, and few studies have represented the immense diversity of microorganisms present in soils. On the other hand, observations of variations in enzyme potentials in response to C or nutrient additions are suggestive of increased enzyme production, but are confounded with changes in enzyme stabilization, degradation, and changes in microbial biomass or community structure (Waldrop et al., 2000). Our understanding of the regulation of microbial enzyme production in soil is being advanced by studies using genomic, transcriptomic, and proteomic tools (Wallenstein and Weintraub, 2008).

In some cases, microorganisms may produce small amounts of extracellular enzymes, regardless of substrate availability, as a speculative sensing mechanism to detect substrate (Klonowska et al., 2002). When the substrate is present, these constitutive enzymes generate reaction products that induce additional enzyme synthesis. Once concentrations of products are sufficient to meet demand, enzyme production is downregulated and returns to low constitutive levels (Chróst, 1991).

This is a process known as quorum sensing and has been well-described for many phytopathogens. For instance, *Erwinia caratovora* (L.R. Jones) Bergy et al. virulence factors are controlled by bacterial-cell density and the local concentration of the self-produced signaling molecule acyl homoserine lactone (Barnard and Salmond, 2007). Gene products in this situation are pectin methyl esterase, pectic lyase, and polygalacturonase that depolymerize the protective coat of the target seed or fruit and facilitate penetration and pathogenesis. Many other effector proteins pass into the host-plant tissue and reduce its resistance to attack. Quorum sensing in the rhizosphere is believed to be an important controlling process for all sorts of catalytic activities (Pang et al., 2009).

Because microbial production of extracellular enzymes is carbon-, nitrogen-, and energy-intensive, microorganisms should produce only enzymes such as polysaccharases when nutrients and soluble C are scarce (Koch, 1985), or to maintain the stoichiometry of microbial biomass (Cleveland and Liptzin, 2007). When a nutrient is available in the soil solution, microorganisms downregulate or suppress their production of the enzymes that acquire that nutrient and thereby reduce the bioenergetic costs of manufacture (Pelletier and Sygush, 1990; Chróst, 1991; Sinsabaugh and Moorhead, 1994). Polysaccharase production is increased when there is a plentiful supply of soluble N (Sinsabaugh et al., 2002), while excess C may increase protease synthesis. Another expression of this control is that extracellular enzyme secretion is usually inversely related to specific growth rate. When particular nutrients are scarce, on the other hand, microorganisms secrete enzymes to liberate those nutrients from organic matter (Harder and Dijkhuizen, 1983). However, this strategy can be successful only if the appropriate organic substrates are present. As a result, the production of some extracellular enzymes may occur only in the presence of a suitable substrate or some other inducer (Allison and Vitousek, 2005). It is important to remember that the inducer molecule may not have to enter the cell to stimulate extracellular enzyme synthesis and release.

Instead, it can bind to cell-wall receptor proteins (sensory kinases) and initiate a sequence that passes the signal into the cell.

The successful depolymerization and subsequent metabolism of complex C and energy sources will be governed by the growth requirements of the attacking microorganisms and, in that context, C/N/P ratios (and yield coefficients) may be rate-limiting. Leaf litter generally has a C/N ratio in the range of 20 to 80:1 and a C/P ratio of around 3000:1. Microorganisms, on the other hand, have C/N ratios of 5 to 10:1 and C/P ratios of 50 to 100:1 (Cleveland and Liptzin, 2007). Therefore, microorganisms degrading plant residues need not only the right complement of penetrative enzymes but also ways to access additional N and P (and all the other elements that make up a microbial cell). One source will be dead microbial cells, although these also are composed of many complex polymers. A high proportion of N in soil will be in the form of proteinaceous compounds which will themselves require the activities of extracellular proteinases before microbial utilization (Geisseler and Horwath, 2008).

Because decomposition is a predominantly extracellular process that may take place remote from the cell, the diffusing enzymes can be destroyed or inactivated long before they locate a substrate. Even if the organic substrate is located and degraded, other microorganisms may intercept the products before the originating cell can benefit. These opportunistic microorganisms may not have invested any resources in extracellular enzyme generation yet will reduce the efficiency of the entire process as far as the actual enzyme producer is concerned (Allison, 2005). However, some microorganisms employ antibiotics and antagonistic enzymes to suppress opportunistic and competitive microorganisms; in the rumen, stomach protozoa may have this function (Modak et al., 2007). Other mechanisms to limit opportunistic enzymes and to increase efficiency of extracellular enzymes are the various forms of mural enzymes that deliver products directly at the cell surface. The coordination of enzyme production by quorum sensing and the spatial aggregation of enzyme producers also may mitigate or reduce competitive interference from opportunistic microorganisms (Ekschmitt et al., 2005).

Conversely, the presence of extracellular enzyme producing microorganisms may enable other microbial groups that do not produce these enzymes to invest energy in other processes that provide community benefits that otherwise might not be possible. This may be part of a complex and poorly understood microbial community cooperation in which the opportunistic microorganisms provide some direct or, more likely, indirect benefit to the extracellular-producing microorganisms or to overall community welfare. Or, it may be that the benefits of a successful extracellular depolymerization far outweigh the disadvantages derived from some of the products being intercepted.

Within a self-supporting microbial community, low levels of continuous enzyme secretion by a small proportion of the population might be effective because rapid upregulation can occur once the substrate is detected and the cells receive the appropriate signal. What proportion of the community can be sacrificed to ensure the success and stability of the remaining components? In this context it would be useful to produce an energy budget that compares energy expenditure to net energy gains and interest within the microbial community and how this is modified subsequently in soil. Game theory analysis, as it may apply to microbial strategies in soils, warrants more attention (Velicer, 2003; Davidson and Surette, 2008) as does intraspecies variation and individual sacrifice for the benefit of the whole (Krug et al., 2008).

2–6 CONCLUSIONS

We have emphasized the role of microbial extracellular enzymes in the degradation of lignocelluloses, but equally detailed and sometimes speculative descriptions could be applied to chitin (Ekschmitt et al., 2005; Duo-Chuan, 2006), pectin (Abbott and Boraston, 2008), mannan (Dhawan and Kaur, 2007), and many other biopolymers. In addition, we have ignored a major component of the overall process of lignocellulolysis and C mineralization—that is the comminuting activities of invertebrates, especially ants (Ikeda-Ohtsubo and Brune, 2009), termites (Douglas, 2009), and earthworms (Nozaki et al., 2009) and the symbiotic activities of the cellulolytic microorganisms they cultivate (Silva et al., 2006) or harbor in their digestive tracts. The contributions of grazing ruminants and their gut microflora (Desvaux, 2005) is also part of the densely complex topic of macromolecule breakdown in soils.

An enhanced knowledge of extracellular enzyme function will have many practical applications including manipulating the soil for bioremediation, biocontrol, plant nutrient generation and availability, and C sequestration. There also are implications for plant pathology, biofuel production, and the impacts of climate change. Protein survival and movement in soils and sediments is a broader issue and may have important human health consequences (Quiquampoix and Burns, 2007).

One of the greatest challenges in soil enzymology is to link the functional and ecological aspects of soil microbial extracellular enzyme activities to organic matter degradation. We are increasingly equipped with the analytical (electrophoretic, chromatographic, mass spectrometric—especially MALDI-TOF/MS), microscopical (fluorescence, scanning probe, atomic and ultrasonic force, confocal laser, scanning probe, differential interference, etc.), molecular (genomics, proteomics, metabolomics, secretomics, metagenomics) (Daniel, 2005), biosensor (Gage et al., 2008), and bioinformatic tools to achieve this objective at both the individual genotype and community level. We must now begin to apply these tools to advance our understanding of the ecology and biochemistry of soil enzymology.

Are the activities of microbial enzymes in soil an example of organized chaos, ongoing selection processes, or the outcome of an advanced and stable community organization? The answers to these fundamental questions and their application to the rational manipulation of soil should emerge within the next decade.

REFERENCES

Abbott, D.W., and B.A. Boraston. 2008. Structural biology of pectin degradation by *Enterobacteriaceae*. Microbiol. Mol. Biol. Rev. 72:301–316.

Alam, M.Z., M.F. Mansor, and K.C.A. Jalal. 2009. Optimization of lignin peroxidase production and stability by *Phanerochaete chrysosporium* using sewage-treatment-plant sludge as substrate in a stirred-tank bioreactor. J. Ind. Microbiol. Biotechnol. 36:757–764.

Alexandre, G., and I.B. Zhulin. 2000. Laccases are widespread in bacteria. Trends Biotechnol. 18:41–42.

Allison, S.D. 2005. Cheaters, diffusion and nutrients constrain decomposition by microbial enzymes in spatially structured environments. Ecol. Lett. 8:626–635.

Allison, S.D., and J.D. Jastrow. 2006. Activities of extracellular enzymes in physically isolated fractions of restored grassland soils. Soil Biol. Biochem. 38:3245–3256.

Allison, S.D., and J.B.H. Martiny. 2008. Resistance, resilience, and redundancy in microbial communities. Proc. Natl. Acad. Sci. USA 105:11512–11519.

Allison, S.D., and P.M. Vitousek. 2005. Responses of extracellular enzymes to simple and complex nutrient inputs. Soil Biol. Biochem. 37:937–944.

Alvarez, V.M., I. von der Weid, L. Seldin, and A.L.S. Santos. 2006. Influence of growth conditions on the production of extracellular proteolytic enzymes in *Paenibacillus peoriae* NRRL BD-62 and *Paenibacillus polymyxa* SCE2. Lett. Appl. Microbiol. 43:625–630.

Aro, N., T. Pakula, and M. Penttila. 2005. Transcriptional regulation of plant cell wall degradation by filamentous fungi. FEMS Microbiol. Rev. 29:719–739.

Asgher, M., H.N. Bhatti, M. Ashraf, and R.L. Legge. 2008. Recent developments in biodegradation of industrial pollutants by white rot fungi and their enzyme system. Biodegradation 19:771–783.

Baldrian, P. 2006. Fungal laccases—occurrence and properties. FEMS Microbiol. Rev. 30:215–242.

Baldrian, P., and V. Valáková. 2008. Degradation of cellulose by basidiomycetous fungi. FEMS Microbiol. Rev. 32:501–521.

Barnard, A.M.L., and G.P.C. Salmond. 2007. Quorum sensing in *Erwinia* species. Anal. Bioanal. Chem. 387:415–423.

Bayer, E.A., J.P. Belaich, Y. Shoham, and R. Lamed. 2004. The cellulosomes: Multienzyme machines for degradation of plant cell wall polysaccharides. Annu. Rev. Microbiol. 58:521–554.

Bayer, E.A., R. Lamed, and M.E. Himmel. 2007. The potential of cellulases and cellulosomes for cellulosic waste management. Curr. Opin. Biotechnol. 18:237–245.

Bayer, E.A., R. Lamed, B.A. White, and H.J. Flint. 2008. From cellulosomes to cellulosomics. Chem. Rec. 8:364–377.

Blum, G., S.R. Mullins, K. Keren, M. Fonovic, C. Jedeszko, M.J. Rice, B.F. Sloane, and M. Bogyo. 2005. Dynamic imaging of protease activity with fluorescently quenched activity-based probes. Nat. Chem. Biol. 1:203–209.

Blum, G., G. von Degenfeld, M.J. Merchant, H.M. Blau, and M. Bogyo. 2007. Noninvasive optical imaging of cysteine protease activity using fluorescently quenched activity-based probes. Nat. Chem. Biol. 3:668–677.

Boraston, A.B., E. Kwan, P. Chiu, R.A.J. Warren, and D.G. Kilburn. 2003. Recognition and hydrolysis of noncrystalline cellulose. J. Biol. Chem. 278:6120–6127.

Bouws, H., A. Wattenberg, and H. Zorn. 2008. Fungal secretomes—nature's toolbox for white biotechnology. Appl. Microbiol. Biotechnol. 80:381–388.

Bradford, M.A., C.A. Davies, S.D. Frey, T.R. Maddox, J.M. Melillo, J.E. Mohan, J.F. Reynolds, K.K. Treseder, and M.D. Wallenstein. 2008. Thermal adaptation of soil microbial respiration to elevated temperature. Ecol. Lett. 11:1316–1327.

Bu, L.T., G.T. Beckham, M.F. Crowley, C.H. Chang, J.F. Matthews, Y.J. Bomble, W.S. Adney, M.E. Himmel, and M.R. Nimlos. 2009. The energy landscape for the interaction of the family 1 carbohydrate-binding module and the cellulose surface is altered by hydrolyzed glycosidic bonds. J. Phys. Chem. B 113:10994–11002.

Burns, R.G. 1978. Soil enzymes. Academic Press, New York.

Burns, R.G. 1982. Enzyme activity in soil: Location and a possible role in microbial ecology. Soil Biol. Biochem. 14:423–427.

Burns, R.G., and R.P. Dick (ed.). 2002. Enzymes in the environment: Activity, ecology, and applications. Marcel Dekker, New York.

Caldwell, B.A. 2005. Enzyme activities as a component of soil biodiversity: A review. Pedobiologia 49:637.

Carreiro, M.M., R.L. Sinsabaugh, D.A. Repert, and D.F. Parkhurst. 2000. Microbial enzyme shifts explain litter decay responses to simulated nitrogen deposition. Ecology 81:2359–2365.

Chen, D.M., A.F.S. Taylor, R.M. Burke, and J.W.G. Cairney. 2001. Identification of genes for lignin peroxidases and manganese peroxidases in ectomycorrhizal fungi. New Phytol. 152:151–158.

Chróst, R.J. 1991. Environmental control of the synthesis and activity of aquatic microbial ectoenzymes. p. 29–59. *In* R.J. Chróst ed. Microbial enzymes in aquatic environments. Springer-Verlag, New York.

Chung, H.G., D.R. Zak, P.B. Reich, and D.S. Ellsworth. 2007. Plant species richness, elevated CO_2, and atmospheric nitrogen deposition alter soil microbial community composition and function. Glob. Change Biol. 13:980–989.

Claus, H. 2003. Laccases and their occurrence in prokaryotes. Arch. Microbiol. 179:145–150.

Cleveland, C.C., and D. Liptzin. 2007. C:N:P stoichiometry in soil: Is there a "Redfield ratio" for the microbial biomass? Biogeochemistry 85:235–252.

Coker, J.A., P.P. Sheridan, J. Loveland-Curtze, K.R. Gutshall, A.J. Auman, and J.E. Brenchley. 2003. Biochemical characterization of a β-galactosidase with a low temperature optimum obtained from an Antarctic *Arthrobacter* isolate. J. Bacteriol. 185:5473–5482.

Conant, R.T., J.M. Steinweg, M.L. Haddix, E.A. Paul, A.F. Plante, and J. Six. 2008. Experimental warming shows that decomposition temperature sensitivity increases with soil organic matter recalcitrance. Ecology 89:2384–2391.

Crecchio, C., M. Curci, M.D.R. Pizzigallo, P. Ricciuti, and P. Ruggiero. 2004. Effects of municipal solid waste compost amendments on soil enzyme activities and bacterial genetic diversity. Soil Biol. Biochem. 36:1595–1605.

Criquet, S., S. Tagger, G. Vogt, G. Iacazio, and J. Le Petit. 1999. Laccase activity of forest litter. Soil Biol. Biochem. 31:1239–1244.

Daniel, R. 2005. The metagenomics of soil. Nat. Rev. Microbiol. 3:470–478.

Daniel, R.M., M.J. Danson, and R. Eisenthal. 2001. The temperature optima of enzymes: A new perspective on an old phenomenon. Trends Biochem. Sci. 26:223–225.

Daniel, R.M., M.J. Danson, R. Eisenthal, C.K. Lee, and M.E. Peterson. 2008. The effect of temperature on enzyme activity: New insights and their implications. Extremophiles 12:51–59.

Davidson, C.J., and M.G. Surette. 2008. Individuality in bacteria. Annu. Rev. Genet. 42:253–268.

Davidson, E.A., and I.A. Janssens. 2006. Temperature sensitivity of soil carbon decomposition and feedbacks to climate change. Nature 440:165.

Davidson, E.A., S.E. Trumbore, and R. Amundson. 2000. Biogeochemistry: Soil warming and organic carbon content. Nature 408:789–790.

Demain, A.L., M. Newcomb, and J.H.D. Wu. 2005. Cellulase, Clostridia, and ethanol. Microbiol. Mol. Biol. Rev. 69:124–154.

Desvaux, M. 2005. The cellulosome of *Clostridium cellulolyticum*. Enzyme Microb. Technol. 37:373–385.

Dhawan, S., and J. Kaur. 2007. Microbial mannanases: An overview of production and applications. Crit. Rev. Biotechnol. 27:197–216.

Di Nardo, C., A. Cinquegrana, S. Papa, A. Fuggi, and A. Fioretto. 2004. Laccase and peroxidase isoenzymes during leaf litter decomposition of *Quercus ilex* in a Mediterranean ecosystem. Soil Biol. Biochem. 36:1539–1544.

Dong, S., D. Brooks, M.D. Jones, and S.J. Grayston. 2007. A method for linking in situ activities of hydrolytic enzymes to associated organisms in forest soils. Soil Biol. Biochem. 39:2414–2419.

Dorodnikov, M., E. Blagodatskaya, S. Blagodatsky, S. Marhan, A. Fangmeier, and Y. Kuzyakov. 2009. Stimulation of microbial extracellular enzyme activities by elevated CO_2 depends on soil aggregate size. Glob. Change Biol. 15:1603–1614.

Douglas, A.E. 2009. The microbial dimension in insect nutritional ecology. Funct. Ecol. 23:38–47.

Dunn, C.G., W.L. Campbell, H. Fram, and A. Hutchins. 1948. Biological and photochemical effects of high energy electrostatically produced Roentgen rays and cathode rays. J. Appl. Phys. 19:605–616.

Duo-Chuan, L. 2006. Review of fungal chitinases. Mycopathologia 161:345–360.

Ekschmitt, K., M.Q. Liu, S. Vetter, O. Fox, and V. Wolters. 2005. Strategies used by soil biota to overcome soil organic matter stability—why is dead organic matter left over in the soil? Geoderma 128:167–176.

Eliasson, P.E., R.E. McMurtrie, D.A. Pepper, M. Stromgren, S. Linder, and G.I. Agren. 2005. The response of heterotrophic CO_2 flux to soil warming. Glob. Change Biol. 11:167–181.

Elsgaard, L., and F.R. Vinther. 2004. Modeling of the fine-scale temperature response of arylsulfatase activity in soil. J. Plant Nutr. Soil Sci. 167:196–201.

Fang, C.M., P. Smith, J.B. Moncrieff, and J.U. Smith. 2005. Similar response of labile and resistant soil organic matter pools to changes in temperature. Nature 433:57–59.

Feldman, M.F., M. Wacker, M. Hernandez, P.G. Hitchen, C.L. Marolda, M. Kowarik, H.R. Morris, A. Dell, M.A. Valvano, and M. Aebi. 2005. Engineering N-linked protein glycosylation with diverse O antigen lipopolysaccharide structures in *Escherichia coli*. Proc. Natl. Acad. Sci. USA 102:3016–3021.

Feller, G., and C. Gerday. 2003. Psychrophilic enzymes: Hot topics in cold adaptation. Nat. Rev. Microbiol. 1:200–208.

Fenner, N., C. Freeman, and B. Reynolds. 2005. Observations of a seasonally shifting thermal optimum in peatland carbon-cycling processes; implications for the global carbon cycle and soil enzyme methodologies. Soil Biol. Biochem. 37:1814.

Fierer, N., J.M. Craine, K. McLauchlan, and J.P. Schimel. 2005. Litter quality and the temperature sensitivity of decomposition. Ecology 86:320–326.

Finzi, A.C., R.L. Sinsabaugh, T.M. Long, and M.P. Osgood. 2006. Microbial community responses to atmospheric carbon dioxide enrichment in a warm-temperate forest. Ecosystems 9:215–226.

Flemming, H.C., and J. Wingender. 2002. What biofilms contain—Proteins, polysaccharides, etc. Chem. Unserer Zeit 36:30–42.

Foster, R. 1985. In situ localization of organic matter in soils. Quaestiones Entomologicae 21:609–633.

Foster, R., and J.K. Martin. 1981. In situ analysis of soil components of biological origin. p. 75–110. *In* E.A. Paul and J.N. Ladd (ed.) Soil biochemistry. Marcel Dekker, New York.

Frankenberger, W.T., and M.A. Tabatabai. 1991a. L-Asparaginase activity of soils. Biol. Fertil. Soils 11:6–12.

Frankenberger, W.T., and M.A. Tabatabai. 1991b. L-Glutaminase activity of soils. Soil Biol. Biochem. 23:869–874.

Gage, D.J., P.M. Herron, C.A. Pinedo, and Z.G. Cardon. 2008. Live reports from the soil grain—the promise and challenge of microbiosensors. Funct. Ecol. 22:983–989.

Gallo, M., R. Amonette, C. Lauber, R.L. Sinsabaugh, and D.R. Zak. 2004. Microbial community structure and oxidative enzyme activity in nitrogen-amended north temperate forest soils. Microb. Ecol. 48:218–229.

Geisseler, D., and W.R. Horwath. 2008. Regulation of extracellular protease activity in soil in response to different sources and concentrations of nitrogen and carbon. Soil Biol. Biochem. 40:3040–3048.

Gianfreda, L., M.A. Rao, and A. Violante. 1995. Formation and activity of urease-tannate complexes affected by aluminum, iron, and manganese. Soil Sci. Soc. Am. J. 59:805–810.

Gold, N.D., and V.J.J. Martin. 2007. Global view of the *Clostridium thermocellum* cellulosome revealed by quantitative proteomic analysis. J. Bacteriol. 189:6787–6795.

Harder, W., and L. Dijkhuizen. 1983. Physiological responses to nutrient limitation. Annu. Rev. Microbiol. 37:1–23.

Hatakka, A. 1994. Lignin-modifying enzymes from selected white-rot fungi: production and role in lignin degradation. FEMS Microbiol. Rev. 13:125–135.

Henry, H.A.L., J.D. Juarez, C.B. Field, and P.M. Vitousek. 2005. Interactive effects of elevated CO_2, N deposition and climate change on extracellular enzyme activity and soil density fractionation in a California annual grassland. Glob. Change Biol. 11:1808–1815.

Hilden, L., and G. Johansson. 2004. Recent developments on cellulases and carbohydrate-binding modules with cellulose affinity. Biotechnol. Lett. 26:1683–1693.

Hope, C.F.A., and R.G. Burns. 1985. The barrier-ring plate technique for studying extracellular enzyme diffusion and microbial-growth in model soil environments. J. Gen. Microbiol. 131:1237–1243.

Huston, A.L., B.B. Krieger-Brockett, and J.W. Deming. 2000. Remarkably low temperature optima for extracellular enzyme activity from Arctic bacteria and sea ice. Environ. Microbiol. 2:383–388.

Ikeda-Ohtsubo, W., and A. Brune. 2009. Cospeciation of termite gut flagellates and their bacterial endosymbionts: *Trichonympha* species and '*Candidatus* Endomicrobium trichonymphae'. Mol. Ecol. 18:332–342.

Irwin, D.C., M. Spezio, L.P. Walker, and D.B. Wilson. 1993. Activity studies of 8 purified cellulases: Specificity, synergism, and binding domain effects. Biotechnol. Bioeng. 42:1002–1013.

Jastrow, J.D., J.E. Amonette, and V.L. Bailey. 2007. Mechanisms controlling soil carbon turnover and their potential application for enhancing carbon sequestration. Clim. Change 80:5–23.

Jervis, E.J., C.A. Haynes, and D.G. Kilburn. 1997. Surface diffusion of cellulases and their isolated binding domains on cellulose. J. Biol. Chem. 272:24016–24023.

Joanisse, G.D., R.L. Bradley, C.M. Preston, and A.D. Munson. 2007. Soil enzyme inhibition by condensed litter tannins may drive ecosystem structure and processes: The case of *Kalmia angustifolia*. New Phytol. 175:535–546.

Jones, C.D., P. Cox, and C. Huntingford. 2003. Uncertainty in climate–carbon-cycle projections associated with the sensitivity of soil respiration to temperature. Tellus Ser. B: Chem. Phys. Meteorol. 55:642–648.

Jones, C., C. McConnell, K. Coleman, P. Cox, P. Falloon, D. Jenkinson, and D. Powlson. 2005. Global climate change and soil carbon stocks; predictions from two contrasting models for the turnover of organic carbon in soil. Glob. Change Biol. 11:154–166.

Kaiser, K., and G. Guggenberger. 2007. Sorptive stabilization of organic matter by microporous goethite: Sorption into small pores vs. surface complexation. Eur. J. Soil Sci. 58:45–59.

Kandeler, E. 1990. Characterization of free and adsorbed phosphatases in soils. Biol. Fertil. Soils 9:199–202.

Kelleher, B.P., and A.J. Simpson. 2006. Humic substances in soils: Are they really chemically distinct? Environ. Sci. Technol. 40:4605–4611.

Kellner, H., P. Luis, and F. Buscot. 2007. Diversity of laccase-like multicopper oxidase genes in *Morchellaceae*: Identification of genes potentially involved in extracellular activities related to plant litter decay. FEMS Microbiol. Ecol. 61:153–163.

Kellner, H., P. Luis, B. Zimdars, B. Kiesel, and F. Buscot. 2008. Diversity of bacterial laccase-like multicopper oxidase genes in forest and grassland Cambisol soil samples. Soil Biol. Biochem. 40:638–648.

Kikot, G.E., R.A. Hours, and T.M. Alconada. 2009. Contribution of cell wall degrading enzymes to pathogenesis of *Fusarium graminearum*: A review. J. Basic Microbiol. 49:231–241.

Kim, E.S., H.J. Lee, W.G. Bang, I.G. Choi, and K.H. Kim. 2009. Functional characterization of a bacterial expansin from *Bacillus subtilis* for enhanced enzymatic hydrolysis of cellulose. Biotechnol. Bioeng. 102:1342–1353.

Kirby, R. 2006. Actinomycetes and lignin degradation. Adv. Appl. Microbiol. 58:125–168.

Kirschbaum, M.U.F. 2004. Soil respiration under prolonged soil warming: Are rate reductions caused by acclimation or substrate loss? Glob. Change Biol. 10:1870–1877.

Klonowska, A., C. Gaudin, A. Fournel, M. Asso, J. Le Petit, M. Giorgi, and T. Tron. 2002. Characterization of a low redox potential laccase from the basidiomycete C30. Eur. J. Biochem. 269:6119–6125.

Knorr, W., I.C. Prentice, J.I. House, and E.A. Holland. 2005. Long-term sensitivity of soil carbon turnover to warming. Nature 433:298.

Koch, A.L. 1985. The macroeconomics of bacterial growth. p. 1–42. *In* M. Fletcher and G.D. Floodgate (ed.) Bacteria in their natural environments. Academic Press, London.

Koch, O., D. Tscherko, and E. Kandeler. 2007. Temperature sensitivity of microbial respiration, nitrogen mineralization, and potential soil enzyme activities in organic alpine soils. Global Biogeochem. Cycles. doi:10.1029/2007GB002983

Kristensen, J.B., R.L. Meyer, B.S. Laursen, S. Shipovskov, F. Besenbacher, and C.H. Poulsen. 2008. Antifouling enzymes and the biochemistry of marine settlement. Biotechnol. Adv. 26:471–481.

Krug, D., G. Zurek, O. Revermann, M. Vos, G.J. Velicer, and R. Muller. 2008. Discovering the hidden secondary metabolome of *Myxococcus xanthus*: A study of intraspecific diversity. Appl. Environ. Microbiol. 74:3058–3068.

Lamed, R., E. Setter, and E.A. Bayer. 1983. Characterization of a cellulose-binding, cellulase-containing complex in *Clostridium thermocellum*. J. Bacteriol. 156:828–836.

Larrainzar, E., F. O'Gara, and J.P. Morrissey. 2005. Applications of autofluorescent proteins for in situ studies in microbial ecology. Annu. Rev. Microbiol. 59:257–277.

Loveland, J., K. Gutshall, J. Kasmir, P. Prema, and J.E. Brenchley. 1994. Characterization of psychrotrophic microorganisms producing β-galactosidase activities. Appl. Environ. Microbiol. 60:12–18.

Luo, Y.Q. 2007. Terrestrial carbon-cycle feedback to climate warming. Annu. Rev. Ecol. Evol. Syst. 38:683–712.

Martinez, D., R.M. Berka, B. Henrissat, M. Saloheimo, M. Arvas, S. Baker, J. Chapman, O. Chertkov, P.M. Coutinho, D. Cullen, E.G.J. Danchin, I.V. Grigoriev, P. Harris, M. Jackson, C.P. Kubicek, C.S. Han, I. Ho, L.F. Larrondo, A.L. de Leon, J.K. Magnuson, S. Merino, M. Misra, B. Nelson, N. Putnam, B. Robbertse, A.A. Salamov, M. Schmoll, A. Terry, N. Thayer, A. Westerholm-Parvinen, C.L. Schoch, J. Yao, R. Barbote, M.A. Nelson, C. Detter, D. Bruce, C.R. Kuske, G. Xie, P. Richardson, D.S. Rokhsar, S.M. Lucas, E.M. Rubin, N. Dunn-Coleman, M. Ward, and T.S. Brettin. 2008. Genome sequencing and analysis of the biomass-degrading fungus *Trichoderma reesei* (syn. *Hypocrea jecorina*). Nat. Biotechnol. 26:553–560.

Marx, M.C., E. Kandeler, M. Wood, N. Wermbter, and S.C. Jarvis. 2005. Exploring the enzymatic landscape: Distribution and kinetics of hydrolytic enzymes in soil particle-size fractions. Soil Biol. Biochem. 37:35–48.

McClaugherty, C.A., and A.E. Linkins. 1990. Temperature responses of enzymes in two forest soils. Soil Biol. Biochem. 22:29–33.

McLaren, A.D., R.A. Luse, and J.J. Skujiņš. 1962. Sterilization of soil by irradiation and some further observations on soil enzyme activity. Soil Sci. Soc. Am. Proc. 26:371–377.

McLaren, A.D., L. Reshetko, and W. Huber. 1957. Sterilization of soil by irradiation with an electron beam, and some observations on soil enzyme activity. Soil Sci. 83:497–502.

Melillo, J.M., P.A. Steudler, J.D. Aber, K. Newkirk, H. Lux, F.P. Bowles, C. Catricala, A. Magill, T. Ahrens, and S. Morrisseau. 2002. Soil warming and carbon-cycle feedbacks to the climate system. Science 298:2173–2176.

Michalet, X., F.F. Pinaud, L.A. Bentolila, J.M. Tsay, S. Doose, J.J. Li, G. Sundaresan, A.M. Wu, S.S. Gambhir, and S. Weiss. 2005. Quantum dots for live cells, in vivo imaging, and diagnostics. Science 307:538–544.

Michel, K., and E. Matzner. 2003. Response of enzyme activities to nitrogen addition in forest floors of different C-to-N ratios. Biol. Fertil. Soils 38:102–109.

Mingardon, F., A. Chanal, C. Tardif, E.A. Bayer, and H.P. Fierobe. 2007. Exploration of new geometries in cellulosome-like chimeras. Appl. Environ. Microbiol. 73:7138–7149.

Modak, T., S. Pradhan, and M. Watve. 2007. Sociobiology of biodegradation and the role of predatory protozoa in biodegrading communities. J. Biosci. 32:775–780.

Morrison, M., P.B. Pope, S.E. Denman, and C.S. McSweeney. 2009. Plant biomass degradation by gut microbiomes: More of the same or something new? Curr. Opin. Biotechnol. 20:358–363.

Nannipieri, P. 2006. Role of stabilised enzymes in microbial ecology and enzyme extraction from soil with potential applications in soil proteomics. p. 75–94 In P. Nannipieri and K. Smalla (ed.) Nucleic acids and proteins in soil. Soil biology. Vol. 8. Springer, Berlin.

Nannipieri, P., and J.M. Bollag. 1991. Use of enzymes to detoxify pesticide-contaminated soils and waters. J. Environ. Qual. 20:510–517.

Nannipieri, P., B. Ceccanti, and D. Bianchi. 1988. Characterization of humus phosphatase complexes extracted from soil. Soil Biol. Biochem. 20:683–691.

Nannipieri, P., B. Ceccanti, S. Cervelli, and P. Sequi. 1978. Stability and kinetic properties of humus urease complexes. Soil Biol. Biochem. 10:143–148.

Nannipieri, P., E. Kandeler, and P. Ruggiero. 2002. Enzyme activities and microbiological and biochemical processes in soil. p. 1–33. In R.G. Burns and R.P. Dick (ed.) Enzymes in the environment. Marcel Dekker, New York.

Niemi, R.M., and M. Vepsalainen. 2005. Stability of the fluorogenic enzyme substrates and pH optima of enzyme activities in different Finnish soils. J. Microbiol. Methods 60:195–205.

Nigmatullin, R., R. Lovitt, C. Wright, M. Linder, T. Nakari-Setala, and A. Gama. 2004. Atomic force microscopy study of cellulose surface interaction controlled by cellulose binding domains. Colloids Surf. B Biointerfaces 35:125–135.

Nozaki, M., C. Miura, Y. Tozawa, and T. Miura. 2009. The contribution of endogenous cellulase to the cellulose digestion in the gut of earthworm (*Pheretima hilgendorfi*: Megascolecidae). Soil Biol. Biochem. 41:762–769.

Nyanhongo, G.S., G. Gubitz, P. Sukyai, C. Leitner, D. Haltrich, and R. Ludwig. 2007. Oxidoreductases from *Trametes* spp. in biotechnology: A wealth of catalytic activity. Food Technol. Biotechnol. 45:250–268.

Osono, T. 2007. Ecology of ligninolytic fungi associated with leaf litter decomposition. Ecol. Res. 22:955–974.

Pang, Y., X. Liu, Y. Ma, L. Chernin, G. Berg, and K. Gao. 2009. Induction of systemic resistance, root colonisation and biocontrol activities of the rhizospheric strain of *Serratia plymuthica* are dependent on N-acyl homoserine lactones. Eur. J. Plant Pathol. 124:261–268.

Parham, J.A., and S.P. Deng. 2000. Detection, quantification and characterization of β-glucosaminidase activity in soil. Soil Biol. Biochem. 32:1183–1190.

Pastor, A., J. Chabert, M. Louwagie, J. Garin, and I. Attree. 2005. PscF is a major component of the *Pseudomonas aeruginosa* type III secretion needle. FEMS Microbiol. Lett. 253:95–101.

Peer, A., S.P. Smith, E.A. Bayer, R. Lamed, and I. Borovok. 2009. Noncellulosomal cohesin- and dockerin-like modules in the three domains of life. FEMS Microbiol. Lett. 291:1–16.

Pelletier, A., and J. Sygush. 1990. Purification and characterization of three chitosanase activities from *Bacillus megaterium* P1. Appl. Environ. Microbiol. 56:844–848.

Pflug, W. 1982. Soil enzymes and clay-minerals: Effect of clay-minerals on the activity of polysaccharide cleaving soil enzymes. Z. Pflanzenernaehr. Bodenk. 145:493–502.

Piccolo, A. 2001. The supramolecular structure of humic substances. Soil Sci. 166:810–832.

Quiquampoix, H., and R.G. Burns. 2007. Interactions between proteins and soil mineral surfaces: Environmental and health consequences. Elements 3:401–406.

Quiquampoix, H., S. Servagent-Noinville, and M.H. Baron. 2002. Enzyme adsorption on soil mineral surfaces and consequences for the catalytic activity. p. 285–306. *In* R.G. Burns and R. Dick (ed.) Enzymes in the environment: Activity, ecology, and applications. Marcel Dekker, New York.

Raut, M.P., S. William, J.K. Bhattacharyya, T. Chakrabarti, and S. Devotta. 2008. Microbial dynamics and enzyme activities during rapid composting of municipal solid waste—A compost maturity analysis perspective. Bioresour. Technol. 99:6512–6519.

Romani, A.M., K. Fund, J. Artigas, T. Schwartz, S. Sabater, and U. Obst. 2008. Relevance of polymeric matrix enzymes during biofilm formation. Microb. Ecol. 56:427–436.

Saiya-Cork, K.R., R.L. Sinsabaugh, and D.R. Zak. 2002. The effects of long term nitrogen deposition on extracellular enzyme activity in an *Acer saccharum* forest soil. Soil Biol. Biochem. 34:1309–1315.

Sanchez-Perez, G., A. Mira, G. Nyiro, L. Pasic, and F. Rodriguez-Valera. 2008. Adapting to environmental changes using specialized paralogs. Trends Genet. 24:154–158.

Schimel, J.P., and M.N. Weintraub. 2003. The implications of exoenzyme activity on microbial carbon and nitrogen limitation in soil: A theoretical model. Soil Biol. Biochem. 35:549–563.

Schwarz, W.H., and V.V. Zverlov. 2006. Protease inhibitors in bacteria: An emerging concept for the regulation of bacterial protein complexes? Mol. Microbiol. 60:1323–1326.

Shackle, V., C. Freeman, and B. Reynolds. 2006. Exogenous enzyme supplements to promote treatment efficiency in constructed wetlands. Sci. Total Environ. 361:18–24.

Silva, A., M. Bacci, F.C. Pagnocca, O.C. Bueno, and M.J.A. Hebling. 2006. Starch metabolism in *Leucoagaricus gongylophorus*, the symbiotic fungus of leaf-cutting ants. Microbiol. Res. 161:299–303.

Sinsabaugh, R.L. 1994. Enzymatic analysis of microbial pattern and process. Biol. Fertil. Soils 17:69–74.

Sinsabaugh, R.L., M.M. Carreiro, and D.A. Repert. 2002. Allocation of extracellular enzymatic activity in relation to litter composition, N deposition, and mass loss. Biogeochemistry 60:1–24.

Sinsabaugh, R.L., and D.L. Moorhead. 1994. Resource-allocation to extracellular enzyme-production: A model for nitrogen and phosphorus control of litter decomposition. Soil Biol. Biochem. 26:1305–1311.

Steinweg, J.M., A.F. Plante, R.T. Conant, E.A. Paul, and D.L. Tanaka. 2008. Patterns of substrate utilization during long-term incubations at different temperatures. Soil Biol. Biochem. 40:2722–2728.

Stursova, M., and R.L. Sinsabaugh. 2008. Stabilization of oxidative enzymes in desert soil may limit organic matter accumulation. Soil Biol. Biochem. 40:550–553.

Suarez, M.B., L. Sanz, M.I. Chamorro, M. Rey, F.J. Gonzalez, A. Llobell, and E. Monte. 2005. Proteomic analysis of secreted proteins from *Trichoderma harzianum:* Identification of a fungal cell wall-induced aspartic protease. Fungal Genet. Biol. 42:924–934.

Tate, R.L. 2002. Microbiology and enzymology of carbon and nitrogen cycling. p. 227–248. *In* R.G. Burns and R. Dick (ed.) Enzymes in the environment: Activity, ecology, and applications. Marcel Dekker, New York.

Trasar-Cepeda, C., F. Gil-Sotres, and M.C. Leiros. 2007. Thermodynamic parameters of enzymes in grassland soils from Galicia, NW Spain. Soil Biol. Biochem. 39:311–319.

Tsumuraya, Y. 1996. Expansins, nonenzymatic proteins promoting cell wall expansion in higher plants. Trends Glycosci. Glycotechnol. 8:443–444.

Valaskova, V., J. Snajdr, B. Bittner, T. Cajthaml, V. Merhautova, M. Hoffichter, and P. Baldrian. 2007. Production of lignocellulose-degrading enzymes and degradation of leaf litter by saprotrophic basidiomycetes isolated from a *Quercus petraea* forest. Soil Biol. Biochem. 39:2651–2660.

Vanden Wymelenberg, A., G. Sabat, M. Mozuch, P.J. Kersten, D. Cullen, and R.A. Blanchette. 2006. Structure, organization, and transcriptional regulation of a family of copper radical oxidase genes in the lignin-degrading basidiomycete *Phanerochaete chrysosporium*. Appl. Environ. Microbiol. 72:4871–4877.

Velicer, G.J. 2003. Social strife in the microbial world. Trends Microbiol. 11:330–337.

Wackett, L.P. 2008. Biomass to fuels via microbial transformations. Curr. Opin. Chem. Biol. 12:187–193.

Wackett, L.P. 2009. Questioning our perceptions about evolution of biodegradative enzymes. Curr. Opin. Microbiol. 12:244–251.

Waldrop, M.P., T.C. Balser, and M.K. Firestone. 2000. Linking microbial community composition to function in a tropical soil. Soil Biol. Biochem. 32:1837–1846.

Wallenstein, M.D., S.K. McMahon, and J.P. Schimel. 2009. Seasonal variation in enzyme activities and temperature sensitivities in Arctic tundra soils. Glob. Change Biol. 15:1631–1639.

Wallenstein, M.D., and M.N. Weintraub. 2008. Emerging tools for measuring and modeling the in situ activity of soil extracellular enzymes. Soil Biol. Biochem. 40:2098–2106.

Whiteside, M.D., K.K. Treseder, and P.R. Atsatt. 2009. The brighter side of soils: Quantum dots track organic nitrogen through fungi and plants. Ecology 90:100–108.

Wilson, D.B. 2008. Three microbial strategies for plant cell wall degradation. Ann. N.Y. Acad. Sci. 1125:289–297.

Wirth, S.J., and G.A. Wolf. 1992. Micro-plate colourimetric assay for *endo* -acting cellulase, xylanase, chitinase, 1,3-β-glucanase and amylase extracted from forest soil horizons. Soil Biol. Biochem. 24:511–519.

Witayakran, S., and A.J. Ragauskas. 2009. Modification of high-lignin softwood kraft pulp with laccase and amino acids. Enzyme Microb. Technol. 44:176–181.

Wong, D.W.S. 2009. Structure and action mechanism of ligninolytic enzymes. Appl. Biochem. Biotechnol. 157:174–209.

Chapter 3

Kinetics of Soil Enzyme Reactions

Warren A. Dick

3–1 INTRODUCTION

For a chemical reaction to occur, the reactants in their ground states must first come together with sufficient energy to form an activated transition state before proceeding to the final product(s) or state. Once sufficient energy has been attained to form the activated transition state, the probability that the chemical reaction will proceed to the formation of product is much increased. The fact that enzymes are so efficient in lowering the energy barrier so that reactions can take place is often accepted without much thought. Yet, a simple exercise illustrates just how amazing enzymes are in catalyzing important functions, even in places such as soil, where they appear in extremely low concentrations.

To estimate the concentration of one common enzyme in soil, alkaline phosphatase, the following assumptions are made. These assumptions are reasonable in that all are based on experimentally derived information.

1. The molecular weight of alkaline phosphatase (from *E. coli*) is 160,000 g mol^{-1}.

2. Measured activity of alkaline phosphatase activity in soil is 14 µg *p*-nitrophenol released h^{-1} g^{-1} soil using *p*-nitrophenyl phosphate as the substrate.

3. The molecular weight of *p*-nitrophenol is 140 g mol^{-1} or 140 µg µmol^{-1}.

4. A unit of enzyme activity is defined as the production of 1 µmol *p*-nitrophenol min^{-1}. Therefore, activity of alkaline phosphatase in 1 g of soil is 0.10 µmol h^{-1} or 0.001667 units.

5. The activity of purified alkaline phosphatase is approximately 3000 units mg^{-1} protein.

With these assumptions the following equation can be developed:

$$\frac{0.001667 \text{ units activity}/\text{g soil}}{3000 \text{ units activity}/\text{mg alkaline phosphatase}} = "x" \text{ mg alkaline phosphatase}/\text{g soil} \quad [1]$$

Solving for x gives us an answer of 5.6 × 10^{-7} mg enzyme g^{-1} soil. On a molar basis, the concentration of alkaline phosphatase in soil would be 3.5 × 10^{-15} mol g^{-1} soil.

Warren A. Dick, Professor, Soil Science, The Ohio State University, Wooster, OH 44691 (dick.5@osu.edu)

doi:10.2136/sssabookser9.c3

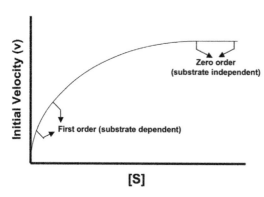

Fig. 3–1. Michaelis–Menten plot relating the initial reaction velocity (v) to the substrate concentration [S].

Enzymes are efficient catalysts because they bring about the formation of an activated transition state at a reduced energy expense. Additional features of an enzyme-catalyzed reaction are the phenomena of specificity and saturation. As the concentration of the substrate increases from zero, the initial velocity of the reaction with respect to substrate concentration is observed to follow first-order kinetics (Fig. 3–1). However, the rate of increase in initial velocity becomes less and less with each unit increase in substrate concentration. Eventually a point is reached where a further increase in initial velocity is no longer observed and the reaction can be described by zero-order kinetics as pertains to substrate concentration. At this point, the reaction rate is no longer a function of both substrate concentration and enzyme concentration but only of enzyme concentration.

3–2 MICHAELIS–MENTEN KINETICS

Early investigators hypothesized that in the initial step of an enzyme-catalyzed reaction, the enzyme and substrate react reversibly to form an enzyme–substrate complex. This hypothesis was very important in the formulation of the general theory of enzyme kinetics and the equation named after Leonor Michaelis and Maud Menten called the Michaelis–Menten equation (Eq. [2]):

$$v = V_{max}[S]/(K_m + [S])$$ [2]

where v is the initial reaction rate at the substrate concentration [S], V_{max} is the maximum rate of reaction and K_m is the Michaelis constant. The relationship between v and [S] can be shown as a hyperbolic graph (Fig. 3–1).

This Michaelis–Menten equation describes simple enzyme kinetics for single substrate enzyme-catalyzed reactions. It has been widely used to characterize extracellular enzyme reactions in soils and to determine the Michaelis constant (K_m) and the maximum rate of reaction (V_{max}).

The development of the Michaelis–Menten equation rests on several assumptions.

1. The rate of an enzyme-catalyzed reaction changes from first-order to zero-order kinetics (Fig. 3–1).

2. Enzyme (E) reversibly binds with substrate (S) to form an intermediate enzyme–substrate (ES) complex that then breaks down to form product (P) (see below). Each reaction is described by a specific rate constant designated k_1, k_2, and k_3:

$$E + S \underset{k_1}{\overset{k_2}{\rightleftharpoons}} ES \overset{k_3}{\longrightarrow} E + P \qquad [3]$$

3. A steady-state equilibrium between the rate of formation and the rate of degradation of ES is rapidly achieved.
4. The concentration of total enzyme $[E_t]$ is defined as the concentration of enzyme in the free state $[E_f]$ and the concentration of the enzyme–substrate complex $[ES]$, i.e., $[E_t] = [E_f] + [ES]$.
5. The rate-limiting step is the decomposition of the enzyme–substrate complex to form the product. The initial velocity of the enzyme-catalyzed reaction is thus proportional to the concentration of the enzyme–substrate complex, i.e., $v \approx [ES]$.
6. The maximum rate of reaction (V_{max}) is attained when the concentration of the enzyme–substrate complex reaches a maximum. This will occur only when all free enzyme becomes complexed with substrate, i.e., saturation of the enzyme with substrate has occurred.

With these assumptions the following Michaelis–Menten equation can be derived: The rate of formation of ES can be expressed as:

$$d[ES]/dt = k_1[E_f][S] \qquad [4]$$

but $[E_f] = [E_t] - [ES]$ so that the rate for formation of ES can be rewritten as:

$$d[ES]/dt = k_1([E_t] - [ES])([S]) \qquad [5]$$

The rate of breakdown of ES is given as:

$$-d[ES]/dt = k_2[ES] + k_3[ES] \qquad [6]$$

Setting the rates equal to each other gives:

$$d[ES]/dt = -d[ES]/dt \text{ or } k_1([E_t] - [ES])([S]) = k_2[ES] + k_3[ES] \qquad [7]$$

Rearranging the above equation yields:

$$[S]([E_t] - [ES])/[ES] = (k_2 + k_3)/k_1 = K_m \qquad [8]$$

Rearranging again yields:

$$[ES] = ([E_t][S]/(K_m + [S]) \qquad [9]$$

When each side of the above equation is multiplied by k_3, the result is the equation:

$$k_3[ES] = k_3([E_t][S]/(K_m + [S]) \qquad [10]$$

But $k_3[ES] = v$ (initial velocity) and $k_3[E_t] = V_{max}$ (maximum rate of reaction). Substituting these equalities yields the classical form of the Michaelis–Menten equation:

$$v = V_{max}[S]/(K_m + [S]) \qquad [11]$$

As can be noted from Eq. [8], the Michaelis constant (K_m) is really a combination of the individual reaction constants k_1, k_2, and k_3 of the enzyme-catalyzed reaction. The Michaelis constant is often used as a way to easily characterize an enzyme's activity in relationship to its substrate concentration. Simple algebra reveals that when the K_m value is set equal to [S]:

$$v = V_{max}[S]/([S] + [S]),$$ [12]

the reaction rate of the enzyme reaches one-half of its maximum rate ($0.5V_{max}$):

$$v = 0.5V_{max}$$ [13]

The K_m value thus provides useful information because the smaller the K_m value, the greater the affinity of the enzyme for its substrate and the lower the concentration of the substrate required to achieve 50% of the maximum rate of reaction. The unit for K_m is the same as that of substrate concentration.

3–3 DETERMINATION OF THE KINETIC CONSTANTS K_m AND V_{MAX}

The values of the K_m and V_{max} constants are easily determined by converting the Michaelis–Menten equation to an equation corresponding to a straight line (i.e., $y = mx + b$) and solving for K_m and V_{max}. There are several linear transformations possible of the Michaelis–Menten equation and these linear transformations can be easily visualized using graphical techniques (Fig. 3–2). The Lineweaver–Burk double-reciprocal equation is the most commonly used linear transformation:

$$1/v = (K_m/V_{max})(1/[S]) + 1/V_{max}$$ [14]

By plotting the inverse of the initial velocity of an enzyme-catalyzed reaction on the y axis against the inverse of the substrate concentration on the x axis, the y-intercept value is $1/V_{max}$, the slope is equal to K_m/V_{max} and the x-intercept is $-1/K_m$.

Other linear transformations are possible and include the Hanes–Woolf transformation:

$$[S]/v = (1/V_{max})([S]) + K_m/V_{max}$$ [15]

and the Eadie–Hofstee transformation:

$$v = -K_m(v/[S]) + V_{max}$$ [16]

Similarly, as for the Lineweaver–Burk equation, the values of the slope, y-intercepts, and x-intercepts of Hanes–Woolf and the Eadie–Hofstee equations can be used to calculate the K_m and V_{max} values.

3–4 COMMENTS RELATED TO MICHAELIS–MENTEN KINETICS

To ensure adherence to Michaelis–Menten kinetics, it is important that care is taken to actually measure soil enzymes under carefully controlled conditions that

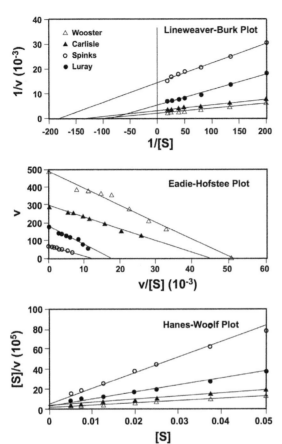

Fig. 3–2. Graphical representation of three possible linear transformations of the Michaelis–Menten equation. These graphs are of myrosinase activity in soils and the reaction velocity (v) is expressed as μg glucose g^{-1} soil 4 h^{-1} and the substrate concentration [S] is in mol L^{-1} (Al-Turki and Dick, 2003).

are zero order for every part of the reaction except that of the enzyme itself. This is because the goal of a soil enzymologist is generally to measure only the concentration of the enzyme in soil. (Note: To review the steps required to develop a valid soil enzyme assay, see Chapter 4). If the enzyme reaction is performed under zero-order kinetics in terms of substrate concentration, it is permissible to compare kinetic constants among soils. This provides a rapid means of assessing the relative amount of an enzyme and the relative enzyme reaction rate in one soil vs. another. For example, comparing V_{max} values among soils immediately informs about the relative concentration of an enzyme among soils. Comparing K_m values provides a rapid assessment of the interaction of an enzyme with its substrate and what substrate concentration in soil would be needed to result in an enzyme having an ecological meaningful level of activity.

The Michaelis constant (K_m) value for soil enzymes is often different from that of a pure enzyme catalyzing the same reaction in a solution contained in a test tube. This is because not all of the assumptions made in the derivation of the Michaelis–Menten equation are strictly met in soils. Thus for soil enzymes, we often speak of the K_m value as the apparent K_m value.

The kinetics of an enzyme reaction in a heterogeneous system such as soil were described by Ruggiero et al. (1996). Factors that affect the kinetic behavior of a soil enzyme include (i) conformational changes in the enzyme; (ii) steric effects in which the accessibility of the active site of the enzyme to its substrate or to inhibitors is altered; (iii) partitioning effects whereby the soil particles that interact with enzymes induce differences in concentrations of substrate, product, hydrogen ions, etc. in the area surrounding the enzyme on the solid support compared with the bulk solution; (iv) diffusional limitations that restrict movement of substrate to and product away from the enzyme; and (v) other microenvironmental effects.

Both an increase or decrease in the value of the Michaelis constant can occur in soil as compared with an enzyme operating in solution under optimal conditions. If the concentration of substrate is less in the microenvironment surrounding the enzyme in soil immobilized on a soil particle compared with in the bulk solution, an increase in K_m will occur. In contrast, the same comparison may result in a decrease in the K_m value in soil if substrate becomes concentrated near the active site of the enzyme due to electrostatic or other types of interactions.

3–5 SIMPLE COMPETITIVE AND NONCOMPETITIVE INHIBITION OF ENZYMES AND ENZYME KINETICS

Inhibitors of single substrate soil enzyme-catalyzed reactions often can be grouped as either simple competitive or simple noncompetitive inhibitors. Both simple competitive and noncompetitive inhibitors can bind reversibly to the enzyme. Inhibitors also can react irreversibly with the enzyme and form a covalent bond with the enzyme (EI) or enzyme–substrate (ESI) complex that then becomes a dead-end complex. These inhibitors are often called suicide inhibitors. The rate at which the EI or ESI complex is formed is called the inactivation rate or k_{inact}. Since formation of EI may compete with formation of ES, binding of inhibitors may be prevented by increasing substrate concentration. The binding and inactivation steps of an irreversible reaction are investigated by incubating the enzyme with inhibitor and assaying the amount of activity remaining over time. The activity will decrease in a time-dependent manner, usually following an exponential decay equation. Fitting these data to a rate equation gives the rate of inactivation at this concentration of inhibitor. This is done at several different concentrations of inhibitor.

Most classical soil enzyme inhibition studies have assumed that inhibition can be reversed. The relationship between substrate concentration [S] and initial rate of reaction (v) for an uninhibited enzyme reaction and for the same reaction in the presence of either a simple competitive or noncompetitive inhibitor is shown in Fig. 3–3.

Competitive inhibitors are those that compete with the substrate for the active site of the enzyme. They often have a chemical similarity to the substrate and, if the inhibitor (I) binds first, form an EI complex. As a result, the substrate cannot bind and the reaction of the enzyme with its substrate will be impeded.

$$E + S \longleftrightarrow ES \longrightarrow E + P$$
$$I -\updownarrow$$
$$EI$$

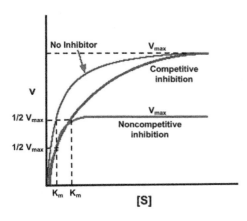

Fig. 3–3. Michaelis–Menten plot relating the initial reaction velocity (v) to the substrate concentration [S] for an enzyme in the presence of either a competitive or noncompetitive inhibitor.

The chance that either the substrate (S) or the inhibitor (I) will bind with the enzyme's active site is a function of their relative concentration and the affinity of the inhibitor to bind to the active site compared with the substrate. If both substrate and inhibitor have an affinity for the enzyme's active site and the binding is reversible, then as the concentration of the inhibitor increases relative to that of the substrate, its chances of binding to the active site of the enzyme will also increase. For example, effective competitive inhibitors for acid or alkaline phosphates are oxyanions similar in many ways to the phosphate ion. Such inhibitors include arsenate, tungstate and molybdate.

A simple noncompetitive inhibitor does not compete with substrate. Instead, a noncompetitive inhibitor reacts either remote from, or very close to, the active site and generally causes a change in the shape of the enzyme. The result is that the substrate can no longer interact with the enzyme and cause a reaction. The effect of a noncompetitive inhibitor may be reversible, but this effect is not influenced by concentration of the substrate, as is the case for a reversible competitive inhibitor.

Noncompetitive inhibitors are those that actually change the effective concentration of the enzyme so that the V_{max} value is decreased. The binding of the substrate to its enzyme does not change so much as the overall amount of enzyme molecules available to bind to the substrate. Merely adding additional amounts of substrate cannot reverse the effect of a simple noncompetitive inhibitor. This type of inhibition is often more ecologically dangerous. Examples would be the presence of mercury in soil because mercury will bind with many of the sulfur atoms in several of the amino acids that make up an enzyme protein. The mercury binding is very strong and essentially disrupts the function of the enzyme. Many heavy metals act as noncompetitive inhibitors in soil enzyme reactions.

Using the Lineweaver–Burk transformation as an example, it can be seen how K_m and V_{max} values change in the presence of a simple competitive inhibitor (Fig. 3–4). Note that the K_m value increases but the V_{max} value remains the same. This is because as the substrate concentration increases, substrate will begin to displace the inhibitor. A point is eventually reached at a saturated substrate concentration where the amount of substrate molecules is so much greater than the amount of inhibitor molecules that all enzyme is essentially bound to its substrate. When this occurs, the maximum rate of reaction (V_{max}) is the same in the presence or absence of an inhibitor at saturated substrate concentrations. The K_m value, how-

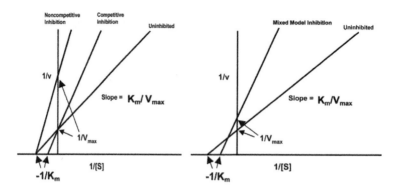

Fig. 3–4. Lineweaver–Burk plot of an uninhibited enzyme reaction and a reaction inhibited by a simple competitive inhibitor, a simple noncompetitive inhibitor, and a mixed-model inhibitor.

ever, increases as it takes a greater amount of substrate to achieve 50% of the V_{max} value in the presence of the inhibitor. The greater the concentration of the inhibitor, the larger the apparent K_m value.

In contrast to simple competitive inhibition, a simple noncompetitive inhibitor causes a decrease in the V_{max} value but does not affect the K_m value (Fig. 3–4). As previously explained, this is because a noncompetitive inhibitor does not affect the binding of the enzyme to its substrate, but effectively reduces the number of enzyme molecules able to participate in a reaction.

A mixed-model inhibitor is one that exhibits both competitive and noncompetitive characteristics. In this case, the lines on the graph for the uninhibited and the inhibited assays meet, not at the y-intercept or the x-intercept value, but somewhere in the open space to the left of the y axis and above the x axis (Fig. 3–4).

Just as the K_m constant characterizes an enzyme's relationship to its substrate, so the K_i constant characterizes the relationship of an inhibitor to the enzyme that it inhibits. Thus the K_i constant reflects the strength of an inhibitor and the smaller the K_i value, the stronger the inhibitor. The K_i constant also is given in units that are similar to substrate concentration.

To determine the value of K_i, a series of reactions is conducted with different amounts of substrate concentration, first without the inhibitor present and then with ever-increasing concentrations of inhibitor. At a minimum, at least two concentrations of inhibitor should be tested but it is better to test four or five concentrations. The reaction without inhibitor will provide values of K_m and V_{max} and these are determined as described previously using the double reciprocal Lineweaver–Burk equation. The inhibited reactions will provide apparent values of K_m and V_{max} (i.e., appK_m and appV_{max}). K_i is calculated for competitive and noncompetitive inhibition using the equations shown below and averaging the calculated values of K_i determined for each concentration of I:

$$K_i = [K_m] [I]/(\text{app}K_m - K_m) \text{ (competitive inhibition)} \qquad [17]$$

$$K_i = [V_{max}] [I]/[\text{app}V_{max} - V_{max}] \text{ (noncompetitive inhibition)} \qquad [18]$$

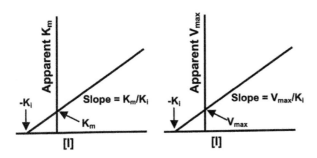

Fig. 3–5. Graphical method to determine K_i values for an enzyme reaction affected by a simple competitive and a simple noncompetitive inhibitor.

An alternative that allows more precision and accuracy is to use graphical methods to determine K_i values; a number of such methods can be found in the literature. The K_i values then can be calculated from the slope value, and the x- and y-intercept values. An example for a simple competitive inhibited reaction is to plot apparent K_m on the y axis and inhibitor concentration on the x axis as shown in Fig. 3–5. The slope value will be equal to K_m/K_i, the y-intercept value will be the K_m value, and the x-intercept value will be $-K_i$. Similarly, for a simple noncompetitive inhibited reaction, apparent V_{max} is plotted on the y axis and inhibitor concentration on the x axis. The slope will equal V_{max}/K_i, the y-intercept will be equal to V_{max}, and the x-intercept value will be $-K_i$.

Although not shown, an alternative graphing procedure is to plot $1/v$ (y axis) against [I] (x axis), with one curve for each substrate concentration. The point at which the lines intersect is the K_i values that should be on the x axis for noncompetitive inhibition and above the x axis for competitive inhibition.

3–6 PARTIAL COMPETITIVE AND PARTIAL NONCOMPETITIVE INHIBITION KINETIC MODELS

Partial competitive inhibition occurs when the substrate and inhibitor bind to the enzyme at different sites to form ES, EI, and ESI complexes, the substrate binds to free enzymes with a greater affinity than to the EI complex, and both the ES and ESI complexes yield product with equal ease (Segal, 1975). The reciprocal form of the initial velocity equation for partial competitive inhibition is:

$$\frac{1}{v} = \frac{K_m(1+[I]/K_i)}{V_{max}(1+[I]/\alpha K_i)}\frac{1}{[S]} + \frac{1}{V_{max}} \tag{19}$$

where α is the factor by which K_m changes when the inhibitor occupies the enzyme. At any inhibitor concentration, the K_m value measured will increase by the factor α $(1 < \alpha < \infty)$ because a portion of the enzyme will exist as EI complex that has a reduced affinity for S $(\alpha K_m > K_m)$. However, increasing the inhibitor concentration to infinity while holding the substrate concentration constant will result in all enzyme being converted to the EI or ESI form and Eq. [18] can be expressed with a variation of the classical Michaelis–Menten equation:

$$v = V_{max}[S]/(\alpha K_m + [S]) \tag{20}$$

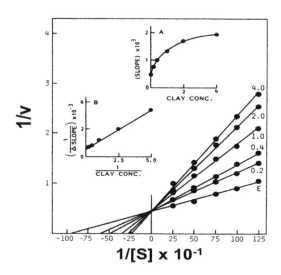

Fig. 3–6. Partial competitive inhibition of acid phosphatase activity in corn root homogenates by kaolinite clay. Initial velocity (v) is expressed as μmol p-nitrophenol released 20 mg^{-1} corn root h^{-1}, [S] is in M, clay concentration is in mg L^{-1} and is shown for each line, and E is the untreated enzyme control (Dick and Tabatabai, 1987).

Because both EI and ESI complexes produce product with equal efficiency, V_{max} value will remain unchanged, but a limiting K_m value is obtained. The inhibition of acid phosphatase and pyrophosphatase by kaolinite clay followed this pattern and is best described by partial competitive inhibition (Fig. 3–6) (Dick and Tabatabai, 1987).

The partial noncompetitive model is a type of inhibition different from simple noncompetitive inhibition in that the EI complex is not a dead-end complex and the ESI complex is not a nonproductive complex. Instead, with partial noncompetitive inhibition, the ESI still can react to form product, but at a slower rate. The reciprocal form of the initial velocity equation for a partial noncompetitive inhibition is:

$$\frac{1}{v} = \frac{K_m(1+[I]/K_i)}{V_{max}(1+\beta[I]/K_i)}\frac{1}{[S]} + \frac{1(1+[I]/K_i)}{V_{max}(1+\beta[I]/K_i)} \tag{21}$$

where K_m is the dissociation constant for the ES complex (because $K_m \approx K_s$ when the rate of breakdown of ES to form product is much less than the rate of breakdown to form the individual E and S molecules), K_i is the dissociation constant for the EI complex, and β is the factor by which the inhibitor (I) decreases the rate of formation of product (P) from the EI complex compared with the rate of product formation from the ES complex. As the inhibitor concentration reaches infinity, Eq. [21] reduces to:

$$v = \beta V_{max}[S]/(K_m + [S]) \tag{22}$$

The V_{max} obtained represents the rate of reaction when all enzyme has been driven to form an ESI complex which can form product, but at a limited rate defined by β ($0 < \beta < 1$).

The partial noncompetitive model predicts that V_{max} values will decrease and that K_m values will remain unchanged in the presence of increasing inhibitor concentrations. The noncompetitive kinetic model was used to explain the results

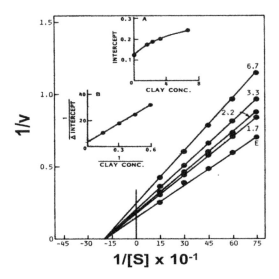

Fig. 3–7. Partial noncompetitive inhibition of inorganic pyrophosphatase activity in corn root homogenates by illite clay. Initial velocity (v) is expressed as μmol inorganic phosphate (Pi) released 50 mg^{-1} corn root h^{-1}, [S] is in M, clay concentration is in mg L^{-1} and is shown for each line, and E is the untreated enzyme control (Dick and Tabatabai, 1987).

obtained when acid phosphatase and pyrophosphatase of corn root homogenates were allowed to react with illite clay (Fig. 3–7) (Dick and Tabatabai, 1987).

The α, β, and K_i constants in the partial competitive and partial noncompetitive kinetic models can be estimated by appropriate techniques (Segal, 1975), which are illustrated in insets A and B of Fig. 3–6 and 3–7. The y-intercept values or the slope values obtained from the double reciprocal plots are first plotted as curves shown in Inset A. However, because these plots are hyperbolic, the limits that would allow α, β, and K_i to be determined are difficult to identify. A re-plot of the data in Inset A as 1/clay concentration vs. 1/change in intercept value or change in slope value converts the relationship between the y-intercept or slope values and [I] (i.e., clay concentration) to a straight line. The line in Inset B of Fig. 3–6 has a slope of $\alpha K_i V_{max}/K_m(\alpha - 1)$, an intercept value on the y axis of $V_{max}/K_m(\alpha - 1)$, and an intercept value on the x axis of $-1/\alpha K_i$. The line in Inset B of Fig. 3–7 has a slope of $K_i V_{max}/(1 - \beta)$, an intercept value on the y axis of $\beta V_{max}/(1 - \beta)$, and an intercept value on the x axis of $-\beta/K_i$. The K_m and V_{max} values used to calculate the other constants are determined from kinetic data obtained when an inhibitor (in this case clay) is not present in the assay mixture (i.e., from the control plot).

A summary of the different kinetic models and graphical techniques to calculate kinetic constants is provided in Table 3–1.

3–7 APPLICATION OF KINETIC MODELS

Kinetic analysis can be used to better understand mechanisms associated with enzyme reactions in soil. For example, the K_i values for acid phosphatase activity in the presence of kaolinite, montmorillonite, and illite were similar, ranging from 0.15 to 0.19 g L^{-1} (Dick and Tabatabai, 1987). However, the K_i for pyrophosphatase in the presence of montmorillonite was 0.74 g L^{-1}, for illite it was 5.4 g L^{-1}, and for kaolinite it was 10.7 g L^{-1}. It can be concluded that acid phosphatase had a greater affinity for clay minerals, except for montmorillonite, than did pyrophosphatase,

Table 3–1. Summary of kinetic constants and graphing techniques to estimate the constants.

Model	Kinetic constants	Graphing technique	Slope value	y-intercept value	x-intercept value
Uninhibited	K_m, V_{max}	Plot $1/v$ on the y axis vs. $1/[S]$ on the x axis	K_m/V_{max}	$1/V_{max}$	$-1/K_m$
Simple competitive inhibition	K_m, K_i, V_{max}	Plot 1/apparent K_m on the y axis vs. $[I]$ on the x axis	K_m/K_i	K_m	$-K_i$
Simple non-competitive inhibition	K_m, K_i, V_{max}	Plot 1/apparent V_{max} on the y axis vs. $[I]$ on the x axis	V_{max}/K_i	V_{max}	$-K_i$
Partial competitive inhibition	$K_m, K_i, V_{max}, \alpha$	Starting from the Lineweaver–Burk graph plot K_m/V_{max} (i.e., the slope) on the y axis and $[I]$ on the x axis. Then replot $1/\Delta$ slope on the y axis and $1/[I]$ on the x axis	$\alpha K_i V_{max}/$ $K_m(\alpha-1)$	$V_{max}/$ $K_m(\alpha-1)$	$-1/\alpha K_i$
Partial non-competitive inhibition	K_m, K_i, V_{max}, β	Starting from the Lineweaver–Burk graph plot $1/V_{max}$ (i.e., the intercept) on the y axis and $[I]$ on the x axis. Then replot $1/\Delta$ intercept on the y axis and $1/[I]$ on the x axis	$K_i V_{max}/$ $K_m(\alpha-1)$	$\beta V_{max}/$ $(1-\beta)$	$-\beta/K_i$

and that pyrophosphatase had a much greater affinity for montmorillonite than for kaolinite.

The kinetic constants of α, β, and K_i were also affected by whether the soil samples were shaken or not during enzyme assay of acid phosphatase and pyrophosphatase (Dick and Tabatabai, 1987). Shaking the samples during incubation caused an increase in the α constant for acid phosphatase and pyrophosphatase bound to kaolinite from 4.6 to 10.6 and from 3.2 to 5.4, respectively. However, the β constant for these two enzymes remained unchanged when samples were shaken during incubation. This suggests that diffusional limitation effects on the catalytic efficiency of the overall enzyme mechanism probably was not very important when kaolinite was present.

In contrast, the K_i constants associated with montmorillonite and illite were greatly decreased when the samples were shaken. For acid phosphatase bound to montmorillonite, the K_i values decreased from 0.15 to 0.086 g L^{-1}; when bound to illite, the decrease was from 0.32 to 0.68 g L^{-1}. Similar changes were observed for pyrophosphatase. Enzymes bound to kaolinite resulted in either no change (acid phosphatase) or a slight decrease (pyrophosphatase) in the K_i value. These K_i data suggest that much of the phosphatase bound to montmorillonite or illite is held loosely, and when the samples are shaken during incubation, the kinetic model indicates only enzymes tightly bound remain on the clay surface. Hughes and Simpson (1978) also reported that only arylsulfatase bound directly to montmorillonite surface was inhibited and that there was also a less tightly bound enzyme layer associated with the clay mineral that was not inhibited by the clay mineral.

3–8 SUMMARY

More attention should be paid to kinetic analysis tools for evaluation of enzyme-catalyzed reactions in soil. At a minimum, reactions should be conducted under zero-order substrate concentrations. Use of inhibitors and appropriate enzyme kinetic models also can be very informative about various functions in soil in which enzymes are intimately involved. This will provide a much deeper understanding of how various biological processes occur and are regulated in soil.

REFERENCES

Al-Turki, A.I., and W.A. Dick. 2003. Myrosinase activity in soil. Soil Sci. Soc. Am. J. 67:139–145.

Dick, W.A., and M.A. Tabatabai. 1987. Kinetics and activities of phosphatase-clay complexes. Soil Sci. 143:5–15.

Hughes, J.D., and G.H. Simpson. 1978. Arylsulfatase-clay interactions: 2. The effect of kaolinite and montmorillonite on arylsulfatase activity. Aust. J. Soil Res. 16:35–40.

Ruggiero, P., J. Dec, and J.-M. Bollag. 1996. Soil as a catalytic system. p. 79–122. In J.-M. Bollag and G. Stotzky (ed.) Soil biochemistry. Vol. 9. Marcel Dekker, New York.

Segal, I.L. 1975. Enzyme kinetics. John Wiley & Sons, New York.

Chapter 4

Development of a Soil Enzyme Reaction Assay

Warren A. Dick

4–1 INTRODUCTION

Soil is not an inert material. A vast array of microorganisms, plants, and animals manufacture, degrade, and/or transform a countless number of organic and inorganic materials in soil. Enzymes catalyze most of these reactions and without their presence, the soil would no longer function properly. In fact, it could be argued that without enzymes in soil, all life on planet earth would eventually be affected.

The reactions mediated by enzymes in soil occur either within the living cell or by extracellular enzymes, i.e., enzymes that are still active but are no longer contained within a viable living plant, animal, or microbial cell. Extracellular enzymes in soil are especially important for the metabolism of macromolecules that often are too big to be transported across membranes into the interior of a cell. The study of these extracellular soil enzymes can yield information that is vital to our understanding of the total soil system.

Enzymes are proteins and like all catalysts, enzymes increase the rate of a chemical reaction without, themselves, undergoing permanent alteration. Enzymes have unique properties, however, that distinguish them from other catalysts. First, they are extremely *efficient* catalysts and are able to increase the speed of a reaction a millionfold or more compared with the same reaction performed in the absence of an enzyme. Second, enzymes are *specific* in the reactions they catalyze. Third, enzymes are subject to *regulation* and thus act to control metabolite concentration and flow. Fourth, enzymes *act under normal physiological conditions* of temperature and pressure.

In this chapter, the factors that affect the activity of an enzyme in soil and need to be considered when developing a valid, standard soil enzyme assay are described. In some instances, examples are taken from a paper that reports the development of an assay to measure myrosinase activity in soil (Al-Turki and Dick, 2003), although there are other examples in the literature that illustrate the same points.

Warren A. Dick, Professor, Soil Science, The Ohio State University, Wooster, OH 44691 (dick.5@osu.edu)

doi:10.2136/sssabookser9.c4

4–2 FACTORS OF A VALID SOIL ENZYME ASSAY

Assessment of the activity of an enzyme in soil is based on measurement of either product appearance or substrate disappearance when soil is incubated with substrate. Other factors that must be considered and controlled during the measurement of the activity of an enzyme in soil include the extraction efficiency of product or substrate from soil, whether the soil is in an air-dried or field-moist state, buffer pH, substrate concentration, amount of soil, time span of reaction, temperature, shaking vs. not shaking during reaction, the stoichiometry of the reaction, the selection of an appropriate analytical procedure, storage or pretreatment of the soil before assay, whether there is a need for cofactors for the reaction to proceed, and the use of a proper control (Table 4–1). All of these factors need to be carefully evaluated and optimized for a variety of soils to provide the most valid soil enzyme assay and to ensure that the only limitation to the reaction rate is enzyme concentration. The goal is to ensure that the reaction rate measured is proportional to, or accurately reports, the enzyme concentration in soil.

The soil enzymologist normally does not actually measure concentration of an enzyme in soil. This would involve the extraction from soil of a specific protein and then subsequent quantification of that protein. This is extremely difficult to do and, in many cases, would be meaningless because, from an ecological perspective, it is the activity of the enzyme that is most important. Instead, the goal of the soil enzymologist is to measure the activity of a specific enzyme-catalyzed reaction in one soil relative to another. This requires standard assay procedures that can be used in laboratories around the world to provide reproducible results. Anything that interferes with this goal will compromise the value of the data obtained. The publication of a well-characterized soil enzyme assay is extremely important as it provides a standard whereby enzyme activity in one soil legitimately can be compared with the activity in another soil. In addition, the results from studies that employ valid enzyme assays improve our understanding of their role in soil, including many important soil processes or functions.

In the sections that follow, each of the factors summarized in Table 4–1 will be discussed in more detail. If these factors and their optimization are stringently developed and tested, the information gained will be useful to enhance our knowledge of known soil enzymes and for the development of any newly discovered enzymes in soil.

4–2.1 Measuring Appearance of Product or Disappearance of Substrate

An enzyme is a catalyst and an enzyme assay is designed to determine the rate at which a reaction proceeds in the presence of the enzyme catalyst, thus obtaining information about the concentration of the enzyme in the system of interest. The two most obvious and common ways to do this are to measure either the disappearance of substrate or the appearance of product over a specified time period. For the soil enzymologist, the system of interest is soil.

In general, the measurement of the appearance of a product is preferred. It is much easier to measure a small change in product against a low or even nonexistent concentration of product in soil than a small change in substrate concentration against the background of a high concentration of substrate. The following example will illustrate this point, even though neither the substrate nor the product are naturally found in soil or only at very low concentrations, and we can equally

Table 4–1. Variables that affect a soil enzyme activity measurement.

Factor	Summary comments
A. Whether to measure appearance of product or disappearance of substrate	Sensitivity of the enzyme assay is generally greatest when appearance of product is measured.
B. Extraction efficiency of product or substrate from soil	An extraction procedure that can accurately quantify changes in substrate or product concentrations must be confirmed before a valid enzyme activity measurement can be made.
C. Whether the soil is in an air-dried or field-moist state	Air-dried soil samples are preferred. However, field-moist samples are sometimes required because air-drying can lead to loss of activity.
D. Buffer pH	Use an appropriate pH buffer at a pH value that gives optimum activity.
E. Substrate concentration	Substrate concentration must be great enough to achieve zero-order kinetics with regard to the substrate concentration.
F. Amount of soil	Choose an amount that provides sufficient activity for detection of enzyme without compromising valid comparison of one soil vs. another.
G. Time span of reaction	Time should be as short as possible while allowing sufficient activity for detection and valid comparison among soils of the enzyme.
H. Temperature	Choose a temperature where activity is high but also where thermal inactivity of the enzyme protein is minimal. Generally, for comparison of soil enzyme activities, a temperature of either 25°C or 37°C is used.
I. Shaking vs. not shaking during enzyme assay	The rate of reaction and K_m are often affected by whether the soil, with substrate, is shaken during incubation. It is important that this factor be standardized to provide valid comparison of activity among soils.
J. The stoichiometry of the reaction	Enzyme reactions, such as urease, will yield a reaction rate based on product appearance that is twice the rate based on substrate disappearance. Units of activity for an enzyme must be standardized.
K. The selection of an appropriate analytical procedure	An analytical procedure that is accurate, precise and inexpensive is preferred.
L. Storage or pretreatment of the soil before assay	Enzyme assays should be conducted as soon as possible after soil sample collection. If this is not possible, soils stored frozen, refrigerated, or air-dried must be tested before and after storage to assess activity changes.
M. The need for cofactors for the reaction to proceed	Enzymes that require a cofactor (e.g., a certain metal) must not be limited in activity due to failure to provide this required cofactor. Otherwise, reaction rate will not be a function of enzyme concentration in soil.
N. The use of a proper control	Negative controls are a necessity. This involves either applying substrate to soil after incubation or using an identical soil that has been treated to completely eliminate all enzyme activity.

measure the same unit concentration of substrate or product. Assume we start with 1000 units of substrate, which must be added to soil in excess to achieve zero-order reaction rate kinetics, and there is a transformation of 1 unit of substrate to 1 unit of product during the assay period. At the conclusion of the assay period, we will be required to consistently measure either a difference of 1000 units vs. 999 units of substrate or a difference of 0 and 1 unit of product. The large background of substrate makes the first option more difficult than the second option and the likelihood of easily detecting a change of one unit of product formed from a background of zero, or near zero, is a much easier task to accomplish.

It must be noted that there are instances in which there are no good analytical methods to extract and detect the product of a soil enzyme-catalyzed reaction. In this case, measurement of substrate disappearance will be required. The following discussion assumes that the appearance of product is to be measured to estimate enzyme activity in soil, although the principles would be similar if disappearance of substrate were to be followed.

4–2.2 Extraction Efficiency

Once the decision has been made to measure the appearance of product as affected by the amount of enzyme in soil, then it is important to conduct tests as to whether it is possible to quantitatively extract the product from the soil.

To begin, one can make a guess of how much product may be created during the course of a soil enzyme reaction. A series of extraction tests is then conducted by adding a known concentration of the product to soil of approximately 100 times less to 100 times greater than what would be expected. The product should be added to soil in the same buffer matrix that will ultimately be used to assay the enzyme in soil. After a short time of equilibration, generally 30 to 60 min is sufficient, extract the product back out of soil. If approximately 100% of the product can be extracted from the soil and analyzed, then we can be assured that what we measure during an enzyme assay is a true reflection of the enzyme activity in soil. In some cases, the extraction efficiency may not be 100% but some consistent fraction. Therefore, if consistently 50% of the product is extracted, one could multiply the extraction results by a correction factor of 2.0 to obtain the true level of enzyme activity. While this may work well for a study of a single soil type and when soil properties are consistent, a correction factor could not be applied to other soils unless it was first verified for each individual soil.

For some products, extraction efficiency may be complicated because they also are readily transformed or taken up by the microbial community. For example, if glucose is the product of an enzyme-catalyzed reaction, we can predict that it will be taken up rapidly and metabolized by microorganisms in soil during the assay incubation period. As such, it is necessary to use a biostatic reagent to suppress microbial utilization of glucose. Toluene is commonly used for this purpose and many enzyme reactions have been tested to assess the effect of a small amount of toluene. For example, myrosinase activity in soil produces glucose and it is readily apparent this enzyme can be assayed accurately in soil by measuring production of glucose only if 0.2 mL of toluene is added per g soil (Table 4–2). If toluene is required, tests also should be conducted to make sure the toluene that is extracted with the product does not interfere with the subsequent analytical procedures used to measure the product. If this occurs, but the toluene does serve

Table 4–2. Recovery of glucose from soils with and without addition of toluene† (Al-Turki and Dick, 2003).

Soil	Glucose added	No toluene		With toluene (0.2 ml)	
		Glucose recovered	Recovery	Glucose recovered	Recovery
	$\mu g\ g^{-1}$ soil	$\mu g\ g^{-1}$ soil	%	$\mu g\ g^{-1}$ soil	%
Wooster	25	2.1 (0.4)‡	8.0	23.1 (1.7)	92.4
	50	2.6 (0.6)	5.1	49.8 (3.4)	99.6
	100	12.3 (1.9)	12.3	98.9 (3.4)	98.9
	250	17.3 (1.9)	6.9	232 (5.3)	92.7
	1000	767.0 (10.8)	76.8	935 (6.7)	93.5
Carlisle	25	1.8 (0.3)	7.2	23.1 (2.1)	92.4
	50	4.1 (0.7)	8.2	46.2 (4.1)	92.2
	100	8.4 (1.7)	8.4	93.4 (6.4)	93.4
	250	14.4 (2.2)	5.8	246 (4.9)	98.4
	1000	781.0 (14.1)	78.1	890 (18.1)	98.0
Luray	25	7.4 (0.8)	29.4	22.9 (1.7)	91.9
	50	16.6 (1.5)	33.1	47.1 (2.1)	94.2
	100	18.9 (1.3)	19.0	96.1 (1.3)	96.1
	250	58. (2.5)	23.2	243 (2.6)	97.1
	1000	714.0 (8.5)	71.5	963 (6.2)	96.3
Spinks	25	18.5 (1.4)	74.2	24.3 (1.6)	97.2
	50	43.4 (1.2)	86.7	49.8 (3.4)	99.7
	100	85.1 (3.7)	85.1	95.9 (2.4)	95.9
	250	226.0 (3.3)	90.6	246 (5.3)	98.4
	1000	912.0 (10.1)	91.2	973 (5.7)	97.4

† The background amount of glucose in soil ranged from 3.8 (Spinks) to 12.9 (Luray) $\mu g\ g^{-1}$ soil when toluene was not added to soil and from 5.8 (Spinks) to 38.8 (Carlisle) $\mu g\ g^{-1}$ soil when toluene was added. This background level was subtracted when calculating recoveries of added glucose.

‡ Values in parentheses are standard deviations.

its intended purpose as a biostatic reagent, attempts could be made to find other analytical procedures that are not affected by the presence of the toluene.

Extractants also can interfere with a soil enzyme reaction assay in other ways. For example, many phosphatase enzyme substrates may be broken down by acid extractants to yield the same product as that of the enzyme itself (Dick and Tabatabai, 1977). When this occurs, there will be an overestimation of activity. An example is the enzyme pyrophosphatase, where a strong acid (i.e., sulfuric acid) is used both to extract the orthophosphate produced during the reaction and to measure the orthophosphate by the molybdenum blue analytical procedure (Dick and Tabatabai, 1978). The sulfuric acid extractant chemically breaks down pyrophosphate to orthophosphate (Dick and Tabatabai, 1977) and it is imperative for the extraction procedure to be rapid for the extractant interference to be overcome.

This can be done by developing an analytical method to quantitatively measure orthophosphate in the presence of pyrophosphate and sulfuric acid (Dick and Tabatabai, 1977, 1978).

4–2.3 Use of Air-Dried or Fresh Soil

Enzyme activities in soil are often affected by air-drying, and oven-drying or drying at elevated temperatures is to be avoided as the high temperatures will inactivate many enzymes. Generally, use of air-dried soils is preferred because they are much easier to store and handle. Use of air-dried soils also generally results in less variation among replicates and makes it much easier to go back to the original soil sample and repeat the enzyme assay. For some enzymes, such as dehydrogenase, field-moist soil samples are almost always used because air-drying can have a great effect on the activity of this enzyme. It is still important, however, to express activity on a soil dry-weight basis. This can be easily done by taking an aliquot of the same soil being used to measure dehydrogenase activity, drying the soil to obtain its moisture content, and then making the appropriate calculation.

To test the effect of air-drying on the activity of a specific enzyme, the same amount of soil on an air-dry basis, removed from the field and in either its field-moist or air-dried states, is assayed. The effect of air-drying may result in an elevated amount of enzyme activity, an activity that is essentially unchanged, or a reduced level of activity (Table 4–3). To compare enzyme activity values from one study to another, it is important to know how the soil was handled before the enzyme assay was conducted.

For comparison purposes of one treatment in the field vs. another, an air-dried sample may still be used even if the activity is affected. It often is not the absolute value of an enzyme's activity that is important but whether the treatment applied in the field increased or decreased that activity. As a general rule, however,

Table 4–3. Effect of air-drying on the activity of various soil enzymes (information taken from Gianfreda and Bollag, 1996; Speir and Ross, 1981).

Enzyme	Change in enzyme activity due to air-drying of soil sample		
	Positive	Negative	Unchanged
Glucosidase	•		
Galactosidase	•		
Sulfatase	•		•
Urease	•		•
Xylanase		•	•
Invertase		•	•
Phosphatase		•	
Amidase		•	
Rhodanese		•	
Amylase		•	
Cellulase		•	
Protease		•	

it is always best to optimize all the enzyme assay factors and use standard enzyme assay protocols if comparisons among studies are to be done.

4–2.4 Buffer pH

Enzymes exhibit optimum activity at specific pH values because changes in pH (i.e., hydrogen ion concentration) cause changes in the ionic state of both the amino acid residues of the enzyme and the substrate molecules. These alterations in charge will affect the conformational structure of the enzyme and thus both its ability to bind with its substrate and its ability to complete catalysis. These effects will be reversible over a narrow pH range, but extremes of acidity or alkalinity often cause permanent denaturation of the enzyme protein. The effect of pH on the activity of an enzyme is most commonly reflected in a bell-shaped curve (see Fig. 4–1 for soil myrosinase) in which a pH value demonstrates maximum activity and that activity drops off as pH either increases or decreases from this optimum pH.

The pH optimum of an enzyme in soil is generally higher than the same enzyme in purified form in solution. This can be illustrated by considering clay particles with a net negative charge. At the surface of the clay particle, and then moving away, a double layer exists where H^+ concentrations are higher than in the bulk soil solution, i.e., the pH at the clay surface will be lower than that in the bulk soil solution (Fig. 4–2). Many enzymes are located in microenvironments in the soil where this double layer exists. Thus, for the optimum pH to be reached in the double layer microenvironment where the enzyme resides, the pH in the bulk soil solution must be one to two units higher than the pH optimum for the same enzyme in a buffer solution without soil. In the biochemical literature, it is the buffer solution pH optimums without soil that generally are reported for an enzyme.

It is important when developing a soil enzyme assay that the assay utilizes a buffer that is efficient in buffering against changes in pH at the pH where optimum activity occurs in soil. The buffer must also not interfere with the extraction and quantitative analysis of the product. When comparing enzyme activities among soils and studies, it is important to standardize the buffer pH to that of the optimum pH. If this is not done, the measured activity may not reflect the actual concentration of the enzyme in soil.

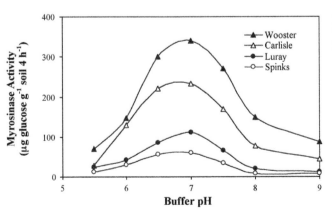

Fig. 4–1. Effect of buffer pH on myrosinase activity in soils (Al-Turki and Dick, 2003).

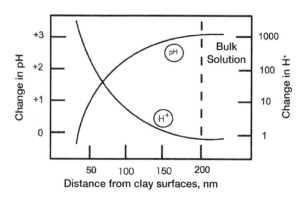

Fig. 4–2. Changes in pH and hydrogen ion concentration at increasing distances from the clay surface.

4–2.5 Substrate Concentration

When assaying enzymes, the substrate concentration must be sufficient to maintain zero-order kinetics with respect to substrate concentration, thus achieving a reaction rate proportional to enzyme concentration (Segel, 1975). Another way of saying this is that the rate of product formation should be directly related or proportional to the enzyme concentration in soil and independent of substrate concentration. This zero-order kinetic standard for substrate concentration must be met for all soils studied. For soil myrosinase, zero-order reaction kinetics was achieved at a concentration of 20 mM sinigrin (substrate) for the soils studied (Fig. 4–3), which included two Alfisols, a Mollisol and a Histosol.

4–2.6 Amount of Soil

A test should be made to determine that the course of the enzyme reaction is linear with respect to the amount of soil. This indicates that soil amount is not a variable in the measuring of the enzyme in soil and that the assay is actually a reflection of the concentration of the enzyme in the soil. Figure 4–4 illustrates the linearity that is desired for a valid enzyme assay. Often, such a test will show the response lines flattening out at high amounts of soil. This may be due to the pH buffer not being strong enough to resist changes in pH as soil amount increases, removal of product or substrate by soil particles affecting the reaction rate, and/or the total amount of enzyme in the larger amount of soil changing the substrate concentration to a point where zero-order substrate concentration kinetics is no longer operative. In Fig. 4–4, the reaction rates are linear for all three soils up to a total of 1.5 g of soil. The choice for amount of soil to be used in this particular myrosinase assay would be a matter of how much sensitivity is desired. The larger the sample size, the more sensitive the enzyme assay will be. The other variables to consider are the amount of reagent needed for a larger sample size and whether small sample sizes are representative of the original soil being assayed.

The zero level of soil serves as one type of negative control. If no activity is observed when soil is not present in the assay, but all other reagents are present, this indicates the reagents are stable during the incubation period and are not being chemically altered. Any product created in the presence of soil is likely due to the enzyme being assayed. A further control would be needed to confirm that

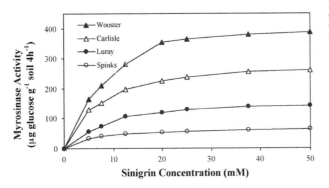

Fig. 4–3. Effect of sinigrin (i.e., substrate) concentration on myrosinase activity in soils.

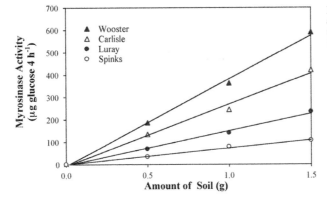

Fig. 4–4. Effect of amount of soil on myrosinase activity (Al-Turki and Dick, 2003).

chemical processes in the soil are not also contributing to the reaction and the use of a proper control will be discussed in the section, Adoption of a Proper Control.

4–2.7 Incubation Time

The effect of incubation time on soil enzyme activity in a valid assay should be as shown in Fig. 4–5. The linear relationship indicates enzyme activity is not affected by microbial assimilation of product and that product formation is directly related to enzyme concentration and not to any other variables in the reaction. The choice of incubation time is often a compromise between increased sensitivity and convenience. Longer incubation times will result in more product and thus a smaller amount of enzyme can be detected; this represents greater sensitivity. However, the longer the incubation time, the more difficult it becomes to control all of the other variables that can affect soil enzyme activity. A short incubation time reduces the potential for microbial proliferation or impacts, chemical reactions occurring, etc. A short incubation time also reduces cost as it allows more reactions to be conducted in a set amount of time. An incubation time is desired that both reduces potential problems and ensures enough product is created to detect low amounts of enzyme in soil.

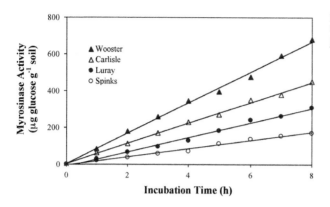

Fig. 4–5. Effect of incubation time on myrosinase activity in soils (Al-Turki and Dick, 2003).

4–2.8 Temperature

Temperature affects an enzyme activity by imparting energy to the reactants and makes it possible for more of the reactants to cross over the energy barrier that exists between the substrate and product. For each 10°C increase in temperature, there is generally a doubling of reaction rate up to a point where any additional temperature rise causes little increase in reaction rate and even further temperature rise will lead to a decrease in activity (Fig. 4–6). This is due to thermal activity becoming so great that it causes inactivation of the enzyme protein because of conformational changes in the protein structure. The temperature at which the enzyme begins to lose its activity is called the temperature of inactivation.

The proper choice of temperature is one that is low enough to avoid any significant inactivation of the enzyme, but still high enough to yield the highest possible rate of reaction (and thus maximize the sensitivity of the reaction). The reality is that biochemists have standardized most enzyme assays at 37°C. The use of 37°C as a standard temperature is based on the fact that this is the normal physiological temperature of a human.

For soil enzyme assays, it is recommended that the temperature for assays also be standardized to allow rapid comparison of results among studies. The two temperatures most commonly used for measuring soil enzyme activity are

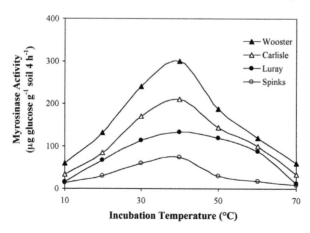

Fig. 4–6. Effect of incubation temperature on myrosinase activity in soils (Al-Turki and Dick, 2003).

25°C and 37°C. The higher temperature will lead to a higher reaction rate and thus increase the overall sensitivity of the enzyme assay. If all soil enzymologists agreed to use either the 25°C or the 37°C temperature, this would represent an advance in standardization of enzyme activity measurements and, presumably, greater progress in soil enzyme research.

4–2.9 Shaking vs. Not Shaking During Enzyme Assay

Shaking of the soil with the buffered substrate often affects the rate of reaction of an enzyme-catalyzed reaction in soil. It also can affect the K_m value. Soil is a complex medium and without shaking, diffusion controls the movement of the substrate to the enzyme and the product away from the enzyme. This issue becomes especially troublesome in soil that is not finely ground and when the enzymes are located in the center of a soil aggregate.

An experiment could be conducted where the identical assay conditions are used but one set of assays is incubated in a static state and one set is incubated with gentle agitation. There will almost certainly be differences in the results obtained. Although there does not seem to be a consensus as to which procedure should be used, my preference would be to incubate under static conditions. The reasoning for this is that it is difficult to standardize shaking conditions as to shaker arm path length and speed. This issue becomes less of a problem if the effect of shaking is small or is a constant proportion of the total enzyme activity across all soils tested. Regardless of which method is selected for incubation, it is imperative that this be mentioned in the description of the assay procedure used.

4–2.10 Stoichiometry of Reaction

Although this may be important in many biochemical systems, in soil enzymology the factor of stoichiometry of reaction is generally ignored. For example, urease enzyme takes one mole of the substrate urea and converts it to two moles of ammonia. Thus, if one were measuring ammonia to follow urease activity, one would actually overestimate the mole reaction rate by a factor of two.

$$urea + H_2O \rightarrow 2NH_3 + CO_2$$

Realistically, most soil enzymologists simply compare reaction rates based on the actual amounts of product created or substrate that disappears. If there is general agreement on this convention, then it is still very informative for comparing activities of an enzyme in one soil vs. another.

Probably more important is to know whether there is some stoichiometric ratio of a cofactor that is required for a reaction to proceed compared with that of the substrate. If this is not considered, it may be that the reaction will become limited due to the presence of too little cofactor.

4–2.11 Selection of an Appropriate Analytical Procedure

The choice of an appropriate analytical procedure for the product of an enzyme reaction can be a reflection of many variables such as sensitivity of the procedure, cost, experience of the researcher with a specific procedure, need for automation, accuracy, precision, etc. Therefore, one cannot make a simple statement of what is the best analytical procedure to use for a specific enzyme assay. Very sensitive assays can be developed using radioactive isotopic or fluorimetric procedures.

As in many scientific disciplines, it seems increasingly desirable for soil enzymologists to collect massive amounts of data as quickly and inexpensively as possible and then use computer programs to mine the data for important scientific information. The trend is to (i) develop analytical methods that require smaller and smaller volumes of reagents and (ii) combine high levels of sensitivity with large throughput. As the field of soil enzymology matures, we likely will see the development of commercial kits that are easy to use. These kits will need to be rigorously tested to assure they can be used in a standardized way so that results obtained in one laboratory or research study can be compared with results from other studies.

4–2.12 Storage or Pretreatment of Soil Before Assay

Soils that are assayed are representative of a larger population of soils that are sampled at a particular point in time and from a particular location. If samples represent a particular field treatment, the same sampling procedures must be followed as used for measurement of other soil properties. Multiple soil cores must be taken, bulked, well mixed, and subsampled. If there is large sample variation, it is imperative that not only the treatments be replicated, but also the enzyme assays themselves. With newer, rapid methods, it is becoming easier to obtain multiple analytical replicates for a single treatment replicate.

If soils must be assayed in their field-moist state, the assays should be made as soon as possible after collection of the samples is completed. If it is known that there will be delay in conducting the soil enzyme assays, a subset of samples should be assayed immediately and then again when the stored samples are measured. If there is no change, then the storage procedures can be assumed to not have affected the results. If there are large differences, either positive or negative, and the differences are not consistent among samples, then the results will be suspect and should not be used. Positive differences could come about if microbial growth occurs or some chemical change occurs during storage that helps to increase the activity of an enzyme, e.g., volatilization of an inhibitor and cell or aggregate breakage due to freeze and thaw action. Negative differences could be a result of the enzyme becoming inactivated due to increased interaction with humic substances, release of inhibitory compounds, or microbial degradation.

In general, use of air-dried soil that has passed through a 2-mm sieve works well for measuring enzyme activities in soil. However, a very common soil pretreatment is to grind the soils to different levels of fineness. Grinding removes much of the heterogeneity of a soil sample, but it can also affect the reaction rate of a soil enzyme by making the substrate more accessible to its enzyme and reducing the diffusion limitations mentioned previously.

There may be occasions when soils are pretreated with an antibiotic or other amendment before conducting an enzyme assay. A common example in soil enzyme assays is the use of toluene. If this pretreatment is not part of a treatment in the actual experimental design, it is important to first test whether it interferes with the enzyme assay being conducted. Measuring enzyme activity in the presence and absence of the pretreatment can easily do this.

4–2.13 Need for Cofactor in Enzyme Assay

Some oxidation–reduction enzymes require coenzymes for activity. Other enzymes require addition of metal cofactors to achieve maximum activity. Generally this information can be obtained from the biochemical literature. Most enzymes in soils for which assays have been developed do not require cofactors for activity. However, some do. An example is soil pyrophosphatase that requires divalent metal ions for activity (Dick and Tabatabai, 1983). There is very little research on how activity of certain enzymes in soil may be limited by failing to optimize the concentration of a cofactor during the enzyme assay procedure.

4–2.14 Adoption of a Proper Control

Controversy often occurs over the proper control that should be used for a soil enzyme assay. The control should correct for any product that exists naturally in the soil but is not created as part of the enzyme reaction or that is created or exists due to chemical and not enzymatic reaction. Both a reagent control and a soil control should be investigated and generally used.

A reagent control or negative enzyme control is needed to determine the stability of the reagents during the time the soil sample is incubated. Measuring the amount of product created in a sample without soil is one type of negative control. If any product is observed, this is assumed to come from impurity in the reagents or to nonenzymatic breakdown of the substrate. This amount of product can be subtracted from the amount of product produced in a normal enzyme assay procedure to correct for activity not associated with enzyme in the soil. If the reagent control is very high, and not reproducible from one sample to another, the enzyme assay will need to be revised or even abandoned.

For the soil control, two identical reactions must be set up. One soil sample receives the substrate at the start of the incubation period and the other after incubation and immediately before the extraction of the product from soil. If any product existed in the soil before the initiation of the enzyme reaction, it can be subtracted.

An alternative control method that combines both the reagent control and soil control is to test whether there is any possible activity in a soil known to have zero enzyme activity. Studies have been conducted to determine the best way to create a zero enzyme activity control soil (Dick and Tabatabai, 2002). For example, the soil can be pretreated by autoclaving or by adding a potent inhibitor to the soil before assay. The goal is to eliminate enzyme activity during the incubation period of the soil enzyme assay, but to maintain all other aspects of the soil matrix. The assumption in this control is that any product measured at the end of the incubation cannot be due to enzyme activity in the soil, but is due to product that existed in the soil before conducting the enzyme assay, a chemical reaction on the substrate caused by the soil, or to inherent chemical instability of the substrate reagents.

Both reagent and soil controls should be rigorously tested before publishing a method to measure enzyme activity in soil.

4–3 SUMMARY

Our knowledge of enzyme activity in soil has rapidly progressed in the past 40 to 50 years and has been closely tied to the development of standard assays. For this to continue, it is imperative that when new enzyme assay methods are developed for existing or newly discovered soil enzymes, they also are rigorously tested and follow standard methodology. When this is done, data from widely diverse soils in the world can be compared and a synthesis of information into new knowledge can occur.

REFERENCES

Al-Turki, A.I., and W.A. Dick. 2003. Myrosinase activity in soil. Soil Sci. Soc. Am. J. 67:139–145.

Dick, W.A., and M.A. Tabatabai. 1977. Determination of orthophosphate in aqueous solutions containing labile organic and inorganic phosphorus compounds. J. Environ. Qual. 6:82–85.

Dick, W.A., and M.A. Tabatabai. 1978. Inorganic pyrophosphatase activity of soils. Soil Biol. Biochem. 10:59–65.

Dick, W.A., and M.A. Tabatabai. 1983. Activation of soil pyrophosphatase by metal ions. Soil Biol. Biochem. 15:359–363.

Dick, W.A., and M.A. Tabatabai. 2002. Enzymes in soil: Research and developments in measuring activities. p. 567–596. In R.G. Burns and R.P. Dick (ed.) Enzymes in the environment. Marcel Dekker, New York.

Gianfreda, L., and J.-M. Bollag. 1996. Influence of natural and anthropogenic factors on enzyme activity in soil. p. 123–194. In G. Stotzky and J.-M. Bollag (ed.) Soil biochemistry. Vol. 9. Marcel Dekker, New York.

Segel, I.H. 1975. Biochemical calculations. 2nd ed. John Wiley & Sons, Chichester, UK.

Speir, T.W., and D.J. Ross. 1981. A comparison of the effects of air-drying and acetone dehydration on soil enzyme activities. Soil Biol. Biochem. 13:225–229.

Chapter 5

Sampling and Pretreatment of Soil before Enzyme Analysis

Nicola Lorenz and Richard P. Dick*

5–1 INTRODUCTION

Soil microbial properties vary widely, both spatially and temporally. Therefore it is important to carefully collect soil samples for enzyme analysis according to a given objective or hypothesis. With a proper sampling design and sample pretreatment protocols, the research outcome will provide accurate, reliable estimates of soil enzyme activities as well as characterization of spatial and temporal variability (Dick et al., 1996). Depending on the objective of the study, it might be necessary to collect soil samples several times on a seasonal basis or for consecutive years to characterize a site or treatment effects sufficiently (Schinner et al., 1996). However, in reality the collection of soil samples is a tradeoff between the gain of useful information and cost of sampling and laboratory analyses (Dick et al., 1996).

This chapter will provide guidelines on soil sampling and pretreatment for enzyme analysis. The following two subsections describe soil sample collection strategies (time of sampling, sampling design, depth) and soil sample handling and pretreatment (soil physical fractionation, sieving, and storage). This chapter builds on previous soil microbial sampling and sample processing guidelines of Dick et al. (1996), Kandeler and Dick (2007), and Wollum II (1994) as well as on more recent findings, and specifically focuses on approaches for soil enzymatic analyses.

5–2 SOIL SAMPLE COLLECTION STRATEGIES AND CONSIDERATIONS

5–2.1 Selection, Analytical, and Sampling Error

Before outlining strategies of soil sample collection, some common but avoidable mistakes are presented during soil sample collection and analysis. Three common sources of error are: sample selection, analytical, and soil sampling errors (Das, 1950), the latter two of which are avoidable. All three sources of error are cumulative and it is important to consider them while developing a sampling plan.

Richard P. Dick, Ohio Eminent Scholar, Professor of Soil Microbial Ecology, The School of Environment & Natural Resources, The Ohio State University, Columbus OH 43210 (Richard.Dick@snr.osu.edu), *corresponding author;
Nicola Lorenz, The School of Environment & Natural Resources, The Ohio State University, Columbus OH 43210 (nicola.lorenz.64@googlemail.com)

doi:10.2136/sssabookser9.c5

Selection error arises when soil samples are taken from spots in the field that do not represent the site being investigated. For example, avoiding areas where it is difficult to obtain a sample, such as rocky or highly vegetated areas, would result in selection error since only the soil in the easily obtained area would be sampled. To overcome selection error, every area in the field should have an equal chance to be selected for sampling, e.g., by using a random sampling design.

Analytical error results from sample handling, storage, preparation, and/or mistakes during measurement. In general, analytical error is small. To reduce this error, it is important to follow the correct pre-analysis handling protocol, to use the best analytical technique possible, and to run at least two analytical replicates.

The major concern is sampling error. Sampling error occurs because the entire soil mass cannot be measured and the collection of soil samples represents an extremely small subset of the entire soil present at a site. To overcome sampling error, the sampling plan should ensure that a high number of soil samples are being collected from one site. This increases the precision and decreases the impact of sampling error in estimating the true mean of a soil property. However, the sampling plan is usually a compromise between precision and cost (Dick et al., 1996).

5–2.2 Time of Sampling

A very important decision for soil research is when to collect soil samples. Temporal or seasonal variability is less important in undisturbed soils for a few soil chemical or physical properties such as total C, texture, or mineralogy. These properties will remain fairly constant in the short term. Unlike treatment effects having a long-term effect on soil C in agricultural soils, soil C is likely to change rapidly (less than a decade) in forest soils after clear-cutting or fire. However, soil microbial properties typically can vary widely over short seasonal periods and between years even when sampled at the same time of year. Although soil enzyme activities can be more stable on a seasonal basis than are other microbial properties measurements (Ndiaye et al., 2000), sampling times should be chosen carefully according to the objectives of the study and should account for seasonal variability.

Environmental factors, particularly moisture and temperature, affect chemical and biological properties in soils (Aon and Colaneri, 2001). One possible strategy is to measure soil enzyme activities at a time of the year when the climate is most stable and when there have been no recent soil disturbances. For example, if the objective is to determine long-term management effects on enzyme activities, times with recent disturbances should be avoided. Thus, in temperate regions, mid-to-late spring or late fall, when there has been no fresh input of organic matter or fertilizer, might be optimal. By doing this, the impact of temporary microbial responses, which can occur during management periods and thus not effectively reflect long-term effects on soil enzyme activity, are reduced. A specific example is fertilizer experiments where Schinner et al. (1996) recommended soil samples for microbial analyses should be collected 2 to 3 mo after fertilization. Similarly, for rhizosphere research it would be important to identify key moments during the life cycle of the plant species of interest and sample accordingly.

One way to overcome temporal variation of soil enzyme activities is to sample several times a year, as appropriate for the question being asked. However, this greatly increases labor and analytical costs. If this approach is used, it is best to collect at least 2 yr of data because of year-to-year variation in enzyme activities. In this case it may be best to choose times each year that are climatically or environmentally similar. However, as mentioned before, the sampling time may need to shift to ensure similar moisture or temperature conditions (e.g., growing degree days) each year.

5–2.3 Presampling Soil Characterization

The primary outcome of a well-designed sampling program should be a collection of samples that represent the horizontal and vertical spatial variability in soil physical and chemical factors on the field site. This includes accounting for the size, shape, slope orientation, and heterogeneity of soil types or properties for the experimental units or field sites. Pre-site characterization is an important exercise to guide experimental and sampling plans because spatial heterogeneity of factors such as soil moisture, clay, and soil organic matter content will affect soil biological properties. As an example of the importance of soil properties, Schutter et al. (2001) found that the relative importance in controlling soil microbial community structure was soil type, the season, and lastly soil management.

A traditional approach for a new field experiment is to use a completely randomized block design to account for spatial heterogeneity in soil characteristics. Alternatively, in the case of pre-existing research sites, sampling schemes that account for soil heterogeneity should be designed. This is particularly important for large-scale experiments that may be addressing pollution gradients or land units (e.g., forest) sites that were established by land managers (typically, the layout would be based on management considerations without scientific foundation or a statistically valid experimental design). Besides reducing the impact of spatial variability, site characterization will facilitate data interpretation by using site-specific data for correlations with soil enzymatic activities or microbial biomass measures. Some general approaches for characterizing the research site are outlined below.

An important first step is to determine the management history of the site. Factors to consider for agricultural soils include: fertilizer type, quantity, and method of application (e.g., banding vs. broadcast); prior tillage; crop rotations; and land leveling or erosion that may have exposed the subsoil (for a detailed discussion of sampling plans for soil testing and/or fertility see James and Wells, 1990). In forest systems, factors such as plant species, previous harvest methods (e.g., clear cut) and dates, or fire history may affect the sampling plan.

The next step is to characterize the soil types present at the sampling site. A very useful resource for obtaining detailed information on soil heterogeneity are local soil surveys (e.g., the U.S. county surveys are available from the Natural Resources Conservation Service, online at http://soils.usda.gov/survey/printed_surveys/ [verified 26 Mar. 2011]). Typically, using printed manuals or the U.S.-based USDA-NRCS web soil survey (WSS, at http://websoilsurvey.nrcs.usda.gov/app/HomePage.htm [verified 26 Mar. 2011]), the soil type(s) and the general soil physical and chemical properties at a given site can be determined. In addition, soil survey manuals give information about the major soil series in the sampling area, descriptions of soil horizons, landscape position of the soil series, and land use categories. However, this is not a substitute for field inspections. In the field, there are no sharp borders between soil types, but rather a gradual transition from one soil type to another. Thus, regional or local soil surveys may not deliver the resolution to identify small changes in soil types at a sampling site.

A low-cost field inspection approach is to visually inspect soil cores taken to a specified depth on a grid of the sampling area and to check for any obvious changes in soil color or texture. However, in forest soils it often is not possible to take cores because the soils are too rocky or contain thick roots or wooden debris. Here, several soil pits need to be dug to evaluate the soil. A more expensive but more informative method is to analyze the samples derived from cores or soil pits taken on a grid for physical (most notably textural analysis) and general chemical (e.g., pH, total organic C and N) properties.

After obtaining soil survey information and chemical and/or physical analyses of the sampling site, a site-specific sampling plan can be developed.

5–3 SAMPLING PROTOCOL
5–3.1 Choosing Sampling Points

Several authors have presented sampling techniques (e.g., Dick et al., 1996; Wollum II, 1994). The discussion here is adapted from Dick et al. (1996) and will give an overview of the most common sampling techniques that are useful for microbial characterization of soils. Some are more labor intensive and require knowledge of the field situation.

For developing a sampling plan, the first decision is whether the site should be randomly sampled (simple random sampling), divided into subsample units (stratified random sample), systematically sampled, selectively sampled (judgment sampling), or whether a composite sample should be collected.

The variance of soil properties can be as high within a relatively short distance as at larger distances. Webster and Oliver (1990) have shown that one-quarter to half of the total variance of many soil properties in areas of 10 to 10,000 ha (25 to 25,000 acres) can occur within a few square meters. This would indicate that unless there is strong evidence for a lack of uniformity at the sampling site (e.g., obvious changes in soil type, topography, land use) a *simple random sampling* is appropriate. A simple random sampling plan is the most straightforward sampling approach. A first step is to establish a grid overlaying the sampling area and assign each square of the grid with a number. Then, random numbers representing squares on the grid are drawn from numbered pieces of paper in a hat or a computer random number generator program. In practice, Webster and Oliver (1990) found that small uniform areas (<0.5 ha or <1.2 acres) can be sampled with as few as 5 to 10 samples and that larger areas gain little in precision when numbers are greater than 25. Others suggest collecting a calculated number of samples that is based on a pre-evaluation of the variability within one site (Dick et al., 1996; Wollum II, 1994). So indeed, the number of samples can be as low as 5 to 10 for uniform sites but high variation within one site can result in sample numbers >40.

Another approach for *random sampling* is to use a serpentine sampling plan. In this method, the distance of the serpentine path is first determined by counting the number of paces from beginning to end of the selected sampling area. Then the distance is divided by the number of samples to be taken to determine the distance between sampling points. To incorporate randomness, the investigator randomly chooses the starting point of the serpentine sampling scheme. To achieve this, 10 numbered pieces of paper can be put in a hat or a random number generator and one number is drawn. The investigator then takes as many paces into the field and takes the first sample at that point. The remaining samples should be taken at the calculated intervals as outlined above (Dick et al., 1996).

If the site clearly is not uniform and subareas can be identified, then the *stratified random sampling* approach should be used (Dick et al., 1996; Wollum II, 1994). This requires a random sample to be collected within each subarea. For collecting random samples, the serpentine random sampling approach can be utilized. Subareas are delineated because there may be unique characteristics that can be readily identified. This could be due to soil series, topography, soil pollution, or obvious changes in soil management practices and vegetation. The advantage of this approach is that it allows the investigator to characterize each subarea and improve the precision of estimating the entire sampling area. The disadvantage is greater labor for sampling and analytical costs.

Systematic sampling or grid sampling ensures that the entire area is sampled from predetermined points (Dick et al., 1996; Wollum II, 1994). These points are selected at regular intervals, which makes sampling faster than random sampling. Systematic sampling is often more precise than simple random sampling, and somewhat more precise than stratified random sampling (Webster and Oliver, 1990). Systematic sampling requires consideration of periodic trends (e.g., banding of fertilizer in rows, equipment traffic patterns). This effect can be reduced by choosing a grid pattern that is unrelated in spacing and orientation to the periodic trend and increasing the number of samples. Systematic sampling is advantageous for geostatistical methods like kriging and for identifying the location of high and low values, which is not possible with random sampling (discussed above) or composite sampling (outlined below). This is an important consideration for soil quality assessment. For example, this approach is effective in identifying exposed subsoil due to land leveling or erosion or to characterize contaminated soils.

Judgment sampling selects an area at a site that is "typical" and avoids areas that are thought to be nonrepresentative of the larger area (Dick et al., 1996; Wollum II, 1994). Judgment sampling is a highly biased approach and is dependent on the expertise of the investigator. This approach may be appropriate for certain situations where the investigator is skilled and/or when there is clear evidence that a representative area can be easily identified. However, in general, this sampling plan is not recommended because the investigator will not be able to assess the accuracy of the results since they are entirely dependent on the sampler's judgment.

Composite sampling is done when a number of field samples are taken to adequately represent the area of consideration and then are thoroughly mixed to form one composite or bulk sample (Dick et al., 1996; Wollum II, 1994). This bulk sample or a well-mixed representative subsample is then analyzed. Composite sampling does not allow the determination of the variance of the mean and consequently it is not possible to estimate precision of the biological results obtained for a given site. However, it is possible to estimate the variance between similar composite samples. Composite sampling can significantly reduce the number of analyses and cost. To be valid, all samples must be drawn from the same soil type; each soil sample must contribute the same amount of soil to the composite sample; and the only objective is to provide an unbiased estimate of the mean. Typically, most soil scientists who are sampling soil for laboratory analyses use a 2.5-cm diameter probe that results in cores to the depth of sampling. Consequently, for many analyses including enzyme assays, taking 25 to 50 such cores and then homogenizing the sample (normally by passing through a 2- or 5-mm sieve) provides about 500 g of a composite soil sample. Composite sampling is an appropriate and low-cost way to integrate spatial variability within a sampling site. The authors have used this method extensively on both experimental field plots and landscape level transects and found it to be an effective means to lower sampling variability and increase the potential to detect land management or environmental impact effects (e.g., Bandick and Dick, 1999 Hinojosa et al., 2004). Composite sampling provides an inexpensive means to obtain average values of soil properties to a sampling depth for the sampling area, but does not provide any information about vertical or horizontal variability.

A better approach might be the *stratified composite sampling,* which combines composite sampling with sampling by subarea (Dick et al., 1996). This includes sampling of discrete subareas or periodic trends such as inside wheel track vs. outside wheel track in row crop fields or compositing samples from the same field under sufficient vs. deficient crop growth and/or nutrition. This results in a reduction of soil analyses but still provides separate estimates of stratified regions that were defined to be distinctly different.

5–3.2 Sample Size and Number

There is a nearly infinite amount of sampling units at most sampling sites that should be collected to characterize the site; however, this is not practical. Thus, the investigator must first decide how many sampling points sufficiently represent a given site. Simple statistics can be used to calculate the number of samples that are required. The sample mean or arithmetic average of a variable is an important statistical measure and is simply calculated as the sum of the observations divided by the number of observations. It is also necessary to have knowledge about the variability or variance (s^2), which is the degree of variation around the mean. Another way to express variance is by calculating the coefficient of variation (CV), which is the standard deviation (s) (square root of variance) divided by the mean and multiplied by 100%. Sample average is useful to minimize information; variance and standard deviation are given to show variation from the average. Ideally, the variation or standard deviation should not exceed 10% of the average. These calculations can be found in any standard statistics book and many calculators or computer spreadsheets have these functions. Further, specific equations for soil sampling are presented in Dick et al. (1996) and Wollum II (1994, p. 8).

Sampling and analytical work is time-consuming and expensive. Thus, the biggest challenge in soil sampling is to reduce the number of soil samples to an acceptable level of sampling and analytical costs. The number of soil samples can be reduced by developing a sampling approach that reduces variability (e.g., composite sampling), accepting a lower level of precision, or by lowering the confidence limit.

A more in-depth approach for describing the spatial relationship of soil properties is *geostatistics,* which will only be introduced here. The reader is referred to detailed publications that fully discuss the applications to soils (Aşkin and Kızılkaya, 2006; Oliver, 1987; Warrick et al., 1986; Webster and Oliver, 1990). Geostatistics is a sequence of methods that analyze and characterize spatially correlated data. Thus, with a systematic grid sampling, the spatial variability of a measurement across the landscape can be characterized and data between sampling points can be predicted. This is more of an "after the fact" approach and may be helpful to guide future sampling in designs at the same or similar sites. However, this approach is relatively impractical for most soil research applications. The common practices for choosing soil sampling points are discussed in Section 5–3.1 above.

5–3.3 Depth of Sampling

Typically, the surface soil will be the most sensitive region for measuring effects of soil use, disturbance, or management on soil enzyme activities. However, it may be important to identify changes with depth. Subsoil layers (30–100 cm) may control water storage and pH changes that may be important in affecting soil enzyme activities. In soils under native conditions, sampling of the surface layer should be based on the depth of horizons (e.g., O or A horizons). In agricultural soils, the surface layer or plow layer may be the most important soil depth to sample.

Research on minimum tillage and no-till agriculture systems has shown that hot spots of biological properties tend to occur in the top 7.5 cm (Dick and Daniel, 1987). For these systems, sampling the surface layer should be spilt to two depths, 0 to 7.5 cm and 7.5 to 15 cm. Pasture or grass-seed production accumulates nutrients near the surface, so a sampling depth of 0 to 10 cm might be most appropriate.

It may be best to sample by horizon for natural or unmanaged ecosystems. To determine an approach, one should consult a soil survey where the horizons are described for a given soil. A survey of the site should be done to verify the horizons in your sampling site by taking a representative number of cores or by digging several soil pits. These cores can be inspected visually or by chemical and/or physical analysis to determine horizon depths in comparison to the soil survey descriptions. Sampling depth can be established for recently logged forest sites, based on the expected root depth of newly planted seedlings (10–15 cm).

5–3.4 Collection of Bulk and Rhizosphere Soil

Typically, most soil scientists who are sampling bulk soil for laboratory analyses use a 2.5-cm probe that results in cores to the depth of sampling. Taking 25 to 50 such cores per area and then homogenizing the sample (normally by passing through a 2- or 5-mm sieve) provides about 500 g of soil. However, for rocky soils, instead of coring probes, 5-cm augers, soil knives, or spades may be necessary to collect samples at each sampling point. *Another caution when sampling for aggregate fractionation is that coring or auguring devices should not be used because the former tends to compress soils and the latter causes significant disruption of aggregates.* Therefore, a shovel or spade should be used by first digging a hole and then taking about 2-cm-thick slices to the desired depth and keeping the soil intact until fractionation.

After the collection of a planned number of replicates, a composite sample is created by mixing the soil cores and transferring the samples to plastic bags. It is advisable to collect more soil than the minimum requirement for planned analyses to allow for sieving losses or further analyses.

A shovel should be used for the collection of rhizosphere soil. The plant should be gently excavated by digging outside the root zone and then gently lifting the plant out of the soil by wedging the shovel beneath the roots. The plant with roots intact, resting on the shovel, should be lifted by hand by grasping the stem near its base. The nonrhizosphere soil should be removed by gently shaking the roots so that soil closely adhering to the roots (rhizosphere soil) stays attached. The stem should then be cut off at the base and discarded. The intact roots and adhering soil should be placed in plastic bags, sealed, and sent back to the laboratory. If available, at least four independent soil rhizosphere replicates should be taken per sampling point, with higher numbers in heterogeneous unmanaged ecosystems.

In general, soil sampling bags should be prelabeled (sample ID, location, date) in the laboratory before sample collection and then laid out at each sampling site before sampling. As sampling proceeds, the labels should be double-checked to ensure the correct sample is being taken and identified. All labels on the bags should be clearly visible and permanent. A field book should be available to note a full sample description for each sample ID (name of the location, GPS coordinates, sampling depth, date). In addition, the field book should be used to record specifics of date of sampling: weather data, disturbances at the sampling site, and any other general observations on the sampling site or sampling process that may be important for data interpretation.

5–3.5 Volumetric or Bulk Density Sampling

Measurement of bulk density can be important not only as a soil physical measurement but can also be used to convert soil measurements from a gravimetric to a volumetric basis. This is an important consideration since soils vary in bulk density and therefore a correct comparison of treatments should be done on a volumetric basis (Ellert and Bettany, 1995). In addition, the unit "per g soil" is not useful to estimate the relevance of a biological parameter on the landscape basis. Considering bulk density, the biological parameter can be expressed as "per cm^2" or "cm^3," which will provide an accurate extrapolation of results to a "per ha or acre" basis to a given soil depth.

5–4 SAMPLE HANDLING AND PRETREATMENT

5–4.1 Sample Transfer from the Field to the Laboratory

For soil enzyme measurement, soil samples should be kept fresh in a sealed plastic bag and brought to the laboratory in a cooler. If samples need to be sieved, they should be stored at 4°C until they are being processed. The soil samples should be analyzed as soon as possible—on the order of days—to most closely attain results that are unchanged from the day of sampling. Overall, most authors have concluded that 4°C storage is adequate for microbial analyses even though moderate alterations were found when compared with fresh soil (e.g., OECD, 1995). This is likely due to the fact that many enzyme activities are associated with extracellular enzymes stabilized in the soil matrix on clay minerals and organic matter (Tietjen and Wetzel, 2003). These enzymes are not associated with viable cells and during 4°C storage they remain protected from denaturation.

For more guidelines and discussion on soil storage before enzyme analysis please refer to Section 5–4.4.

5–4.2 Soil Physical Fractionation

Soil enzymes are spatially distributed in the soil matrix at microscales and therefore physical fractionation techniques are useful tools to study enzyme activities

within their microenvironment (Kandeler and Dick, 2007). There are three fractionation methods for studying microscale distributions of soil properties, which
for soil enzyme activity were reviewed in detail by Kandeler and Dick (2007).
Depending on the research objective, soil enzyme activities might be studied (i)
across soil aggregate size fractions (Six et al., 2004), (ii) on particulate organic
matter (POM), or (iii) among particle-size fractions (PSF). When conducting soil
aggregate fractionation that includes sieving for fractions >2 mm, it is important
not to sieve the soil before fractionation. Furthermore, soil should be used directly
from the field by taking an intact portion of the soil sample for fractionating. Soil
aggregates can be obtained by a wet-sieving method as described by Fansler et
al. (2005). Here, sieve sizes of 2 mm, 1 mm, 250 μm, and 53μm are stacked with
the biggest size on the top where the soil is loaded on the 2-mm sieve. The soil
is washed and shaken by rotation using distilled water. The rotation of the sieve
stack is 4 cm at a rate of 50 × min^{-1} for 2 min.

Particulate organic matter (POM) can be separated from the soil using a
flotation method described by Magid and Kjærgaard (2001). Soil is dispersed in
5% NaCl, shaken by hand, and allowed to stand for 45 min. The sample is then
poured gradually onto a sieve (~100-mm mesh size) and washed with tap water.
Subsequently, the aggregates degrade due to the flushing and the water passing
the sieve becomes clear. The material retained on the sieve is then transferred
into a plastic bucket. Tap water is added and the bucket is swirled, enabling the
POM to be separated from the mineral material phase by flotation–decantation.
The POM yielded by this process can be separated further into size fractions
from 100 to 400 μm, 400 to 800 μm, 800 to 2,000 μm, and >2000 μm using appropriate sieves.

Soil particle-size fractions (PSF) for enzymatic analysis can be obtained by
low-energy sonication of soil water suspensions, followed by wet sieving and
centrifugation to yield sand, silt, and clay fractions as described in Stemmer et
al. (1998). Accordingly, 30 g of field-moist soil is dispersed in 80 mL of deionized water by using an ultrasound probe (50 Js^{-1} for 2 min). Macroaggregates are
hereby disrupted, while organic matter physically protected within microaggregates is not released. Coarse and medium sand (2000–250 μm) and fine sand
(250–63 μm) can be separated by manual wet sieving. Silt-sized particles (63–2
μm) are separated from the clay fraction (2–0.1 μm) by centrifugation at 150 g for
2 min. The pellet will contain the silt fraction that needs to be separated from the
clay by further washing steps, specifically by addition of fresh water and repeated
centrifugation until the supernatant is clear. The centrifugation can be performed
at 20°C; however, a refrigerated (4°C) centrifuge is optimal. Subsequently, the fractions should be stored at 4°C prior to enzyme analysis. Using the PSF technique,
Stemmer et al. (1998) have shown that tillage systems influenced the spatial distribution of soil enzyme activities. In several studies, authors have used the assay
described by Stemmer et al. (1998) to analyze enzyme activities and other biological properties in particle-size fractions (e.g., Kandeler et al., 1999a; Kandeler et al.,
1999b; Kandeler et al., 2000; Poll et al., 2003).

Another PSF method is described in Filip et al. (2000), which separates up to
seven fractions each for organic matter and mineral soil. Using the approach of
Filip et al. (2000), PSFs of >5 mm, 5 to 2 mm, 2 to 1 mm, 1 to 0.5 mm, 0.5 to 0.2 mm,
0.2 to 0.1 mm, and 0.1 to 0.05 mm are obtained. The organic matter in the fractions
is separated from the mineral soil by considering differences in sedimentation

velocity. Using this method, Filip et al. (2000) have measured β-glucosidase, protease, and β-acetylglucosaminidase activities in PSFs.

5–4.3 Soil Sieving

Several factors should be considered before sieving soil samples. It is important that the sample is completely mixed before sieving when composite samples are taken. Dry soils that do not readily pass through the sieve can be crushed carefully before sieving. Sometimes soils are too wet. In this case, soils can be air-dried in the refrigerator or cold room to a moisture at which sieving is easier. However, it is important that the sample does not completely dry out or be subject to high temperatures during partial drying or sieving. Wetland soils (Gleysols) or lake sediments should be sieved at the actual water content since drying changes the redox potential and this does not reflect the situation in the field.

The sample should be carefully passed through a 2- to 5-mm sieve to remove roots and stones from the soil sample. In general, a sieve mesh size of 2 mm is recommended; however, a 4- or 5-mm sieve is advised for soils with a high organic matter or clay content. *It is essential that the whole sample, or a representative aliquot, passes through the sieve. If aggregates are discarded because they are difficult to force through the sieve, the sample will be biased because not the entire soil sample will be represented after sieving (Dick et al., 1996).* When rhizosphere soil is sieved, the sample (roots with the soil attached) should be shaken gently onto the sieve. In addition, the attached soil should be removed from the roots using forceps. This is a time-consuming task but it is an important step and care should be taken to remove all visible roots from the soil sample. Finally, it is important to clean the sieve and work area between samples to avoid cross-contamination.

5–4.4 Soil Storage

The objective for measuring enzyme activities will, to some extent, dictate the method of storage. If the goal is to use enzyme activities to assess soil quality, and ultimately to be adopted for commercial applications, then the most important factor is temporal stability and a period of storage that has low variability. On the other hand, if the activities are being conducted for research purposes, then the goal is to use a storage method that best maintains the soil between sampling and analysis, which normally means as soon as possible.

5–4.4.1 Air-Drying

A potential advantage of analyzing soil enzymes rather than other soil microbial properties for assessing soil quality is that enzyme activities have been shown to preserve the ranking of field management treatments comparing air-dried soils and field-fresh soils (Bandick and Dick, 1999). This is particularly useful for long-term monitoring projects where relative changes over time, e.g., due to land management, are of interest. In this case, the goal typically is to have a practical means to detect effects of disturbance, inputs, or other management practices on soils, rather than trying to closely relate the enzyme activities to microbial communities or functions under field conditions for research purposes.

It seems probable that air-drying has the biggest impact on enzymes associated with the viable microbial population, with less effect on the catalytic fraction stabilized in the soil matrix. This could be an advantage for soil quality testing, as air-drying would reduce the impact of conditions that affect the highly variable microbial component relative to enzyme activity. Therefore, enzyme activity after air-drying likely better reflects the true long-term trajectory of a given management system on soil quality. Because air-drying stabilizes enzyme activity, it greatly facilitates handling the sample and allows for later analyses (unlike most other microbial measures that must be done as soon as possible after sampling). Supporting the idea of soil air-drying to facilitate commercial adoption of enzyme assays is the report by Bandick and Dick (1999). They found that although some enzyme activities changed during air-drying, the ranking of field management treatments was the same between air-dried soils and field-fresh soils. In addition, Hinojosa et al. (2004) pointed out while analyzing heavy metal polluted soils, rewetted field-dry soils increased enzyme activities and improved discrimination between heavy metal polluted and unpolluted soils. Tables 1 to 3 show that air-drying did not affect some soil enzymes: urease (Lee et al., 2007; Kandeler and Gerber, 1988; Zantua and Bremner, 1977); β-glucosidase (Lee et al., 2007; Bandick and Dick, 1999); invertase (Bandick and Dick, 1999); cellulase (Lee et al., 2007); acid phosphatase (Lee et al., 2007); arylsulfatase (Bandick and Dick, 1999; Zornoza et al., 2006). Urease has proven to remain stable for periods for up to 2 yr after air-drying at room temperature (Zantua and Bremner, 1977; Pettit et al., 1976)

Finally, several other reports have shown that enzyme activities increased after air-drying the soil (urease for Kandeler and Gerber, 1988; McGarity and Myers, 1967; Palma and Conti, 1990; Speir and Ross, 1981; arylsulfatase for Tabatabai and Bremner, 1970). Eivazi and Tabatabai found that acid phosphatase (1977) and α- and β-glucosidase, as well as α- and β-galactosidase (1990), were higher in air-dried soils compared with field-fresh soil (Tables 1–3).

Overall, the literature shows no clear or consistent air-drying effect that is consistent among different soil types and enzyme activities (Tables 1–3). There is no indication that enzyme activities in soils with higher clay and organic matter content are less affected by air-drying.

Several studies have shown that soil drying and rewetting before enzyme analyses is the least desirable soil pretreatment because enzyme activities are drastically reduced compared with fresh soil and therefore it should be avoided (Lee et al., 2007; Pancholy and Rice, 1972; Rao et al., 2003).

5–4.4.2 Cold Storage

Normally the goal for research purposes is to have enzyme activities reflect the true state of the ability of the soil to perform a given enzymatic reaction under in situ conditions. In this case, cold storage may be the best approach. For enzyme analysis, it is best if the samples are maintained at ambient or cool temperatures (4°C) during transport to the lab. As a general rule, enzyme analyses should be made within the least amount of time necessary (days to a few weeks). However, in most cases, soil storage is necessary. Although clay mineralogy and humic acids are known to be main drivers of enzyme stability (Allison, 2006), in certain cases soil storage at different temperatures has been shown to affect soil enzyme activities.

In the study by Lee et al. (2007), storage at 4°C for 28 d resulted in enzyme activities that were generally statistically indistinguishable from those of fresh soil across three soil types. Some authors have concluded that 4°C storage is adequate for microbial analyses even though moderate alterations were found when compared with fresh soil (OECD, 1995). However, there also are negative reports about

Table 5–1. Effect of soil drying (+ rew = rewetted and 14 d incubation) on soil urease (U) and β-glucosaminidase (β-Gm) enzyme activities in relation to field-moist soil observed in different studies.

Enzyme	Treatment: effect	N	pH	C	Sand	Silt	Clay	Reference
						%		
U	dry + rew: no effect	1	5.8†	3.38‡	23.4	64.4	12.2	Lee et al., 2007
U	dry + rew: no effect	1	5.4†	2.05‡	20.7	37.2	42.1	Lee et al., 2007
U	dry: no effect	4	5.2 – 7.4§	1.5 – 19.2¶	5 – 72	17 – 61	8 – 31	Kandeler & Gerber, 1988
U	dry 2 y: no effect	6	5 – 8†	0.3 – 5.9#	5 – 53	31 – 71	13 – 41	Zantua & Bremner, 1977
U	dry: decreased 17%	1	5.95†	2.65††	–	–	–	Bandick & Dick, 1999
U	dry + rew: decreased 40%	1	6.1†	2.83‡	86	10.9	3.1	Lee et al., 2007
U	dry: increased 11 – 33%	9	5.6 – 7.2†	1.7 – 10.4††	–	–	–	Speir & Ross, 1981
U	dry: increased 16%	5	5.3 – 5.7†	1.8 – 3.2#	–	–	–	McGarity & Myers, 1967
U	dry: increased 20%	1	7.0†	2.95††	–	–	–	Bandick & Dick, 1999
U	dry: increased ±30%	22	5 – 7.7†	0.6 – 4#	–	–	–	Palma & Conti, 1990
U	dry: increased 47 – 110%	3	5.6 – 7.1§	1.4 – 4.5¶	3 – 75	8 – 81	8 – 16	Kandeler & Gerber, 1988
β-Gm	dry: decreased 12%	1	6.5†	1.14††	17.5	52.5	30	Parham & Deng, 2000
β-Gm	dry + rew: decreased 20%	1	5.4†	2.05‡	20.7	37.2	42.1	Lee et al., 2007
β-Gm	dry: decreased 22%	1	5.9†	0.72††	55	32.5	12.5	Parham & Deng, 2000
β-Gm	dry + rew: decreased 40%	1	5.8†	3.38‡	23.4	64.4	12.2	Lee et al., 2007
β-Gm	dry + rew: decreased 40%	1	6.1†	2.83‡	86	10.9	3.1	Lee et al., 2007
β-Gm	dry: decreased 60 – 66%	2	5.3 – 5.9†	2.6 – 3††	–	–	33 – 62	Turner & Romero, 2010

† pH in water.
‡ C_t dry combustion (960°C).
§ pH in 0.01 M $CaCl_2$.
¶ chromic titration.
# C_{org} by wet potassium dichromate oxidation.
†† C_{org} by dry combustion (no temperature given).

Table 5–2. Effect of soil drying (+ rew = rewetted and 14 d incubation) on soil β-glucosidase (β-Gs), invertase (Inv.) and cellulase (Cel.) enzyme activities in relation to field-moist soil observed in different studies.

Enzyme	Treatment: effect	N	pH†	C	Sand	Silt	Clay	Reference
						—%—		
β-Gs	dry + rew: no effect	1	6.1	2.83‡	86	10.9	3.1	Lee et al., 2007
β-Gs	dry + rew: no effect	1	5.8	3.38‡	23.4	64.4	12.2	Lee et al., 2007
β-Gs	dry: no effect	1	7.0	2.95§	–	–	–	Bandick & Dick, 1999
β-Gs	dry: decreased 10%	1	5.95	2.65§	–	–	–	Bandick & Dick, 1999
β-Gs	dry + rew: decreased 20%	1	5.4	2.05‡	20.7	37.2	42.1	Lee et al., 2007
β-Gs	dry: decreased 50 – 60%	1	5.3 – 5.9	2.6 – 3§	–	–	33 – 62	Turner & Romero, 2010
β-Gs	dry: increased 40 – 63%	5	6.1 – 7.8	1.8 – 5.5	16 – 39	37 – 50	23 – 34	Eivazi & Tabatabai, 1990
Inv.	dry: no effect	1	5.95	2.65§	–	–	–	Bandick & Dick, 1999
Inv.	dry: decreased 20%	1	7.0	2.95§	–	–	–	Bandick & Dick, 1999
Inv.	dry: decreased 22 – 43%	9	5.6 – 7.2	1.7 – 10.4§	–	–	–	Speir & Ross, 1981
Inv.	dry: decreased 36%	13	5.3 – 8.3	1.3 – 9.2§	–	–	–	Ross, 1965
Cel.	dry + rew: no effect	1	5.8	3.38‡	23.4	64.4	12.2	Lee et al., 2007
Cel.	dry: decreased 40%	1	7.0	2.95§	–	–	–	Bandick & Dick, 1999
Cel.	dry: decreased 41 – 80%	9	5.6 – 7.2	1.7 – 10.4§	–	–	–	Speir & Ross, 1981
Cel.	dry + rew: decreased 50%	1	6.1	2.83‡	86	10.9	3.1	Lee et al., 2007
Cel.	dry: decreased 60%	1	5.95	2.65§	–	–	–	Bandick & Dick, 1999
Cel.	dry + rew: increased 28%	1	5.4	2.05‡	20.7	37.2	42.1	Lee et al., 2007

† pH in water.

‡ C_t dry combustion (960°C).

§ C_{org} by dry combustion (no temperature given).

Table 5–3. Effect of soil drying (+ rew = rewetted and 14 d incubation) on soil alkaline phosphatase (AkP), acid phosphatase (AcP), and arylsulfatase (Aryl.) enzyme activities in relation to field-moist soil observed in different studies.

Enzyme	Treatment: effect	N	pH†	C	Sand	Silt	Clay	Reference
						—%—		
AkP	dry: decreased 4 – 7%	2	7.6	4.6‡	21	47	32	Eivazi & Tabatabai, 1977
AkP	dry + rew: decreased 10%	1	5.8	3.38§	23.4	64.4	12.2	Lee et al., 2007
AkP	dry: decreased 17 – 47%	9	5.6 – 7.2	1.7 – 10.4¶	–	–	–	Speir & Ross, 1981
AkP	dry + rew: decreased 30%	1	5.4	2.05§	20.7	37.2	42.1	Lee et al., 2007
AkP	dry + rew: decreased 30%	1	6.1	2.83§	86	10.9	3.1	Lee et al., 2007
AkP	dry: decreased 46 – 53%	3	6.1	2.6‡	34	40	26	Eivazi & Tabatabai, 1977
AcP	dry + rew: no effect	1	5.8	3.38§	23.4	64.4	12.2	Lee et al., 2007
AcP	dry + rew: decreased 10%	1	6.1	2.83§	86	10.9	3.1	Lee et al., 2007
AcP	dry: increased 4 – 10%	3	6.1	2.6‡	34	40	26	Eivazi & Tabatabai, 1977
AcP	dry + rew: increased 10%	1	5.4	2.05§	20.7	37.2	42.1	Lee et al., 2007
AcP	dry: increased 42 – 54%	2	7.6	4.6‡	21	47	32	Eivazi & Tabatabai, 1977
Aryl.	dry: no effect	1	5.95	2.65¶	–	–	–	Bandick & Dick, 1999
Aryl.	dry: no effect	1	7.0	2.95¶	–	–	–	Bandick & Dick, 1999
Aryl.	dry: no effect	3	6.1 – 6.7	3.6 – 6.3¶	–	–	–	Speir & Ross, 1981
Aryl.	dry: decreased 4 – 19%	5	5.6 – 7.2	1.8 – 10.4¶	–	–	–	Speir & Ross, 1981
Aryl.	dry + rew: decreased 20%	1	6.1	2.83§	86	10.9	3.1	Lee et al., 2007
Aryl.	dry + rew: decreased 20%	1	5.4	2.05§	20.7	37.2	42.1	Lee et al., 2007
Aryl.	dry + rew: decreased 30%	1	5.8	3.38§	23.4	64.4	12.2	Lee et al., 2007
Aryl.	dry: increased 43%	13	5.9 – 8	1.4 – 5.9‡	1 – 58	30 – 74	12 – 45	Tabatabai & Bremner, 1970

† pH in water.
‡ C_{org} by wet potassium dichromate oxidation.
§ C_t by dry combustion (960°C).
¶ C_{org} by dry combustion (no temperature given).

the effects of cold storage on enzyme activities. For example, alkaline phosphatase and dehydrogenase activities were reduced after cold storage at temperatures ranging from +5°C to −12°C (Vieira and Nahas, 1998). When using −20°C for soil storage, soil aggregates have been shown to be degraded due to physical disruption, which tends to increase enzyme activities as shown by McGarity and Myers (1967) and Tabatabai and Bremner (1970). Overall, the most consistent and recommended method of soil storage for enzyme analysis is cold storage at 4°C for a maximum of 1 mo (Gianfreda and Bollag, 1996; Schinner et al., 1996; Lee et al., 2007; Turner and Romero, 2010). In addition, it might be acceptable to freeze the soil at −20°C if the analyses cannot be completed within 1 mo (Schinner et al., 1996; Stenberg et al., 1998). This was supported by Turner and Romero (2010), who found that freezing a tropical soil at −35°C closely maintained enzyme activities better than ambient (22°C) or cold storage (4°C).

In conclusion, the most consistent recommendation we could find across the literature was to store soil at 4°C before enzyme analysis. However, there were soil types, depending on particular enzyme assays, that deviated from this outcome. Therefore, before launching a new research or monitoring project, if resources and time are available, it would be best to test several soil storage methods for the enzyme assays of interest.

REFERENCES

Allison, S.D. 2006. Soil minerals and humic acids alter enzyme stability: Implications for ecosystem processes. Biogeochemistry 81:361–373.

Aon, M.A., and A.C. Colaneri. 2001. II. Temporal and spatial evolution of enzymatic activities and physico-chemical properties in an agricultural soil. Appl. Soil Ecol. 18:255–270.

Aşkin, T., and R. Kızılkaya. 2006. Assessing spatial variability of soil enzyme activities in pasture topsoils using geostatistics. Eur. J. Soil Biol. 42:230–237.

Bandick, A., and R.P. Dick. 1999. Field management effects on soil enzyme activities. Soil Biol. Biochem. 31:1471–1479.

Das, A.C. 1950. Two-dimensional systematic sampling and associated stratified and random sampling. Sankhya 10:95–108.

Dick, R.P., D.R. Thomas, and J.J. Halverson. 1996. Standardized methods, sampling, and sample pretreatment. p. 107–122. In J.W. Doran and A.J. Jones (ed.) Methods of assessing soil quality. SSSA Spec. Publ. 49. SSSA, Madison, WI.

Dick, W.A., and T.C. Daniel. 1987. Soil chemical and biological properties as affected by conservation tillage: Environmental impacts. In T.J. Logan et al. (ed.) Effects of conservation tillage on groundwater quality: Nitrates and pesticides. Lewis Publishers, Chelsea, MI.

Eivazi, F., and M.A. Tabatabai. 1977. Phosphatases in soils. Soil Biol. Biochem. 9:167–172.

Eivazi, F., and M.A. Tabatabai. 1990. Factors affecting glucosidase and galactosidase activities in soils. Soil Biol. Biochem. 22:891–897.

Ellert, B.H., and J.R. Bettany. 1995. Calculation of organic matter and nutrients stored in soils under contrasting management regimes. Can. J. Soil Sci. 75:529–538.

Fansler, S.J., J.L. Smith, H. Bolton, Jr., and V.L. Bailey. 2005. Distribution of two C cycle enzymes in soil aggregates of a prairie chronosequence. Biol. Fertil. Soils 42:17–23.

Filip, Z., S. Kanazawa, and J. Berthelin. 2000. Distribution of microorganisms, biomass ATP, and enzyme activities in organic and mineral particles of a long-term wastewater irrigated soil. J. Plant Nutr. Soil Sci. 163:143–150.

Gianfreda, L., and J.M. Bollag. 1996. Influence of natural and anthropogenic factors on enzyme activity in soil. p. 123–193. In G. Stotzky and J.M. Bollag (ed.) Soil biochemistry. Vol. 9. Marcel Dekker, New York.

Hinojosa, M.B., J.A. Carriera, R. García-Ruíz, and R.P. Dick. 2004. Soil moisture pre-treatments effects on enzyme activities as indicators of heavy metal–contaminated and reclaimed soils. Soil Biol. Biochem. 36:1559–1568.

James, D.W., and K.L. Wells. 1990. Soil sample collection and handling: Technique based on source and degree of field variability. p. 25–44. *In* R.L. Westerman (ed.) Soil testing and plant analysis. SSSA, Madison, WI.

Kandeler, E., and R.P. Dick. 2007. Soil enzymes: Spatial distribution and function in agro-ecosystems. p. 263–287. *In* G. Benckiser and S. Schnell (ed.) Biodiversity in agricultural production systems. CRC/Taylor & Francis, Boca Raton, FL.

Kandeler, E., and H. Gerber. 1988. Short-term assay of soil urease activity using colorimetric determination of ammonium. Biol. Fertil. Soils 6:68–72.

Kandeler, E., S. Palli, M. Stemmer, and M.H. Gerzabek. 1999b. Tillage changes microbial biomass and enzyme activities in particle-size fractions of a Haplic Chernozem. Soil Biol. Biochem. 31:1253–1264.

Kandeler, E., M. Stemmer, and E.M. Klimanek. 1999a. Response of soil microbial biomass, urease and xylanase within particle size fractions to long-term soil management. Soil Biol. Biochem. 31:261–273.

Kandeler, E., D. Tscherko, K.D. Bruce, M. Stemmer, P.J. Hobbs, R.D. Bardgett, and W. Amelung. 2000. The structure and function of the soil microbial community in micro-habitats of a heavy metal polluted soil. Biol. Fertil. Soils 32:390–400.

Lee, Y.B., N. Lorenz, L. Kincaid Dick, and R.P. Dick. 2007. Cold storage and pretreatment incubation effects on soil microbial properties. Soil Sci. Soc. Am. J. 71:1299–1305.

Magid, J., and C. Kjærgaard. 2001. Recovering decomposing plant residues from the particulate soil organic matter fraction: Size versus density separation. Biol. Fertil. Soils 33:252–257.

McGarity, J.W., and M.G. Myers. 1967. A survey of urease activity in soils of Northern New South Wales. Plant Soil 27:217–238.

Ndiaye, E.L., J.M. Sandeno, D. McGrath, and R.P. Dick. 2000. Integrative biological indicators for detecting change in soil quality. Am. J. Altern. Agric. 15:26–36.

OECD (Organization for Economic and Cooperative Development). 1995. OECD guideline for the testing of chemicals. Final Rep. OECD Worksh. on Selection of Soils/Sediments, Belgirate, Italy. 18–20 Jan. 1995. OECD, Paris.

Oliver, M.A. 1987. Geostatistics and its application to soil science. Soil Use Manage. 3:8–20.

Palma, R.M., and M.E. Conti. 1990. Urease activity in argentine soils: Field studies and influence of sample treatment. Soil Biol. Biochem. 22:105–108.

Pancholy, S.K., and E.L. Rice. 1972. Effect of storage conditions on activities of urease, invertase, amylase, and dehydrogenase in soil. Soil Sci. Soc. Am. Proc. 36:536–537.

Parham, J.A., and S.P. Deng. 2000. Detection, quantification and characterization of β-glucosaminidase activity in soil. Soil Biol. Biochem. 32:1183–1190.

Pettit, N.M., A.R.J. Smith, R.B. Freedman, and R.G. Burns. 1976. Soil urease: Activity, stability and kinetic properties. Soil Biol. Biochem. 8:479–484.

Poll, C., A. Thiede, N. Wermbter, A. Sessitsch, and E. Kandeler. 2003. Micro-scale distribution of microorganisms and microbial enzyme activities in a soil with long-term organic amendment. Eur. J. Soil Sci. 54:715–724.

Rao, M.A., F. Sanino, G. Nocerino, E. Puglisi, and L. Gianfreda. 2003. Effect of air-drying treatment on enzymatic activities of soils affects by anthropogenic activities. Biol. Fertil. Soils 38:327–332.

Ross, D.J. 1965. Effect of air-dry, refrigerated, and frozen storage on activities of enzymes hydrolizing sucrose and starch in soils. J. Soil Sci. 16:86–94.

Schinner, F., R. Öhlinger, E. Kandeler, and R. Margesin (ed.). 1996. Methods in soil biology. Springer, Berlin and London.

Schutter, M., J. Sandeno, and R.P. Dick. 2001. Seasonal, soil type, and alternative management influences on microbial communities of vegetable cropping systems. Biol. Fertil. Soils 34:397–410.

Six, J., H. Bossuyt, S. Degryze, and K. Denef. 2004. A history of research on the link between (micro)aggregates, soil biota, and soil organic matter dynamics. Soil Tillage Res. 79:7–31.

Speir, T.W., and D.J. Ross. 1981. A comparison of the effects of air-drying and acetone dehydration on soil enzyme activities. Soil Biol. Biochem. 13:225–229.

Stemmer, M., M.H. Gerzabek, and E. Kandeler. 1998. Organic matter and enzyme activity in particle-size fractions of soils obtained after low-energy sonication. Soil Biol. Biochem. 30:9–17.

Stenberg, B., M. Johansson, M. Pell, K. Sjödahl-Sevensson, J. Stenström, and L. Torstensson. 1998. Microbial biomass and activities in soil as affected by frozen and cold storage. Soil Biol. Biochem. 30:393–402.

Tabatabai, M.A., and J.M. Bremner. 1970. Factors affecting soil arylsulfatase activity. Soil Sci. Soc. Am. J. 34:427–429.

Tietjen, T., and R.G. Wetzel. 2003. Extracellular enzyme-clay mineral complexes: Enzyme adsorption, alteration of enzyme activity, and protection from photodegradation. Aquat. Ecol. 37:331–339.

Turner, B.L., and T.E. Romero. 2010. Stability of hydrolytic enzyme activity and microbial phosphorus during storage of tropical rain forest soils. Soil Biol. Biochem. 42:459–465.

Vieira, F.C.S., and E. Nahas. 1998. Microbial activity of dark red laterosol samples stored at different temperatures. Rev. Microbiol. Vol. 29. No. 3. doi:10.1590/S0001–37141998000300002

Warrick, A.W., D.E. Myers, and D.E. Neilsen. 1986. Geostatistical methods applied to soil science. p. 53–82. In A. Klute (ed.) Methods of soil analysis. Part 1. Physical and mineralogical methods. 2nd ed. SSSA, Madison WI.

Webster, R., and M.A. Oliver. 1990. Statistical methods in soil and land resource survey. p. 44–46. Oxford Univ. Press, Oxford, UK,

Wollum, A.G., II. 1994. Soil sampling for microbiological analysis. p. 1–14. In J.M. Bigham (ed.) Methods of soil analysis. Part 2. Microbiological and biochemical properties. SSSA Book Ser. 5. SSSA, Madison, WI.

Zantua, M.I., and J.M. Bremner. 1977. Stability of urease in soils. Soil Biol. Biochem. 9:135–140.

Zornoza, R., C. Guerrero, J. Mataix-Solera, V. Arcenegui, F. García-Orenes, and J. Mataix-Beneyto. 2006. Assessing air-drying and rewetting pre-treatment effect on some soil enzyme activities under Mediterranean conditions. Soil Biol. Biochem. 38:2125–2134.

Chapter 6

Soil Oxidoreductases and FDA Hydrolysis

Jennifer A. Prosser, Tom W. Speir,* and Diane E. Stott

6-1 GENERAL INTRODUCTION

The oxidoreductases (EC 1) comprise the largest enzyme group and consist of enzymes that catalyze reactions in which one substrate is oxidized (the donor) while another is reduced (the acceptor) (Dixon and Webb, 1979). In common with all redox reactions, the reaction mechanism involves electron transfer, expressed in a simplistic representation as:

$$A^- + B \rightarrow A + B^- \tag{1}$$

However, the observed reaction usually involves the transfer of two hydrogen atoms from the donor to the acceptor (dehydrogenation) and, consequently, most of the enzymes are called dehydrogenases. The entire dehydrogenase-catalyzed reaction system is an enzyme donor–acceptor complex, located inside the cell, and does not involve ions or electrons reacting in solution (Dixon and Webb, 1979). Where molecular oxygen (O_2) is the acceptor, the O_2 is reduced to water and the enzymes are termed oxidases. Here again, the reaction mechanism is not straightforward, and may involve the transfer of one or two H atoms (or one or two electrons) to the O_2. For one particular oxidase enzyme, laccase, which catalyzes a one-electron transfer, it appears that the enzyme may operate like a battery, storing electrons from individual oxidation reactions to reduce molecular oxygen, intermediates of which remain bound to the enzyme complex (Thurston, 1994). One or both atoms from O_2 can also be incorporated into the substrate being oxidized and the enzymes that catalyze these transformations all fall into EC 1.13. and 1.14. and are termed oxygenases. There are many other oxidoreductase acceptors, including hydrogen peroxide (H_2O_2), which acts as both donor and acceptor for the enzyme catalase, and as acceptor for peroxidases.

We propose the use of an assay for fluorescein diacetate (FDA) hydrolysis as an alternative indicator of overall enzymatic activity in soil. FDA hydrolysis is mediated by a number of different enzymes, including lipases, proteases, and

Jennifer A. Prosser, ESR Ltd., 34 Kenepuru Dr., Porirua 5240, New Zealand (jennifer.prosser@esr.cri.nz); **Tom W. Speir**, ESR Ltd., 34 Kenepuru Dr., Porirua 5240, New Zealand (tom.speir@esr.cri.nz), *corresponding author; **Diane E. Stott**, USDA-ARS, National Soil Erosion Research Laboratory, 275 S. Russell St., West Lafayette, IN 47907 (diane.stott@ars.usda.gov).

doi:10.2136/sssabookser9.c6

esterases (Rotman and Papermas, 1966), and appears to be ubiquitous among fungi and bacteria that are the primary decomposers in the soil (Lundgren, 1981). Organisms that degrade plant and other organic residues have been identified as major contributors to soil enzyme activity (Speir and Ross, 1976; Speir, 1977). Therefore, since FDA hydrolysis is a wide-ranging assay of soil hydrolase activities, it should represent microbial activity and have an important role in soil microbial ecology (Dick et al., 1996).

6–2 DEHYDROGENASE

6–2.1 Introduction

The oxidoreductase class contains close to 600 known enzymes, more than half of which are dehydrogenases that use the metabolic cofactors NAD^+ or $NADP^+$ as the acceptor molecule (Dixon and Webb, 1979). In oxidative catabolism of organic compounds, these cofactors, having been reduced, transfer the two electrons to another compound and are thus available for further oxidative reactions. In this way, they serve as electron transport intermediaries in the dehydrogenase-catalyzed reactions of the respiratory chain. For a more complete description of the metabolic role of dehydrogenases, see Alef (1995).

In soil enzymology, "dehydrogenase activity" comprises the cumulative activities of many microbial dehydrogenases involved in the oxidation of a multitude of organic molecules during microbial respiration, with the terminal acceptor being molecular oxygen. It is considered that the biochemical properties of dehydrogenases are such that the free "enzyme" cannot function in soil and that all activity is intracellular (Skujiņš, 1967). Soil dehydrogenase was originally assumed to estimate microbial respiration and metabolic activity due to the direct involvement of dehydrogenases in these processes (Skujiņš, 1967), but subsequently, many studies have found no relationship with other measures of respiration and activity (Nannipieri et al., 1990).

In the assay methods used for soil dehydrogenase, the hydrogen atoms and electrons from the oxidation reactions are transferred to an electron acceptor competitor of $NAD(P)^+$, thereby forming a reduced product that accumulates because it is unable to take part in any further catabolic reactions. The lack of correlation of soil dehydrogenase activity with other measures of respiration and activity (Skujiņš, 1976; Frankenberger and Dick, 1983) may be due to the variety of experimental conditions used by different research groups, and/or the low competitive ability of the chosen electron acceptor, compared with O_2, to capture the electrons (Nannipieri et al., 1990). Howard (1972) reviewed the literature and presented data that provided a potential explanation of these results. Using oxygen uptake to calculate theoretical dehydrogenase activity, he showed that the observed dehydrogenase activity was substantially more than theoretical dehydrogenase activity. He hypothesized that extracellular phenol oxidases may also carry out the dehydrogenase reaction and that this causes overestimation of dehydrogenase activity. Another problem with this assay is that Cu can interfere with the analytical procedure so that soils with high concentrations of soil solution Cu, or that have received Cu-contaminated amendments, will show artificially low dehydrogenase activity levels (Chander and Brookes, 1991).

6–2.2 Principle

For the assay of soil dehydrogenase activity, there have been two principal electron acceptors used, 2,3,5-triphenyltetrazolium chloride (TTC) and 2-(4-iodophenyl)-3-(4-nitrophenyl)-5-phenyl-2*H*-tetrazolium chloride (INT). There has, arguably, been more critical analysis of soil dehydrogenase methodology than of any other method in soil biochemistry, and both of these compounds have been found to have a number of deficiencies. However, the consensus appears to favor INT as the preferred electron acceptor for the following reasons: i) it is a better competitor with O_2 than other electron acceptors, ii) it is less toxic to microorganisms, iii) it is more rapidly reduced than TTC, iv) it performs in both aerobic and anaerobic conditions, and v) it is sensitive over a wide range of temperatures (Benefield et al., 1977; Trevors et al., 1982; Trevors,1984a; Friedel et al., 1994; Mosher et al., 2003). It has been noted that approximately 10% of the evolved hydrogen is transferred to INT (Benefield et al., 1977), compared with only 2 to 3% with TTC (Öhlinger, 1996b). Although there are suggestions that TTC is more sensitive than INT when assayed under optimum conditions (Gong, 1997), this work has not been widely discussed. Consequently, the method suggested in this chapter is based on the use of INT as the electron acceptor for dehydrogenase activity. In the reaction, water-soluble INT is enzymatically reduced to form the water-insoluble purple compound, iodonitrotetrazolium formazan (INTF), which is extracted using an organic solvent and determined colorimetrically (Shaw and Burns, 2006). The method is based on that originally described by Benefield et al. (1977), with adaptations by von Mersi and Schinner (1991), von Mersi (1996), and Shaw and Burns (2006).

6–2.3 Assay Method (Benefield et al., 1977; von Mersi and Schinner, 1991; von Mersi, 1996; and Shaw and Burns, 2006)

6–2.3.1 Apparatus

- Volumetric flasks, 1 L, 250 mL, 100 mL
- Platform shaker or ultrasonic bath
- 20-mL glass tubes with ethanol-resistant caps
- Shaking incubator set at 37°C
- Vortex mixer
- Spectrophotometer
- Spectrophotometer cuvettes

6–2.3.2 Reagents

- Dilute hydrochloric acid, 3 M: Mix 100 mL of concentrated hydrochloric acid (HCl, 37%) with 300 mL of deionized water.
- THAM buffer, 1 M, pH 7: Weigh 30.28 g of tris(hydroxymethyl)aminomethane $[NH_2C(CH_2OH)_3]$ into a 250-mL volumetric flask and dissolve in 200 mL of deionized water. Adjust to pH 7 using HCl (dropwise), and bring to 250 mL with deionized water.
- INT solution, 0.2% (w/v): In a 100-mL volumetric flask, mix 200 mg of 2-(4-iodophenyl)-3-(4-nitrophenyl)-5-phenyl-2*H*-tetrazolium chloride $(C_{19}H_{13}ClIN_5O_2)$ with 2 mL of N,N-dimethylformamide $[HCON(CH_3)_2]$ and

shake vigorously. Bring the volume to 90 mL with THAM buffer (1 M, pH 7) and shake or place in an ultrasonic bath until dissolved. Dissolution is slow and should be performed in the dark. When dissolved, bring the solution to 100 mL. Prepare the reagent daily and store in the dark. Four replicates are required per sample and each replicate uses 3.5 mL of this solution; ensure enough is prepared for the required number of analyses.

- Extractant: Mix together 100 mL of N,N-dimethylformamide with 100 mL of ethanol (97%).

- Standard stock solution, 100 mg INTF mL^{-1}: Weigh 10 mg of iodonitrotetra-zolium formazan ($C_{19}H_{14}IN_5O_2$) into a 100-mL volumetric flask. Dissolve in 80 mL of extractant solution and bring to 100 mL with extractant solution.

- Working standard solutions: Pipette 0, 0.25, 0.5, 1, 2, 3, and 5 mL of INTF standard stock solution into seven 20-mL tubes and bring to 13.5 mL with extractant solution. These standards correspond to 0, 25, 50, 100, 200, 300, and 500 µg of INTF in 13.5 mL of solution.

6–2.3.3 Procedure

Sterilize a subsample of each soil to be used in the assay by autoclaving fresh soil for 30 min at 120°C. Weigh 1 g (±0.003) fresh soil (sieved <2mm and thoroughly mixed) into three replicate tubes and 1 g autoclaved soil into a fourth to serve as a sterile control. Add 3.5 mL of INT solution to all tubes. Close tubes and shake briefly. Incubate for 2 h at 37°C in the dark. Immediately after incubation add 10 mL of extractant solution to all tubes. Maintain tubes at room temperature in the dark for 1 h, shaking vigorously for a few seconds every 20 min (vortex mixer). Filter tube contents in a dark room, and measure INTF in the samples using a spectrophotometer (464 nm), using the 0 µg INTF standard as the blank. If dilutions of the colored product are required, these must be performed using the extraction solution.

6–2.3.4 Calculations

Dehydrogenase activity is expressed as µg of INTF g^{-1} dry soil h^{-1}. Using the calibration curve, INTF concentrations can be determined from the corresponding absorbance value at 464 nm. Activity is calculated as follows:

$$\text{Dehydrogenase activity} = (S - C)/(2 \times DM) \text{ µg INTF g}^{-1} \text{ dry soil h}^{-1} \qquad [2]$$

where S is mean concentration of INTF (µg) in the sample, C is concentration of INTF (µg) in the sterile control, and DM = dry mass of soil aliquot (determined by oven-drying soil at 105°C).

Note: to express activity on a molar basis, use the following conversion

$$1 \text{ µg INTF g}^{-1} \text{ dry soil h}^{-1} = 2 \text{ nmol INTF g}^{-1} \text{ dry soil h}^{-1} \qquad [3]$$

6–2.4 Comments

Soil should be analyzed fresh and as soon after collection as possible (<1 mo), as results have shown that dehydrogenase activity is adversely affected by air-drying or storage, even at 4°C (Ross, 1970).

INT is very sensitive to light and should be kept in the dark at all times (von Mersi and Schinner, 1991). INT has a solubility limit of 0.3% (w/v) in water (Shaw and Burns, 2006) and, consequently, Shaw and Burns, (2006) recommend a lower concentration than used by von Mersi and Schinner (1991). It is important to ensure that all of the INT has dissolved before use.

As it is possible for chemical reduction of INT to occur in soil, it is suggested that autoclaved soil be used as a sterile control (von Mersi and Schinner, 1991).

Many variations of incubation pH (with and without buffers), incubation temperature, substrate concentration, and extractant, are found in the literature for this method for determining dehydrogenase activity (Shaw and Burns, 2006). The pH optimum for INT reduction has been determined to be pH 7.7 in phosphate buffer, with an effective range of pH 6.8 to 8.0 (Benefield et al., 1977). In addition, higher dehydrogenase activity has been reported in the presence of buffers at pH 7.6 than in water, with a notable decrease in activity below pH 6.6 or above pH 9.5 (Trevors, 1984a). Subsequently, THAM (tris(hydroxymethyl-)amino-methane) buffer has been shown to be the most suitable (Friedel et al., 1994), and the buffer pH and concentration we recommend are those used by von Mersi and Schinner (1991) who showed optimal activity at pH 7.0 to 7.5.

Studies have shown that dehydrogenase activity increases linearly with increasing temperature over a wide temperature range (5–70°C) (Trevors, 1984a). Shaw and Burns (2006) recommend an incubation time of 48 h at 25°C. Since incubation times of this length can lead to microbial growth, we have chosen to minimize the incubation time by increasing the incubation temperature and consequently have chosen conditions (2 h at 37°C) close to those used by von Mersi and Schinner (1991).

There is evidence that the presence of Cu in soil can have an adverse effect on the measurement of dehydrogenase activity by directly affecting the absorbance of the colored product (Chander and Brookes, 1991; Obbard, 2001). This has been noted when using either TTC or INT as substrates, although the effect is smaller with INT (Obbard, 2001).

Although INT has been shown to be a better competitor than TTC for liberated electrons, 90% of the observed O_2 uptake is still not accounted for, indicating that both substrates are poor competitors for O_2 in dehydrogenase reactions (Gong, 1997; Benefield et al., 1977). This highlights the need for further soil dehydrogenase method development, especially to find a more efficient electron acceptor.

6–3 FLUORESCEIN DIACETATE HYDROLYSIS
6–3.1 Introduction

Fluorescein diacetate [3′,6′-diacetylfluorescein (FDA)] can be used to measure microbial activity in soils (Brunius, 1980; Lundgren, 1981; Schnürer and Rosswall, 1982). It is a relatively nonpolar compound that is presumed to diffuse easily through cell membranes (Rotman and Papermas, 1966), where it can be hydrolyzed, releasing fluorescein, by several nonspecific enzymes such as proteases, lipases, and esterases. Within the soil environment, FDA can also be hydrolyzed by extracellular enzymes produced by the soil microbial population.

The product of this enzymatic reaction is fluorescein, which can be visualized within cells by fluorescence microscopy (Lundgren, 1981). Fluorescein released in soil also can be measured by spectrophotometry (Swisher and Carroll, 1980; Schnürer and Rosswall, 1982; Adam and Duncan, 2001; Green et al., 2006).

The Schnürer and Rosswall method (1982), which frequently has been used to determine FDA hydrolysis activity in the soil, was developed for pure microbial cultures, and was not originally optimized for soil samples. Pure culture work does not take into account the influence of the soil itself on the methodology, which may adsorb substrate or introduce interfering factors such as Fe or Al. Additionally, the fluorescein diacetate immobilized within a soil environment would come from many microbial and plant species. This method has been modified over time by various research groups (Perucci et al., 1999; Adam and Duncan, 2001; Green et al., 2006).

6–3.2 Principle

FDA hydrolysis is based on the following enzyme-mediated reaction:

$$[4]$$

Fluorescein Diacetate   Fluorescein

The product, fluorescein, is extracted and measured by spectrophotometry. Green et al. (2006) thoroughly tested this method for optimal pH, substrate concentration, incubation time and temperature, and shaking during the assay, as well as presenting basic kinetics of this composite enzyme activity.

This method has been tested for precision using eight soils representing two Alfisols, an Aridisol, two Mollisols, two Ultisols, and a Vertisol, with clay contents ranging from 11 to 57% and C contents from 4.7 to 24.8 g kg^{-1}. The percent standard deviation averages 3.2% and ranges from 2.2 to 4.4% (Green et al., 2006). It is important to thoroughly mix the soil sample before weighing out the subsamples to achieve an acceptable standard deviation between replicates.

The Adam and Duncan (2001) method uses field-moist soil, while the Green et al. (2006) method uses air-dried soil. In a recent study (Stott, unpublished data, 2007) that compared FDA hydrolytic activity of soils that were incubated in the field-moist or air-dried condition, there were no significant differences between the two conditions. Thus, for ease of soil preparation and storage, air-dried soil is used in this method.

The method presented here is that developed by Green et al. (2006) with a single modification: THAM buffer is used rather than the original phosphate buffer. As is apparent in the Green et al. (2006) paper, the sodium phosphate buffer, which was the same used by Schnürer and Rosswall (1982), did not always adequately maintain the incubation solution pH. Our laboratory tested THAM buffer with this assay, using a variety of soils, and we found that the incubation solutions were maintained consistently at pH 7.6 (Stott, unpublished data, 2007). Therefore we are recommending this alteration to the published method.

6–3.3 Assay Method (Green et al., 2006)

6–3.3.1 Apparatus

- Volumetric flasks, 1L, 500 mL, 50 mL
- 125-mL Erlenmeyer flasks with No. 5 stoppers
- Glass funnels with Whatman No. 2 filter paper
- Incubator set to 37°C
- Spectrophotometer set to 490-nm wavelength
- Spectrophotometer cuvettes

6–3.3.2 Reagents

1. Acetone, reagent grade.
2. Sulfuric acid (H_2SO_4), 3.0 M: In a 500-mL volumetric flask, add about 300 mL of deionized water and slowly add 83.4 mL concentrated H_2SO_4. Adjust to final volume with deionized water.
3. THAM buffer [tris (hydroxymethyl) aminomethane; $NH_2C(CH_2OH)_3$], 0.1 M, pH 7.6: In a 1-L volumetric flask, add about 800 mL deionized water and 12.1 g THAM. Adjust to pH 7.6 with 3.0 M H_2SO_4 (dropwise). Bring to volume with deionized water.
4. Fluorescein diacetate (FDA) lipase substrate solution ($C_{24}H_{16}O_7$), 4.9 mM FDA: Dissolve 0.020 g FDA into 10 mL acetone. Make fresh solution just before use as this substrate hydrolyzes quickly.
5. Fluorescein standard stock solution, 602 mM: In a 50-mL volumetric flask, add 0.010 g fluorescein ($C_{20}H_{12}O_5$) into 10 mL acetone. Adjust to final volume with THAM buffer.
6. Working standard solutions: Pipette 0, 0.15, 0.50, 1.5, and 2.5 mL of the fluorescein standard stock solution into 50-mL volumetric flasks. Bring to volume with THAM buffer (pH 7.6) and then add 2.5 mL acetone to match the matrix of the sample. These standards correspond to 0, 0.03, 0.10, 0.30, and 0.50 mg of fluorescein in a sample.

6–3.3.3 Procedure

Weigh 1.000 g (±0.0005) of air-dried soil (sieved <2 mm) into four 125-mL Erlenmeyer flasks. Three soil replicates and one nonsubstrate control are required per sample. Add 50 mL of THAM buffer (0.1M, pH 7.6) to each flask. Add 0.50 mL FDA substrate solution to the three replicate flasks. This results in a final concentration of 47.6 µM FDA in the incubation solution. Add 0.50 mL acetone to the control flask. A blank should also be run per batch (THAM buffer with substrate, no soil). Swirl flasks for a few seconds to mix the contents. Stopper the flasks and incubate them at 37°C for 3 h.

At the end of 3 h, remove the samples from the incubator and quickly add 2 mL acetone to the three replicate flasks and control flask; swirl suspensions to terminate FDA hydrolysis. To the control flask, add the substrate at this time. Mix the solution thoroughly and filter through a Whatman No. 2 filter paper. Transfer the filtrate to a cuvette and measure the absorbance on a spectrophotometer set at a wavelength of 490 nm. If the color intensity of the filtrate exceeds the range of the standard curve, dilute the filtrate with THAM buffer until the spectrometer reading falls within the limits of the calibration graph.

6–3.3.4 Calculations

Fluorescein diacetate hydrolytic activity is expressed as mg fluorescein released kg^{-1} oven-dried soil 3 h^{-1}. Using the calibration curve, fluorescein concentrations can be determined from the corresponding absorbance value at 490 nm. Activity is calculated as follows:

FDA hydrolytic activity =
$(S – C)/(DM)$ mg fluorescein kg^{-1} oven-dried soil $3h^{-1}$ incubation [5]

where S is mean concentration of fluorescein (mg) in the sample, C is concentration of fluorescein (mg) in the control, and DM is dry mass of soil aliquot (determined by oven-drying soil at 105°C).

To report activity on a 1-h basis (mg fluorescein kg^{-1} oven-dried soil h^{-1} incubation), divide the result by 3.

6–3.4 Comments

Clean glassware is critical, and rinsing glassware in an acid bath followed by deionized water is recommended.

Acetone is used both as a solvent for the FDA and to stop the hydrolysis reaction. The small amount of acetone used as a solvent is diluted in the incubation solution and does not appear to inhibit the reaction. Use of 2 mL acetone to halt the reaction does not result in complete stoppage of the hydrolysis reaction, but does slow it down sufficiently. Spectrophotometer readings taken within 30 min of reaction stoppage show no significant differences within that time span. After 30 min, noticeable and significant differences are seen in the spectrophotometer data (Green et al., 2006).

Increases in FDA hydrolytic activity are essentially linear between 25 and 50°C, but activity falls sharply at incubation temperatures >60°C (Green et al., 2006). Additionally, the effects of shaking during incubation were explored, and shaking was shown to reduce the level of activity compared with static incubation.

Soil with high clay content may be difficult to filter. If so, centrifuging can be used as an alternative. Mix the incubation solution thoroughly, and transfer approximately 30 mL of the soil suspension to a 50-mL centrifuge tube. Centrifuge at 8800 g for 5 min. Make sure to centrifuge the three replicates and the control at the same time. Proceed with measuring the resulting color as per the protocol above (Green et al., 2006).

Soils with high organic matter and/or clay contents may have such high microbial activity that substrate availability might be limiting. It is, however, expedient not to increase the final substrate concentration in the incubation solution (47.6 µM FDA), because with higher concentrations a precipitate begins to form (Green et al., 2006). For soils such as these, it is best to reduce the amount of soil by half or alternatively, to decrease the incubation time. Be sure to make the corresponding changes in the calculations of activity.

6–4 PHENOL OXIDASE (TYROSINASE)

6–4.1 Introduction

Phenol oxidases (diphenol oxidase, polyphenol oxidase, catechol oxidase) are Cu-containing enzymes that are widely distributed in plants, fungi, and bacteria

(Durán et al., 2002). These enzymes have very low substrate specificity, which overlaps with that of the laccases (EC 1.10.3.2). Substrate oxidation by phenol oxidases generates a free radical, which is converted to a quinone, either spontaneously or by further enzyme catalysis. The quinone and the free radical also can undergo nonenzymatic coupling reactions leading to polymerization (Durán et al., 2002). Tyrosinase (EC 1.14.18.1), a monophenol monooxygenase, is a specific phenol oxidase that catalyses reactions in which a second hydroxyl group is inserted in the *ortho* position into a monophenol (e.g., tyrosine). It is then able to further catalyze oxidation of the diphenol to a quinone.

Laccases have been assayed in forest litter using syringaldazine, a substrate that is specific for this enzyme (Harkin and Obst, 1973), but only after extraction and concentration of the extract (Criquet et al., 1999). We have found no evidence of any direct soil enzyme assay that is specific for laccase.

6–4.2 Principle

The following method for the determination of phenol oxidase activity uses the phenolic amino acid L-3,4-dihydroxy phenylalanine (L-DOPA) as the substrate and is based on the method of Pind et al. (1994). The unstable enzyme reaction product is spontaneously converted to the red pigment, dopachrome (2-carboxy 2,3-dihydroindole-5,6-quinone), which has a molar extinction coefficient of 3700 at 475 nm (Mason, 1948), and its absorbance is determined on a spectrophotometer. It is almost certain that L-DOPA is also a substrate for tyrosinase and catechol oxidase (E.C 1.10.3.1), and it may also be a substrate for laccase, so the enzyme activity assayed in soil should probably be regarded as a general phenol oxidase. We have incorporated minor modifications from Sinsabaugh et al. (1999), Williams et al. (2000), and Iyyemperumal and Shi (2008). These are discussed in the Comments section following the method.

6–4.3 Assay Method (Pind et al., 1994; Sinsabaugh et al.,1999; Williams et al., 2000; Iyyemperumal and Shi, 2008)

6–4.3.1 Apparatus

- Volumetric flasks, 1 L, 100 mL
- Incubator shaker set at 25°C
- 10-mL Nalgene centrifuge tubes
- Shaking incubator
- Refrigerated centrifuge (up to 12,000 rpm)
- Spectrophotometer
- Spectrophotometer cuvettes

6–4.3.2 Reagents

1. Acetate buffer, 50 mM, pH 5.0: Add 4.356 g sodium acetate tri-hydrate ($CH_3COONa·3H_2O$) to a 1-L volumetric flask and bring to approximately 800 mL with deionized water. Mix to dissolve. Add 1.08 mL glacial acetic acid (CH_3COOH), mix and adjust volume to 1 L with deionized water. Check the pH and adjust if necessary with glacial acetic acid dropwise.
2. L-3,4-dihydroxy phenylalanine (L-DOPA), 10 mM: Weigh 0.197 g L-DOPA

$[(HO)_2C_6H_3CH_2CH(NH_2)CO_2H]$ into a 100-mL volumetric flask. Add 50 mL 50 mM acetate buffer and dissolve. Adjust to 100 mL with 50 mM acetate buffer. This reagent is unstable and must be made up fresh daily.

6–4.3.3 Procedure

Weigh 0.5 g ($\pm$0.003) of soil (sieved <2 mm and thoroughly mixed) into four replicate 10-mL centrifuge tubes (three sample replicates and one nonsubstrate control). Add 3 mL acetate buffer and gently mix on a platform shaker for 10 min. Add 2 mL of 10 mM L-DOPA to three replicate tubes and 2 mL acetate buffer to the control. Rapidly swirl to mix. Place tubes in a shaking incubator (100 rev min^{-1}, 25°C) for 30 min. The incubation time should be predetermined to ensure linear oxidation of L-DOPA as this may vary for different soil types. Immediately centrifuge at 12,000 rpm at 5°C for 5 min to terminate the reaction. Filter through GF/C filter paper and measure the absorbance of the red product (dopachrome) in the filtrate at 475 nm using deionized water to zero the spectrophotometer.

6–4.3.4 Calculations

The unstable enzyme reaction product is spontaneously converted to the red pigment, dopachrome (2-carboxy2,3-dihydroindole-5,6-quinone), which has a molar extinction coefficient of 3700 at 475 nm (Mason, 1948). This can be used to directly determine the concentration of dopachrome in the filtrate. Note that in the method described by Pind et al. (1994), the authors erroneously ascribe to Mason (1948) a molar extinction coefficient of 37,000 for dopachrome.

Phenol oxidase activity is expressed as μmol product formed (measured as dopachrome) g^{-1} dry soil h^{-1} and this is calculated as follows:

$$\text{Phenol oxidase activity} = OD \times (5 + M_{H2O})/(EC \times 10^3 \times T_h \times DM)$$
μmol product formed g^{-1} dry soil h^{-1} [6]

where OD is sample absorbance minus control absorbance, M_{H2O} is mass of water in the moist soil aliquot and $(5 + M_{H2O})/10^3$ converts μM to μmol in the assay volume, EC is micromolar extinction coefficient for dopachrome (3.7×10^{-3}), T_h is incubation time in hours, and DM is dry mass of soil aliquot (determined by oven-drying soil at 105°C).

6–4.4 Comments

It is recommended that the soil be assayed in a field-moist condition whenever possible.

There are many variations throughout the literature of the above method for analysis of phenol oxidase. Some have suggested making up the L-DOPA substrate in water (e.g., Toberman et al., 2008) instead of acetate buffer. In general, the use of water instead of buffer in enzyme assays is thought to more closely mimic conditions that would be found in the field, while the use of a buffer is thought to yield the maximum potential enzyme activity (assuming that the assay is conducted at the pH optimum for the enzyme). We recommend that a buffered system is always used; otherwise, cross-laboratory comparisons cannot be made. In this method, the assay is conducted at pH 5, even though the optimum for the enzyme was originally determined by Pind et al. (1994) to be pH 8. Pind et al. (1994) do not explain their choice of pH, but this (pH 5) has been adopted in all subsequent papers using this method. Pind et al. (1994) do, however, stress that absorbance of the dopachrome solution must be read at pH 5 to avoid artifacts from pH-induced shifts in the absorbance spectrum.

Temperatures for incubation can be chosen to either give optimal conditions for enzyme activity (Kourtev et al., 2002) or to more closely mimic field conditions (Pind et al., 1994), and consequently there is a range of temperatures in the literature for phenol oxidase activity (11°C to 25°C). Since comparison of soil enzyme activities between laboratories is also impossible if the assays are conducted at different temperatures, we recommend 25°C, the upper temperature in this range.

The length of incubation time also varies greatly in the literature. Incubation times for phenol oxidase activity using L-DOPA range from 1 min to 1 h depending on the type of soil being analyzed (Pind et al., 1994; Sinsabaugh et al., 1999; Toberman et al., 2008; Iyyemperumal and Shi, 2008). Soils with high organic matter content may need much shorter incubation times than the 30 min outlined in this method. For example, Pind et al. (1994) found that L-DOPA oxidation was linear for only 5 min in incubations of peat soils and, subsequently, incubated samples for 3 min only. Therefore, it is best to first determine the optimal incubation duration for each soil before conducting the assay.

It has been suggested that incubation times should be no longer than 1 h due to the instability of L-DOPA in the presence of oxygen and the potential for reaction rates that are not linear through time (Sinsabaugh et al., 1999).

Pind et al. (1994) do not include a nonsoil control to evaluate abiotic L-DOPA oxidation as part of their method. However, the 96-well microplate adaptation of this method includes incubation of buffer and L-DOPA together (Saiya-Cork et al., 2002), and the adaptation of Matocha et al. (2004) includes incubation with autoclave-sterilized soil. In neither of these adaptations do the authors comment on the extent of the abiotic reaction, but Sinsabaugh et al. (1999) suggested that unless a nonsoil control is run alongside the samples, incubation times should be kept as short as possible. We suggest that abiotic oxidation be checked as part of the preliminary work to establish assay conditions and that a suitable control be included if warranted.

6–5 PEROXIDASE

6–5.1 Introduction

The peroxidases (EC 1.11.1.) comprise a group of Fe-containing enzymes that use H_2O_2 as the acceptor molecule and which are generally very substrate specific (Dixon and Webb, 1979). However, one particular member, simply called peroxidase (EC 1.11.1.7), catalyzes the oxidation of many organic molecules, including the substrates for phenol oxidase and laccase, and therefore is also involved in lignin degradation (Iyyemperumal and Shi, 2008).

Peroxidase catalyzes a one-electron oxidation, producing a free radical and forming an enzyme intermediate. This intermediate then reacts with a second substrate molecule, forming another free radical and regenerating the enzyme (Gianfreda and Bollag, 2002). The enzyme is susceptible to side reactions that can temporarily reduce its efficiency or can lead to permanent inactivation (Gianfreda and Bollag, 2002). We consider it unlikely that peroxidases can function naturally as free enzymes in soil, because very little, if any, H_2O_2 could exist outside cells: i) because of its reactivity and ii) because of the efficiency of the enzyme catalase (see Section 6–7). Consequently, the assay method described below may overestimate the potential role of peroxidases, because addition of H_2O_2 will allow estimation of both intracellular and extracellular activities.

6–5.2 Principle

The following method based on that of Sinsabaugh et al. (1999) and adapted by Iyyemperumal and Shi (2008), uses L-DOPA (see Section 6–4.2) and hydrogen peroxide as substrates. The procedure used is the same as that outlined in this chapter for phenol oxidase (Section 6–4), except that it involves an additional set of replicates receiving H_2O_2.

Peroxidase activity in soil involves determining both phenol oxidase activity and total oxidase activity using the same procedure, but with the addition of H_2O_2 for the latter activity. Peroxidase activity is thus defined as the difference between the catalysis due to O_2 plus H_2O_2 and that due to O_2 alone (Kourtev et al., 2002).

6–5.3 Assay Method (Sinsabaugh et al., 1999; and Iyyemperumal and Shi, 2008)

6–5.3.1 Apparatus

- Volumetric flasks, 1 L, 100 mL
- Platform shaker
- 10-mL centrifuge tubes
- Shaking incubator set at 25°C
- Refrigerated centrifuge (up to 12,000 rpm)
- Spectrophotometer
- Spectrophotometer cuvettes

6–5.3.2 Reagents

1. Acetate buffer, 50 mM, pH 5.0: Add 4.356 g sodium acetate tri-hydrate ($CH_3COONa \cdot 3H_2O$) to a 1-L volumetric flask and bring to approximately 800 mL with deionized water. Mix to dissolve. Add 1.08 mL glacial acetic acid (CH_3COOH), mix and adjust volume to 1 L with deionized water. Check the pH and adjust if necessary with glacial acetic acid dropwise.

2. L-3,4-dihydroxy phenylalanine (L-DOPA), 10 mM: Weigh 1.97 g L-DOPA [$(HO)_2C_6H_3CH_2CH(NH_2)CO_2H$] into a 100-mL volumetric flask. Add 50 mL 50 mM acetate buffer and dissolve. Adjust to 100 mL with 50 mM acetate buffer. This reagent must be made up fresh daily as the solution is unstable.

3. Hydrogen peroxide solution, 0.3% w/v: Dilute 1 mL commercial hydrogen peroxide (H_2O_2, 30%) in 100 mL deionized water. This reagent must be prepared immediately before use.

6–5.3.3 Procedure

Weigh 0.5 g (± 0.003) of soil (sieved <2mm and thoroughly mixed) into eight replicate 10-mL centrifuge tubes (three sample replicates and one nonsubstrate control for phenol oxidase and the same for total oxidase). Add 3.2 mL acetate buffer to the phenol oxidase tubes and 3 mL acetate buffer to the peroxidase tubes, then gently mix on a platform shaker for 10 min. For each activity, add 2 mL of 10 mM L-DOPA to three replicate tubes and 2 mL acetate buffer to the control. Then immediately add 0.2 mL of 0.3% H_2O_2 to the four total oxidase tubes (i.e., including the control). Rapidly swirl to mix. Place tubes in a shaking incubator (100 rev min^{-1}, 25°C) for 30 min. (As noted in Section 6–4.4, the incubation time should be predetermined to ensure linear oxidation of L-DOPA—this may vary for different soil types.) Immediately centrifuge at 12,000 rpm at 5°C for 5 min to terminate the reaction. Filter through GF/C filter paper and measure the absorbance of the filtrate at 475 nm using deionized water to zero the spectrophotometer.

6–5.3.4 Calculations

Peroxidase activity is determined by calculating the difference between the dopachrome absorption at 475 nm in the assay including H_2O_2 and that in the assay without H_2O_2 (phenol oxidase).

Peroxidase activity is expressed as μmol product formed (measured as dopachrome) g^{-1} dry soil h^{-1} and this is calculated as follows:

$$\text{Peroxidase activity} = OD_{Diff} \times (5.2 + M_{H2O})/(EC \times 10^3 \times T_h \times DM) \qquad [7]$$
μmol product formed g^{-1} dry soil h^{-1}

where OD_{Diff} is mean absorbance including H_2O_2– mean absorbance without H_2O_2, M_{H2O} is mass of water in the moist soil aliquot and $(5.2 + M_{H2O})/10^3$ converts μM to μmol in the assay volume, EC is micromolar extinction coefficient for dopachrome (3.7×10^{-3}), T_h is incubation time in hours, and DM is dry mass of soil aliquot (determined by oven-drying soil at 105°C).

6–5.4 Comments

It is recommended that the soil be assayed in a field-moist condition whenever possible.

To ensure linearity of L-DOPA oxidation, the incubation time must be optimized for peroxidase plus phenol oxidase and the same time period be used for both assays, even if this is suboptimal for phenol oxidase. If phenol oxidase activity is to be reported in addition to peroxidase, it may be necessary to repeat phenol oxidase determinations alone using a longer incubation period.

Peroxidase can be inactivated by H_2O_2 (Gianfreda and Bollag, 2002), so it is important to keep the incubation as short as is practicable.

As discussed above for phenol oxidase, the method proposed by Pind et al. (1994) does not include a nonsoil control, even though it is known that L-DOPA is oxidized chemically by O_2 (and probably also by H_2O_2). We suggest that abiotic oxidation be checked as part of the preliminary work to establish assay conditions and that a suitable control be included if warranted.

Johnsen and Jacobsen (2008) claim that the L-DOPA method for determining peroxidase activity does not discriminate well between peroxidase and phenol oxidase and also point out that L-DOPA is teratogenic, mutagenic, and carcinogenic. They proposed an alternative that appears to be specific for peroxidase using 3,3′,5,5′-tetramethylbenzidine (TMB) as substrate. However, this new method has not been sufficiently validated to supplant those using L-DOPA, still the most commonly used and cited in the literature.

6–6 CATECHOL OXIDASE

6–6.1 Introduction

Catechol oxidase (EC 1.10.3.1) is a Cu-containing enzyme that catalyzes the oxidation of o-diphenols such as catechol, to yield a quinone and water. In soil, catechol oxidase plays a role in the formation of humic substances during decay of plant and animal remains, and in the breakdown of organic contaminants (Perucci et al., 2000). However, because of the overlapping substrate specificities of phenol oxidases and other phenol-oxidizing enzymes (catechol oxidase and tyrosinase [EC 1.14.18.1]), it is not possible to determine what enzymes are catalyzing this reaction. In reality, the oxidation of catechol is almost certainly catalyzed by the same enzymes measured above as phenol oxidase. Therefore, in this method, we define catechol oxidase activity as the enzymatic oxidation of catechol.

6–6.2 Principle

The method outlined in this chapter is that of Perucci et al. (2000), who called the enzyme o-diphenol oxidase and used catechol as substrate. The enzyme activity is determined by spectrophotometric analysis of a red product [4-(N-proline)-o-benzoquinone] derived from the reaction between the enzyme reaction product, o-benzoquinone, and proline (Perucci et al., 2000). The method is based on colorimetric reactions investigated by Jackson and Kendal (1949) and Mason and Peterson (1955), modified for use with soil extracts by Mayaudon et al. (1973) and for whole soil by Perucci et al. (2000).

6–6.3 Assay Method (Jackson and Kendal, 1949; Mason and Peterson, 1955; and Perucci et al., 2000)

6–6.3.1 Apparatus

- Volumetric flasks, 1 L, 100 mL
- 10-mL centrifuge tubes
- Shaking incubator set at 30°C
- Ice bath
- Refrigerated centrifuge (up to 5000 rpm)
- Spectrophotometer
- Spectrophotometer cuvettes

6–6.3.2 Reagents

1. Phosphate buffer, 0.1 M, pH 6.5: Combine 9.52 g sodium phosphate monobasic ($NaH_2PO_4 \cdot H_2O$) and 8.45 g sodium phosphate dibasic heptahydrate ($Na_2HPO_4 \cdot 7H_2O$) in a 1-L volumetric flask. Add approx. 600 mL deionized water, swirl to dissolve, and bring to 1 L with deionized water.

2. Catechol, 0.2 M: Weigh 2.2 g of catechol ($C_6H_4(OH)_2$) into a 100-mL volumetric flask, dissolve in phosphate buffer (0.1 M, pH 6.5), and bring to 100 mL with phosphate buffer.

3. Proline, 0.2 M: Weigh 2.3 g of proline ($C_5H_9NO_2$) into a 100-mL volumetric flask, dissolve in phosphate buffer (0.1 M, pH 6.5) and bring to 100 mL with phosphate buffer.

4. Ethanol, standard analytical grade (97%).

6–6.3.3 Procedure

Sterilize a subsample of each soil to be used in the assay by autoclaving fresh soil for 30 min at 120°C. Oxygenate subsamples of the proline and catechol solutions for 3 min and incubate for 10 min at 30°C. Take five 10-mL centrifuge tubes (three sample replicates, one nonsubstrate control, and one sterile-soil control). Weigh 1 g ($\pm$0.003) of soil, (sieved <2mm and thoroughly mixed) into four tubes and 1 g of sterilized soil into the fifth tube. Add 2 mL phosphate buffer and 1.5 mL proline reagent to all five tubes. Immediately add 1.5 mL of catechol reagent to the three sample replicates and the sterile-soil control, and 1.5 mL phosphate buffer to the remaining tube (nonsubstrate control). Rapidly swirl to mix. Incubate tubes for 10 min in a shaking incubator at 30°C. Following incubation, immediately add 5 mL ethanol to each tube and rapidly cool in an ice bath to terminate the reaction. Centrifuge the tubes at 5000 g at 4°C for 5 min and read the supernatant at 525 nm. Zero the spectrophotometer against deionized water.

6–6.3.4 Calculations

The red product formed by the stoichiometric reaction of the enzyme reaction product, benzoquinone, with proline [4-(N-proline)-o-benzoquinone] has a molar extinction coefficient of 5000 at 525 nm (Yamaguchi et al., 1970).

Catechol oxidase activity is expressed as µmol product formed [measured as 4-(N-proline)-o-benzoquinone] g^{-1} dry soil h^{-1} and this is calculated as follows:

$$\text{Catechol oxidase activity} = OD \times (10 + M_{H2O})/(EC \times 10^3 \times T_h \times DM) \qquad [8]$$
$$\text{µmol product formed } g^{-1} \text{ dry soil } h^{-1}$$

where OD is sample absorbance minus control absorbance (taking account of both controls), M_{H2O} is mass of water in the moist soil aliquot and $(10 + M_{H2O})/10^3$ converts µM to µmol in the assay volume, EC is micromolar extinction coefficient for 4-(N-proline)-o-benzoquinone (5.0×10^{-3}), T_h is incubation time in hours, and DM is dry mass of soil aliquot (determined by oven-drying soil at 105°C).

6–6.4 Comments

It is recommended that the soil be assayed in a field-moist condition whenever possible.

The red compound was found to be stable from pH 5.5 to 8 and at temperatures between 10 and 50°C for at least 2 h (Perucci et al., 2000).

Optimum pH and temperatures for catechol oxidase activity were also determined (Perucci et al., 2000), with optimum activity observed between pH 6.5 and 7.5 and 30 to 37°C. For the assay, the lower regions of these ranges were chosen, due to reduced stability of catechol at higher pH and temperature.

It should be noted that significant abiotic oxidation of catechol also occurs in soils, especially organic soils, and Perucci et al. (2000) suggested running sterile-soil controls at the same time as test samples. We have incorporated this treatment into the method, replacing the original "nonsoil" control with a "sterile-soil" control. It may be necessary to add a further "sterile-soil without substrate" control to account for increased absorbance at 525 nm due to organic matter solubilized by the autoclaving process.

The enzyme-catalyzed oxidation of catechol is suboptimal at a substrate concentration of 0.2 M. However, at higher substrate concentrations, abiotic oxidation rates increase, affecting the sensitivity of the assay (Perucci et al., 2000).

6–7 CATALASE

6–7.1 Introduction

Catalase (EC 1.11.1.6), a specific peroxidase, is the most active of all enzymes, with one enzyme molecule able to destroy many millions of H_2O_2 molecules per second. It is found in all aerobic organisms, including soil bacteria and fungi, and in most facultative anaerobes (Weetall et al., 1965; Guwy et al., 1999; Isobe et al., 2006), where its role is to relieve oxidative stress by converting H_2O_2, produced during normal aerobic metabolism, to water.

In soil, catalase has been assumed to be purely an intracellular enzyme (Schinner et al., 1996), and this assumption still persists (Trasar-Cepeda et al., 1999, 2007; Pérez-de-Mora et al., 2008) even though there are a number of studies that have demonstrated the presence of active extracellular catalase in soil (Stotzky, 1974; Perez Mateos and Gonzalez Carcedo, 1985, 1987). It was the first enzyme studied in soil (Skujiņš, 1976) and it continues to be assayed as an indicator of aerobic microbial activity and of soil fertility (García and Hernández, 1997).

Methods for determination of catalase activity in soil are based on the addition of H_2O_2 to a soil sample and subsequent measurement of the rate of O_2 release or of residual H_2O_2. This is usually by manometric techniques (U-tube or Scheibler apparatus) (Weetall et al., 1965; Beck, 1971) or gas chromatography for O_2 production (Trevors, 1984b), or by titration with $KMnO_4$ (Johnson and Temple, 1964) or colorimetrically (Trasar-Cepeda et al., 1999) for residual H_2O_2 determination.

6–7.2 Principle

The catalyzed reaction involves disproportionation of H_2O_2, represented as:

$$2H_2O_2 \rightarrow 2H_2O + O_2 \qquad [9]$$

Both titration and gas analysis require complex equipment, and can be time-consuming. (Trasar-Cepeda et al., 1999); for this reason, the method outlined in this chapter is based on the colorimetric method of Trasar-Cepeda et al. (1999). Soil catalase activity is determined by estimating residual H_2O_2 by a peroxidase-catalyzed oxidative coupling between 4-aminoantipyrine (4-AAP) and phenol. The method is based on the automated procedure described by Holz (1986) and produces a colored product that absorbs at 505 nm (Trasar-Cepeda et al., 1999).

6–7.3 Assay Method (Holz, 1986; and Trasar-Cepeda et al., 1999)

6–7.3.1 Apparatus

- Volumetric flasks, 1L, 100 mL, 50 mL
- 50-mL Ehrlenmeyer flasks or 30- to 50-mL glass bottles
- Rotary shaker
- Glass tubes, 15 mL
- Shaking incubator set at 20°C

- Vortex mixer
- Spectrophotometer
- Spectrophotometer cuvettes

6–7.3.2 Reagents

1. Phosphate buffer, 110 mM, pH 7.4: Dissolve 15.18 g of monosodium phosphate monohydrate ($NaH_2PO_4.H_2O$) in 800 mL deionized water in a 1-L volumetric flask. Adjust the pH to 7.4 dropwise with 10 M NaOH. Allow the solution to cool, check the pH, and adjust to 1 L with deionized water.

2. Phosphate buffer, 1 M, pH 8.0: Dissolve 138 g of monosodium phosphate monohydrate ($NaH_2PO_4.H_2O$) in 800 mL deionized water in a 1-L volumetric flask. Adjust the pH to 8.0 drop-wise with 10 M NaOH. Allow the solution to cool, check the pH, and adjust to 1 L with deionized water.

3. Sulfuric acid, 3 M: Carefully and slowly add 30 g concentrated sulfuric acid (98%) to 70 mL deionized water and bring to 100 mL when cool.

4. 4-Aminoantipyrine (4-AAP), 30 mM: Dissolve 0.305 g of 4-AAP ($C_{11}H_{13}N_3O$) in 30 mL deionized water in a 50-mL volumetric flask, bring to 50 mL and store at 4°C.

5. Ammonium sulfate, 3.9 M: Dissolve 51.5 g ammonium sulfate (($NH_4)_2SO_4$) in 80 mL deionized water in a 100-mL volumetric flask, bring to 100 mL.

6. Peroxidase (POD): Take one vial of 10,000 U peroxidase (Boehringer Mannheim product No. 127 361) and add 5 mL 3.9 M ammonium sulfate. Store at 4°C.

7. Phenol stock: Melt phenol at 50 to 60°C and equilibrate with an equal volume of the 1 M phosphate buffer (pH 8.0). Shake the mixture and leave it to separate into two phases. Discard the aqueous phase. Re-equilibrate the organic phase with an equal volume of the 110 mM phosphate buffer (pH 7.4). Again discard the aqueous phase leaving a 2- to 3-mm layer of the aqueous phase above the phenol.

8. Phenol, 5.3% v/v: Measure 5.3 mL stock phenol (organic phase) into a 100-mL volumetric flask. Dilute to 100 mL with deionized water.

9. Indicator reagent: Mix together in the following order 48 mL 110 mM phosphate buffer (pH 7.4), 1 mL 30 mM 4-APP, 1 mL 5.3% phenol solution, and 30 μL POD.

10. Hydrogen peroxide solution, 0.3% w/v: Dilute 1 mL commercial H_2O_2 (30%) to 100 mL with deionized water. Must be prepared immediately before use.

11. Stock standard: 0.3% H_2O_2 working solution previously prepared for the catalase assay is used as a stock standard.

12. Working standards: Prepare six dilutions in deionized water of H_2O_2 working solution in a range from 0 (blank) to 0.3%.

6–7.3.3 Procedure

Weigh 0.2 g (±0.003) of soil (sieved <2 mm and thoroughly mixed) into four glass vessels (volume at least 30 mL, 50 mL recommended; three soil replicates and one nonsubstrate control). Add 16 mL of deionized water to each and shake for 30 min on a rotary shaker at 30 rev min⁻¹. Add 2 mL of 0.3% H_2O_2 solution (2 mL deionized water for the control) to each suspended sample and shake for a further 10 min at 20°C. Add 2 mL 3 M H_2SO_4 to terminate the reaction and filter through filter paper (pore size 20–25 μm).

Concurrently, add 1 mL of each standard (0 to 0.3% H_2O_2) and 1 mL 3 M H_2SO_4 to 8 mL deionized water and filter as above.

Pipette 100-μL aliquots of the assay filtrates and of the standards filtrates into 15-mL glass tubes and neutralize by the addition of 900 μL of 110 mM phosphate buffer. Add 9 mL of indicator reagent to each tube and mix by vortex. Leave to stand for 5 min and read sample absorbance at 505 nm. Compare sample absorbances against those of the standard solutions by interpolation from the standard curve.

6–7.3.4 Calculations

The concentration of H_2O_2 in the working stock solution (0.3%) is 100 mM (assuming a density of 1.13 for 30% H_2O_2). The H_2O_2 in the standards and the enzyme assays is diluted tenfold before removal of the 100-μL aliquot; consequently, the final H_2O_2 concentrations in the standards and assays will range from 0 to 10 mM. As all subsequent steps are the same for both assays and standards, absorbances of the standard can be directly related to mM concentrations on the standard curve and assay concentrations can be read directly from the curve. These represent the concentration of residual H_2O_2 in the assay.

Catalase activity is reported as mmol H_2O_2 consumed g^{-1} dry soil h^{-1} and this is calculated as follows:

$$\text{Catalase activity} = \{[H_2O_2 \text{ top standard (mM)}] - [H_2O_2 \text{ residual (mM)}]\} \times 6/(50 \times DM) \text{ mmol } H_2O_2 \text{ consumed } g^{-1} \text{ dry soil } h^{-1} \qquad [10]$$

where "6" converts the 10-min assay to 1 h, "50" converts mM concentrations to mmol in the assay volume (20 mL), and DM is dry mass of soil aliquot (determined by oven-drying soil at 105°C).

6–7.4 Comments

It is recommended that the soil be assayed in a field-moist condition whenever possible.

Trasar-Cepeda et al. (1999) do not specify incubation vessels, but Johnson and Temple (1964) used Ehrlenmeyer flasks. We have also scaled down to 40% of the specified soil mass and reagent volumes. Depending on the configuration of the orbital shaker, we suggest either 50-mL Ehrlenmeyer flasks or 30- to 50-mL glass bottles.

Investigations into the stability of the colored product have shown that the color at 505 nm is stable for up to 1 h and slowly declines thereafter (Trasar-Cepeda et al., 1999).

There are instances where buffers have been used in the determination of soil catalase activity, e.g., the phosphate buffer, pH 6.8, proposed by Beck (1971), used in the methods of Alef and Nannipieri (1995) and Öhlinger (1996a). However, the method of Johnson and Temple (1964), on which Trasar-Cepeda et al. (1999) is based, uses deionized water instead of buffer. We have chosen to follow the original method because we are unsure how altering the pH during the assay will affect the formation of the colored product. However, it has been noted that catalase is so active that it is insensitive to pH (Weetall et al., 1965; Trevors, 1984b) and temperature (Trasar-Cepeda et al., 2007), with the only limitation to its catalytic activity probably being the rate of diffusion of H_2O_2 to the enzyme.

Elements such as manganese and iron in soils may have the ability to chemically catalyze H_2O_2 breakdown (Trevors, 1984b); this should be considered when assaying soils containing high concentrations of these elements.

ACKNOWLEDGMENTS

We thank Liz Shaw, The University of Reading, UK and Richard Burns, The University of Queensland, Australia, for assistance with the dehydrogenase introduction and comments. We also acknowledge the helpful advice received from the late Dr. Des Ross, Landcare Research Emeritus Scientist, Palmerston North New Zealand, who passed away during the writing of this chapter.

REFERENCES

Adam, G., and H. Duncan. 2001. Development of a sensitive and rapid method for the measurement of total microbial activity using fluorescein diacetate (FDA) in a range of soils. Soil Biol. Biochem. 33:943–951.

Alef, K. 1995. Dehydrogenase activity. p. 228–231. In K. Alef and P. Nannipieri (ed.) Methods in applied soil microbiology and biochemistry. Academic Press Ltd, London.

Alef, K., and P. Nannipieri. 1995. Catalase activity. p. 362–363. In K. Alef and P. Nannipieri (ed.) Methods in applied soil microbiology and biochemistry. Academic Press Ltd, London.

Beck, T. 1971. Die Messung der Katalaseaktivität von Böden. Z. Pflanzenernaehr. Bodenkd. 130:68–81.

Benefield, C.B., P.J.A. Howard, and D.M. Howard. 1977. The estimation of dehydrogenase activity in soil. Soil Biol. Biochem. 9:67–70.

Brunius, G. 1980. Technical aspects of the use of 3′,6′-diacetyl fluorescein for vital fluorescent staining of bacteria. Curr. Microbiol. 4:321–323.

Chander, K., and P.C. Brookes. 1991. Is the dehydrogenase assay invalid as a method to estimate microbial activity in copper-contaminated soils. Soil Biol. Biochem. 23:909–915.

Criquet, S., S. Tagger, G. Vogt, G. Iacazio, and J. Le Petit. 1999. Laccase activity of forest litter. Soil Biol. Biochem. 31:1239–1244.

Dick, R.P., D.P. Breakwell, and R.F. Turco. 1996. Soil enzyme activities and biodiversity measurements as integrative microbiological indicators. p. 247–271. In J.W. Doran and A.J. Jones (ed.) Methods for assessing soil quality. SSSA, Madison, WI.

Dixon, M., and E.C. Webb. 1979. Enzymes. 3rd ed. Longman Group Ltd., London.

Durán, N., M.A. Rosa, A. D'Annibale, and L. Gianfreda. 2002. Applications of laccases and tyrosinases (phenoloxidases) immobilized on different supports: A review. Enzyme Microb. Technol. 31:907–931.

Frankenberger, W.T., Jr., and W.A. Dick. 1983. Relationships between enzyme activities and microbial growth and activity indices in soil. Soil Sci. Soc. Am. J. 47:945–951.

Friedel, J.K., K. Mölter, and W.R. Fischer. 1994. Comparison and improvement of methods for determining soil dehydrogenase activity by using triphenyltetrazolium chloride and iodonitrotetrazolium chloride. Biol. Fertil. Soils 18:291–296.

García, C., and T. Hernández. 1997. Biological and biochemical indicators in derelict soils subject to erosion. Soil Biol. Biochem. 29:171–177.

Gianfreda, L., and J.-M. Bollag. 2002. Isolated enzymes for the transformation and detoxification of organic pollutants. p. 495–538. In R.G. Burns and R.P. Dick (ed.) Enzymes in the environment: Activity, ecology, and applications. Marcel Dekker, New York.

Gong, P. 1997. Dehydrogenase activity in soil: A comparison between the TTC and INT assay under their optimum conditions. Soil Biol. Biochem. 29:211–214.

Green, V.S., D.E. Stott, and M. Diack. 2006. Assay for fluorescein diacetate hydrolytic activity: Optimization for soil samples. Soil Biol. Biochem. 38:693–701.

Guwy, A.J., S.R. Martin, F.R. Hawkes, and D.L. Hawkes. 1999. Catalase activity measurements in suspended aerobic biomass and soil samples. Enzyme Microb. Technol. 25:669–676.

Harkin, J.M., and J.R. Obst. 1973. Syringaldazine, an effective reagent for detecting laccase and peroxidase in fungi. Experientia 29:381–387.

Holz, F. 1986. Automatisierte, photometrische Bestimmung der Aktivität von Bodenenzymen durch Anwendung (enzymatisch)-oxidativer Kupplungsreaktionen im Durchfluß. I. Mitteilung: Die Bestimmung der Katalaseaktivität. Landwirtsch. Forsch. 39:139–153.

Howard, P.J.A. 1972. Problems in the estimation of biological activity in soil. Oikos 23:235–240.

Isobe, K., N. Inoue, Y. Takamatsu, K. Kamada, and N. Wakao. 2006. Production of catalase by fungi growing at low pH and high temperature. J. Biosci. Bioeng. 101:73–76.

Iyyemperumal, K., and W. Shi. 2008. Soil enzyme activities in two forage systems following application of different rates of swine lagoon effluent or ammonium nitrate. Appl. Soil Ecol. 38:128–136.

Jackson, H., and L.P. Kendal. 1949. The oxidation of catechol and homocatechol by tyrosinase in the presence of amino-acids. Biochem. J. 44:477–487.

Johnsen, A.R., and O.S. Jacobsen. 2008. A quick and sensitive method for the quantification of peroxidase activity of organic surface soil from forests. Soil Biol. Biochem. 40:814–821.

Johnson, J.L., and K.L. Temple. 1964. Some variables affecting the measurement of "catalase activity" in soil. Soil Sci. Soc. Am. Proc. 28:207–209.

Kourtev, P.S., J.G. Ehrenfeld, and W.Z. Huang. 2002. Enzyme activities during litter decomposition of two exotic and two native plant species in hardwood forests of New Jersey. Soil Biol. Biochem. 34:1207–1218.

Lundgren, B. 1981. Fluorescein diacetate as a stain of metabolically active bacteria in soil. Oikos 36:17–22.

Mason, H.S. 1948. The chemistry of melanin. III. Mechanism of the oxidation of dihydroxyphenylalanine by tyrosinase. J. Biol. Chem. 172:83–99.

Mason, H.S., and E.W. Peterson. 1955. The reaction of quinones with protamine and nucleoprotamine: N-terminal proline. J. Biol. Chem. 212:485–493.

Matocha, C.J., G.R. Haszler, and J.H. Grove. 2004. Nitrogen fertilization suppresses soil phenol oxidase enzyme activity in no-tillage systems. Soil Sci. 169:708–714.

Mayaudon, J., M. El Halfawi, and M.A. Chalvignac. 1973. Propriétés des diphénol oxydases extraites des sols. Soil Biol. Biochem. 5:369–383.

Mosher, J.J., B.S. Levison, and C.G. Johnston. 2003. A simplified dehydrogenase enzyme assay in contaminated sediment using 2-(p-iodophenyl)-3-(p-nitrophenyl)-5-phenyltetrazolium chloride. J. Microbiol. Methods 5:411–415.

Nannipieri, P., S. Grego, and B. Ceccanti. 1990. Ecological significance of the biological activity of soil. p. 293–395. In J.-M. Bollag and G. Stotzky (ed.) Soil biochemistry. Vol. 6. Marcel Dekker, New York.

Obbard, J.P. 2001. Measurement of dehydrogenase activity using 2-p-iodophenyl-3-p-nitrophenyl-5-phenyltetrazolium chloride (INT) in the presence of copper. Biol. Fertil. Soils 33:328–330.

Öhlinger, R. 1996a. Catalase activity. p. 237–240. In F. Schinner, R. Öhlinger, E. Kandeler, and R. Margesin (ed.) Methods in soil biology. Springer-Verlag, Berlin.

Öhlinger, R. 1996b. Dehydrogenase activity with the substrate TTC. p. 241–243. In F. Schinner, R. Öhlinger, E. Kandeler, and R. Margesin (ed.) Methods in soil biology. Springer-Verlag, Berlin.

Perez Mateos, M., and S. Gonzalez Carcedo. 1985. Effect of fractionation on location of enzyme activities in soil structural units. Biol. Fertil. Soils 1:153–159.

Perez Mateos, M., and S. Gonzalez Carcedo. 1987. Effect of fractionation on the enzymatic state and behaviour of enzyme activities in different structural soil units. Biol. Fertil. Soils 4:151–154.

Pérez-de-Mora, A., E. Madejón, F. Cabrera, F. Buegger, R. Fuß, K. Pritsch, and M. Schloter. 2008. Long-term impact of acid resin waste deposits on soil quality of forest areas. II. Biological indicators. Sci. Total Environ. 406:99–107.

Perucci, P., C. Casucci, and S. Dumontet. 2000. An improved method to evaluate the o-diphenol oxidase activity of soil. Soil Biol. Biochem. 32:1927–1933.

Perucci, P., C. Vischetti, and F. Battistoni. 1999. Rimsulfuron in a silty clay loam soil: Effects upon microbiological and biochemical properties under varying microcosm conditions. Soil Biol. Biochem. 31:195–204.

Pind, A., C. Freeman, and M.A. Lock. 1994. Enzymic degradation of phenolic materials in peatlands—measurement of phenol oxidase activity. Plant Soil 159:227–231.

Ross, D.J. 1970. Effects of storage on dehydrogenase activities of soils. Soil Biol. Biochem. 2:55–61.

Rotman, B., and B.W. Papermas. 1966. Membrane properties of living mammalian cells as studied by enzymatic hydrolysis of fluorogenic esters. Proc. Natl. Acad. Sci. USA 55:134–141.

Saiya-Cork, K.R., R.L. Sinsaugh, and D.R. Zak. 2002. The effects of long term nitrogen deposition on extracellular enzyme activity in an Acer saccharum forest soil. Soil Biol. Biochem. 34:1309–1315.

Schinner, F., R. Öhlinger, E. Kandeler, and R. Margesin. 1996. Methods in soil biology. Springer-Verlag, Berlin.

Schnürer, J., and T. Rosswall. 1982. Fluorescein diacetate hydrolysis as a measure of total microbial activity in soil and litter. Appl. Environ. Microbiol. 43:1256–1261.

Shaw, L.J., and R.G. Burns. 2006. Enzyme activity profiles and soil quality. p. 158–171. In J. Bloem, D.W. Hopkins, and A. Benedetti (ed.) Microbiological methods for assessing soil quality. CABI Publishing, Oxfordshire, UK.

Sinsaugh, R.L., M.J. Klug, H.P. Collins, P.E. Yeager, and S.O. Petersen. 1999. Characterizing soil microbial communities. p. 318–348. In G.P. Robertson, D.C. Coleman, C.S. Bledsoe, and P. Sollins (ed.) Standard soil methods for long-term ecological research. Oxford Univ. Press, New York.

Skujiņš, J. 1967. Enzymes in soil. p. 371–414. In A.D. McLaren and G.H. Peterson (ed.) Soil biochemistry. Vol. 1. Marcel Dekker, New York.

Skujiņš, J. 1976. Extracellular enzymes in soil. CRC Crit. Rev. Microbiol. 4:383–421.

Speir, T.W. 1977. Studies on a climosequence of soils in tussock grasslands. II. Urease, phosphatase, and sulfatase activities of topsoils and their relationships with other properties including plant available sulfur. N. Z. J. Sci. 20:159–166.

Speir, T.W., and D.J. Ross. 1976. Studies on a climosequence of soils in tussock grasslands. 9. Influence of age of Chionochloa rigida leaves on enzyme activities. N. Z. J. Sci. 19:389–396.

Stotzky, G. 1974. Activity, ecology and population dynamics of microorganisms in soil. p. 57–135. In A. Laskin and H. Lechvalier (ed.) Microbial ecology. CRC Press, Cleveland.

Swisher, R., and G.C. Carroll. 1980. Fluorescein diacetate hydrolysis as an estimator of microbial biomass on coniferous needle surfaces. Microb. Ecol. 6:217–226.

Thurston, C.F. 1994. The structure and function of fungal laccases. Microbiology 140:19–26.

Toberman, H., C.D. Evans, C. Freeman, N. Fenner, M. White, B.A. Emmett, and R.R.E. Artz. 2008. Summer drought effects upon soil and litter extracellular phenol oxidase activity and soluble carbon release in an upland Calluna heathland. Soil Biol. Biochem. 40:1519–1532.

Trasar-Cepeda, C., F. Camiña, M.C. Leirós, and F. Gil-Sotres. 1999. An improved method to measure catalase activity in soils. Soil Biol. Biochem. 31:483–485.

Trasar-Cepeda, C., F. Gil-Sotres, and M.C. Leirós. 2007. Thermodynamic parameters of enzymes in grassland soils from Galicia, NW Spain. Soil Biol. Biochem. 39:311–319.

Trevors, J.T. 1984a. Effect of substrate concentration, inorganic nitrogen, O_2 concentration, temperature and pH on dehydrogenase in soil. Plant Soil 77:285–293.

Trevors, J.T. 1984b. Rapid gas chromatographic method to measure oxidoreductase (catalase) activity in soil. Soil Biol. Biochem. 16:525–526.

Trevors, J.T., C.I. Mayfield, and W.E. Inniss. 1982. Measurement of electron transport system (ETS) activity in soil. Microb. Ecol. 8:163–168.

von Mersi, W. 1996. Dehydrogenase activity with the substrate INT. p. 243–245. In F. Schinner, R. Öhlinger, E. Kandeler, and R. Margesin (ed.) Methods in soil biology. Springer-Verlag, Berlin.

von Mersi, W., and F. Schinner. 1991. An improved and accurate method for determining the dehydrogenase activity of soils with iodonitrotetrazolium chloride. Biol. Fertil. Soils 11:216–220.

Weetall, H.H., N. Weliky, and S.P. Vango. 1965. Detection of micro-organisms in soil by their catalytic activity. Nature 206:1019–1021.

Williams, C.J., E.A. Shingara, and J.B. Yavitt. 2000. Phenol oxidase activity in peatlands in New York State: Response to summer drought and peat type. Wetlands 20:416–421.

Yamaguchi, M., P.M. Hwang, and J.D. Campbell. 1970. Latent o-diphenol oxidase in mushrooms (*Agaricus bisporus*). Can. J. Biochem. 28:198–202.

Chapter 7

Sulfur Cycle Enzymes

Susanne Klose,* Serdar Bilen, M. Ali Tabatabai, and Warren A. Dick

7–1 INTRODUCTION

Sulfur (S) is the 8th most abundant element in the solar atmosphere and the 14th most abundant element in the earth's crust (Dick, 1992). Sulfur cycles among the lithosphere, pedosphere, biosphere, atmosphere, and hydrosphere and also within each of these environmental components. All organisms require S for protein and enzyme synthesis and it is a constituent of the amino acids methionine and cysteine. Therefore, all organisms participate in the biological S cycle. In the past, fertilizers and atmospheric inputs supplied the soils with adequate amounts of S; more recently, areas of S deficiency are becoming widespread throughout the world. In agroecosystems, an insufficient S supply can affect yield and quality of crops.

Sulfur occurs in soil in inorganic and organic forms, with the organic fraction accounting for 90 to 98% of the total S present in most surface soils of humid and semihumid regions (Scherer, 2001; Tabatabai, 2005). Although soil organic S compounds are still poorly characterized, they are commonly divided in two main groups: organic sulfate S and carbon-bonded S. Information about the chemical characteristics of these groups has been derived mainly from the comparative reactivity of soil S with different reducing reagents (Tabatabai, 2005). The organic sulfate-S fraction (e.g., ester sulfates and some organic sulfites) is reducible by hydriodic acid (HI) and accounts for 30 to 75% of the organic S in surface soils worldwide (Germida, 2005; Haneklaus et al., 2007). The high variation is a result of climatic factors, organic inputs, and land use practice. Although the exact form of ester sulfates in soil is unclear, they are probably arylsulfates and polysaccharide sulfates such as adenosine-5'-phosphosulfate, chlorine sulfates, glucose sulfates, glucosinolates and sulfamates. The carbon-bonded S fraction represents the difference between total soil organic S and the organic sulfate-S fraction and may constitute up to 30% of the organic S in soil. It includes mainly S-containing amino acids (e.g., cysteine, cystine, and methionine), but also heterocyclic compounds (e.g., biotine, thiamin) and disulfides (Freney, 1986).

Susanne Klose, Chiquita Brands North America, James Lugg Research Center, 607 Brunken Ave., Salinas, CA 93901 (sklose@chiquita.com), *corresponding author;
Serdar Bilen, Atatürk University, Faculty of Agriculture, Department of Soil Science, Erzurum, Turkey (sbilen@atauni.edu.tr);
M. Ali Tabatabai, Department of Agronomy, Iowa State University, 2403 Agronomy Hall, Ames, IA 50011 (malit@iastate.edu);
Warren A. Dick, School of Environment and Natural Resources, The Ohio State University, 101A Hayden Hall, 1680 Madison Ave., Wooster, OH 44691 (dick.5@osu.edu).

doi:10.2136/sssabookser9.c7

Inorganic S forms are generally much less abundant in most soils than is organic S and, depending on the soil conditions, they may include sulfide (S^{2-}), polysulfide (S_n^{2-}), elemental S (S^0), sulfite (SO_3^{2-}), thiosulfate ($S_2O_3^{2-}$), trithionate ($S_3O_6^{2-}$), tetrathionate ($S_4O_6^{2-}$), and sulfates (SO_4^{2-}) (Bohn et al., 1986).

Soil S is continuously cycled between organic and inorganic S forms. Organic S compounds are unavailable to plants and must be converted by biochemical or microbial mineralization to inorganic SO_4^{2-} for plant uptake (Castellano and Dick, 1991). The major reactions that characterize S transformations in the environment are oxidation, reduction, mineralization, immobilization, and volatilization. Because S atoms can exist in a valence state ranging from –2 for sulfides to +6 for sulfates, a large number of oxidation–reduction reactions are possible. In many instances, mineralization–immobilization reactions and volatilization reactions are really special cases of oxidation–reduction reactions of S compounds in the environment. Basic information about the nature, quantity, and distribution of different S compounds in soil, and the transformation processes in which they participate, is of fundamental interest for a better understanding of the significance of S in various natural and managed ecosystems.

Many S transformation reactions are catalyzed by enzymes released into the soil environment by microorganisms, plant roots, and soil fauna. The biogeochemical cycling of organic and gaseous S compounds in soil catalyzed by some of these enzymes is summarized in Fig. 7–1 and Table 7–1. A number identifies each enzyme-catalyzed reaction in Fig. 7–1. This number is also reflected in Table 7–1 where the common name of the enzyme that catalyzes the reaction is provided along with the Enzyme Commission number.

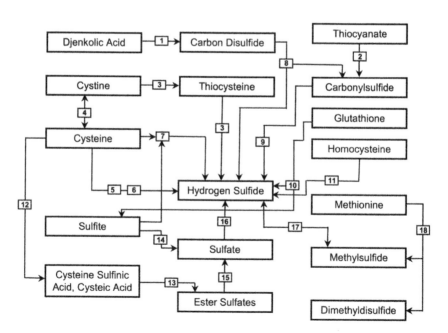

Fig. 7–1. Schematic diagram integrating biological reactions involving organic S and gaseous S compounds in soil. The numbers refer to specific enzyme-catalyzed reactions that are identified in Table 7–1.

Table 7–1. Summary of biological reactions involving organic and gaseous S compounds in soil (key to Fig. 7–1).

Reaction Key	Enzymes involved in reactions	EC number
1	S-alkylcysteine lyase	4.4.1.6
2	Thiocyanate hydrolase	3.5.5.8
3	Cystathionine lyases (γ and β)	4.4.1.1, 4.4.1.8
4	Cystine reductase (reversible)	1.8.1.6
5	Cysteine desulfhydrase	4.4.1.15
6	β-cyanoalinine synthase	4.4.1.9
7	Cysteine lyase	4.4.1.10
8	Conversion of carbon disulfide to hydrogen sulfide and carbonyl sulfide	–†
9	Carbonic anhydrase ($COS + H_2O = CO_2 + H_2S$)	4.2.1.1
10	Thiosulfate–thiol sulfurtransferase (with $S_2O_3^{2-}$)	2.8.1.3
11	Homocysteine desulfhydrase	4.4.1.2
12	Cysteine dioxygenase	1.13.11.20
13	Formation of ester sulfates	–†
14	Sulfite oxidase	1.8.3.1
15	Arylsulfatase	3.1.6.1
16	Sulfate reductase—multiple pathways and enzyme steps	Not applicable
17	Methanethiol oxidase	1.8.3.4
18	β-cystathionase	4.4.1.8

† Enzyme Commission number not assigned.

The Enzyme Commission (EC) number, or the enzyme common name, can be used to search the "BRENDA" web site (http://www.brenda-enzymes.org/ [verified 12 April 2011]). This site is a comprehensive enzyme information system site that can provide much additional information such as that related to nomenclature, reaction and specificity, organism-related information, enzyme stability, enzyme structure, and enzyme isolation and purification. Another web site used to construct Fig. 7–1 (and also Fig. 7–2) is the MetaCyc site for hydrogen sulfide (http://biocyc.org/ [verified 12 April 2011]) and related pages, where one can search S-related enzyme reactions.

Although most of the reactions shown in Fig. 7–1 are described in the BRENDA web site, there are a few exceptions. The conversion of carbonyl sulfide to hydrogen sulfide is described in some detail by Kato et al. (2008). The reactions involving transformation of cysteine (cystine) are taken largely from Freney (1967). The formation of volatile S gases from methionine is described by Lee et al. (2007).

The biogeochemical cycling of inorganic S compounds in soil is summarized in Fig. 7–2 and Table 7–2 (http://www.brenda-enzymes.org/ [verified 12 April 2011]; http://biocyc.org/ [verified 12 April 2011]). Similar to Fig. 7–1, a number identifies each enzyme-catalyzed reaction in Fig. 7–2 and this number is also

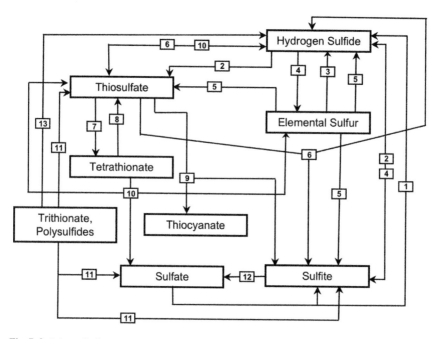

Fig. 7–2. Schematic diagram integrating biological reactions involving inorganic S compounds in soil. The numbers refer to specific enzyme-catalyzed reactions that are identified in Table 7–2.

Table 7–2. Summary of biological reactions involving inorganic S compounds in soil (key to Fig. 7–2).

Reaction Number	Enzymes involved in reactions	EC number
1	Sulfate reductase—multiple pathways and enzyme steps	Not applicable
2	Sulfite reductase	1.8.1.2, 1.8.7.1, 1.8.99.1
3	Sulfur reductase	1.97.1.3
4	Sulfide oxidase	1.8.99.1, 1.97.1.3
5	Sulfur oxygenase/reductase (mixed redox reaction)	1.13.11.18, 1.13.11.5
6	Thiosulfate reductase	2.8.1.5
7	Thiosulfate oxidase	1.8.2.2, 1.8.5.2
8	Tetrathionate synthase	1.8.2.2, 1.8.5.2
9	Rhodanese	2.8.1.1
10	Tetrathionate hydrolase	–†
11	Trithionate hydrolase, triothionate oxidoreductase	3.12.1.1, 1.8.99.3
12	Sulfite oxidase	1.8.2.1, 1.8.3.1
13	Polysulfide reductase (sulfur reductase)	1.97.1.3

† Enzyme Commission number not assigned.

reflected in Table 7–2, where the common name of the enzyme that catalyzes the reaction is provided along with the EC number.

This chapter will not attempt to provide an exhaustive description of all the enzyme-catalyzed reactions involved in the S cycle in soils. First, it is currently impossible because many of the enzymes have not been studied within a soil or environmental context. For example, among the many enzymes involved in the degradation of ester sulfates (i.e., arylsulfatases, alkylsulfatases, steroid sulfatases, glucosulfatases, chondrosulfatases, and mycosulfatases), only arylsulfatase activity has been studied intensively in soils. Second, many of the S transformation processes have been studied on an overall functional basis rather than as discrete biochemical reactions. For example, hydrogen sulfide evolution from soil or elemental S oxidation in soil can be studied without an understanding of the biochemical or enzymatic processes that mediate these reactions. Third, even if there is a basic understanding of the biochemical pathway(s) involved in a specific S transformation process, a valid assay often has not yet been developed and tested. For information about how to develop a soil enzyme assay, see Chapter 4.

7–2 CYSTATHIONINE LYASE

7–2.1 Introduction

More than 90% of the total S in soil is in organic forms. This organic fraction is often divided into two different groups. The ester sulfate fraction is composed of molecules that have a C–O–S linkage. The carbon-bonded fraction is primarily made up of the amino acids cysteine (cystine) and methionine generally found as components of proteins. Cysteine and methionine together can comprise between 11 and 31% of the total soil organic S (Freney et al., 1972; Scott et al., 1981). The ratio of cysteine to methionine was found to be 2:1 in four mineral soils from Scotland (Scott et al., 1981).

The transformation and mineralization of cysteine to sulfate involves intermediates such as cysteine sulfinic acid and cysteic acid (Fig. 7–1). The transformation of cysteine and cystine to produce hydrogen sulfide (H_2S) involves several different pathways, as have been proposed from studies involving microorganisms or partially purified enzymes. Cavallini et al. (1962a, b) present evidence suggesting the desulfuration of cysteine is catalyzed by a single enzyme. However, others provide evidence that two distinct enzymes with different substrate specificities exist (Guarneros and Ortega, 1970; Collins and Monty, 1973; Kredich et al., 1973). Confusion occurs because of the rapid autoxidation of cysteine to cystine. The Enzyme Commission has assigned the desulfurization reaction of cysteine (cystine) to both cystathionine γ-lyase (EC 4.4.1.1) and cystathionine β-lyase (EC 4.4.1.8) (Table 7–1).

7–2.2 Principle

The various pathways of hydrogen sulfide (H_2S) production from the amino acids cysteine and cystine can involve several enzymes (Fig. 7–1 and Table 7–1). One mechanism studied in soils, and reported by Morra and Dick (1985), is catalyzed by enzymes called cystathionine lyases. These enzymes belong to the family of lyases and specifically to the class of carbon-sulfur lyases. The reaction catalyzed by cystathionine lyase causes cystine cleavage to form thiocysteine, NH_3, and pyruvate (Yamanishi and Tuboi, 1981) and can be described as follows:

$$COOH(NH_2)CH_2S\text{-}SCH_2(NH_2)COOH + H_2O \rightarrow$$
$$O\ HSSHCH_2(NH_2)COOH + NH_3 + CH_3CCOOH$$
[1]

Cystine + water $\rightarrow$ thiocysteine + ammonia + pyruvate   [2]

The substrate of this reaction is cystine that is rapidly produced in soils from the amino acid cysteine, and presumably being catalyzed by trace amounts of iron, copper, and manganese (Freney, 1967). To demonstrate the presence of cystathionine lyase in soil, Morra and Dick (1985) utilized the information reported in the literature that thiocysteine reacts nonenzymatically with cysteine to yield H_2S and cystine in a pathway that can be described as follows:

$$HSSHCH_2(NH_2)COOH + SHCH_2(NH_2)COOH \rightarrow$$
$$H_2S + COOH(NH_2)CH_2S\text{-}SCH_2(NH_2)COOH$$
[3]

Thiocysteine + cysteine $\rightarrow$ hydrogen sulfide + cystine   [4]

After incubation of soil with cysteine (cystine), the amount of thiocysteine produced can be determined by measuring the increase in the amount of H_2S after an additional treatment with a larger amount of cysteine. A method to rapidly and sensitively measure H_2S evolved from soil is thus needed to assay cystathionine lyase in soil and has been provided by Morra and Dick (1985, 1989).

7–2 Assay Method (Morra and Dick, 1985 and 1989)

7–2.3.1 Apparatus

- Erlenmeyer flask (25 mL) containing a center well to hold a pleated filter paper (Fig. 7–3). The center well can be created using a small plastic vessel suspended from a wire that passes through the flask stopper. The size of the center well must be sufficient to comfortably hold 0.5 mL of solution and a folded (pleated) filter paper. The center well must be easily detached so that it can be moved into a second 50-mL Erlenmeyer flask.
- Small plastic vessel to serve as a center well
- Whatman No. 42 filter paper (4.0 by 2.5 cm) folded (pleated) to allow it to be placed, on end, in the center well
- Incubator
- Spectrophotometer

7–2.3.2 Reagents

1. Sodium azide (0.5 M): Weigh 3.25 sodium azide (NaN_3) into a volumetric flask. Adjust final volume to 100 mL.
2. Zinc acetate (1.0 M): Dissolve 110 g of zinc acetate [$Zn(CH_3COO)\cdot 2H_2O$] in about 400 mL of water. Adjust final volume to 500 mL. Filter the solution through Whatman No. 42 filter paper if turbid.

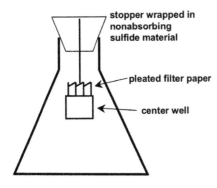

stopper wrapped in
nonabsorbing
sulfide material

pleated filter paper

center well

Fig. 7–3. Schematic illustration of the apparatus used to measure H_2S evolved from soil.

3. Aminodimethylaniline: Dissolve 1 g of p-aminodimethylaniline sulfate in about 700 mL of water in 1-L glass bottle. Place the bottle in a pan of cold water and slowly add 200 mL of concentrated reagent-grade sulfuric acid. This acid should be added in small portions, mixing the solution and allowing the mixture to cool after the addition of each portion. Cool and adjust to 1 L.

4. Ferric ammonium sulfate: To 25 g of ferric ammonium sulfate $[Fe_2(SO_4)_3 \cdot (NH_4)_2SO_4 \cdot 24H_2O]$, add 5 mL of concentrated reagent grade sulfuric acid and 195 mL of water. Allow reagent to stand at room temperature for several days to ensure complete dissolution before using.

5. L-cysteine HCl (10 mM) substrate solution, pH 6.0: Weigh 0.176 g L-cysteine HCl·H$_2$O into a 100-mL volumetric flask. Adjust pH to 6.0 with HCl and then bring final volume to 100 mL.

6. Citric acid (0.4 M) and sodium phosphate (0.8 M) buffer, pH 6.0: Weigh 117.6 g sodium citrate $[HOC(COONa)(CH_2COONa)_2 \cdot 2H_2O]$ and 214.5 g sodium phosphate $(NaHP_4 \cdot 7H_2O)$ into a 1-L volumetric flask. Add approximately 800 mL of water to dissolve the reagents and then adjust the pH to 6.0 using HCl. Bring the final volume to 1000 mL.

7. L-cysteine HCl (100 mM) reagent solution, pH 6.0: Weigh 17.6 g L-cysteine HCl·H$_2$O into a 1-L volumetric flask containing approximately 800 mL of 0.4-M citric acid–0.8-M phosphate buffer solution. Adjust pH to 6.0 with HCl and then bring final volume to 1000 mL.

8. Sulfide antioxidant buffer (SAOB): Mix 80 g NaOH, 320 g sodium salicylate, and 72 g ascorbic acid and bring the volume to 1 L with water.

9. Sulfide standards: Dissolve $Na_2S \cdot 9H_2O$ in 12.5% SAOB into various 100 volumetric flasks so that each flask contains 0, 2, 5, 15, 25, 35, and 45 mg S L^{-1}. Once the standards are created using the SAOB, they must be refrigerated and made fresh on a regular basis. If color intensity is too great at the higher concentrations, the standard curve can be expanded by using more standards at lower concentrations.

7–2.3.3 Procedure

Place 3 g of air-dried soil (<2 mm) into a 25-mL Erlenmeyer flask. Add 4.9 mL of 10 mM L-cysteine HCl (substrate) adjusted to pH 6.0 with NaOH, followed by 0.1 mL of deionized water. Mix the contents in the flask and then cap the flask with a stopper from which a center well is suspended so that the center well is above the soil surface and within the headspace of the flask (see Fig. 7–3). The center well should be previously prepared to contain 0.5 mL 1 M zinc acetate and a piece of pleated filter paper (Whatman No. 42, 4.0 cm by 2.5 cm). Incubate for 24 h at 37°C. After incubation, add 0.1 mL NaN₃ (0.5 M) to stop the reaction. Then add 5 mL of 100 mM L-cysteine in a 0.4-M citric acid–0.8-M sodium phosphate ($Na_2HP_4\cdot7H_2O$) buffer adjusted to pH 6.0. Immediately stopper the flask and set the assay flask in the incubator for at least 20 min to allow the H_2S released to be trapped by the zinc acetate in the well and on the pleated filter paper. Remove the center well and add the entire content containing the zinc acetate solution and the pleated filter paper into a 50-mL Erlenmeyer flask containing 40 mL of water. Add 4 mL p-aminodimethylaniline sulfate, stopper the flask, and mix the contents thoroughly. Remove the stopper and add 1 mL ferric ammonium sulfate solution, bring to 50-mL volume, stopper the flask again, and mix. Measure the absorbance of the methylene blue color formed by using a spectrophotometer adjusted to a wavelength of 670 nm. Controls are conducted by adding 0.1 mL of NaN₃ (0.5 M) before addition of the 10 mM L-cysteine substrate. Activity of the cystathionine lyase reaction is reported as mg H_2S produced kg⁻¹ soil h⁻¹.

The amount of H_2S evolved during reaction is quantified by using a standard curve created from sulfide standards. To prepare this curve, add 1 mL sulfide standard prepared from $Na_2S\cdot H_2O$ and containing 0, 2, 5, 15, 25, 35, and 45 mg S L⁻¹ to a 50-mL Erlenmeyer flask containing 40 mL of water. Add 4 mL p-aminodimethylaniline sulfate and 1 mL ferric ammonium sulfate solution as described previously and adjust final volume to 50 mL. Measure the absorbance of the methylene blue color formed by using a spectrophotometer adjusted to a wavelength of 670 nm. Note: at first the solution is red but it changes to blue as the dye is formed. The final methylene blue color formed is stable for at least several days if stored in the dark, but direct sunlight causes rapid fading.

7–2.4 Comments

The coupling of the cystathionine lyase reaction with the strictly chemical reaction of cysteine with thiocysteine is one mechanism by which S may be lost from natural systems. Other compounds with free sulfhydryl groups also may react with thiocysteine to produce hydrogen sulfide. The overall stoichiometry of the reaction shows that cystine is continually replaced during reaction, so that it can once again undergo reaction, and there is a net consumption of cysteine.

In environments that are neither highly aerobic nor anaerobic, both cysteine and cystine may be present. If active cystathionine lyase is also present in soil, sulfur as H_2S may be lost. The biological nature of the reaction was demonstrated when the production of thiocysteine was inhibited when soils were treated with 0.1 mL of 0.5 M sodium azide or steam sterilized for 1 h at 121°C (Morra and Dick, 1985). Other evidence that cystathionine lyase is active in the formation of thiocysteine was obtained when propargylglycine was included in the assay procedure. Propargylglycine is known to be a "suicide" inhibitor of cystathionine lyase causing complete inactivation of the enzyme (Abeles and Walsh, 1973; Abeles, 1983). Because a suicide inhibitor is initially chemically nonreactive and is modified only by the target enzyme, reactions with other enzymes are unlikely, making suicide inhibitors highly specific. Thiocysteine produced in soils was greatly decreased in the presence of propargylglycine with inhibitions ranging from 58 to 91% (Morra and Dick, 1985) and 90 to 98% (Morra and Dick, 1989). Almost complete inhibition of cystathionine lyase also was found to occur in the presence of toluene and formaldehyde, which is important to note because toluene is commonly used to prevent microbial proliferation during enzyme assays, and formaldehyde is known to be a potent inhibitor of rhodanese activity, another S enzyme in soil (Tabatabai and Singh, 1976).

7–3 ARYLSULFATASE

7–3.1 Introduction

Ester sulfates represent an important fraction of total organic S in soil (30 to 75%), and are considered to be the most labile form of organic S in soil (Scherer, 2001; Tabatabai, 2005). Additionally, there is evidence that ester sulfates are of a more transitory nature than is C-bonded S and serve as a temporary sink of SO_4^{2-} in soil, and thus are an important source of plant-available S.

Sulfatases (sulfohydrolases, EC 3.1.6) are enzymes of the esterase class that catalyze the hydrolysis of ester sulfates, where the linkage with sulfate is in the form of R–O–S and R represents a diverse group of organic moieties. Several types of sulfatases occur in nature, including arylsulfatases, alkylsulfatases, steroid sulfatases, glucosulfatases, chondrosulfatases, and mycosulfatases (Germida et al., 1992; Haneklaus et al., 2007). However, most of the studies have been directed to arylsulfatase, which was the first detected sulfatase in nature (Fitzgerald, 1978). Arylsulfatase (arylsulfate sulfohydrolase, EC 3.1.6.1) (Fig. 7–1, Table 7–1) catalyzes the hydrolysis of ester-bonded S by fission of the O–S linkage in a pathway that can be described as follows:

$$R \cdot OSO_3^- + H_2O \rightarrow R \cdot OH + H^+ + SO_4^{2-} \qquad [5]$$

$$\text{Arylsulfate} + \text{water} \rightarrow \text{phenol} + \text{hydrogen ion} + \text{sulfate} \qquad [6]$$

The reaction is irreversible, and there is no evidence that any acceptor other than water can be used or that any metal ion is required for its catalytic function. Arylsulfatase is found in plants, fungi, bacteria, and animals, but microorganisms are believed to be the main source for this enzyme in soil (Fitzgerald, 1978; Germida et al., 1992). Microbial sulfatases can be grouped into constitutive, inducible, and derepressible enzymes (Germida et al., 1992). Arylsulfatase activity was first detected in soils at Iowa State University by Tabatabai and Bremner (1970a), and since then, has been described in a wide range of agricultural and forest soils, marine, and fresh water sediments, salt marshes, arctic, and wetland soils worldwide (Farrell et al., 1994; Tabatabai, 1994; Deng and Tabatabai, 1997; Bandick and Dick, 1999; Klose et al., 1999, 2003, 2004, 2006; Prietzel, 2001; Acosta-Martinez et al., 2003, 2007a,b, 2008; Li and Sarah, 2003; Dinesh et al., 2004; Wang and Lu, 2006). These studies suggest that arylsulfatase activity is linked to the S supply, because the biochemical mineralization relies on the release of SO_4^{2-} from the ester sulfate fraction through enzymatic hydrolysis, and the majority of the mineralizable soil organic S is in the ester sulfate form (Speir and Ross, 1978; Germida et al., 1992; Germida, 2005; Haneklaus et al., 2007).

The arylsulfatase activity assay involves the quantitative determination of p-nitrophenol released when a soil sample is incubated with buffered p-nitrophenyl sulfate solution and toluene at 37°C for 1 h (Tabatabai and Bremner, 1970a). This method has been used in almost all subsequent studies with no or only minor modifications (Klose and Tabatabai, 1999; Klose et al., 2003, 2004, 2006; Li and Sarah, 2003; Vong et al., 2003; Gianfreda et al., 2005; Acosta-Martinez et al., 2007a,b; 2008; Chen et al., 2008b). The pH optimum of this enzyme ranged between 5.5 and 6.2 (Tabatabai and Bremner, 1970b; Germida et al., 1992, Tabatabai, 1994). A recent study showed that soil arylsulfatase activity increased with increase in temperature up to 57°C, but decreased thereafter (Trasar-Cepeda et al., 2007). Arylsulfatase activity is inactivated at temperatures ranging from 60 to 70°C (Tabatabai and Bremner, 1971). The K_m values of this soil enzyme range from 0.2 to 5.7 mM, and will change during incubation with shaking; usually the K_m values are lower and more uniform among soils during incubation with shaking if compared with static incubation (Tabatabai and Bremner, 1971; Thornton and McLaren, 1975).

Arylsulfatase activity was significantly correlated with soil organic matter content (Tabatabai, 1994; Klose et al., 1999; Knights et al., 2001) and decreased with soil depth reflecting the organic matter content (Tabatabai, 1994). Recent studies suggest that arylsulfatase activity is driven by the quantity and quality of available C rather than total organic C (DeLuca and Keeney, 1994; Dedourge et al., 2004). Studies showed that several trace elements inhibit arylsulfatase activity and that the inhibition by MoO_4^{2-}, WO_4^{2-}, AsO_4^{3-}, and PO_4^{3-} shows competitive kinetics. At concentrations of 25 µmol g^{-1} of soil, common soil anions such as NO_2^-, NO_3^-, Cl$^-$, and SO_4^{2-} are not inhibitors of this enzyme (Al-Khafaji and Tabatabai, 1979). In contrast, arylsulfatase activity was inversely affected by elevated concentrations of SO_4^{2-} (Prietzel, 2001; Wang and Lu, 2006), heavy metals (Gianfreda and Bollag, 1996; Kandeler et al., 1996), soil amendments with urban and industrial waste (Gianfreda et al., 2005), and soil fumigation with methyl bromide and alternative fumigants (Klose and Ajwa, 2004; Klose et. al., 2006). Arylsulfatase activity was affected by soil order (Acosta-Martinez et al., 2007b), season (Castellano and Dick, 1991; Prietzel, 2001), plant cover, and soil management (Farrell et al., 1994; Klose et al., 1999; Dinesh et al., 2004; Acosta-Martinez et al., 2007a, 2008), was unaffected by long-term soil irrigation with reclaimed wastewater (Chen et al., 2008b), but increased in forest soils subjected to long-term lignite-derived atmospheric depositions (Klose et al., 2003, 2004).

Air-drying of field-moist soils resulted in an increase in arylsulfatase activity in some studies (Tabatabai and Bremner, 1970b; Wang and Lu, 2006), and in an activity decrease in other studies (Bandick and Dick, 1999). A possible explanation of this increase is that rewetting of air-dried soil causes a breakdown of aggregates, thus increasing the accessibility of arylsulfatase to the substrate. Other factors that may affect arylsulfatase activity in soils are summarized by Speir and Ross (1978) and Germida et al. (1992).

7–3.2 Principle

The method is based on spectrophotometric determination of p-nitrophenol released by arylsulfatase activity when soil is incubated with buffered (pH 5.8) potassium p-nitrophenyl sulfate solution and toluene. Usually the soil–buffer–substrate mixture is incubated at 37°C for 1 h. The p-nitrophenol released is extracted by filtration after the addition of $CaCl_2$ and NaOH reagents. The spectrophotometric method used for determination of *p*-nitrophenol depends on the fact that alkaline solutions (e.g., achieved by addition of 0.5 M NaOH) of this phenol have a yellow color (acid solutions of *p*-nitrophenol and acid and alkaline solutions of *p*-nitrophenyl sulfate are colorless). The $CaCl_2$ is added to prevent dispersion of clay and extraction of soil organic matter during the treatment with NaOH. Dispersion of clay complicates filtration, and the dark-yellow-brown-colored organic matter extracted with NaOH interferes with the colorimetric determination of *p*-nitrophenol.

7–3.3 Assay Method (Tabatabai and Bremner, 1970a)

Note that this assay on organic soils or organic forest horizons should be modified as outlined in section 7–7.

7–3.3.1 Apparatus

- Erlenmeyer flasks (50 mL) fitted with No. 2 stoppers
- Whatman No. 2 or 2v (folded) filter paper (Thermo Fisher Scientific, Inc.)
- Incubator
- Funnels
- Spectrophotometer

7–3.3.2 Reagents

1. Toluene, Fisher certified reagent (Fisher Scientific Co., Chicago).
2. Acetate buffer, 0.5 M, pH 5.8: Dissolve 68 g of sodium acetate trihydrate in about 700 mL of di-ionized (DI) water, add 1.70 mL of glacial acetic acid (99%), and dilute the volume to 1 L with DI water.
3. *p*-Nitrophenyl sulfate solution, 0.05 M: Dissolve 0.614 g of potassium *p*-nitrophenyl sulfate (Sigma-Aldrich Co., St. Louis, MO) in about 40 mL of acetate buffer, and dilute the solution to 50 mL with buffer. Store this solution in a refrigerator.
4. Calcium chloride ($CaCl_2$), 0.5 M: Dissolve 73.5 g $CaCl_2 \cdot 2H_2O$ in about 700 mL DI water and dilute the volume to 1 L with DI water.
5. Sodium hydroxide (NaOH), 0.5 M: Dissolve 20 g NaOH in about 700 mL DI water, and dilute the volume to 1 L with DI water.
6. Standard *p*-nitrophenol solution: Dissolve 1.0 g *p*-nitrophenol in about 700 mL DI water, and dilute the solution to 1 L with water. Store the solution in a refrigerator.

7–3.3.3 Procedure

Place 1 g of field-moist soil (<2 mm) into a 50-mL Erlenmeyer flask, add 0.25 mL of toluene, 4 mL of acetate buffer, and 1 mL of *p*-nitrophenyl sulfate solution (0.05 M) and swirl the flask for a few seconds to mix the contents. Stopper the flask and place it in an incubator at 37°C. After 1 h, remove the stopper, add 1 mL $CaCl_2$ (0.5 M) and 4 mL NaOH (0.5 M), swirl the flask for a few seconds, and filter the soil suspension through a Whatman No. 2 (or 2v) filter paper. Measure the yellow color intensity of the filtrate with a spectrophotometer adjusted to a wavelength of 420 nm. Calculate the p-nitrophenol concentration of the filtrate by reference to a calibration graph prepared from p-nitrophenol standards containing 0, 10, 20, 30, 40, and 50 mg p-nitrophenol L^{-1}. To prepare this graph, dilute 1 mL of the standard p-nitrophenol solution to 100 mL in a volumetric flask and mix the solution thoroughly. Then pipette 0-, 1-, 2-, 3-, 4-, and 5-mL aliquots of this diluted standard solution into 50-mL Erlenmeyer flasks, adjust the volume to 5 mL by addition of DI water, add 1 mL $CaCl_2$ (0.5 M) and 4 mL NaOH (0.5 M), mix, filter, and measure the yellow color intensity of the filtrate with a spectrophotometer adjusted to a wavelength of 420 nm.

Controls should be performed with each soil analyzed to account for color not derived from p-nitrophenol released by arylsulfatase activity. Controls are conducted by adding 1 mL of p-nitrophenyl sulfate solution (0.05 M) after addition of the 1 mL $CaCl_2$ (0.5 M) and 4 mL NaOH (0.5 M) and immediately before filtration of the soil suspension. The p-nitrophenol yellow color formed is stable for at least several hours if stored in the dark, but direct sunlight causes rapid fading.

7–3.4 Comments

Although several compounds (e.g., potassium phenyl sulfate, potassium nitrocatechol sulfate, potassium phenolphthalein sulfate, tyrosine sulfate) can serve as substrates for arylsulfatase activity, it is difficult to extract the phenolic compounds released by enzymatic hydrolysis of these substrates (Tabatabai and Dick, 2002). Furthermore, in contrast with the p-nitrophenyl sulfate method, these substrates have not been tested on a wide range of soils. The p-nitrophenyl sulfate procedure results in a quantitative (99–100%) recovery of the p-nitrophenol added to soils. The addition of $CaCl_2$ (0.5 M) is necessary to prevent dispersion of clay and extraction of soil organic matter following the NaOH treatment of the soil suspension. In organic soils or organic layers of forest soils, NaOH should be replaced with THAM buffer (pH 12) to prevent extraction of organic matter and humic substances, which would interfere with the measurement of the p-nitrophenol yellow color. For information about how to conduct enzyme assays based on the determination of p-nitrophenol yellow color in organic or forest soils see Section 7–7.

Controls account for the presence of trace amounts of p-nitrophenol in some commercial samples of p-nitrophenyl sulfate and for the extraction of trace amounts of colored soil material by the $CaCl_2$–NaOH treatment. No chemical hydrolysis of p-nitrophenyl sulfate is detected under the described assay conditions.

Arylsulfatase activity values are usually lower if deionized water is used instead of the buffer solution (i.e., acetate buffer pH 5.8). The reason for this difference is presumably the higher arylsulfatase activities obtained at optimum soil pH (i.e., pH 5.5 to 6.2) for this enzyme if compared with activity values at the actual pH of the soil (Tabatabai and Bremner, 1970b; Germida et al., 1992).

The substrate concentration in the incubated mixture (10 mM) recommended is much higher (5- to 10-fold) than the K_m values of arylsulfatase in most soils. If desired, this substrate concentration can be changed to meet the objective of the assay.

Choice of buffer and buffer pH in the method described is based on studies of Tabatabai and Bremner (1970a) revealing that the amount of p-nitrophenol released by incubation of soil with buffered p-nitrophenyl sulfate solution is considerably higher in acetate buffer than in other buffers, and that maximal enzymatic hydrolysis of p-nitrophenyl sulfate occurs at pH 6.2 of the acetate buffer. However, because the buffering poise of the acetate buffer is greater at pH 5.8 than at 6.2, and because the amount of p-nitrophenol released at pH 5.8 is not markedly different from that released at pH 6.2, a buffer pH of 5.8 is recommended.

Arylsulfatase activity values are not affected by the amount of toluene added (0.1–1.0 mL). Toluene, however, increases the arylsulfatase activity in various soils up to 31% during static incubation due to its plasmolytic action, which causes an increased diffusion of substrates and products across the cell membrane (Frankenberger and Johanson, 1986; Tabatabai, 1994; Klose and Tabatabai, 1999; Elsgaard et al., 2002). However, toluene had little effect on arylsulfatase activity during incubation with shaking, and shaking was shown to increase the precision of the assay in treatments without toluene (Elsgaard et al., 2002). Furthermore, to simplify the arylsulfatase assay, sample extraction was successfully done by centrifugation (at 5000 r min^{-1} for 10 min) rather than by filtration (Elsgaard et al., 2002).

7–4 RHODANESE

7–4.1 Introduction

Elemental S is an important S fertilizer. However, S^0 must first be converted to SO_4^{2-} by oxidation before it can contribute to plant uptake of S. Nor and Tabatabai (1975) detected thiosulfate as one of the intermediates during S^0 oxidation.

Rhodanese (thiosulfate-cyanide sulfurtransferase, EC 2.8.1.1) (Fig. 7–2, Table 7–2) is the enzyme that catalyzes the formation of thiocyanate and sulfite from thiosulfate and cyanide in a pathway that can be summarized as follows:

$$S_2O_3^{2-} + CN^- \rightarrow SCN^- + SO_3^{2-} \qquad [7]$$

Thiosulfate + cyanide $\rightarrow$ thiocyanate + sulfite $\qquad [8]$

Rhodanese activity was first detected in animals by Lang (1933), and is currently regarded as one of the mechanisms evolved for cyanide detoxification (Wróbel et al., 2009). Rhodaneses are ubiquitous enzymes, occurring in tissues of various plants, animals, and human (Billaut-Laden et al., 2006; Glatz et al., 1999). The presence of rhodanese activity has been reported for some bacteria, including *Escherichia coli* , *Thiobacillus* spp., and *Chromatium* spp. (Smith and Lascelles, 1969; Tabita et al., 1969; Alexander and Volini, 1987; Ray et al., 2000) and for some fungi, including *Fusarium* spp. and *Trichoderma* spp. (Ezzi and Lynch, 2002; Ezzi et al., 2003). Although rhodanese activity was first detected in Iowa soils more than 30 yr ago (Tabatabai and Singh, 1976), there are very few reports on rhodanese activity in soils from other regions. Rhodanese has been proposed to play an important role in the S cycle because it converts $S_2O_3^{2-}$ to SO_3^{2-}, both of which are intermediates produced during oxidation of elemental S (S^o) in soils (Nor and Tabatabai, 1977; Deng and Dick, 1990; Dick and Deng, 1991, Germida et al., 1992). Germida et al. (1992) reported an increase in rhodanese activity during S^0 oxidation in aerobic soils. However, studies by Ray et al. (1985) and Deng and Dick (1990) revealed that an inconsistent, or no, relationship could be established between rhodanese activity and oxidation of S^0 in flooded rice soils or in Oregon soils with varying physicochemical properties, respectively. Deng and Dick (1990) concluded that other enzymes and alternative pathways in soil are more important for S^0 oxidation than is rhodanese. A more recent study found a positive relationship between rhodanese activity and the concentration of free sulfuric amino acids (i.e., secondary metabolites) in soil, suggesting a potential role of rhodanese activity for cropping system evaluations (Szajdak, 1996).

The procedure of assaying rhodanese activity in soils involves the quantitative determination of the SCN^- produced when a soil sample is incubated with buffered $S_2O_3^{2-}$ and CN^- solutions and toluene at 37°C for 1 h (Tabatabai and Singh, 1976). Kinetic studies support a hypothesis where the overall catalysis occurs in two steps in which rhodanese cycles between two intermediates (Cipollone et al., 2006). Tabatabai and Singh (1979) calculated the kinetic and thermodynamic parameters of rhodanese activity in five Iowa soils. The K_m values for soil rhodanese activity with $S_2O_3^{2-}$ and CN^- as substrates ranged from 1.2 to 10.3 (avg. 5.46 for $S_2O_3^{2-}$) mM and from 2.5 to 10.2 (avg. 5.81 for CN^-) mM. The V_{max} values ranged from 0.51 to 1.43 mM SCN^- produced per gram of soil per hour. The activation energy values ranged from 21.6 to 34 kJ mol^{-1} for temperatures from 10 to 60°C, and the Q_{10} under these conditions ranged from 1.25 to 1.45 (avg. 1.37). These kinetic studies showed similar K_m values of $S_2O_3^{2-}$ and CN^- for the reaction catalyzed by soil rhodanese and by rhodanese from other biological systems (Tabatabai and Singh, 1979).

Rhodanese activity is correlated with total organic C (Singh and Tabatabai, 1978) and microbial biomass C in soils, while no relationship was observed with soil pH (Dick and Deng, 1991). Rhodanese activity is modulated by soil pretreatment, various inorganic salts (Singh and Tabatabai, 1978), phosphate ions and divalent anions that have been shown to interact with the active site (Alexander and Volini, 1987; Bordo et al., 2001), soil texture, temperature, and soil water potential (Ray et al., 1985; Deng and Dick, 1990) as well as season (Szajdak, 1996). Higher rhodanese activity values were found in rhizosphere soils compared with nonrhizosphere soils (Ray et al., 1985), and in soil under continuous rye cropping compared with soil under crop rotation (Szajdak, 1996). Furthermore, rhodanese activity was significantly increased in soils subjected to atmospheric S depositions (Press et al., 1985). Studies on the effect of S fertilizers revealed contrasting results, ranging from an increase in rhodanese activity after long-term application of S^0 fertilizers (Lawrence and Germida, 1988), to a decrease in the activity after soil amendment with S^0 fertilizers, K_2SO_4, or Na_4SO_4 (Deng and Dick, 1990). The latter authors suggested that rhodanese activity in these soils is suppressed by SO_4^{2-}, i.e., the end product of S^0 oxidation.

Rhodanese activity in soils was decreased by air-drying and, generally, by increasing storage time in both air-dried (stored at room temperature) and field-moist soils (stored at 4°C) (Singh and Tabatabai, 1978; Deng and Dick, 1990).

7–4.2 Principle

The method described is based on spectrophotometric determination of SCN^- production due to rhodanese activity when a soil sample is incubated with buffered (pH 6.0) $Na_2S_2O_3$ and KCN solutions and toluene. The soil–buffer–substrate mixture is incubated at 37°C for 1 h. The SCN^- produced is extracted by filtration after the addition of $CaSO_4^-$ formaldehyde solution. The method to measure

SCN^- production in soil is based on the reaction of SCN^- with Fe^{3+} in an acidic medium forming a colored Fe–SCN complex that is stable for at least 1 h. The formaldehyde present in the extractant stops rhodanese activity and prevents the formation of a blue-colored $Fe-S_2O_3^-$ complex.

No SCN^- formation is expected if $Na_2S_2O_3$ and KCN solutions are incubated without soil under the described assay conditions. However, copper ions catalyze the reaction of $S_2O_3^{2-}$ and CN^- resulting in the formation of SCN^- (Nor and Tabatabai, 1975, 1976). Therefore, autoclaved soil samples should be included as controls in the assay for rhodanese activity to account for potential SCN^- formation through a chemical pathway.

7–4.3 Assay Method (Tabatabai and Singh, 1976)

7–4.3.1 Apparatus

- Erlenmeyer flasks (50 mL) fitted with No. 2 stoppers
- Whatman No. 2 or 2v (folded) filter paper (Thermo Fisher Scientific, Inc.)
- Autoclave
- Incubator
- Funnels
- Spectrophotometer

7–4.3.2 Reagents

1. Toluene, Fisher certified reagent (Fisher Scientific Co., Chicago).
2. Tris(hydroxymethyl)aminomethane–sulfuric acid (THAM–H_2SO_4) buffer (0.05 M, pH 6.0): Dissolve 6.1 g of THAM (Fisher certified reagent, Fisher Scientific Co., Chicago) in about 600 mL of DI water, bring the pH of the solution to 6.0 by titration with approximately 0.2 M H_2SO_4, and dilute with DI water to 1 L.
3. Sodium thiosulfate ($Na_2S_2O_3$) solution (0.1 M): Dissolve 24.9 g of $Na_2S_2O_3$ and 6.1 g of tris(hydroxymethyl)aminomethane (THAM) in about 600 mL of DI water, bring the pH of the solution to 6.0 by titration with approximately 0.2N sulfuric acid (H_2SO_4) and dilute with DI water to 1 L.
4. Potassium cyanide (KCN) solution (0.1 M): Dissolve 6.5 g of KCN and 6.1 g of tris(hydroxymethyl)aminomethane (THAM) in about 600 mL of DI water, bring the pH of the solution to 6.0 by titration with approximately 0.2 N sulfuric acid (H_2SO_4), and dilute with DI water to 1 L.
5. Formaldehyde solution (37%), (Fisher certified reagent).
6. Calcium sulfate ($CaSO_4$)–formaldehyde solution: Dissolve 2.2 g of $CaSO_4\cdot2H_2O$ in about 900 mL of DI water, and adjust the volume to 1 L with DI water. Mix 100 mL of formaldehyde with 900 mL of the calcium sulfate solution.
7. Ferric nitrate–nitric acid solution [0.25 M Fe $(NO_3)_3\cdot9H_2O$-3.1 M HNO_3]: Dissolve 50 g of ferric nitrate nanohydrate in 100 mL of concentrated HNO_3 (analytical reagent grade, sp. grade 1.42), add this solution to 300 mL of DI water, and adjust the volume to 500 mL with DI water.
8. Standard thiocyanate stock solution (20 mM): Dissolve 1.621 g of sodium thiocyanate (NaSCN) in about 800 mL of DI water, and dilute the solution to 1 L with DI water. One liter of this solution contains 20 mM of NaSCN.

7–4.3.3 Procedure

Place 4 g of soil (<2 mm) in a 50-mL Erlenmeyer flask, add 0.5 mL toluene, 8 mL THAM–H_2SO_4 buffer (0.05 M, pH 6.0), 1 mL $Na_2S_2O_3$ (0.1 M), and 1 mL KCN (0.1 M), and swirl the flask for a few seconds to mix the contents. Stopper the flask and place it in an incubator at 37°C. After 1 h, remove the stopper, add 10 mL $CaSO_4$–formaldehyde solution, swirl the flask for a few seconds, and filter the soil suspension through a Whatman No. 2 (or 2v) filter paper. Pipette 5 mL filtrate into a test tube, add 1 mL ferric nitrate reagent, and measure the intensity of the reddish-brown color of the filtrate with a spectrophotometer adjusted to a wavelength of 460 nm. Calculate the thiocyanate concentration of the filtrate by reference to a calibration graph prepared from thiocyanate standard solutions containing 0, 0.2, 0.4, 0.6, 0.8, and 1.0 mM of thiocyanate. To prepare this graph, dilute 0, 2, 4, 6, 8, and 10 mL of the standard thiocyanate stock solution to 1 L in volumetric flasks, and mix the solutions thoroughly. Then pipette 5-mL aliquots of these diluted standards into test tubes, and proceed as described for analysis of the thiocyanate in the soil filtrate. If the color intensity of the filtrate exceeds the limit of the calibration graph, take a smaller aliquot of the filtrate and adjust the volume to 5 mL with DI water before adding 1 mL ferric nitrate reagent until the spectrophotometer reading falls within the limits of the graph.

Controls should be performed with each soil analyzed to account for color not derived from thiocyanate produced by rhodanese activity. To perform controls, follow the procedure described for the assay of rhodanese activity but use autoclaved soil samples and add the CN^- solution after incubation together with the addition of $CaSO_4$–formaldehyde solution immediately before filtration of the soil suspension.

7–4.4 Comments

Because of the significance of rhodanese to human and animal toxicology, semi-automated assay methods have been developed for rhodanese activity for these biological systems, including spectrofluorimetric methods, the use of ion-selective electrodes, and sequential capillary zone electrophoresis (Glatz et al., 1999). In contrast, studies on rhodanese activity in soil are limited and entirely based on the method described here with no, or only minor, modifications (Ray et al., 1985; Deng and Dick, 1990; Dick and Deng, 1991; Szajdak, 1996). A few authors addressed the contribution of rhodanese to overall S oxidation in soil (Deng and Dick, 1990; Dick and Deng, 1991). These authors concluded that the reaction catalyzed by rhodanese may not be the rate-limiting step in S^0 oxidation in selected Oregon and Chinese soils, and postulated the existence of several S^0 oxidation pathways in these soils. Although S deficiencies have become common in recent years in worldwide cropping systems (McGrath and Zhao, 1995; Niknahad Gharmakher et al., 2009), there has been little conceptualization to assist in developing effective S fertilization strategies for these cropping systems. The relationship between S^0 oxidation and rhodanese activity is unknown for a large number of soil types and regions. A sound understanding of S transformation processes and factors controlling these is essential for predicting soil S supply and S fertilizer management in agroecosystems.

7–5 SULFUR OXIDATION

7–5.1 Introduction

Inorganic S exists in a number of oxidation states in soils ranging from –2 in sulfides (S^{2-}) and its derivates, to 0 in elemental S (S^0), +4 in sulfites (SO_3^{2-}), and +6 in sulfates (SO_4^{2-}), and undergoes similar types of chemical reactions and biological transformations as nitrogen, including oxidation and reduction reactions. Major forms of inorganic S in soil include S^0, sulfides (S^{2-}), and SO_4^{2-} (Tabatabai, 2005).

In recent years, deficiencies of S in crops have increased worldwide due to the use of high-analysis low-S fertilizers, low S returns with farmyard manure, high-yielding varieties and intensive agriculture, declining use of S-containing fungicides, and reduced atmospheric S depositions (Chen et al., 2008a). This has increased the interest in the use of S-containing fertilizers and by-products from industrial processes in agriculture.

Elemental S is an important fertilizer, which must be oxidized to plant-available SO_4^{2-}. Sulfur oxidation in soils is a result of chemical and microbial reactions, but microbial reactions dominate the process (Germida, 2005). The nature of the biological oxidation of S in soils has not been clearly delineated, but the following sequence of reactions has been suggested, although some of the products may result from chemical reactions:

$$S^0 \rightarrow S_2O_3^{2-} \rightarrow S_4O_6^{2-} \rightarrow SO_4^{2-} \qquad [9]$$

Elemental S $\rightarrow$ thiosulfate $\rightarrow$ tetrathionate $\rightarrow$ sulfate $\qquad [10]$

Biological oxidation of S can be performed by a diverse group of microorganisms capable of oxidizing S under aerobic or anaerobic conditions and light or dark conditions (Laishley and Bryant, 1987). These can be divided into three main groups: (i) chemoautotrophs (lithotrophs), including members of the genus *Acidithiobacillus*, previously known as *Thiobacillus* (Kelly and Wood, 2000); (ii) photoautotrophs, including species of purple and green S bacteria; and (iii) chemoheterotrophs (organotrophs), including species of the genera *Bacillus*, *Arthrobacter*, *Flavobacterium*, *Achromobacter* and *Pseudomonas* (Germida, 2005; Pisz, 2008). Heterotrophic S oxidizers were found to be the most abundant S oxidizers in agricultural soils (Lawrence and Germida, 1991). The same authors reported that Canadian agricultural soils supplemented with S^0 beads revealed a shift in microbial populations toward $S_2O_3^{2-}$–oxidizing autotrophs, while S^0–oxidizing autotrophs, such as *Acidithiobacilli*, were not detected.

The biochemical processes involved in S oxidation differ among different S-oxidizing organisms, but they all involve one of two main pathways, the polythionate ($S_xO_6^{2-}$) or the nonpolythionate pathway (Kelly et al., 1997; Druschel et al., 2003). The oxidation of $S_2O_3^{2-}$ through the nonpolythionate pathway results in the formation of SO_4^{2-} via SO_3^{2-}, catalyzed by either a sulfite-oxidizing enzyme (Fig. 7–2, Table 7–2) or adenosine phosphosulfate reductase (APS) (Suzuki et al., 1992). This reaction is the major energy-generating reaction for the autotroph organisms *Thiobacillus denitrificans* and *T. thioparus*. Alternatively, a second nonpolythionate pathway exists in some facultative autotrophs, such as *Paracoccus* (or *Thiobacillus*) *versutus*, *P. denitrificans*, and possibly *T. novellas* and *Xanthobacter* spp., in which $S_2O_3^{2-}$ is oxidized by the periplasmatic thiosulfate-oxidizing multienzyme system

(TOMES) (Fig. 7–2, Table 7–2), which produces SO_4^{2-} without detectable intermediates (Kelly et al., 1997). The polythionate pathway involves the combination of two molecules of $S_2O_3^{2-}$ to form $S_4O_6^{2-}$ (Kelly et al., 1997). Cleavage of $S_4O_6^{2-}$ by tetrathionate hydrolase (Fig. 7–2, Table 7–2) produces $S_2O_3^{2-}$, S, and SO_4^{2-} (Meulenberg et al., 1993; De Jong et al., 1997). Under sulfite-accumulating conditions, $S_4O_6^{2-}$ converts to $S_2O_3^{2-}$ and $S_2O_6^{2-}$, which can be hydrolyzed by trithionate hydrolase (Fig. 7–2, Table 7–2) to $S_2O_3^{2-}$ and SO_4^{2-} (Suzuki, 1999). Microorganisms utilizing the polythionate pathway include autotrophic bacteria species in the genera *Thiobacillus* and *Acidithiobacillus*.

The influence of S^0 applications on S-oxidizing *Thiobacilli* and heterotrophic microorganisms has been studied comprehensively and a wide spectrum of S-oxidizing microorganisms was identified in soils, including 273 different bacteria, such as *Arthrobacter*, *Bacillus*, *Micrococcus*, and *Pseudomonas*, some actinomycetes, and 70 fungi (Germida, 2005). These microorganisms may (i) oxidize S and produce mainly $S_2O_3^{2-}$, (ii) oxidize S and produce mainly SO_4^{2-}, and (iii) oxidize $S_2O_3^{2-}$ and produce SO_4^{2-}. Although the pathways by which these inorganic S compounds are formed are still largely unknown, the evaluation of the concentrations and distribution of these compounds in various soil–plant systems is important for a better understanding of soil–crop interactions (Haneklaus et al., 2007). Additionally, $S_2O_3^{2-}$, also a known nitrification inhibitor, may inhibit germination and growth of certain plant species (Audus and Quastel, 1947). These authors found that $S_2O_3^{2-}$ at a concentration of 250 mg L^{-1} S inhibited root growth of garden pea (*Pisum sativum* L.) by 50%, while $S_2O_3^{2-}$ applied as ammonium thiosulfate did not affect growth and yield of Bermudagrass (*Cynodon dactylon* L. Pers.) (Sloan and Anderson, 1998). Thiosulfate also inhibits NH_4^+ oxidation by heterotrophic nitrifiers, and thus contributes to the production of NO and N_2O with known adverse effects on the atmospheric chemistry and climate (Saad et al., 1996). Tetrathionate has more recently been shown to be the actual inhibitor of soil urease rather than $S_2O_3^{2-}$ itself (Sullivan and Havlin, 1992). Turnover of these inorganic S compounds in soil may also result in the production of small amounts of volatile organic S compounds, such as carbon disulfide (CS_2), dimethyl sulfide (CH_3SCH_3), or dimethyl disulfide (CH_3SSCH_3) (Saad et al., 1996; Yang et al., 1996).

Studies by Nor and Tabatabai (1977) showed that the amount of $S_2O_3^{2-}$, $S_4O_6^{2-}$, and SO_4^{2-} produced during incubation (5, 15, and 30°C) of S^0–amended soils under aerobic conditions over 70 d varied with time and soil type. They showed that $S_2O_3^{2-}$ was produced within the first few days of incubation and that SO_4^{2-} accumulated in some soils. The rate of S oxidation increased with increasing temperature and S^0 application rate (50, 100, and 200 mg of S^0 kg^{-1} soil). For 100 mg of S^0 kg^{-1} soil, the S^0 oxidation rate in 10 Iowa surface soils ranged from 39 to 75 mg of S^0 kg^{-1} soil after incubation at 30°C for 70 d. Changes in soil pH were small among the three S^0 application rates (50 to 200 mg of S^0 kg^{-1} soil). The S^0 oxidation rate was lower in acid soils than in alkaline soils and in air-dried soils than in field-moist soils (Nor and Tabatabai, 1977).

In S^0–amended soils, S oxidation rates depend on soil type (particularly clay content), soil organic matter content, soil pH, soil moisture and temperature, total microbial biomass, population size and activity of S-oxidizing microorganisms, as well as S^0 particle size and application rate (Lawrence and Germida, 1988; Watkinson and Bolan, 1998; Haneklaus et al., 2007). This has implications for the applicability of S^0 oxidation rates for soils obtained under laboratory conditions into prediction of S^0 oxidation rates in the field. Since direct contact between microorganisms and S^0 particles is necessary for S oxidation, the S^0 oxidation rate is directly correlated to the surface area of the S^0 particles. Previous studies have shown that the use of S^0 particles in the range of 0.106 to 0.150 mm can minimize the effect of S^0 particle size on S^0 oxidation rate (Janzen and Bettany, 1987a). The effect of soil moisture on S^0 oxidation is highly dependent on the soil type, with optimum water potentials of −0.02 MPa in silty clay or silt loam soils, −0.01 MPa in a sandy loam soil, and −0.1 MPa in a sandy soil (Deng and Dick, 1990). Optimum temperature for S^0 oxidation ranged from 25 to 30°C under aerobic conditions (Waksman and Joffe, 1922; Li and Caldwell, 1966; Nor and Tabatabai, 1977; Janzen and Bettany, 1987b; Deng and Dick, 1990). Li et al. (2005) reported that repeated applications of S^0 increased the oxidation rate due to a higher number of S-oxidizing microbial populations and thus a substantially higher oxidation potential.

Rate of S^0 oxidation in soils can be determined by measuring the decrease in S^0 concentrations using titration following S^0 extraction from soil with organic solvents and reduction to H_2S or directly by high performance liquid chromatography (HPLC) or inductively coupled plasma atomic emission spectroscopy (ICP–AES) (Zhao et al., 1996). In most studies, however, S^0 oxidation rate is determined by measuring SO_4^{2-} production during a defined time period after S^0 application to soils. Few studies have been conducted that differentiate oxidation rates of S^0 products differing in size and shape. Janzen and Bettany (1987a) suggested that S^0 oxidation rate should be expressed per unit of S^0 surface area rather than per mass basis, because oxidation rate is influenced by S^0 particle size. Their model further accounted for the progressive decrease in S^0 particle size during oxidation. The decline in surface area over time can be mathematically determined if it is assumed that the S^0 particles are relatively uniform spheres. Janzen and Bettany (1987a) derived the following equations to calculate the proportion of S^0 oxidized and the rate constant:

$$m/m_0 = 1 - [1 - 2\,kt/g\,D_0]^3 \qquad [11]$$

where m/m_o is the proportion of S^0 oxidized as a function of a rate constant (k), time (t), density (g), and initial particle diameter (D). Rearrangement of Eq. [11] gives:

$$k = \{1 - [1 - m/m_0]^{1/3}\} \cdot gD_0/2t \qquad [12]$$

The rate constant calculated from Eq. [12] has the desired units (g S m^{-1} d^{-1}) and is independent of both particle diameter and time.

7–5.2 Principle

Estimation of S^0 oxidation rate in soils involves determination of SO_4^{2-} produced when soil is incubated with S^0 under specific conditions. The variables that should be considered include particle size of S^0, time, and temperature of incubation. These variables must be selected so that sufficient SO_4^{2-} is produced to ensure accurate analytical determination, SO_4^{2-} production from organic S mineralization is minimized, acidification of the soil sample is prevented, and S^0 is applied accurately (Janzen and Bettany, 1987a).

Any particle size of S^0 can be used in this assay, but generally S^0 oxidation rate increases with decreasing particle size. If very fine (e.g., <150 mesh, 106 μm) S^0 is used, it can be diluted with the similar-size glass beads to facilitate weighing (Nor and Tabatabai, 1977). Also, relatively coarse (0.199 mm) S^0 can be used successfully for estimation of S^0 oxidation rate in soils (Janzen and Bettany, 1987a; Deng and Dick, 1990). According to Janzen and Bettany (1987a), this relatively coarse material has the following advantages over finer particle sizes: (i) it can be commercially produced with small particle size variations; (ii) its particle size can be accurately confirmed by microscopic analysis; and (iii) because of its lower specific surface area, coarse S^0 can be applied at higher rates without adversely affecting the oxidation rate due to soil acidification. Therefore, this method is convenient and precise for even small soil samples.

7–5.3 Assay Method (Nor and Tabatabai, 1977; Janzen and Bettany, 1987a; Deng and Dick, 1990, Dick and Deng, 1991)

7–5.3.1 Apparatus

- Extraction bottles (Nalgene bottles fitted with caps or French square bottles fitted with No. 7 stoppers) (250 mL)
- Whatman No. 42 filter papers (Thermo Fisher Scientific, Inc.)
- GA-8 membrane filters (47-mm diameter, 0.20-mm pore size, Gelman Instrument Co., Ann Arbor, MI)
- Vacuum filter system
- Incubator
- Funnels
- pH glass electrode
- Spectrophotometer or gas chromatograph

7–5.3.2 Reagents

1. Elemental S: Sublimed S^0 (J.T. Baker Chemical Co., Phillipsburg, NJ)
2. Elemental S^0 powder (0.119 mm): Finely grind S^0 particles and collect the particles that passed through a 0.150-mm sieve, but were retained on a 0.106-mm sieve. Wet-sieve sieved S^0 powder in ethanol to ensure complete removal of fine particles adhering to 0.119-mm particles. Rinse the ethanol-washed S^0 several times with deionized water and dry overnight at 65°C.
3. Glass beads (0.100 mm): Can be obtained from Fisher Scientific (Thermo Fisher Scientific, Inc.).
4. Calcium phosphate monohydrate solution [$Ca(H_2PO_4)_2 \cdot H_2O$] (500 mg P L^{-1}): Dissolve 2.02 g of $Ca(H_2PO_4)_2 \cdot H_2O$ in about 700 mL of deionized water, and bring to volume of 1L with deionized water.

7–5.3.3 Procedure

Amend a 20-g sample of field-moist soil (<2 mm) with S^0 in a 250-mL extraction bottle to achieve the desired S^0 concentration (for example, 500 mg S^0 kg^{-1} soil), and adjust soil moisture content to 60% of water-holding capacity. For convenience, the required amount of S^0 (0.199-mm or finer particle size) can be mixed with glass beads of similar particle size. Close the extraction bottle and incubate for the time and temperature necessary to produce detectable concentrations of inorganic S compounds. It is recommended to incubate the S^0–amended soil samples for at least 6 d at a temperature between 25 and 30°C (Deng and Dick, 1990). Open and aerate the bottle every 2 d to maintain aerobic incubation conditions. After incubation, remove stopper and add 50 mL water (soil/water ratio 1:2.5), and mix thoroughly with the soil. After 45 min measure the pH of the soil by a glass electrode. Add an additional 100 mL of 500 mg P L^{-1} as $Ca(H_2PO_4)_2$ to each bottle, stopper the bottle, and shake it on an end-to-end shaker for 1h. Filter the soil extractant through a Whatman No. 42 filter paper, followed by filtration through a GA-8 membrane filter under vacuum for a complete removal of any particulate matter in the soil extract. Analyze this filtrate for $S_2O_3^{2-}$ and $S_4O_6^{2-}$ by the spectrophotometric method described by Nor and Tabatabai (1976), or total inorganic S ($S_2O_3^{2-}$, $S_4O_6^{2-}$, and SO_4^{2-}) by the methylene blue method (Tabatabai, 1996), or for SO_4^{2-} by ion chromatographic method (Tabatabai and Frankenberger, 1996). The SO_4^{2-} production in a S^0–amended soil during a specific time period can be calculated by subtracting the sum of ($S_2O_3^{2-}$ + $S_4O_6^{2-}$) from the total inorganic S ($S_2O_3^{2-}$ + $S_4O_6^{2-}$ + SO_4^{2-}).

7–5.4 Comments

In an attempt to further advance the model by Janzen and Bettany (1987a) to predict S^0 oxidation rates in soils, McCaskill and Blair (1989) proposed a model based on particle size, soil temperature, and moisture. However, both models require knowledge of the soils' potential to oxidize S^0. The S^0 used in most studies with the goal to predict or compare S^0 oxidation rates between different soils has been finely ground and sieved to a uniform particle size, which is not representative of most commercial S^0 products. Slaton et al. (2001) characterized the oxidation rate of several commercial S^0 sources, including soil pH and EC changes due to S^0 oxidation in an alkaline silt loam soil. Their results suggest that commercial S^0 products have different oxidation rates, which needs to be considered in developing recommendations to growers.

Air-drying of field-moist soils reduced the rate of S^0 oxidation, but the effect of air-drying decreased with increasing incubation time. Results of the effect of autoclaving on S^0 oxidation in soil were inconclusive and deserve further investigation to differentiate between chemical and microbial reactions during this process in soil.

He et al. (1994) measured S oxidation and distribution in four different soils using labeled S^0 (^{35}S-S^0, sulfur flowers), and reported that the half-life ($t_{1/2}$) of the ^{35}S-S^0 ranged from 12 to 176 d in a paddy soil and an Ultisol, respectively. By the end of the 56-d experiment, generally 54 to 70% of the oxidized labeled S in the soil was present in the SO_4^{2-} fraction, while the remaining oxidized labeled S was presumed to have been converted into organic S fractions.

More recently, a noninvasive technique, i.e., S K-edge x-ray absorption near-edge structure spectroscopy (XANES), was used to identify different oxidation states of organic S compounds in humic substances and their relative quantities in a variety of geochemical samples (Xia et al., 1998). The potential of this method to evaluate the distribution and dynamic of S in terrestrial ecosystems deserves further investigation.

7–6 SULFATE REDUCTION

7–6.1 Introduction

Although sulfate (SO_4^{2-}) is the most common, stable form of inorganic S in well-drained and aerated soils, SO_4^{2-} reduction is a major component of the S cycle in soils exposed to waterlogging or periodic flooding, especially where sufficient amounts of organic matter are available, as well as in salt marshes and anaerobic lake sediments (Germida, 2005). Under these conditions, SO_4^{2-} can be reduced to sulfides (S^{2-}), sulfites (SO_3^{2-}), thiosulfates ($S_2O_3^{2-}$), polysulfides (S_n^{2-}), and elemental S (S^0). Sulfate reduction in soils is a result of chemical and microbial reactions, the latter respiring a variety of inorganic S compounds as anaerobic terminal electron acceptors. The term "sulfate-reducing bacteria" is conventionally reserved for groups of microorganisms that conduct assimilatory or dissimilatory SO_4^{2-} reduction. In this process, SO_4^{2-} acts as an oxidizing agent for the degradation of organic matter, similar to O_2 in common respiration processes. A small amount of reduced S is assimilated by the microorganisms to meet their S requirements (e.g., assimilatory SO_4^{2-} reduction). However, the majority of SO_4^{2-} is reduced by microorganisms to meet their energy requirements (e.g., dissimilatory SO_4^{2-} reduction) in a pathway that can be described as follows:

$$4\,H_2 + SO_4^{2-} \rightarrow 4\,H_2O + S^{2-} \qquad\qquad [13]$$

$$\text{Hydrogen + sulfate} \rightarrow \text{water + sulfide} \qquad\qquad [14]$$

Dissimilatory SO_4^{2-} reduction generates large quantities of biogenic S gases—such as hydrogen sulfide (H_2S), dimethylsulfide (DMDS), carbon disulfide (CS_2), and dimethyldisulfide (DMDS)—which are released into the environment (Postgate, 1979; Germida, 2005). Dissimilatory SO_4^{2-} reduction is commonly viewed as a strictly anaerobic process, and under these conditions, mediated by bacteria of the genera *Desulfobacter*, *Desulfobulbus*, *Desulfococcus*, *Desulfonema*, *Desulfosarcina*, *Desulfotomaculum*, and *Desulfovibrio* (Trudinger, 1986; Germida, 2005). Studies by Canfield and Des Marais (1991) measuring bacterial SO_4^{2-} reduction and dissolved O_2 in hypersaline bacterial mats from Baja California, Mexico, revealed that SO_4^{2-} reduction occurred within the well-oxygenated photosynthetic zone of the mats, which suggests that this process also occurs under aerobic conditions.

Accurate determination of reduced inorganic S species in soils is often difficult because of their rapid oxidation on exposure to aerobic conditions, their low concentration, the presence of interfering compounds, and the limitations in current analytical techniques for soil analysis. Gilboa-Garber (1971) proposed a method for direct spectrophotometric determination of inorganic S^{2-} in biological materials, involving the colorimetric determination of S^{2-} as methylene blue complex after extraction with Zn acetate. Methylene blue is formed during the reaction of H_2S with an acid solution of *p*-aminodimethylaniline in the presence of Fe^{3+}. Howarth et al. (1983) modified this method for S^{2-} determination in pore waters of core samples of a salt marsh. Smittenberg et al. (1951) proposed a method for determination of S^{2-}, SO_3^{2-}, $S_2O_3^{2-}$, S_n^{2-}, S^0, and organically bound S in soils, involving digestion of a soil sample with 6 M HCl for determining monosulfidic S species, and digestion with Sn in HCl for determining total oxidizable S. The H_2S evolved is determined spectrophotometrically as methylene blue resulting in the determination of acid-volatile sulfides (AVS). Accuracy of these methods is low, because digestion of soil with HCl does not release S^{2-} in metal sulfides (acid-insoluble fraction), and thus underestimates S^{2-} pools. Digestion with Sn in HCl releases AVS from organic S forms, and thus overestimates the total oxidizable S (Melville et al., 1971). Allen and Parkes (1995) conducted a detailed study to determine recovery of different metal sulfides and forms of reduced S from biological materials and sediments by various digestion procedures. Their study indicates that hot AVS digestion, and cold followed by hot chromous chloride digestion produced best recoveries.

Sulfate reduction in salt marsh sediments was measured involving the use of $^{35}SO_4^{2-}$ as a tracer and the digestion of reduced S compounds (e.g., pyrite and S^0) with Cr(II) to H_2S (Howarth and Merkel, 1984). The reduction of reduced S compounds with Cr(II) improved selectivity and sensitivity of the method relative to oxidation of these S compounds to SO_4^{2-} with aqua regia (e.g., HNO_3 plus HCl). The Cr(II) reduction method has been employed by researchers for measuring chromium-labile sulfides (CLS) in marine and freshwaters sediments for over 15 yr, but was shown to be inappropriate for measuring CLS in oxic freshwaters in which SO_4^{2-} concentrations are large relative to the dissolved metal sulfides. Under these conditions SO_4^{2-} is reduced by Cr(II), which represents a significant interference (Mylon et al., 2002).

A polarographic method for determination of the reduced S species, including $S_2SO_3^{2-}$, S^{2-}, HS^-, and S_n^{2-} in marine pore water samples based on differential pulse determination using a mercury electrode was first proposed by Luther et al. (1985). These authors considered the S_n^{2-} to be composed of one S atom and in the -2 oxidation state, S^{2-}, and the remaining $(n - 1)$ S atoms in the zero-valent (0) oxidation state. The number of S atoms in each S species is measurable by this technique. Although this method also enables the measurement of $S_2SO_3^{2-}$ and other polythionates, these reduced S species were not detectable in pore water samples. Furthermore, the authors showed that salt marsh and subtidal pore water profiles contain significant concentrations of $S_2SO_3^{2-}$, HS^-, and S_n^{2-}. This technique has been successfully used in measuring seasonal cycling of S and Fe in pore water samples of a Delaware salt marsh (Luther and Church, 1988), and more recently, S^{2-} concentration and distribution in laminated sediment samples from the Santa Barbara Basin (Bernhard et al., 2003).

A highly sensitive HPLC method was developed for the measurement of reduced inorganic S compounds in small sample volumes of less than 50 μL (Rethmeier et al., 1997). In this method, the combination of fluorescence labeling of reduced inorganic S compounds such as S^{2-}, SO_3^{2-}, and $S_2O_3^{2-}$ with monobromobimane followed by an extraction of S^0 by chloroform treatment enabled the detection of all these S species as well as SO_4^{2-} (remaining aqueous phase) in the same sample. While the derivatized S compounds were detected by fluorescence emission at 480 nm, S^0 is identified by UV absorption at 263 nm. Sulfate in the remaining aqueous phase is detected by HPLC with indirect UV detection at 254 nm. Detection ranges for the different S compounds examined were as follows: S^{2-} (5 μM to 1.5 mM), SO_3^{2-} (5 μM to 1.0 mM), $S_2O_3^{2-}$ (1 μM to 1.5 mM), S^0 (2 μM to 32 mM) and SO_4^{2-} (5 μM to >1 mM).

Although the methods for detection of sulfate reduction are used mainly in studies of marine and freshwater sediments, and forms and concentration of reduced S compounds in these environments are markedly different from those in soils (Bashkin and Howarth, 2002), they will be described here because of their potential use in studies on soils developed from marine sediments, paddy soils, and soils exposed to periodic flooding.

7–6.2 Principles

Estimation of sulfate reduction involves addition of SO_4^{2-} into a homogenized or undisturbed sediment core or soil sample, reduction with Cr(II) in an acid solution [Cr(II) reduction, e.g., single-step method], and spectrophotometric determination of reduced S compounds as H_2S released after formation of the methylene blue complex. For information about how the methylene blue complex is formed, see Section 7–5. If $^{34}SO_4^{2-}$ is used as a tracer, rate of incorporated ^{34}S into reduced S compounds (e.g., H_2S) is measured using a scintillator. The single-step method results in the chromium-labile sulfide (CLS) pool, which comprises H_2S, S^0, FeS, and FeS_2. Alternately, sulfate reduction can be measured by a two-step method in which acid volatile sulfides (AVS) (e.g., H_2S + FeS) and the remaining CLS (S^0 + FeS_2) are sequentially reduced from the sample. If stable isotope technology is used, the fraction of $^{34}SO_4^{2-}$ reduced during incubation is calculated from the sum of ^{34}S in the AVS and CLS pools. The single-step method is simpler and faster than the sequential reduction of AVS and CLS pools. Because the Cr(II) reduction (single-step) method showed higher selectivity and accuracy, and is less time-consuming than the sequential reduction (two-step) method, the Cr(II) reduction method will be described here.

7–6.3 Assay Method (Fossing and Jørgensen, 1989)

7–6.3.1 Apparatus

- Modified Johnson–Nishita apparatus (Tabatabai, 1996), the apparatus described by Canfield et al. (1986), or that described by Zhabina and Volkov (1978)
- Scintillation counter

7–6.3.2 Reagents

1. Zinc acetate [Zn(OAc)$_2$], 20%, and zinc acetate [Zn(OAc)], 5% in 1% acetic acid.
2. Nitrogen gas.
3. Aminodimethylaniline solution: Dissolve 2 g p-aminodimethylaniline sulfate (Eastman Kodak Co., Rochester, NY) in about 1500 mL di-ionized (DI) water in a 2-L volumetric flask. Place the flask in a pan of cold water, and slowly add 400 mL concentrated reagent-grade sulfuric acid (H_2SO_4). This acid should be added in small portions, mixing the solution and allowing it to cool after each portion. Cool and adjust volume to 2 L with DI water.
4. Ferric ammonium sulfate solution: To 25 g ferric ammonium sulfate [Fe$_2$(SO$_4$)$_3$ (NH$_4$)$_2$SO$_4$· 24 H$_2$O], add 5 mL concentrated reagent-grade sulfuric acid and 195 mL DI water. Allow to stand at room temperature until dissolved. (It can take several days before all ferric ammonium sulfate is dissolved.)
5. Concentrated HCl: HCl, 0.5 M and HCl, 12 M.
6. Ethanol (95%).
7. Methanol.

8. Antifoam (Sigma Aldrich Corp., St. Louis, MO).

9. Scintillation fluid (Dynagel, Baker Chemical).

10. Chromic chloride hexahydrate ($CrCl_3.6H_2O$): Prepare the highly reactive Cr(II) solution from the more stable Cr(III) by percolating 1 M $CrCl_3.6H_2O$ in 0.5 M HCl through a Jones reductor with amalgamated Zn granules:

$$2\,Cr^{3+} + Zn \rightarrow 2\,Cr^{2+} + Zn^{2+} \tag{15}$$

An efficient reduction is verified by a color change from dark green [Cr(III)] to bright blue [Cr(II)]. For construction of the Jones reductor, see Kolthoff and Sandell (1963, p. 569). The Jones reductor can be constructed from a glass column (40 cm long, 1.5 cm i.d.) with an integral sinter at the bottom and stopcocks in both ends (Fossing and Jørgensen, 1989). Wash the granular Zn (0.3–1.5-mm grain size) three times with 1 M HCl and twice with DI water. Amalgamate the Zn for a few minutes with a saturated solution of $HgCl_2$ (0.25 M), transfer to the glass column, and wash it with three column volumes of 0.5 M HCl. Remove the reduced Cr solution by suction into a large polyethylene syringe in which it can be stored under reduced conditions (bright blue) for several weeks. Otherwise, the Cr(II) solution should be freshly prepared every 2 to 3 d and stored in a ground-glass stoppered bottle (Canfield et al., 1986).

An alternate and faster method of producing large volumes of Cr(II) solution is recommended by Fossing and Jørgensen (1989). In this method, fill a glass bottle with 1 M HCl-rinsed "mossy zinc" (Aldrich Chemicals, Milwaukee, WI) and then fill the bottle with the Cr(III) solution (1 M $CrCl_3.6H_2O$ in 0.5 M HCl) under continuous flow of N_2. The mossy zinc does not need to be amalgamated. Chromium(III) is reduced to Cr(II) within 10 min and is kept under N_2 until it is drawn into syringes through an outlet at the bottom of the bottle. After use, the mossy Zn can be regenerated by washing with 1M HCl.

7–6.3.3 Procedure

Sulfate reduction can be measured in undisturbed sediment core (3 cm in diam.) by the core injection technique. In this method, a volume of 2 μL carrier-free $^{35}SO_4^{2-}$ (70 kBq) is injected at a specific depth into replicated sediment cores from each location. The sediment core is incubated for 18 to 24 h at room temperature (23°C), then stepwise extruded from the tube and cut into segments and transferred to 20 mL Zn acetate [20% Zn(OAc)] and frozen to terminate the reaction and fix the sulfides. Then reduced S compounds are distilled in the form of H_2S from the sediment–Zn(OAc) mixture, trapped as described below; ^{35}S radioactivity in H_2S and precipitated ZnS is determined. Sulfate concentration in pore water of sediments or soil samples can be determined turbidimetrically in replicated parallel core samples as described by Tabatabai (1974).

Segments from each depth interval are pooled from the replicated cores and homogenized. The homogenized sediment is centrifuged and $^{35}SO_4^{2-}$ radioactivity is determined in a subsample of the supernatant. The sediment pellet is washed twice with a salt solution (e.g., 0.1 M NaCl) to remove $^{35}SO_4^{2-}$. The washed sediment is homogenized and 1 to 2 g are transferred into a distillation flask (modified Johnson–Nishita apparatus, or apparatus described by Canfield et al. [1986] or by Zhabina and Volkov [1978]) and mixed with 5 mL DI water and 5 mL methanol.

The distillation flask is degassed for 20 min with N_2, 16 mL $CrCl_2$ (1M) in HCl (0.5 M) and 8 mL HCl (12 M) are added, and the sediment slurry is gently boiled for 40 min. During this distillation, the total reduced inorganic S is dissolved and distilled into $Zn(OAc)_2$, sequential traps containing 10 mL $Zn(OAc)_2$ (5%) buffered with 0.1% acetate and with a drop of antifoam.

More than 98% of the distilled H_2S is recovered as ZnS in the first trap. After distillation, the two traps are pooled, 5 mL are subsampled and mixed with 5 mL of scintillation fluid, and ^{35}S radioactivity is determined. The concentration of the CLS fraction is determined spectrophotometrically at 670 nm in an aliquot of the trapping solution after adding 2 mL ferric ammonium sulfate solution, 10 mL of p-aminodimethylaniline solution, and adjusting the volume to 100 mL with DI water.

The sulfate reduction rate is calculated according to the following equation:

$$SRR = \frac{[SO_4^{2-}] \, a \times 24 \times 1.06}{(A + a) \, h} \text{ nmol } SO_4^{2-} \text{ cm}^{-3} \text{ d}^{-1} \qquad [16]$$

where a is the total radioactivity of ZnS, A is the total radioactivity of $^{35}SO_4^{2-}$ after incubation, h is the incubation time in hours, $[SO_4^{2-}]$ is the sulfate concentration in nmol per cm^3 sediment, and 1.06 is a correction factor for the expected isotope fractionation (Jørgensen and Fenchel, 1974; Fossing and Jørgensen, 1989).

7–6.4 Comments

Comparison of the two methods by Fossing and Jørgensen (1989) showed that the single-step method resulted in greater (4–50%) SO_4^{2-} reduction rates than the two-step method. The difference was largest when the sediment had been dried after acid reduction (acid-volatile sulfate) and before Cr(II) reduction (chromium-labile sulfate). These authors also showed that SO_4^{2-} reduction rate was between 8 and 87% greater in the Cr(II) reduction (single-step) method if compared with the SO_4^{2-} reduction rate following the first step (acid reduction) of the two-step method. Recovery of $^{35}SO_4^{2-}$ reduced and incorporated into the chromium-labile sulfate pool was, on average, 10 to 15% in different marine sediments and salt marshes (for review, see Fossing and Jørgensen, 1989), but the percentage varied widely, depending on the type of sediment and the S chemistry. Because the pathways involved in ^{35}S incorporation into S^0 and FeS_2 during short-term incubation periods are unclear, only the ^{35}S incorporated into the entire chromium-labile sulfate pool gives a correct measure of the total SO_4^{2-} reduction rate.

More recently, a three-step sequential digestion procedure, after prior elemental S extraction, on ZnOAc-treated samples was used to distinguish three different reduced S species, namely, acid volatile sulfate, pyrite S, and S^0 (elemental S) from sediments (Duan et al., 1997). Modifications also have been made to the process of distillation and collection of released H_2S to retrieve enough precipitated S for later isotopic analysis. Pyrite S, as defined by this extraction scheme, is divided into a less mature fraction, as distilled by cold Cr reduction, and a mature fraction, by hot Cr(II) reduction. This method could be used as an index of the degree of the maturity of iron sulfide phases in recent sedimentary environments.

7–7 MODIFICATION OF *p*-NITROPHENOL-BASED ENZYME ACTIVITIES FOR ORGANIC SOILS OR FOREST ORGANIC HORIZONS (KLOSE ET AL., 2003)

7–7.1 Introduction

Enzyme assays that produce *p*-nitrophenol as the end product result in a yellow color that is determined colorimetrically. The standard solution used to kill the reaction and develop the yellow color is 0.5 M NaOH. However, in soil high in organic matter (e.g., Histosols) or in forest organic horizons, this can cause the release of humic acids that will develop a yellow color and interfere with the analysis of *p*-nitrophenol. Furthermore, high organic matter content in the reaction vessel can limit the availability of substrates to enzymes. The alternative method is outlined below for *p*-nitrophenol-based methods presented. If there is concern that soils are high in organic matter, a preliminary extraction may follow the standard procedure without adding the substrate and then the degree of color development is determined. If this has high colorimetric readings, then the procedure outlined here should be followed.

7–7.2 Principle

The modified method is for organic soils and organic horizons of forest soils for any enzyme assay that measures *p*-nitrophenol as the end product. The volume of the buffer–substrate solution is increased from 4 to 8 mL to ensure a thorough mixing of the soil–buffer–substrate suspension when working with soils that have high levels of extractable humic acids. The *p*-nitrophenol released is extracted by filtration after the addition of 0.05 M $CaCl_2$ and 0.1 M THAM buffer (pH 12). The THAM buffer (pH 12) is used instead of the 0.5 M NaOH to avoid the extraction of humic acids from organic soils. The latter interferes with colorimetrical determination of *p*-nitrophenol.

7–7.3 Assay Method

7–7.3.1 Apparatus

- Erlenmeyer flasks (50 mL), rubber stoppers (No. 2)
- Incubator (forced-air incubator, Isotemp with temperature uniformity ±0.5°C)
- Standard bench-top pH meter
- UV/Vis-Spectrophotometer (wavelength range 190–1100 nm)
- Whatman No. 2 or No. 2v filter paper

7–7.3.2 Reagents

1. Toluene, Fisher certified reagent (Fisher Scientific Co., Chicago).
2. Buffer stock solutions: Prepare as described for reagents in standard procedure.
3. pH-adjusted buffer solutions: Prepare as described for reagents in standard procedure.
4. *p*-Nitrophenyl substrate solutions, 0.05 M: Prepare as described for reagents in standard procedure.

5. Calcium chloride ($CaCl_2$), 0.5 M: Prepare as described for reagents in standard procedure.

6. Tris(hydroxymethyl)aminomethane (THAM) buffers, 0.1 M, pH 12: Dissolve 12.2 g of THAM in about 800 mL of water and adjust the pH of the solution to 12.

7. Standard p-nitrophenol solution: Prepare as described for reagents in standard procedure.

7–3.3.3 Procedure

Place a 2.5- or 1.25-g oven-dry equivalent sample of organic soils or organic horizons (<5 mm) in a 50-mL Erlenmeyer flask, add 0.2 mL of toluene, 8 mL of enzyme-specific buffer solution, 1 mL of p-nitrophenyl substrate solution made in the same buffer, and swirl the flask for a few seconds to mix the contents. Stopper the flask and place it in an incubator at 37°C. After 1 h, remove the stopper, add 1 mL of 0.5 M $CaCl_2$ and 4 mL of 0.1 M THAM buffer (pH 12), swirl the flask for a few seconds, and filter the soil suspension through a Whatman No. 2 or No. 2v filter paper. Measure the yellow color intensity of the filtrate with a UV/Vis-Spectrophotometer. Calculate the p-nitrophenol content of the filtrate by reference to a calibration graph prepared from results obtained with standards containing 0, 10, 20, 30, 40, and 50 μg of p-nitrophenol. To prepare these standards, dilute 1 mL of the standard p-nitrophenol solution to 100 mL in a volumetric flask and mix the solution thoroughly. Then pipette 0-, 1-, 2-, 3-, 4-, and 5-mL aliquots of this diluted standard solution into a 50-mL Erlenmeyer flask, adjust the volume to 9 mL by addition of water, and proceed as described for p-nitrophenol analysis of the incubated soil sample (i.e., add 1 mL of 0.5 M $CaCl_2$ and 4 mL of the 0.1 M THAM [pH 12], mix, and filter the suspension). If the color intensity of the filtrate exceeds that of 50 μg of the p-nitrophenol standard, an aliquot of the filtrate should be diluted with water until the spectrophotometer reading falls within the limits of the calibration graph. To account for the generally higher yellow color intensity of the filtrates from organic soils and organic horizons of forest soils relative to mineral soils, concentrations of standard solutions could be increased by 10-fold for enzyme assays in organic soils.

Controls should be performed with each soil and organic horizon to allow for color not derived from p-nitrophenol released by enzymatic reaction. To perform controls, follow the procedure as described above, but make the addition of the 1 mL p-nitrophenyl substrate solution after the addition of 0.5 M $CaCl_2$ and 4 mL of 0.1 M THAM (pH 12) (i.e., immediately before filtration of the soil suspension).

REFERENCES

Abeles, R.H. 1983. Suicide enzyme inactivators. Chem. Eng. News 61:48–56. doi:10.1021/cen-v061n038.p048

Abeles, R.H., and C.T. Walsh. 1973. Acetylenic enzyme inactivation. Inactivation of γ-cystathionase, in vitro and in vivo, by propargylglycine. J. Am. Chem. Soc. 95:6124–6125. doi:10.1021/ja00799a053

Acosta-Martinez, V., D. Acosta-Mercado, D. Sotomayor-Ramirez, and L. Cruz-Rodriguez. 2008. Microbial communities and enzymatic activities under different management in semiarid soils. Appl. Soil Ecol. 38:249–260. doi:10.1016/j.apsoil.2007.10.012

Acosta-Martinez, V., L. Cruz, D. Sotomayor-Ramirez, and L. Perez-Alegria. 2007a. Enzyme activities as affected by soil properties and land use in a tropical watershed. Appl. Soil Ecol. 35:25–45. doi:10.1016/j.apsoil.2006.05.013

Acosta-Martinez, V., S. Klose, and T.M. Zobeck. 2003. Enzyme activities in semiarid soils under conservation reserve program, native rangeland, and cropland. J. Plant Nutr. Soil Sci. 166:699–707. doi:10.1002/jpln.200321215

Acosta-Martinez, V., M.M. Mikha, and M.F. Vigil. 2007b Microbial communities and enzyme activities in soils under alternative crop rotations compared to wheat-fallow for the Central Great Plains. Appl. Soil Ecol. 37:41–52. doi:10.1016/j.apsoil.2007.03.009

Alexander, K., and M. Volini. 1987. Properties of an *Escherichia coli* rhodanese. J. Biol. Chem. 262:6595–6604.

Al-Khafaji, A.A., and M.A. Tabatabai. 1979. Effects of trace elements on arylsulfatase activity in soils. Soil Sci. 127:129–133. doi:10.1097/00010694-197903000-00001

Allen, R.E., and R.J. Parkes. 1995. Digestion procedures for determining reduced sulfur species in bacterial cultures and in ancient and recent sediments. p. 243–257. *In* M.A. Vairavamurthy, M.A.A. Schoonen, T.I. Eglinton, G.W. Luther, and B. Manowitz, (ed.) Geochemical transformation of sedimentary sulfur. ACS Symp. Series. Vol. 612. Am. Chem. Soc., Washington, DC.

Audus, L.J., and J.H. Quastel. 1947. Selective toxic action of Thiosulfate on plants. Nature 60:264–265.

Bandick, A.K., and R.P. Dick. 1999. Field management effects on soil enzyme activities. Soil Biol. Biochem. 31:1471–1479. doi:10.1016/S0038-0717(99)00051-6

Bashkin, V.N., and R.W. Howarth. 2002. Biogeochemistry cycle of sulfur. p. 134–139. *In* V.N. Bashkin and R.W. Howarth (ed.) Modern biogeochemistry. Kluwer Academic Publishers, Dordrecht, The Netherlands.

Bernhard, J.M., P.T. Visscher, and S.S. Bowser. 2003. Submillimeter life positions of bacteria, protists, and metazoans in laminated sediments of the Santa Barbara Basin. Limnol. Oceanogr. 48:813–828. doi:10.4319/lo.2003.48.2.0813

Billaut-Laden, I., D. Allorge, A. Crunelle-Thibaut, E. Rat, C. Cauffiez, D. Chevalier, N. Houdret, J.M. Lo-Guidice, and F. Broly. 2006. Evidence for functional genetic polymorphism of the human thiosulfate sulfurtransferase (Rhodanese), a cyanide and H_2S detoxification enzyme. Toxicology 225:1–11. doi:10.1016/j.tox.2006.04.054

Bohn, H.L., N.J. Barrow, S.S.S. Rajan, and R.L. Parfitt. 1986. Reactions of inorganic sulfur in soils. p. 233–249. *In* M.A. Tabatabai (ed.) Sulfur in agriculture. Agron. Monogr. 27. ASA, CSSA, and SSSA, Madison, WI.

Bordo, D., F. Forlani, A. Spallarossa, R. Colnaghi, A. Carpen, M. Bolognesi, and S. Pagani. 2001. A persulfurated cysteine promotes active site reactivity in *Azotobacter vinelandii* rhodanese. Biol. Chem. 382:1245–1252. doi:10.1515/BC.2001.155

Canfield, D.E., and D.J. Des Marais. 1991. Aerobic sulfate reduction in microbial mats. Science 251:1471–1473. doi:10.1126/science.11538266

Canfield, D.E., R. Raiswell, J.T. Westrich, C.M. Reaves, and R.A. Berner. 1986. The use of chromium reduction in the analysis of reduced inorganic sulfur in sediments and shales. Chem. Geol. 54:149–155. doi:10.1016/0009-2541(86)90078-1

Castellano, S.D., and R.P. Dick. 1991. Cropping and sulfur fertilization influence on sulfur transformations in soil. Soil Sci. Soc. Am. J. 55:114–121. doi:10.2136/sssaj1991.03615995005500010020x

Cavallini, D., B. Mondovi, C. DeMarco, and A. Scioscia-Santoro. 1962a. The mechanism of desulphydration of cysteine. Enzymologia 24:263–266.

Cavallini, D., B. Mondovi, C. DeMarco, and A. Scioscia-Santoro. 1962b. Inhibitory effect of mercaptoethanol and hypotaurine on the desulfhydration of cysteine by cystathionase. Arch. Biochem. Biophys. 96:456–457. doi:10.1016/0003-9861(62)90436-8

Chen, L., D. Kost, and W.A. Dick. 2008a. Flue gas desulfurization products as sulfur sources for corn. Soil Sci. Soc. Am. J. 72:1464–1470. doi:10.2136/sssaj2007.0221

Chen, W., L. Wu, W.T. Frankenberger, Jr., and A. Chang. 2008b. Soil enzyme activities of long-term reclaimed wastewater-irrigated soils. J. Environ. Qual. 37:36–42.

Cipollone, R., P. Ascenzi, E. Frangipani, and P. Visca. 2006. Cyanide detoxification by recombinant bacterial rhodanese. Chemosphere 63:942–949. doi:10.1016/j.chemosphere.2005.09.048

Collins, J.M., and K.J. Monty. 1973. The cysteine desulfhydrase of *Salmonella typhimurium*. J. Biol. Chem. 248:5943–5949.

De Jong, G.A.H., W. Hazeu, P. Bos, and J.G. Kuenen. 1997. Isolation of the tetra-thionate hydrolase from *Thiobacillus acidophilus*. Eur. J. Biochem. 243:678–683. doi:10.1111/j.1432-1033.1997.00678.x

Dedourge, O., P.C. Vong, F. Lasserre-Joulin, E. Beniri, and A. Guckert. 2004. Effects of glu-cose and rhizodeposits (with or without cysteine-S) on immobilized-^{35}S, microbial biomass-^{35}S and arylsulfatase activity in a calcareous and an acid brown soil. Eur. J. Soil Sci. 55:649–656. doi:10.1111/j.1365-2389.2004.00645.x

DeLuca, T.H., and D.R. Keeney. 1994. Soluble carbon and nitrogen pools of prairie and cultivated soils: Seasonal variation. Soil Sci. Soc. Am. J. 58:835–840. doi:10.2136/sssaj1994.03615995005800030029x

Deng, S.P., and R.P. Dick. 1990. Sulfur oxidation and rhodanese activity in soils. Soil Sci. 150:552–560. doi:10.1097/00010694-199008000-00009

Deng, S.P., and M.A. Tabatabai. 1997. Effect of tillage and residue management on enzyme activities in soils. III. Phosphatase and arylsulfatase. Biol. Fertil. Soils 24:141–146. doi:10.1007/s003740050222

Dick, R.P., and S.P. Deng. 1991. Multivariate factor analysis of sulfur oxidation and rhoda-nese activity in soils. Biogeochemistry 12:87–101. doi:10.1007/BF00001808

Dick, W.A. 1992. Sulfur cycle. p. 123–133. *In* J. Lederberg (ed.) Encyclopedia of Microbiology, Volume 4. Academic Press, San Diego, CA.

Dinesh, R., M.A. Suryanarayana, S. Ghoshal Chaudhuri, and T.E. Sheeja. 2004. Long-term influence of leguminous cover crops on the biochemical properties of a sandy loam Fluventic Sulfaquent in a humid tropical region of India. Soil Tillage Res. 77:69–77. doi:10.1016/j.still.2003.11.001

Druschel, G.K., R.J. Hamers, and J.F. Banfield. 2003. Kinetics and mechanism of polythionate oxidation to sulfate at low pH by O_2 and Fe^{3+}. Geochim. Cosmochim. Acta 67:4457–4469. doi:10.1016/S0016-7037(03)00388-0

Duan, W.-M., M.L. Coleman, and K. Pye. 1997. Determination of reduced sulphur spe-cies—an evaluation and modified technique. Chem. Geol. 141:185–194. doi:10.1016/S0009-2541(97)00062-4

Elsgaard, L., G. Hastrup Anderson, and J. Erikson. 2002. Measurement of arylsulphatase activity in agricultural soils using a simplified assay. Soil Biol. Biochem. 34:79–82. doi:10.1016/S0038-0717(01)00157-2

Ezzi, M.I., and J.M. Lynch. 2002. Cyanide catabolizing enzymes in *Trichoderma* spp. Enzyme Microb. Technol. 31:1042–1047. doi:10.1016/S0141-0229(02)00238-7

Ezzi, M.I., J.A. Pascual, B.J. Gould, and J.M. Lynch. 2003. Characterization of the rhoda-nese enzyme in *Trichoderma* spp. Enzyme Microb. Technol. 32:629–634. doi:10.1016/S0141-0229(03)00021-8

Farrell, R.E., V.V.S.R. Gupta, and J.J. Germida. 1994. Effects of cultivation on the activity and kinetics of arylsulfatase in Sakatchewan soils. Soil Biol. Biochem. 26:1033–1040. doi:10.1016/0038-0717(94)90118-X

Fitzgerald, J.W. 1978. Naturally occurring organosulfur compounds in soil. p. 391–443. *In* J.O. Nriagu (ed.) Sulfur in the environment. Part 2. John Wiley & Sons, New York.

Fossing, H., and B.B. Jørgensen. 1989. Measurement of bacterial sulfate reduction in sedi-ments: Evaluation of a single-step chromium reduction method. Biogeochemistry 8:205–222. doi:10.1007/BF00002889

Frankenberger, W.T., Jr., and J.B. Johanson. 1986. Use of plasmolytic agents and antiseptics in soil enzyme assays. Soil Biol. Biochem. 18:209–213. doi:10.1016/0038-0717(86)90029-5

Freney, J.R. 1967. Sulfur-containing organics. p. 229–259. *In* A.D. McLaren and G.H. Peterson (ed.) Soil biochemistry. Marcel Dekker, New York.

Freney, J.R. 1986. Forms and reactions of organic sulfur compounds in soils. p. 207–232. *In* M.A. Tabatabai (ed.) Sulfur in agriculture. Agron. Monogr. 27. ASA, CSSA, SSSA, Madison, WI.

Freney, J.R., F.J. Stevenson, and A.H. Beavers. 1972. Sulfur-containing amino acids in soils hydrolysates. Soil Sci. 114:468–476. doi:10.1097/00010694-197212000-00010

Germida, J.J. 2005. Transformations of sulfur. p. 433–462. *In* D.M. Sylvia, J.J. Fuhrmann, P.G. Hartel, and D.A. Zuberer (ed.) Principles and applications of soil microbiology. 2nd ed. Pearson Education, New Jersey.

Germida, J.J., M. Wainwright, and V.V.S.R. Gupta. 1992. Biochemistry of sulfur cycling in soil. p. 1–38. *In* G. Stotzky and J.M. Bollag (ed.) Soil biochemistry. Vol. 7. Marcel Dekker, New York.

Gianfreda, L., and J.-M. Bollag. 1996. Influence of natural and anthropogenic factors on enzyme activity in soil. p. 123–193. *In* G. Stotzky and J.-M. Bollag (ed.) Soil biochemistry. Vol. 9. Marcel Dekker, New York.

Gianfreda, L., M.A. Rao, A. Piotrowska, G. Palumbo, and C. Colombo. 2005. Soil enzyme activities as affected by anthropogenic alterations: Intensive agricultural practices and organic pollution. Sci. Total Environ. 341:265–279. doi:10.1016/j.scitotenv.2004.10.005

Gilboa-Garber, N. 1971. Direct spectrophotometric determination of inorganic sulfide in biological materials and in other complex mixtures. Anal. Biochem. 43:129–133. doi:10.1016/0003-2697(71)90116-3

Glatz, Z., P. Bouchal, O. Janiczek, M. Mandl, and P. Česková. 1999. Determination of rhodanese enzyme activity by capillary zone electrophoresis. J. Chromatogr. A 838:139–148. doi:10.1016/S0021-9673(98)00972-8

Guarneros, G., and M.V. Ortega. 1970. Cysteine desulfhydrase activities of *Salmonella typhimurium* and *Escherichia coli*. Biochim. Biophys. Acta 198:132–142.

Haneklaus, S., E. Bloem, and E. Schnug. 2007. Sulfur interactions in crop ecosystems. p. 17–58. *In* M.J. Hawkesford and L.J. De Kok (ed.) Sulfur in plants: an ecological perspective. Springer Publishing Company, Berlin.

He, Z.-L., A.G. O'Donnell, J. Wu, and J.K. Syers. 1994. Oxidation and transformation of elemental sulphur in soils. J. Sci. Food Agric. 65:59–65. doi:10.1002/jsfa.2740650110

Howarth, R.W., A. Giblin, J. Gate, B.J. Peterson, and G.W. Luther, III. 1983. Reduced sulfur compounds in the pore waters of a New England salt marsh. Environ. Biogeochem. Ecol. Bull. 35:135–152.

Howarth, R.W., and S. Merkel. 1984. Pyrite formation and the measurement of sulfate reduction in salt marsh sediments. Limnol. Oceanogr. 29:598–608. doi:10.4319/lo.1984.29.3.0598

Janzen, H.H., and J.R. Bettany. 1987a. Measurement of sulfur oxidation in soils. Soil Sci. 143:444–452. doi:10.1097/00010694-198706000-00008

Janzen, H.H., and J.R. Bettany. 1987b. The effect of temperature and water potential on sulfur oxidation in soils. Soil Sci. 144:81–89. doi:10.1097/00010694-198708000-00001

Jørgensen, B.B., and T. Fenchel. 1974. The sulfur cycle of a marine model system. Mar. Biol. 24:189–201. doi:10.1007/BF00391893

Kandeler, E., C. Kampichler, and O. Horak. 1996. Influence of heavy metals on the functional diversity of soil microbial communities. Biol. Fertil. Soils 23:299–306. doi:10.1007/BF00335958

Kato, H., M. Saito, Y. Nagahata, and Y. Katayama. 2008. Degradation of ambient carbonyl sulfide by *Mycobacterium* spp. in soil. Microbiology 154:249–255. doi:10.1099/mic.0.2007/011213-0

Kelly, D.P., J.K. Shergill, W.-P. Lu, and A.P. Wood. 1997. Oxidative metabolism of inorganic sulfur compounds by bacteria. Antonie van Leeuwenhoek 71:95–107. doi:10.1023/A:1000135707181

Kelly, D.P., and A.P. Wood. 2000. Reclassification of some species of *Thiobacillus* to the newly designated genera *Acidithiobacillus* gen. nov., *Halothiobacillus* gen. nov. and *Thermithiobacillus* gen. nov. Int. J. Syst. Evol. Microbiol. 50:511–516.

Klose, S., V. Acosta-Martinez, and H.A. Ajwa. 2006. Microbial community composition and enzyme activities in a sandy loam soil after fumigation with methyl bromide or alternative biocides. Soil Biol. Biochem. 38:1243–1254. doi:10.1016/j.soilbio.2005.09.025

Klose, S., and H.A. Ajwa. 2004. Enzyme activities and microbial biomass in agricultural soils fumigated with methyl bromide alternatives. Soil Biol. Biochem. 36:1625–1635. doi:10.1016/j.soilbio.2004.07.009

Klose, S., J.M. Moore, and M.A. Tabatabai. 1999. Arylsulfatase activity of the microbial biomass in soils affected by cropping systems. Biol. Fertil. Soils 29:46–54. doi:10.1007/s003740050523

Klose, S., and M.A. Tabatabai. 1999. Arylsulfatase activity of the microbial biomass in soil. Soil Sci. Soc. Am. J. 63:569–574. doi:10.2136/sssaj1999.03615995006300030020x

Klose, S., K.D. Wernecke, and F. Makeschin. 2003. Microbial biomass and enzyme activities in coniferous forest soils under long-term fly ash pollution. Biol. Fertil. Soils 38:32–44. doi:10.1007/s00374-003-0615-4

Klose, S., K.D. Wernecke, and F. Makeschin. 2004. Microbial activities in forest soils exposed to chronic depositions from a lignite power plant. Soil Biol. Biochem. 36:1913–1923. doi:10.1016/j.soilbio.2004.05.011

Knights, J.S., F.J. Zhao, S.P. McGrath, and N. Magan. 2001. Long-term effects of land use and fertilizer treatments on sulphur transformations in soils from Broadbalk experiment. Soil Biol. Biochem. 33:1797–1804. doi:10.1016/S0038-0717(01)00106-7

Kolthoff, I.M., and E.B. Sandell. 1963. Textbook of quantitative inorganic analysis. 3rd ed. Macmillan, New York.

Kredich, N.M., L.J. Foote, and B.S. Kennan. 1973. The stoichiometry and kinetics of the inducible cysteine desulfhydrase from Salmonella typhimurium. J. Biol. Chem. 248:6187–6196.

Laishley, E.J., and R. Bryant. 1987. Critical review of inorganic sulphur microbiology with particular reference to Alberta soils. University of Calgary, Calgary, Alberta.

Lang, K. 1933. Die Rhodanbildung im Tierkörper. Biochem. Z. 259:243–256.

Lawrence, J.R., and J.J. Germida. 1988. Relationship between microbial biomass and elemental sulfur oxidation in agricultural soils. Soil Sci. Soc. Am. J. 52:672–677. doi:10.2136/sssaj1988.03615995005200030014x

Lawrence, J.R., and J.J. Germida. 1991. Enumeration of sulfur-oxidizing populations in Saskatchewan agricultural soils. Can. J. Soil Sci. 71:127–136.

Lee, W.J., D.S. Banavara, J.E. Hughes, J.K. Christiansen, J.L. Steele, J.R. Broadbent, and S.A. Rankin. 2007. Role of cystathionine beta-lyase in catabolism of amino acids to sulfur volatiles by genetic variants of Lactobacillus helveticus CNRZ 32. Appl. Environ. Microbiol. 73:3034–3039. doi:10.1128/AEM.02290-06

Li, P., and A.L. Caldwell. 1966. The oxidation of elemental sulfur in soil. Soil Sci. Soc. Am. Proc. 30:370–372. doi:10.2136/sssaj1966.03615995003000030021x

Li, S.T., B. Lin, and W. Zhou. 2005. Effects of previous elemental sulfur applications on oxidation of additional applied elemental sulfur in soils. Biol. Fertil. Soils 42:146–152. doi:10.1007/s00374-005-0003-3

Li, X., and P. Sarah. 2003. Arylsulfatase activity of soil microbial biomass along a Mediterranean-arid transect. Soil Biol. Biochem. 35:925–934. doi:10.1016/S0038-0717(03)00143-3

Luther, G.W., III, and T.M. Church. 1988. Seasonal cycling of sulfur and iron in porewaters of a Delaware salt marsh. Mar. Chem. 23:295–309. doi:10.1016/0304-4203(88)90100-4

Luther, G.W., III, E.A. Giblin, and R. Varsolona. 1985. Polarographic analysis of sulfur species in marine porewaters. Limnol. Oceanogr. 30:727–736. doi:10.4319/lo.1985.30.4.0727

McCaskill, M.R., and G.J. Blair. 1989. A model for the release of sulfur from elemental S and superphosphate. Fert. Res. 19:77–84. doi:10.1007/BF01054678

McGrath, S.P., and F.J. Zhao. 1995. A risk assessment of sulphur deficiency in cereals using soil and atmospheric deposition data. Soil Use Manage. 11:110–114. doi:10.1111/j.1475-2743.1995.tb00507.x

Melville, G.E., J.R. Freney, and C.H. Williams. 1971. Reduction of organic sulfur compounds in soil with tin and hydrochloric acid. Soil Sci. 112:245–248. doi:10.1097/00010694-197110000-00005

Meulenberg, R., E.J. Scheer, J.T. Pronk, W. Hazeu, P. Bos, and J.G. Kuenen. 1993. Metabolism of tetrathionate in Thiobacillus acidophilus. FEMS Microbiol. Lett. 112:167–172. doi:10.1111/j.1574-6968.1993.tb06443.x

Morra, M.J., and W.A. Dick. 1985. Production of thiocysteine (sulfide) in cysteine amended soils. Soil Sci. Soc. Am. J. 49:882–886. doi:10.2136/sssaj1985.03615995004900040018x

Morra, M.J., and W.A. Dick. 1989. Hydrogen sulfide production from cysteine (cystine) in soil. Soil Sci. Soc. Am. J. 53:440–444. doi:10.2136/sssaj1989.03615995005300020021x

Mylon, S., H. Hu, and G. Benoit. 2002. Unsuitability of Cr(II) reduction for the measurement of sulfides in oxic water samples. Anal. Chem. 74:661–663. doi:10.1021/ac010924k

Niknahad Gharmakher, H.N., J.M. Machet, N. Beaudoin, and S. Recous. 2009. Estimation of sulfur mineralization and relationships with nitrogen and carbon in soils. Biol. Fertil. Soils 45:297–304. doi:10.1007/s00374-008-0332-0

Nor, Y.M., and M.A. Tabatabai. 1975. Colorimetric determination of microgram quantities of thiosulfate and tetrathionate. Anal. Lett. 8:537–547.

Nor, Y.M., and M.A. Tabatabai. 1976. Extraction and colorimetric determination of thiosulfate and tetrathionate in soils. Soil Sci. 122:171–178. doi:10.1097/00010694-197609000-00008

Nor, Y.M., and M.A. Tabatabai. 1977. Oxidation of elemental sulfur in soils. Soil Sci. Soc. Am. J. 41:736–741. doi:10.2136/sssaj1977.03615995004100040025x

Pisz, J. 2008. Characterization of extremophilic sulfur oxidizing microbial communities inhabiting the sulfur blocks of Alberta's oil sands. MS thesis. Dep. of Soil Sci. Univ. of Saskatchewan, Saskatoon, Canada.

Postgate, J.R. 1979. The sulphate-reducing bacteria. Cambridge Univ. Press, Cambridge.

Press, M.C., J. Henderson, and J.A. Lee. 1985. Arylsulfatase activity in peat in relation to acid deposition. Soil Biol. Biochem. 17:99–103. doi:10.1016/0038-0717(85)90096-3

Prietzel, J. 2001. Arylsulfatase activities in soils of the black forest, Germany—seasonal variation and effect of $(NH_4)_2SO_4$ fertilization. Soil Biol. Biochem. 33:1317–1328. doi:10.1016/S0038-0717(01)00037-2

Ray, C.R., N. Behera, and N. Sethunathan. 1985. Rhodanese activity of flooded and non-flooded soils. Soil Biol. Biochem. 17:159–162. doi:10.1016/0038-0717(85)90108-7

Ray, W.K., G. Zeng, M.B. Potters, A.M. Mansuri, and T.J. Larson. 2000. Characterization of a 12-kilodalton rhodanese encoded by glpE of Escherichia coli and its interaction with thioredoxin. J. Bacteriol. 182:2277–2284. doi:10.1128/JB.182.8.2277-2284.2000

Rethmeier, J., A. Rabenstein, M. Langer, and U. Fischer. 1997. Detection of traces of oxidized and reduced sulfur compounds in small samples by combination of different high-performance liquid chromatography methods. J. Chromatogr. A 760:295–302. doi:10.1016/S0021-9673(96)00809-6

Saad, O.A.L.O., S. Lehmann, and R. Conrad. 1996. Influence of thiosulfate on nitrification, denitrification, and production of nitric oxide and nitrous oxide in soil. Biol. Fertil. Soils 21:152–159. doi:10.1007/BF00335927

Scherer, H.W. 2001. Sulphur in crop production—invited paper. Eur. J. Agron. 14:81–111. doi:10.1016/S1161-0301(00)00082-4

Scott, N.M., W. Bick, and H.A. Anderson. 1981. The measurement of sulphur-containing amino acids in some Scottish soils. J. Sci. Food Agric. 32:21–24. doi:10.1002/jsfa.2740320105

Singh, B.B., and M.A. Tabatabai. 1978. Factors affecting rhodanese activity in soils. Soil Sci. 125:337–342. doi:10.1097/00010694-197806000-00001

Slaton, N.A., R.J. Norman, and J.T. Gilmour. 2001. Oxidation rates of commercial elemental sulfur products applied to an alkaline silt loam from Arkansas. Soil Sci. Soc. Am. J. 65:239–243. doi:10.2136/sssaj2001.651239x

Sloan, J.J., and W.B. Anderson. 1998. Influence of calcium chloride and ammonium Thiosulfate on bermudagrass uptake of urea nitrogen. Commun. Soil Sci. Plant Anal. 29:435–446. doi:10.1080/00103629809369956

Smith, A.J., and J. Lascelles. 1969. Thiosulphate metabolism and rhodanese in Chromatium sp. Strain D. J. Gen. Microbiol. 15:357–370.

Smittenberg, J., G.W. Harmsen, A. Quispel, and D. Otzen. 1951. Rapid methods for determining different types of sulphur compounds in soil. Plant Soil 3:353–360. doi:10.1007/BF01394032

Speir, T.W., and D.J. Ross. 1978. Soil phosphatase and sulfatases. p. 197–250. In R.G. Burns (ed.) Soil enzymes. Academic Press, New York.

Sullivan, D.M., and J.L. Havlin. 1992. Mechanism of thiosulfate inhibition of urea hydrolysis in soils: Tetrathionate as a urease inhibitor. Soil Sci. Soc. Am. J. 56:957–960. doi:10.2136/sssaj1992.03615995005600030045x

Suzuki, I. 1999. Oxidation of inorganic sulfur compounds: Chemical and enzymatic reactions. Can. J. Microbiol. 45:97–105. doi:10.1139/cjm-45-2-97

Suzuki, I., C.W. Chan, and K. Takeuchi. 1992. Oxidation of elemental sulfur to sulfite by Thiobacillus thiooxidans cells. Appl. Environ. Microbiol. 58:3767–3769.

Szajdak, L. 1996. Impact of crop rotation and phonological periods on rhodanese activity and free sulfuric amino acids concentrations in soils under continuous rye cropping and crop rotation. Environ. Int. 22:563–569. doi:10.1016/0160-4120(96)00053-0

Tabatabai, M.A. 1974. Determination of sulphate in water samples. Sulphur Inst. J. 10:11–13.

Tabatabai, M.A. 1994. Soil Enzymes. p. 775–833. In R.W. Weaver, S. Angle, P. Bottomley and D. Bezdiecek (ed.) Methods of soil analysis. Part 2. Microbial and biochemical properties. SSSA Book Ser. 5. ASA and SSSA, Madison, WI.

Tabatabai, M.A. 1996. Sulfur. p. 921–960. In D.L. Sparks, A.L. Page, P.A. Helmke, R.H. Loeppert, P.N. Soltanpour, M.A. Tabatabai, C.T. Johnson and M.E. Summer (ed.) Methods of soil analysis. Part 3. Chemical methods. SSSA Book Ser. 5. ASA and SSSA, Madison, WI.

Tabatabai, M.A. 2005. Chemistry of sulfur in soils. p. 193–226. In M.A. Tabatabai and D.L. Sparks (ed.) Chemical processes in soils. SSSA Book Ser. 8. ASA and SSSA, Madison, WI.

Tabatabai, M.A., and J.M. Bremner. 1970a. Arylsulfatase activity of soils. Soil Sci. Soc. Am. Proc. 34:225–229. doi:10.2136/sssaj1970.03615995003400020016x

Tabatabai, M.A., and J.M. Bremner. 1970b. Factors affecting soil arylsulfatase activity. Soil Sci. Soc. Am. Proc. 34:427–429. doi:10.2136/sssaj1970.03615995003400030023x

Tabatabai, M.A., and J.M. Bremner. 1971. Michaelis constant of soil enzymes. Soil Biol. Biochem. 3:317–323. doi:10.1016/0038-0717(71)90041-1

Tabatabai, M.A., and W.A. Dick. 2002. Soil enzymes: Research and developments in measuring activities. p. 567–596. In R.G. Burns and R.P. Dick (ed.) Enzymes in the environment: Activity, ecology and applications. Marcel Dekker, New York.

Tabatabai, M.A., and W.T. Frankenberger. 1996. Liquid Chromatography. p. 225–245. In D.L. Sparks, A.L. Page, P.A. Helmke, R.H. Loeppert, P.N. Soltanpour, M.A. Tabatabai, C.T. Johnson and M.E. Summer (ed.) Methods of soil analysis. Part 3. Chemical methods. SSSA Book Ser. 5. ASA and SSSA, Madison, WI.

Tabatabai, M.A., and B.B. Singh. 1976. Rhodanese activity of soils. Soil Sci. Soc. Am. J. 40:381–385. doi:10.2136/sssaj1976.03615995004000030023x

Tabatabai, M.A., and B.B. Singh. 1979. Kinetic parameters of the rhodanese reaction in soils. Soil Biol. Biochem. 11:9–12. doi:10.1016/0038-0717(79)90111-1

Tabita, R., M. Silver, and D.G. Lundgren. 1969. The rhodanese enzyme of Ferrobacillus ferrooxidans (Thiobacillus ferrooxidans). Can. J. Biochem. 47:1141–1145. doi:10.1139/o69-184

Thornton, J.I., and A.D. McLaren. 1975. Enzymatic characterization of soil evidence. J. Forensic Sci. 20:674–692.

Trasar-Cepeda, C., F. Gil-Sotres, and M.C. Leirós. 2007. Thermodynamic parameters of enzymes in grassland soils from Galicia, NW Spain. Soil Biol. Biochem. 39:311–319. doi:10.1016/j.soilbio.2006.08.002

Trudinger, P.A. 1986. Chemistry of the sulfur cycle. p. 1–22. In M.A. Tabatabai (ed.) Sulfur in agriculture. Agron. Monogr. 27. ASA, CSSA, SSSA, Madison, WI.

Vong, P.-C., O. Dedourgw, F. Lasserre-Joulin, and A. Guckert. 2003. Immobilized-S, microbial biomass-S and soil arylsulfatase activity in the rhizosphere soil of rape and barley as affected by labile substrate C and N additions. Soil Biol. Biochem. 35:1651–1661. doi:10.1016/j.soilbio.2003.08.012

Waksman, S.A., and J.S. Joffe. 1922. Oxidation of sulfur in the soil. J. Bacteriol. 7:231–256.

Wang, X.C., and Q. Lu. 2006. Effect of waterlogged and aerobic incubation on enzyme activities in paddy soil. Pedosphere 16:532–539. doi:10.1016/S1002-0160(06)60085-4

Watkinson, J.H., and N.S. Bolan. 1998. Modeling the rate of elemental sulfur oxidation in soils. p. 135–170. In D.G. Maynard ed. Sulfur in the environment. Marcel Dekker, New York.

Wróbel, M., I. Lewandowska, P. Bronowicka-Adamska, and A. Paszewski. 2009. The level of sulfane sulfur in the fungus Aspergillus nidulans wild type and mutant strains. Amino Acids 37:565–571. doi:10.1007/s00726-008-0175-x

Xia, K., F. Weesner, W.F. Bleam, P.R. Bloom, U.L. Skyllberg, and P.A. Helmke. 1998. XANES studies of oxidation states of sulfur in aquatic and soil humic substances. Soil Sci. Soc. Am. J. 62:1240–1246. doi:10.2136/sssaj1998.03615995006200050014x

Yamanishi, T., and S. Tuboi. 1981. The mechanism of the L-cystine cleavage reaction catalyzed by rat liver γ-cystathionase. J. Biochem. 89:1913–1921.

Yang, Z., K. Kanda, H. Tsuruta, and K. Minami. 1996. Measurement of biogenic sulfur gases emission from some Chinese and Japanese soils. Atmos. Environ. 30:2399–2405. doi:10.1016/1352-2310(95)00247-2

Zhabina, N.N., and I.I. Volkov. 1978. A method for determination of various sulfur compounds in sea sediments and rocks. p. 735–746. *In* W.E. Krumbein (ed.) Environmental biogeochemistry and geomicrobiology. Vol. 3. Ann Arbor Sci., Ann Arbor, MI.

Zhao, F.J., S.Y. Loke, A.R. Crosland, and S.P. McGrath. 1996. Method to determine elemental sulphur in soils applied to measure sulphur oxidation. Soil Biol. Biochem. 28:1083–1087. doi:10.1016/0038-0717(96)00073-9

Chapter 8

Phosphorus Cycle Enzymes

Verónica Acosta-Martínez* and M. Ali Tabatabai

8–1 INTRODUCTION

The phosphorus (P) cycling in soil depends on immobilization, mineralization, and redistribution of P, which is controlled by physical–chemical properties (i.e., P sorption by colloidal surfaces and precipitation reactions) as much as it depends on phosphatase-mediated reactions (Stewart and Tiessen, 1987). Phosphorus is the second-most limiting nutrient after nitrogen (N) in agricultural production worldwide, and it is one of the three nutrients generally added to soils in fertilizers, as large amounts of P are required by plants. Phosphorus is essential for all organisms in the transfer of energy from one reaction to another reaction within cells. Adequate P availability for plants stimulates early plant growth and maturity while excessive P concentration in soil can have negative impacts on the whole environmental quality, as it can be toxic to plants and detrimental to natural waters (i.e., eutrophication). Total P concentration in soils can range from 200 to 5000 mg P kg^{-1} soil (Lindsay, 1979), and is distributed within organic and inorganic forms (Kuo, 1996). Phosphatases play a major role in the transformation of soil organic P forms, which constitute a significant percentage (at least 50%) of soil total P (Dalal, 1977), and otherwise would be immobile and structurally unavailable for plant uptake. The activities of phosphatases mediate the transformation and recycling of P forms in soil, plus P forms applied in fertilizers and manure, into free phosphates (PO_4^{3-}) that can be taken up by plants and soil microorganisms.

The presence of phosphatases in soil and their role in soil organic matter dynamics was detected very early in the soil enzymology field (Rogers, 1942) following two important observations: (i) mineral phosphate was released from soil when adding glycerolphosphate or nucleic acid and toluene, and (ii) orthophosphate was liberated more rapidly from glycerophosphate than from nucleic acids. Phosphatases participate in the hydrolysis of phosphate monoesters, which represent up to one-third of the soil organic P, including inositol phosphates, sugar phosphates, and mononucleotides (Grindel and Zyrin, 1965; McKercher and Anderson, 1968; Veinot and Thomas, 1972; Baker, 1977; Turner and Newman, 2005). These enzymes are also responsible for the transformation of other organic P (phosphate esters) from plants and microbial biomass, including phospholipids

V. Acosta-Martinez, USDA-ARS, Cropping Systems Research Laboratory, 3810 4th Street, Lubbock, TX 79415 (veronica.acosta-martinez@ars.usda.gov), *corresponding author;
M. Ali Tabatabai, 2403 Agronomy Hall, Dep. of Agronomy, Iowa State University, Ames, IA 50011-1010 (malit@iastate.edu).

doi:10.2136/sssabookser9.c8

(Kowalenko and McKercher, 1970), glycerol phosphates, phosphatidyl choline (Hance and Anderson, 1963), and nucleic acids (Anderson, 1970). Microbial biomass P represents a low proportion (<1 or 2%) of the total soil P (Torsvik and Goksor, 1978), but it is the most active labile P pool in soil P cycling. Other organic P sources in soil are the phosphonates (C–P bond) (Newman and Tate, 1980; Tate and Newman, 1982; Ogner, 1983), polyphosphates (Anderson and Russell, 1969; Ghonsikar and Miller, 1973; Pepper et al., 1976) and other nonorthophosphates (Beever and Burns, 1980). Phosphatases also must play an important role in the mineralization and transformation of a large proportion of soil organic P that remains uncharacterized in soil (Stewart and Tiessen, 1987), which may be orthophosphates loosely absorbed to soil organic matter (Carloni and Garcia Lopez de Sa, 1978) or bound through metal bridges (Dormaar, 1963; Grindel and Zyrin, 1965; Harter, 1969; Fares et al., 1974).

Phosphatases are the general group of enzymes that catalyze the hydrolysis of both esters and anhydrides of H_3PO_4 (Schmidt and Laskowski, 1961). The enzyme commission (EC) of the International Union of Biochemistry has classified the phosphatases into phosphomonoesterases (EC 3.1.3), phosphodiesterases (EC 3.1.4), and phosphotriesterases (EC 3.1.5) based on the number of ester bonds of the respective substrate, and also into enzymes acting on phosphoryl-containing anhydrides (EC 3.6.1) or those acting on P–N bonds (EC 3.9) (Florkin and Stotz, 1964). Soil assay protocols are currently available to determine the activities of five soil phosphatases such as phosphomonoesterases (acid phosphatase, EC 3.1.3.2 and alkaline phosphatase, EC 3.1.3.1), phosphodiesterase (EC 3.1.4.1), trimetaphosphatase (EC 3.6.1.2) and inorganic pyrophosphatase (EC 3.6.1.1). These phosphatases are known to play important roles in organic P mineralization and transformation into inorganic phosphate (HPO_4^{2-}, $H_2PO_4^-$) that affects soil biogeochemical cycling and plant nutrition (Schmidt and Laskowski, 1961; Speir and Ross, 1978).

Phosphatases are present in different locations in soils. McLaren et al. (1962) and Skujiņš et al. (1962) observed that at least half of phosphatase activity was still present in soil after sterilization with electrons. Bowman et al. (1967) further concluded that phosphatases already in soil before an irradiation treatment or the phosphatases released from lysed microbial cells (intracellular) could explain the still-existent (limited) intensity of mineralization after irradiation. Burns (1978, 1982) explained that microbial communities contribute to the total enzyme activity detected in soil by both intracellular and extracellular enzyme pools. The distinction between these enzyme pools is that extracellular phosphatases are those released into soil from active or nonproliferating cells (i.e., spores, cysts, seeds, endospores), attached to dead cells or cell debris and absorbed to clay and humic colloids. Many efforts have been made to distinguish between the different pools of enzyme activities in soils, but the specific percent of distribution of phosphatases among the intracellular and extracellular pools has not been reported to date. Furthermore, the distribution of intracellular vs. extracellular locations varies depending on the specific phosphatase.

Phosphatase activities are sensitive indicators of land management and ecological stress or restoration of soil ecosystems related to P cycling (Dick, 1994; Dick, 1997). Changes in the activities of phosphatases have been sensitive to cropping systems and land use (Emmerling et al., 2002; Acosta-Martínez et al., 2004) and tillage (Deng and Tabatabai, 1997; Eivazi and Bayan, 2001). Furthermore, significant amounts of P enter soil P cycling with application of organic amendments (i.e., about 8.23 × 10^7 kg of P from animal manures in the U.S.) (Wodzinski and Ullah, 1996), and the fate and transformation of these P sources in the soil environment depend on the phosphatases (Eivazi and Bayan, 2001; Parham et al., 2002; Acosta-Martínez and Harmel, 2006). Recent studies have emphasized the importance of understanding the factors affecting phosphatase activities for modeling soil organic P turnover in soil (Turner and Haygarth, 2005). To accomplish this, the measurement of various phosphatase activities is necessary to obtain a better understanding of P cycling of soil as enzyme activities are substrate specific, and one enzyme activity can provide information about only a single reaction of soil (Nannipieri et al., 1990). This chapter will describe for phosphomonoesterases, phosphodiesterase, inorganic pyrophosphatase, and trimetaphosphatase: the reactions, location in soil, kinetics, correlation with other soil properties, ecological significance in soil quality and ecosystem functioning, and the assay protocols to determine their activities.

8–2 PHOSPHOMONOESTERASES

8–2.1 Introduction

The phosphomonoesterases are the most studied phosphatases in soil. Phosphomonoesterases, such as acid phosphatase (orthophosphoric monoester phosphohydrolase, EC 3.1.3.2) and alkaline phosphatase (orthophosphoric monoester phosphohydrolase, EC 3.1.3.1), are classified according to their optimum pH activities, which vary toward acid and alkaline ranges, respectively. The phosphomonoesterases are known to hydrolyze a variety of phosphomonoesters in soil, including β-glycerophosphate, phenylphosphate, β-naphthyl phosphate, and p-nitrophenyl phosphate. The general equation of the hydrolysis of phosphomonoesters into orthophosphates catalyzed by acid and alkaline phosphatases is:

$$O=\overset{\displaystyle OH}{\underset{\displaystyle OR}{P}}-OH + H_2O \xrightarrow{\text{Acid or Alkaline Phosphatase}} O=\overset{\displaystyle OH}{\underset{\displaystyle OH}{P}}-OH + R\text{-}OH \qquad [1]$$

where R represents either alcohol or phenol groups or nucleosides (Privat de Garilhe, 1967).

In the soil ecosystem, acid phosphatase is predominant in acidic soils whereas alkaline phosphatase is predominant in alkaline soils (Eivazi and Tabatabai, 1977; Juma and Tabatabai, 1977, 1978). The inverse relationship between phosphatase activity and soil pH suggests that either the rate of synthesis and release of this enzyme by soil microbial communities or the stability of this enzyme are strongly related to soil pH (Tabatabai, 1994). It appears that in contrast to other soil enzymes, these phosphatases are affected more significantly by changes in soil pH due to management independent of organic matter content and level of disturbance. Acosta-Martínez and Harmel (2006) reported that alkaline phosphatase activity, different from other enzyme activities, did not consistently increase with increasing poultry-litter application rates in a cultivated soil, and it only increased at the highest application rates. They attributed these results to the fact that soil pH was not affected by the different rates of poultry litter. Studies in a limed soil, which showed no changes in organic matter content, showed an increase in alkaline phosphatase activity with increasing soil pH while a decrease in acid phosphatase activity with increasing soil pH (Acosta-Martínez and Tabatabai, 2000; Ekenler and Tabatabai, 2003). Although not necessary, measurement of both acid and alkaline phosphatases in soil can provide further information on the total phosphatase activity involved in hydrolytic reactions of phosphomonoesters in soil. Dick and Tabatabai (1993) have suggested that an adequate level of pH for crop growth could be defined as the pH at which a proper acid phosphatase/alkaline phosphatase activity ratio occurs. Also, the liming of a soil could be considered sufficient only when alkaline phosphatase activity increases to a specified level.

Several studies have suggested that phosphomonoesterases (alkaline phosphatase and acid phosphatase) are mainly derived from soil microbial communities, but that acid phosphatase is also released by plants (Estermann and McLaren, 1961; Beever and Burns, 1980; Dick et al., 1983; Juma and Tabatabai, 1988a,b,c). It is also believed that within the soil microbial community, fungal populations represent a major source of soil phosphomonoesterases. The role of mycorrhiza populations as a source of phosphatases in soil has been confirmed by studies by Acosta-Martínez et al. (2003, 2008b) with fatty acid methyl esters (FAME) techniques applied to soils under cropping systems that include peanuts, which are able to support arbuscular mycorrhizal fungal (AMF) associations. These studies reported higher abundance of a fatty acid suggested as mycorrhiza fatty acid indicator (i.e., 18:1ω9c), in agreement with higher alkaline phosphatase activity, in soils under peanut- and cotton-based cropping systems compared with cotton monocropping. According to Dighton (1983), mycorrhiza fungi must release the phosphatases into soil to participate in the reactions where organic P is mineralized and phosphorus is released in a plant-available form (i.e., HPO_4^{2-}) for root uptake. Dighton (1983) concluded that since sheathing mycorrhizas have no intracellular hyphal penetration into the root, HPO_4^{2-} released by phosphatase activity would be available only if released into the soil solution adjacent to the root surface or into the intercellular spaces of the Hartig net.

The kinetic parameters of phosphomonoesterases in soil are different from those in enzymes extracted from plants, animals, or microorganisms and also different from the kinetic parameters of other soil enzymes (Dick and Tabatabai, 1987). Michaelis constants of soil acid phosphatase have been reported in several studies (Tabatabai and Bremner, 1971; Cervelli et al., 1973; Brams and McLaren, 1974; Thornton and McLaren, 1975; Eivazi and Tabatabai, 1977; Dick and Tabatabai, 1984). The apparent K_m values of acid phosphatase in soils range from 1.3 to 4.5 mM, whereas values of alkaline phosphatase range from 0.4 to 4.9 mM. Evaluation of protein concentrations in soil revealed that the catalytic efficiency varies among the phosphomonoesterases as alkaline phosphatase seems to have greater catalytic efficiency than acid phosphatase (Klose and Tabatabai, 2002). Studies have shown that all heavy metals and trace elements inhibit phosphomonoesterases in soils, and the degree of inhibition is related to the soil and type and concentration of the trace elements used (Tyler, 1974, 1976a, b; Juma and Tabatabai, 1977). Kinetic studies indicate that orthophosphate is a competitive inhibitor of acid and alkaline phosphatases in soils (Juma and Tabatabai, 1978). Generally, increased production of phosphatase enzymes by plant roots and microorganisms is observed when P is limiting. Consequently, an increase in phosphatase activity may reflect a high demand for P while a decrease in phosphatase activity in soil may also be limited by the amount of its own hydrolyzable substrate (i.e., the forms of organic P).

Phosphomonoesterases are the most widely studied phosphatases to evaluate the effects of land management on soil P cycling. Studies have reported increases in these phosphatases in soil due to addition of poultry litter (Acosta-Martínez and Harmel, 2006), municipal solid waste (Pascual et al., 1999), manure (Parham et al., 2002), and swine lagoon effluents (Iyyemperumal and Shi, 2008). Several studies have reported higher activities of phosphomonoesterases under pasture compared with cultivated soils due to the increase in substrate availability and richness under the more dense root system, better surface cover, and lack of tillage of pasture (Acosta-Martínez et al., 2007, 2008a). Acosta-Martínez et al. (2007) found that acid phosphatase activity was greater in Oxisols and Ultisols than in Inceptisols, and they decreased due to land use in the order of forest = pasture > agriculture. Peshakov et al. (1984) tried to define optimum NPK fertilizer rates as those giving maximum activities of several enzymes including the phosphatase activities. A study with mine soils found phosphatase activities increased with time, as did respiration rate, so that activity levels in the top 10 cm approached those of native soils after 20 yr (Stroo and Jencks, 1982). They attributed the ecosystem recovery of P cycling to the accumulation of organic matter and N. Similarly, another study revealed the importance of these phosphatases for P cycling, as the activity of these phosphatases was reduced after a short-term (2-yr) cessation of fertilizing, as well as cessation of liming and grazing in a reseeded upland grassland soil (Bardgett and Leemans, 1995). Studies by Deng and Tabatabai (1997) reported lowest activity of phosphomonoesterases under no-till soil where residues were removed and under moldboard plowed soil treatments, compared with no-till soil with residues on the soil surface.

8–2.2 Principle

Several methods have been proposed for estimation of the activities of phosphomonoesterase in soils (Skujinš, 1967). The basic difference is in the substrate used and consequently in the technique employed in measuring the product of hydrolysis of the substrate by phosphatases (Dick and Tabatabai, 1978b). This has included the substrates of phenyl phosphate (Kroll and Kramer, 1955), β-glycerophosphate (Skujinš et al., 1962), and β-naphthol with β-naphthyl phosphate (Ramirez-Martinez and McLaren, 1966). All of these methods suffer from poor recovery of products or interference by co-extracted and confounding soil compounds.

Of the various methods available for assay of phosphatase activity in soils, the method developed by Tabatabai and Bremner (1969) is the most rapid, accurate, and precise and is described here. In brief, this method involves colorimetric estimation of the p-nitrophenol released when soil is incubated with toluene and buffered sodium p-nitrophenyl phosphate solution pH 6.5 or pH 11, for acid phosphatase or alkaline phosphate activity, respectively. Treatment of $CaCl_2$–NaOH (pH 10) after incubation is used to extract the p-nitrophenol released by phosphatase activity, which develops the stable color used to estimate this phenol and gives quantitative recovery of p-nitrophenol added to soils. The colorimetric procedure used for estimation of p-nitrophenol is based on the fact that alkaline solutions of this phenol have a yellow color while acid solutions of p-nitrophenol or acid and alkaline solutions of p-nitrophenyl phosphate are colorless. The described procedure for extraction of p-nitrophenol after incubation for assay of acid and alkaline phosphatases serves (i) to stop phosphatase activity, (ii) to develop the yellow color used to estimate this phenol, and (iii) to give quantitative recovery of p-nitrophenol from soils.

8–2.3 Assay Method (Acid and Alkaline Phosphatases) (Tabatabai and Bremner, 1969; Eivazi and Tabatabai, 1977)

8–2.3.1 Apparatus

- Incubation flasks, Erlenmeyer flasks (preferably 25 mL or 50 mL, fitted with No. 1 or 2 stoppers, respectively)
- Incubator
- Glass funnels
- Test tubes
- Spectrophotometer that can be adjusted to a wavelength between 400 and 410 nm. It is also possible to use a Klett–Summerson photoelectric colorimeter fitted with a blue (No. 42) filter.

8–2.3.2 Reagents

1. Toluene, Fisher certified reagent (Fisher Scientific Co., Chicago).

2. Modified universal buffer (MUB) stock solution: Dissolve 12.1 g of tris(hydroxymethyl)aminomethane (THAM), 11.6 g of maleic acid, 14.0 g of citric acid, and 6.3 g of boric acid (H_3BO_3) in 488 mL of 1 N sodium hydroxide (NaOH) and dilute the solution to 1 L with water. Store it in a refrigerator.

3. Modified universal buffer, pH 6.5 and 11: Place 200 mL of MUB stock solution in a 500-mL beaker containing a magnetic stirring bar, and place the beaker on a magnetic stirrer. Titrate the solution to pH 6.5 with 0.1 M hydrochloric acid (HCl), and adjust the volume to 1 L with water. Titrate another 200 mL of the MUB stock solution to pH 11 by using 0.1 M NaOH, and adjust the volume to 1 L with water.

4. p-Nitrophenyl phosphate solution, 0.05 M: Dissolve 0.840 g of disodium p-nitrophenyl phosphate tetrahydrate (Sigma 104, Sigma Chemical Co., St. Louis, MO) in about 40 mL of MUB pH 6.5 (for assay of acid phosphatase) or pH 11 (for assay of alkaline phosphatase), and dilute the solution to 50 with MUB of the same pH. Store the solution in a refrigerator.

5. Calcium chloride ($CaCl_2$), 0.5 M: Dissolve 73.5 g of $CaCl_2 \cdot 2H_2O$ in about 700 mL of water, and dilute the volume to 1 L with water.

6. Sodium hydroxide (NaOH), 0.5 M: Dissolve 20 g of NaOH in about 700 mL of water, and dilute the volume to 1 L with water.

7. Standard p-nitrophenol solution: Dissolve 1.0 g of p-nitrophenol in about 700 mL of water and dilute the solution to 1 L with water. Store the solution in a refrigerator.

8–2.3.3 Procedure

Place 1 g of soil (<5mm) in a 25-mL (or 50-mL) Erlenmeyer flask, add 0.2 mL of toluene, 4 mL of MUB (pH 6.5 for assay of acid phosphatase or pH 11 for assay of alkaline phosphatase), 1 mL of p-nitrophenyl phosphate solution made in the same buffer, and swirl the flask for a few seconds to mix the contents. Stopper the flask, and place it in an incubator at 37°C. After 1 h, remove the stopper, add 1 mL of 0.5 M CaCl$_2$ and then 4 mL of 0.5 M NaOH, swirl the flask for a few seconds, and filter the soil suspension through a Whatman No. 2v folded filter paper. Measure the yellow color intensity of the filtrate with a spectrophotometer at 400 to 410 nm. If the color intensity of the filtrate exceeds that of 50 µg of the p-nitrophenol standard, an aliquot of the filtrate should be diluted with water until the colorimeter reading falls within the limits of the calibration graph, which is described in the next section.

A soil control should be performed with each soil analyzed to allow for color not derived from p-nitrophenol released by phosphatase activity. To perform controls, follow the same procedure as for a sample, but make the addition of 1 mL of p-nitrophenyl phosphate solution after the additions of 0.5 M CaCl$_2$ and 4 mL of 0.5 M NaOH (i.e., immediately before filtration of the soil suspension).

8–2.3.4 Calibration Curve for p-Nitrophenol Standard

Calculate the p-nitrophenol content of the filtrate by reference to a calibration graph plotted from the results obtained with standards containing 0, 10, 20, 30, 40, and 50 µg of p-nitrophenol. To prepare this graph, dilute 1 mL of the standard p-nitrophenol solution to 100 mL in a volumetric flask and mix the solution thoroughly. Then pipette 0-, 1-, 2-, 3-, 4-, and 5-mL aliquots of this diluted standard solution into Erlenmeyer flasks (25 or 50 mL), adjust the volume to 5 mL by addition of water (i.e., 5, 4, 3, 2, 1, and 0 mL, respectively), and proceed as described in the enzyme assay protocol after incubation of the soil sample (i.e., add 1 mL of 0.5 M CaCl$_2$ and 4 mL of 0.5 M NaOH, mix, and filter the resultant suspension). Measure the yellow color intensity of the filtrate with a spectrophotometer at the same wavelength as for the soil enzyme assay (i.e., 400–410 nm), and prepare a calibration curve (p-nitrophenol concentration vs. absorbance).

8–2.4 Comments

For a detailed description of alkaline and acid phosphates activities as affected by soil conditions (fresh vs. air-dried) and several common cold storage temperatures (i.e., 4°C, –20°C, and –80°C), it is recommended to see Lee et al. (2007). According to several studies, including Lee et al. (2007), the effects of storage treatments on the phosphomonoesterases are complex and can vary as a function of the soil type. A decrease in the activity of these phosphatases is generally observed in the air-dried soil compared with the field-moist soil. However, the soil enzyme activity levels were similar among most of the storing conditions (air-dried, 4°C or –20°C) (Lee et al., 2007).

The solutions of the substrates used for assay of phosphomonoesterases are stable for several days if stored in a refrigerator. The compounds used for assay of these enzymes are artificial substrates; they are not expected to be found in soils. The dry substrates should be stored in a freezer. The standard p-nitrophenol solution is stable for a few weeks if stored in a refrigerator. The substrate concentrations in the incubation mixtures during the assay of the activities of phosphomonoesterases are about 5 to 10 times greater than the K_m values determined for these soil enzymes. If necessary, these substrate concentrations can be changed to meet the objectives of the assay.

It is necessary to add CaCl$_2$ before the addition of NaOH to prevent dispersion of clay and extraction of soil organic matter. The use of 0.5 M NaOH stops the reaction catalyzed by phosphomonoesterases. Dispersion of clay complicates filtration, and the filtrate shows a dark color due to extraction of organic matter, which interferes with colorimetric analysis for p-nitrophenol. The procedures described give quantitative recovery of p-nitrophenol added to soils.

At least two laboratory replicates are necessary to be analyzed per sample. The use of a soil control is so designed that it allows for detection of trace amounts of p-nitrophenol in some commercial samples of p-nitrophenyl phosphate and for extraction of trace amounts of colored soil material by the CaCl$_2$–NaOH treatment used for extraction of p-nitrophenol in the assay of phosphomonoesterase. No chemical hydrolysis of p-nitrophenyl phosphate is detected under the conditions of the assay procedure described. It can be possible to select representative soil controls per batch, rather than running a soil control for each sample, if periodic samplings are conducted for the same soil over time.

Early studies reported that toluene had little effect on the activity of purified phosphatases (Frankenberger and Johanson, 1986). Other studies reported that toluene increased the activities of acid and alkaline phosphatases by a range of 8 to 18% in 10 surface soils (Klose and Tabatabai, 2002). It is important to acknowledge that due to environmental health and safety considerations, the elimination of toluene in the assay protocol described can be a personal choice. It is possible to assume that the comparison of treatment effects (among samples) will not be affected by eliminating toluene. It is also possible to assume that microbial growth is really insignificant without toluene addition in an hourly assay. Elimination of toluene in the assays of phosphatase activities can avoid the need to conduct these assays under a hood and reduces the hazard level of the wastes produced and thus the environmental impacts on their handling and disposal.

8–3 PHOSPHODIESTERASE

8–3.1 Introduction

Phosphodiesterase (orthophosphoricdiester phosphohydrolase, EC 3.1.4.1) is the second- most studied phosphatase in soils after the phosphomonoesterases. The activity of phosphodiesterase has been detected in various plants, animals, and microorganisms (Browman and Tabatabai, 1978). This enzyme is best known for being involved in the degradation of nucleic acids (Pearson et al., 1941; Rogers, 1942; Razzell and Khorana, 1959) and phospholipids (Cosgrove, 1967), which constitute a main component of fresh organic P inputs in soil. Phosphodiesterase catalyzes the overall reaction (Reaction [2]) of the type:

$$O=\overset{\displaystyle OH}{\underset{\displaystyle OR_2}{P}}-OR_1 + H_2O \xrightarrow{\text{Phosphodiesterase}} O=\overset{\displaystyle OH}{\underset{\displaystyle OR_2}{P}}-OH + R_1\text{-OH} \qquad [2]$$

where R_1 and R_2 represent either alcohol or phenol groups or nucleosides (Privat de Garilhe, 1967).

Kinetics studies indicate that the apparent K_m values of phosphodiesterase in soils range from 1.3 to 2.0 mM, the Q_{10} is 1.7, and the average activation energy is 37 kJ mol^{-1} (Browman and Tabatabai, 1978). This enzyme can be inhibited by orthophosphate, EDTA, and citrate (5 mM solution concentrations).

The ecological role of phosphodiesterase (Reaction [3]) and phosphomonoesterases (Reaction [4]) is tightly linked in the soil P cycling as the reactions they catalyze occur in series that ultimately lead to free phosphates that can be taken up by plants and soil microorganisms:

$$O=\overset{\displaystyle OH}{\underset{\displaystyle OR_2}{P}}-OR_1 + H_2O \xrightarrow{\text{Phosphodiesterase}} O=\overset{\displaystyle OH}{\underset{\displaystyle OR_2}{P}}-OH + R_1\text{-OH} + H_2O \xrightarrow[\text{Phosphatase}]{\text{Acid or Alkaline}} O=\overset{\displaystyle OH}{\underset{\displaystyle OH}{P}}-OH + R_2\text{-OH} \quad [3,4]$$

where R1 and R2 represent either alcohol or phenol groups or nucleosides (Privat de Garilhe, 1967).

Studies have reported up to twofold increases in the activity of phosphodiesterase with increasing soil pH from 4.5 to 7 due to liming (Acosta-Martínez and Tabatabai, 2000; Ekenler and Tabatabai, 2003). Parham et al. (2002) reported an increase in phosphodiesterase activity of twofold in manure-treated soils compared with nontreated soils, and the activity was the highest even at a depth of 20 to 30 cm in the manure-treated soils. Phosphodiesterase activity was increased in no-tilled soils compared with the tilled counterparts (Deng and Tabatabai, 1997), while no significant effects in tillage were detected for this phosphatase in other soils (Ekenler and Tabatabai, 2003). Margesin and Schinner (1994) detected phosphodiesterase activity in only two of five forest soils that were within an alkaline range while this enzyme activity was not detected in strongly acidic forest soils. A similar trend was found by Turner and Haygarth (2005) for a pasture; they reported that acidic pasture soils showed lower phosphodiesterase activity and a high concentration of labile organic P, whereas the reverse was true in more neutral soils. These researchers concluded in their study that since most of the organic P inputs to soil are phosphate diesters, it seems that phosphodiesterase activity regulates labile organic P turnover in pasture soils. Turner and Newman (2005) also highlighted the greater emphasis that should be given to understanding the role of phosphate diesters and thus, phosphodiesterase activity in P cycling of wetlands.

8–3.2 Principle

The principle of the assay for phosphodiesterase activity in soils is similar to the assay of acid and alkaline phosphatases in which p-nitrophenol released is extracted and determined colorimetrically. The two procedures differ, however, in that 0.1 M THAM buffer pH 12 is used for extraction of the p-nitrophenol released in the assay of phosphodiesterase activity instead of the 0.5 M NaOH used in the assay of phosphomonoesterases. The reason for this difference is that the substrate of phosphodiesterase, bis-p-nitrophenyl phosphate (BPNP), is not stable in NaOH solutions. The BPNP is hydrolyzed with time in the presence of NaOH. The $CaCl_2$–THAM treatment described for extraction of the p-nitrophenol released in the assay of phosphodiesterase serves the same purpose as that of $CaCl_2$–NaOH used for extraction of this product in the assay of phosphomonoesterases.

8–3.3 Assay Method (Browman and Tabatabai, 1978)

8–3.3.1 Apparatus

- Erlenmeyer flasks (choice of 25 or 50 mL, fitted with stoppers No. 1 or 2, respectively)
- Incubator
- Spectrophotometer described in 8–2.3.1

8–3.3.2 Reagents

1. Toluene, Fisher certified reagent (Fisher Scientific Co., Chicago).
2. Tris(hydroxymethyl)aminomethane (THAM) buffer, 0.05 M, pH 8.0: Dissolve 6.1 g of THAM (Fisher certified reagent, Fisher Scientific Co., Chicago) in about 800 mL of water, adjust the pH to 8.0 by titration with approximately 0.1 M H_2SO_4, and dilute the solution to 1 L with water.
3. Bis-p-nitrophenyl phosphate (BPNP) solution, 0.05 M: Dissolve 0.906 g of bis-p-nitrophenyl phosphate sodium salt in about 40 mL of THAM buffer pH 8.0, and dilute the volume to 50 mL with buffer. Store the solution in a refrigerator.
4. Calcium chloride ($CaCl_2$) solution, 0.5 M: Prepare as described for Reagent 5 in 8–2.3.2.
5. Tris(hydroxymethyl)aminomethane–sodium hydroxide (THAM–NaOH) extractant solution, 0.1 M, pH 12: Dissolve 12.2 g of THAM in about 800 mL of water, adjust the pH of the solution to 12 by titration with 0.5 M NaOH, and dilute the volume to 1 L with water.
6. Tris(hydroxymethyl)aminomethane (THAM) diluent, 0.1 M, pH about 10: Dissolve 12.2 g of THAM (Fisher certified reagent, Fisher Scientific Co., Chicago) in about 800 mL of water, and adjust the volume to 1 L with water.
7. Standard p-nitrophenol solution: Prepare as described for Reagent 7 in 8–2.3.2.

8–3.3.3 Procedure

Place 1 g of soil (<5 mm) in a 25-mL (or 50-mL) Erlenmeyer flask, and add 0.2 mL of toluene, 4 mL of THAM buffer pH 8.0, and 1 mL of BPNP solution. Swirl the flask for a few seconds to mix the contents. Stopper the flask and incubate it at 37°C. After 1 h, remove the stopper, add 1 mL of 0.5 M $CaCl_2$ and then 4 mL of THAM–NaOH (pH 12) extractant solution, swirl the flask for a few seconds, and filter the suspension through a Whatman No. 2v folded filter paper. Measure the yellow color intensity of the filtrate with a spectrophotometer at 400 to 410 nm. Calculate the p-nitrophenol content of the filtrate by reference to a calibration graph prepared as described in section 8–2.3.4. If the color intensity of the filtrate exceeds that of 50 μg of p-nitrophenol standard, an aliquot of the filtrate should be diluted with 0.1 M THAM pH about 10 until the colorimeter reading falls within the limits of the calibration graph.

A soil control should be performed with each soil analyzed to allow for color not derived from *p*-nitrophenol released by phosphodiesterase activity. To perform controls, follow the procedure described for assay of phosphodiesterase activity, but make the addition of 1 mL of BPNP solution after the addition of 0.5 M CaCl$_2$ and 4 mL of 0.1 M THAM–NaOH pH 12 (i.e., immediately before filtration of the soil suspension).

8–3.4 Comments

The substrate solution used for assay of phosphodiesterase activity is stable for several days if stored in a refrigerator. The compound used for assay of this enzyme activity is an artificial substrate, which is not expected to be found in soils. In addition, the substrate concentration in the incubation mixtures during the assay is about 5 to 10 times greater than the K_m values determined for this soil enzyme. If necessary, the substrate concentration can be changed to meet the objectives of the assay. The dry substrates should be stored in a freezer. The standard *p*-nitrophenol solution is stable for a few weeks if stored in a refrigerator.

It is necessary to add CaCl$_2$ before the addition of THAM–NaOH pH 12 to prevent dispersion of clay and extraction of soil organic matter. It is necessary to use THAM–NaOH (pH 12) to stop the reaction catalyzed by phosphodiesterase. Dispersion of clay complicates filtration, and the filtrate shows a dark color due to extraction of organic matter, which interferes with colorimetric analysis for *p*-nitrophenol. The procedures described give quantitative recovery of *p*-nitrophenol added to soils.

At least two laboratory replicates need to be analyzed per sample. The use of a soil control is so designed that it allows for detection of trace amounts of *p*-nitrophenol in some commercial samples of *p*-nitrophenyl phosphate and for extraction of trace amounts of colored soil material by the CaCl$_2$–THAM–NaOH treatments used for extraction of *p*-nitrophenol in the assay. No chemical hydrolysis of *p*-nitrophenyl phosphate is detected under the conditions of the assay procedure described. It is possible to select representative soil controls per batch, rather than running a soil control for each sample, if periodic samplings are conducted for the same soil over time.

As described earlier for the phosphomonoesterases, the elimination of toluene in the assay protocols described can be a personal choice. It is possible to assume that the comparison of treatment effects among samples will not be affected by eliminating toluene. It is also possible to assume that microbial growth is really insignificant in an hourly assay without toluene addition. Elimination of toluene in the assays of phosphatases activities can avoid the need to conduct these assays under a hood and reduces the hazard level of the wastes produced and thus, the environmental impacts on their handling and disposal.

The assay for phosphodiesterase should not be performed over an extended period of time because the risk of hydrolysis of the second product (*p*-nitrophenyl phosphate) produced from the action of phosphodiesterase on BPNP increases as incubation time increases. This is especially true in the absence of buffer (Browman and Tabatabai, 1978). This risk is minimal, however, under the conditions of the procedure described.

8–4 INORGANIC PYROPHOSPHATASE

8–4.1 Introduction

Inorganic pyrophosphatase activity has been reported in early soil enzymology records (Gilliam and Sample, 1968; Hashimoto et al., 1969; Hossner and Phillips, 1971) because its substrate, pyrophosphate, is used as a fertilizer. Similar to the other phosphatases described, inorganic pyrophosphatase is widely distributed in nature. Its presence has been reported in bacteria, insects, mammalian tissues, and plants (Feder, 1973). However, very limited information is available on the activity of this phosphatase, and it was not until 1983 that an accurate method became available for its assay (Dick and Tabatabai, 1983). Inorganic pyrophosphatase (pyrophosphate phosphohydrolase, EC 3.6.1.1) catalyzes the hydrolysis of pyrophosphate to orthophosphate. The overall reaction is:

$$^{-}O-\overset{\overset{\displaystyle O}{\|}}{\underset{\underset{\displaystyle O}{|}}{P}}-O-\overset{\overset{\displaystyle O}{\|}}{\underset{\underset{\displaystyle O}{|}}{P}}-O^{-} \; + \; H_2O \xrightarrow{\quad \text{Inorganic Pyrophosphatase} \quad} 2HPO_4^{2-} \qquad\qquad [5]$$

Formaldehyde, fluoride, oxalate, and carbonate inhibit the activity of this enzyme in soils (Dick and Tabatabai, 1978a). Studies by Stott et al. (1985) showed that pyrophosphatase activity in soils is inhibited by many metals and that AsO_4^{3-}, BO_3^{2-}, MoO_4^{2-}, PO_4^{3-}, VO^{2+}, and WO_4^{2-} are competitive inhibitors of this enzyme in soils. At low concentrations, however, many metals are activators of soil pyrophosphatase (Dick and Tabatabai, 1983). Kinetic studies with surface soils have shown that the apparent K_m values of pyrophosphatase in soils range from 20 to 51 mM, and the activation energy values range from 32 to 43 kJ mol^{-1} (Dick and Tabatabai, 1978a).

The activity of inorganic pyrophosphatase was found to have increased in manure-treated soils (Parham et al., 2002) in agreement with increases in soil pH values and microbial activity and diversity (Sun et al., 2004). Although the importance of pyrophosphatase activity in the transformation of P forms added to soils in organic amendments is obvious, few studies related to organic amendments in soil have included this enzyme within the evaluation of soil phosphatase activities.

8–4.2 Principle

The information available indicates that pyrophosphatase activity in soils is optimum at a buffer pH of 8.0. The assay of inorganic pyrophosphatase activity is based on determination of the orthophosphate released when soil is incubated with buffered (pH 8.0) pyrophosphate solution. Generally, there are three problems associated with the measurement of orthophosphate released by enzymatic hydrolysis of pyrophosphate during the pyrophosphatase assay: (i) the orthophosphate released may be sorbed by the soil constituents and therefore not extracted, (ii) orthophosphate may continue to be hydrolyzed from the substrate (pyrophosphate) after extraction from the soil for reasons other than the enzyme (e.g., low pH), and (iii) the presence of pyrophosphate may interfere with the measurement of orthophosphate. All of these problems must be overcome in any method used to assay the pyrophosphatase activity in soils. In the method described, 0.5 M H_2SO_4 is used to extract orthophosphate. This reagent gives quantitative recovery of orthophosphate added to soils. The colorimetric method used for determination of the orthophosphate extracted in the presence of pyrophosphate is specific for orthophosphate. This method involves a rapid formation of heteropoly blue by the reaction of orthophosphate with molybdate ions in the presence of ascorbic acid–trichloroacetic acid reagent and complexation of the excess molybdate ions by a citrate–arsenite reagent to prevent further formation of blue color from the orthophosphate derived from hydrolysis of the substrate, pyrophosphate (Dick and Tabatabai, 1977b).

8–4.3 Assay Method (Dick and Tabatabai, 1977b, 1978a)

8–4.3.1 Apparatus

- Centrifuge tubes, 50 mL, plastic
- Incubator
- Shaker, end-to-end shaker
- Centrifuge, high speed, 12000 rpm (17390 g)
- Spectrophotometer that can be adjusted to a wavelength of 700 nm

8–4.3.2 Reagents

1. Sodium hydroxide (NaOH), 1 M: Dissolve 40 g of NaOH in about 700 mL of water, and dilute the volume to 1 L with water.
2. Modified universal buffer (MUB) stock solution: Prepared as described for Reagent 2 in section 8–2.3.2.
3. Pyrophosphate solution, 50 mM: Dissolve 2.23 g of sodium pyrophosphate decahydrate ($Na_4P_2O_7 \cdot 10H_2O$) (Matheson Coleman and Bell Manufacturing Chemists, Norwood, OH) in 20 mL of MUB stock solution, titrate the solution to pH 8 with 0.1 M HCl, and dilute the volume to 100 mL with water. This reagent should be prepared daily.
4. Modified universal buffer (MUB) working solution, pH 8.0: Titrate 200 mL of the MUB stock solution to pH 8.0 with 0.1 M HCl, and dilute the volume to 1 L with water.
5. Sulfuric acid (H_2SO_4), 0.5 M: Add 250 mL of concentrated H_2SO_4 to 8 L of water, and dilute the volume to 9 L with water.
6. Ascorbic acid (0.1 M)–trichloroacetic acid (0.5 M) solution (Reagent A): Dissolve 8.8 g of ascorbic acid and 41 g of trichloroacetic acid (Fisher certified reagent) in about 400 mL of water, and dilute the volume to 500 mL with water. This reagent should be prepared daily.
7. Ammonium molybdate tetrahydrate [$(NH_4)_6Mo_7O_{24} \cdot 4H_2O$] (0.015 M) solution (Reagent B): Dissolve 9.3 g of ammonium molybdate (J.T. Backer Chemical Co., Phillipsburg, NJ) in about 450 mL of water, and adjust the volume to 500 mL with water.
8. Sodium citrate (0.15 M)–sodium arsenite (0.3 M)–acetic acid (7.5%) solution (Reagent C): Dissolve 44.1 g of sodium citrate and 39 g of sodium arsenite in

about 800 mL of water, add 75 mL of glacial acetic acid (99.9%), and adjust the volume to 1 L with water.

9. Standard phosphate stock solution: Dissolve 0.4390 g of potassium dihydrogen phosphate (KH_2PO_4) in about 700 mL of water, and dilute the volume to 1 L with water. This solution contains 100 μg of PO_4^{3-}–P mL^{-1}.

8–4.3.3 Procedure

Place 1 g of soil (<5 mm) in a 50-mL plastic centrifuge tube, add 3 mL of 50 mM pyrophosphate solution, and swirl the tube for a few seconds to mix the contents. Stopper the tube and incubate it at 37°C. After 5 h, remove the stopper and immediately add 3 mL of MUB pH 8 and 25 mL of 0.5 M H_2SO_4 and shake the tubes horizontally in a reciprocal shaker for 3 min. Centrifuge the soil suspension for 30 s at 17390 g, and immediately take an aliquot of the supernatant for PO_4^{3-}–P analysis. To analyze for the PO_4^{3-}–P released by inorganic pyrophosphatase, pipette a 1-mL aliquot of the supernatant solution into a 25-mL volumetric flask containing 10 mL of Reagent A, immediately add 2 mL of Reagent B and 5 mL of Reagent C, and adjust the volume with water. After 15 min, measure the absorbance of the heteropoly blue color developed using a spectrophotometer adjusted to a wavelength of 700 nm.

A soil control should be included with each soil analyzed to allow for PO_4^{3-}–P not derived from pyrophosphate through pyrophosphatase activity. To perform controls, add 3 mL of MUB pH 8.0 to 1 g of soil and incubate for 5 h. After incubation, add 3 mL of 50 mM pyrophosphate solution, immediately add 25 mL of 0.5 M H_2SO_4, and then continue the assay treating the control as the other samples.

8–4.3.4 Calibration Curve for Inorganic Phosphate (Pi)

To calculate the PO_4^{3-}–P content in soils, use a calibration graph with absorbance vs. standards containing 0, 5, 10, 15, 20, and 25 μg of PO_4^{3-}–P. To prepare this graph, pipette 0-, 5-, 10-, 15-, 20-, and 25-mL aliquots of the standard PO_4^{3-}–P stock solution into 100-mL volumetric flasks, make up the volumes with water, and mix thoroughly. Then analyze 1-mL aliquots of these diluted standards as described for analysis of PO_4^{3-}–P in the aliquot of the incubated sample (i.e., add to the 25-mL volumetric flask containing 10 mL of Reagent A, immediately add 2 mL of Reagent B and 5 mL of Reagent C, and adjust the volume with water). Measure the absorbance of the heteropoly blue color developed using a spectrophotometer adjusted to a wavelength of 700 nm.

8–4.4 Comments

When assaying for pyrophosphatase activity, it is important that the extraction and analysis of the orthophosphate released from pyrophosphatase be performed immediately, because pyrophosphate hydrolyzes slowly with time in the presence of the extractant (0.5 M H_2SO_4). Also, the steps involved in the determination of the orthophosphate in the presence of pyrophosphate should be adhered to. The reagents described should be tested for orthophosphate, because the presence of orthophosphate in any of the reagents may give a high value for the controls.

8–5 TRIMETAPHOSPHATASE

8–5.1 Introduction

Trimetaphosphatase (EC 3.6.1.2) occurs in all living organisms (Berg and Gordon, 1960), and it has been found to be widely distributed in nature including yeast (Kornberg, 1956; Mattenheimer, 1956; Meyerhof et al., 1953), intestinal mucosa of mammals (Berg and Gordon, 1960; Ivey and Shaver, 1977), plants (Stossel et al., 1981; Pierpoint, 1957), and soils (Rotini and Carloni, 1953; Blanchar and Hossner, 1969). This enzyme is important for P cycling because it participates in the hydrolysis of a cyclic polyphosphate that is not sorbed in soil, known as trimetaphosphate (TMP). Trimetaphosphate also can be hydrolyzed chemically by acidic or basic solutions, especially in the presence of certain cations (Van Wazer, 1958; Berg and Gordon, 1960). Trimetaphosphate ($P_3O_9^{3-}$) is hydrolyzed by a series of biochemical reactions to yield phosphates that are sorbed by soils such as triphosphate ($H_2P_3O_{10}^{3-}$), pyrophosphate ($H_2P_2O_7^{2-}$), and orthophosphate (i.e., $2H_2PO_4^-$), which are catalyzed by trimetaphosphatase (Reaction [6]), triphosphatase (Reaction [7]), and pyrophosphatase (Reaction [8]), respectively (Tabatabai and Dick, 1979; Busman and Tabatabai, 1984):

$$\underset{+\,H_2O}{P_3O_9^{-3} \xrightarrow{\hspace{1cm}}} \underset{+\,H_2O}{\overset{\text{Trimetaphosphatase}}{H_2P_3O_{10}^{-3} \xrightarrow{\hspace{1cm}}}} H_2PO_4^- + H_2P_2O_7^{-2} \underset{+\,H_2O}{\overset{\text{Pyrophosphatase}}{\xrightarrow{\hspace{1cm}}}} 2H_2PO_4^- \quad [6,7,8]$$

Among the phosphatases involved in TMP hydrolysis, trimetaphosphatase is the least investigated enzyme while pyrophosphatase is the most thoroughly studied. Studies by Busman and Tabatabai (1985a, b) with several surface soils (up to 28 different soils) showed that the activity of trimetaphosphatase is correlated with the organic C, total N, and clay content. Although TMP is part of many polyphosphate-based fertilizers and trimetaphosphatase activity is crucial to TMP hydrolysis leading to plant available P compounds such as orthophosphates, Dick and Tabatabai (1986) found TMP added to soils (500 µg g^{-1}) had rapid rates of hydrolysis (>75% decrease in 14 d under aerobic conditions) but was still somewhat slower than noncyclic polyphosphate oligomers with chain lengths ranging from 2 to 65 P units. This study showed that phosphatases were important in driving TMP hydrolysis in soils.

Although this enzyme activity was not investigated in the study by Torres-Dorante et al. (2006), their study reported very low poly-P concentrations in soil over time following the application of polyphosphate-based fertilizers, which is evidence of the presence of trimetaphosphatase activity in soils. In addition, the reported lower poly-P concentrations in a silty-loam soil than in a sandy soil by Torres-Dorante et al. (2006) can be attributed to the higher enzyme activities generally found in higher clay and organic matter content soils compared with sandy soils.

8–5.2 Principle

Orthophosphate can be measured to determine trimetaphosphatase activity because of the difficulty of measuring the primary product of TMP hydrolysis, triphosphate, in the presence of other phosphates. However, the dependence of the formation of orthophosphate from TMP on other enzymes (triphosphatase and pyrophosphatase) further complicates the interpretation of the results. In addition, determination of the orthophosphate released is subject to interference by the presence of the other phosphate compounds (Dick and Tabatabai, 1977b).

The method described for assay of trimetaphosphatase activity in soils is based on the precipitation of other phosphates in a soil extract with $BaCl_2$ and determining TMP as total P remaining in the solution (Busman and Tabatabai, 1984, 1985a, b). Since TMP is also chemically hydrolyzed by metal ions such as Ca^{2+} and Mg^{2+}, the method also allows determination of the rate of nonenzymatic (chemical) hydrolysis of TMP in soils. Formaldehyde and EDTA will inhibit growth, but toluene has no effect on trimetaphosphatase activity in soils. The method involves incubating 1 g of soil with 3 mL buffered (100 mM of Tris) 25 mM of TMP at 37°C for 5 h followed by determination of the unaltered substrate. Steam-sterilized soil is used as a control to account for the amount of trimetaphosphate hydrolyzed nonenzymatically.

8–5.3 Assay Method (Dick and Tabatabai, 1977b; Busman and Tabatabai, 1984, 1985a, b)

8–5.3.1 Apparatus

- Centrifuge tubes, 50 mL, plastic
- Incubator, ordinary incubator, or temperature-controlled water bath
- Centrifuge, high speed, 12,000 rpm (17390 g)
- Spectrophotometer or colorimeter

8–5.3.2 Reagents

1. Trimetaphosphate (TMP) solution (25 mM): Dissolve 1.53 g of $Na_3P_3O_9$ (practical grade, Sigma Chemical Co., St. Louis, MO) and 2.42 g of tris(hydroxymethyl)aminomethane (THAM) in about 100 mL of water, adjust the pH to 8.0 with 0.2 M of HCl, and adjust the volume to 200 mL with water.

2. Trimetaphosphate stock solution (100 mM): Dissolve 3.06 g of $Na_3P_3O_9$ (practical grade, Sigma Chemical Co., St. Louis, MO) in about 80 mL of water, and adjust the volume to 100 mL with water.

3. Trimetaphosphate working solution: Dissolve 1.21 g of THAM in about 25 mL of water, add 1, 5, 10, 25, or 50 mL of the TMP stock solution (to make 1, 5, 10, 25, or 50 mM of TMP, respectively), adjust the pH to 8.0 with 0.2 M of HCl, and adjust the volume to 100 mL with water.

4. Barium chloride ($BaCl_2$) solution (100 mM): Dissolve 24.4 g of $BaCl_2$ in about 800 mL of water, and adjust the volume to 1 L with water.

5. Ascorbic acid (0.1 M)–trichloroacetic acid (0.5 M) reagent (Reagent A), ammonium molybdate tetrahydrate reagent (Reagent B), sodium citrate (0.15 M)–sodium arsenite (0.3 M)–acetic acid (7.5%) reagent (Reagent C), and standard phosphate stock solution: Prepare as described in section 8–4.3.2.

6. Ammonium molybdate-antimony potassium tartrate solution: Prepare one solution in which 12 g of ammonium molybdate (J. T. Baker Chemical Co.,

Phillipsburg, NJ) are dissolved in 250 mL of water. Prepare another solution where 0.2908 g of antimony potassium tartrate (General Chemical Co., New York, NY) are dissolved in 100 mL of water. Add both solutions to 1 L of 4 M HCl, dilute to 2 L with water, and mix the solution thoroughly. Store this reagent in a Pyrex glass bottle in a dark, cool place. This solution is described in Dick and Tabatabai (1977b), but using 2.5 M H_2SO_4.

7. Murphy–Riley reagent (modified): Dissolve 1.056 g of ascorbic acid (J.T. Baker Chemical Co., Phillipsburg, NJ) in 200 mL of reagent 6 and mix the solution thoroughly. This reagent should be prepared daily.

8–5.3.3 Procedure

Place 1 g of soil (<5 mm air-dried) in a 50-mL plastic centrifuge tube, add 3 mL of 25 mM of TMP solution, swirl the tube for a few seconds to mix the contents. Stopper the tube and incubate at 37°C. After 5 h, remove the stopper, mix on a vortex stirrer and during mixing add 10 mL of 100 mM of $BaCl_2$ slowly to precipitate other phosphate compounds. Centrifuge the tube at 12,000 × g for 5 min, and immediately remove aliquots of the supernatant for determination of orthophosphate and TMP by placing an aliquot containing <25 µg of orthophosphate-P into a 25-mL volumetric flask and another aliquot (1–5 mL) containing <1 mg P into a 50-mL volumetric flask, respectively. Determine orthophosphate-P (PO_4^{3-}–P) as described in section 8–4.3.3 and prepare a calibration curve as described in 8–4.3.4.

Acidify the aliquot removed for determination of TMP with sufficient 2 M of HCl to bring the concentration to 1 M with respect to HCl and heat the flask on a steam plate (85°C) for 1 h to hydrolyze the TMP to orthophosphate. After cooling to room temperature, adjust the volume to 50 mL with water. Determine the orthophosphate content of this flask by analyzing an aliquot by the Murphy–Riley reagent after it is neutralized with 1 M of NaOH as described by Dick and Tabatabai (1977a).

Two types of controls should be performed. In one control, the above procedure should be followed with steam-sterilized soil. A second control should be performed with each set of assays to determine precisely the amount of TMP added to the soils. To perform this control, follow the above procedures, but without soil.

The amount of TMP-P hydrolyzed enzymatically is calculated by subtracting the amount of TMP-P remaining in the assay tube (i.e., the tube containing nonsterilized soil) from the amount remaining in the control tube (i.e., the tube containing steam-sterilized soil). The amount of TMP-P hydrolyzed nonenzymatically is calculated by subtracting the amount of TMP-P remaining in the control tube (i.e., the tube containing sterilized soil) from the amount of P in the control tube containing no soil.

8–5.4 Comments

For the assay of trimetaphosphatase, a concentration of 25 mM of TMP was chosen because at least 20 mM is necessary for enzyme saturation and much higher concentrations would be detrimental to the precision of the assay (i.e., with little TMP hydrolyzed a high amount would remain to be determined). Tests performed on several soils showed that TMP can be recovered quantitatively by the method described and that the enzymatic and nonenzymatic hydrolysis of TMP are proportional to time of incubation (up to 5 h) and to the amount of soil used (up to 1.5 g of soil) (Busman and Tabatabai, 1984, 1985a). Other studies showed that treatment of soils with formaldehyde, EDTA, mercuric chloride, sodium nitrate, ascorbic acid, and orthophosphate inhibited trimetaphosphatase activity in soils. The activity of this enzyme is noncompetitively inhibited by EDTA and PO_4^{3-} analogs (i.e., MoO_4^{2-}, WO_4^{2-}, VO^{2+}, AsO_4^{3-} and $B_4O_7^{2-}$). Treatment of soils with formaldehyde, EDTA, ascorbic acid, orthophosphate, pyrophosphate, or triphosphate decreases the nonenzymatic hydrolysis of TMP, whereas treatment of soils with K_2SO_4, NH_4Cl, $CaCl_2$, and $MgCl_2$ increases the rate of nonenzymatic hydrolysis (Busman and Tabatabai, 1985b).

Storing field-moist soil samples at 5°C was an effective method for preserving the trimetaphosphatase activity of soils for at least 3 mo (Busman and Tabatabai, 1985b). Storage of air-dried soil samples for 3 mo retained >58 % of the initial trimetaphosphatase activity of freshly sampled field-moist soils.

REFERENCES

Acosta-Martínez, V., D. Acosta-Mercado, D. Sotomayor-Ramírez, and L. Cruz-Rodríguez. 2008a. Microbial communities and enzymatic activities under different management in semiarid soils. Appl. Soil Ecol. 38:249–260. doi:10.1016/j.apsoil.2007.10.012

Acosta-Martínez, V., L. Cruz, D. Sotomayor-Ramírez, and L. Pérez-Alegría. 2007. Enzyme activities as affected by soil properties and land use in a tropical watershed. Appl. Soil Ecol. 35:35–45. doi:10.1016/j.apsoil.2006.05.012

Acosta-Martínez, V., and R.D. Harmel. 2006. Soil microbial communities and enzyme activities under various poultry litter application rates. J. Environ. Qual. 35:1309–1318. doi:10.2134/jeq2005.0470

Acosta-Martínez, V., D. Rowland, R.B. Sorensen, and K.M. Yeater. 2008b. Microbial community structure and functionality under peanut-based cropping systems in a sandy soil. Biol. Fertil. Soils 44:681–692. doi:10.1007/s00374-007-0251-5

Acosta-Martínez, V., and M.A. Tabatabai. 2000. Enzyme activities in a limed agricultural soil. Biol. Fertil. Soils 31:85–91. doi:10.1007/s003740050628

Acosta-Martínez, V., D.R. Upchurch, A.M. Schubert, D. Porter, and T. Wheeler. 2003. Early impacts of cotton and peanut cropping systems on selected soil chemical, physical, microbial and biochemical properties. Biol. Fertil. Soils 40:44–54. doi:10.1007/s00374-004-0745-3

Acosta-Martínez, V., T.M. Zobeck, and V. Allen. 2004. Soil microbial, chemical and physical properties in continuous cotton and integrated crop–livestock systems. Soil Sci. Soc. Am. J. 68:1875–1884. doi:10.2136/sssaj2004.1875

Anderson, G. 1970. The isolation of nucleoside diphosphates form alkaline extracts of soil. Soil Sci. 21:96–104. doi:10.1111/j.1365-2389.1970.tb01156.x

Anderson, G., and J.D. Russell. 1969. Identification of inorganic polyphosphates in alkaline extracts of soil. J. Sci. Food Agric. 20:78–81. doi:10.1002/jsfa.2740200204

Baker, R.T. 1977. Humic acid-associated organic phosphate. N.Z. J. Sci. 20:439–441.

Bardgett, R.D., and D.K. Leemans. 1995. The short- term effects of cessation of fertilizer applications, liming, and grazing on microbial biomass and activity in a reseeded upland grassland soil. Biol. Fertil. Soils 19:148–154. doi:10.1007/BF00336151

Beever, R.E., and D.J.W. Burns. 1980. Phosphorus uptake, storage and utilization by fungi. Adv. Bot. Res. 8:128–129.

Berg, G.G., and L.H. Gordon. 1960. Presence of trimetaphosphatase in the intestinal mucosa and properties of the enzyme. J. Histochem. Cytochem. 8:85–91.

Blanchar, R.W., and L.R. Hossner. 1969. Hydrolysis and sorption reactions of orthophosphate, pyrophosphate, tripolyphosphate, and trimetaphosphate anions added to an Elliot soil. Soil Sci. Soc. Am. Proc. 33:141–144. doi:10.2136/sssaj1969.03615995003300010037x

Bowman, B.T., R.L. Thomas, and D.E. Elrick. 1967. The movement of phytic acid in soil cores. Soil Sci. Soc. Am. Proc. 31:477–481. doi:10.2136/sssaj1967.03615995003100040018x

Brams, W.H., and A.D. McLaren. 1974. Phosphatase reactions in a column of soil. Soil Biol. Biochem. 6:183–189. doi:10.1016/0038-0717(74)90025-X

Browman, M.G., and M.A. Tabatabai. 1978. Phosphodiesterase activity of soils. Soil Sci. Soc. Am. J. 42:284–290. doi:10.2136/sssaj1978.03615995004200020016x

Burns, R.G. 1978. Soil enzymes. p. 295–340. Academic Press, New York.

Burns, R.G. 1982. Enzyme activity in soil: Location and a possible role in microbial ecology. Soil Biol. Biochem. 14:423–427. doi:10.1016/0038-0717(82)90099-2

Busman, L.M., and M.A. Tabatabai. 1984. Determination of trimetaphosphate added to soils. Commun. Soil Sci. Plant Anal. 15:1257–1268. doi:10.1080/00103628409367555

Busman, L.M., and M.A. Tabatabai. 1985a. Hydrolysis of trimetaphosphate in soils. Soil Sci. Soc. Am. J. 49:630–636. doi:10.2136/sssaj1985.03615995004900030021x

Busman, L.M., and M.A. Tabatabai. 1985b. Factors affecting enzymic and non-enzymic hydrolysis of trimetaphosphate in soils. Soil Sci. 140:421–428. doi:10.1097/00010694-198512000-00004

Carloni, L., and M.E. Garcia Lopez de Sa. 1978. Ricerche sul fosoro organic in alcuni terreni toscani. Agric. Ital. 107:151–158.

Cervelli, S., P. Nannipieri, B. Ceccanti, and P. Sequi. 1973. Michaelis constant of soil acid phosphatase. Soil Biol. Biochem. 5:841–845. doi:10.1016/0038-0717(73)90029-1

Cosgrove, D.J. 1967. Metabolism of organic phosphate in soil. p. 216–288. In A.D. McLaren and G.H. Peterson (ed.) Soil biochemistry. Vol. 1. Marcel Dekker, New York.

Dalal, R.C. 1977. Soil organic phosphorus. Adv. Agron. 29:83–117. doi:10.1016/S0065-2113(08)60216-3

Deng, S.P., and M.A. Tabatabai. 1997. Effect of tillage and residue management on enzyme activities in soils: Phosphatases and arylsulfatase. Biol. Fertil. Soils 24:141–146. doi:10.1007/s003740050222

Dick, R.P. 1994. Soil enzyme activities as indicators of soil quality. p. 107–124. In J.W. Doran, D.C. Coleman, D.F. Bezdicek, and B.A. Stewart (ed.) Defining soil quality for a sustainable environment. SSSA, Madison, WI.

Dick, R.P. 1997. Soil enzyme activities as integrative indicators of soil health. p. 121–153. In C.E. Pankhurst, B.M. Doube, and V.V.S.R. Gupta (ed.) Biological indicators of soil health. CABI, Wallingford, New York.

Dick, R.P., and M.A. Tabatabai. 1986. Hydrolysis of polyphosphates in soils. Soil Sci. 142:132–140. doi:10.1097/00010694-198609000-00002

Dick, W.A., N.G. Juma, and M.A. Tabatabai. 1983. Effects of soils on acid phosphatase and inorganic pyrophosphate of corn roots. Soil Sci. 136:19–25. doi:10.1097/00010694-198307000-00003

Dick, W.A., and M.A. Tabatabai. 1977a. An alkaline oxidation method for determination of total phosphorus in soils. Soil Sci. Soc. Am. J. 41:511–514. doi:10.2136/sssaj1977.03615995004100030015x

Dick, W.A., and M.A. Tabatabai. 1977b. Determination of orthophosphate in aqueous solutions containing labile organic and inorganic phosphorus compounds. J. Environ. Qual. 6:82–85. doi:10.2134/jeq1977.00472425000600010018x

Dick, W.A., and M.A. Tabatabai. 1978a. Inorganic pyrophosphatase activity in soils. Soil Biol. Biochem. 10:58–65. doi:10.1016/0038-0717(78)90011-1

Dick, W.A., and M.A. Tabatabai. 1978b. Hydrolysis of organic and inorganic phosphorus compounds added to soils. Geoderma 21:175–182. doi:10.1016/0016-7061(78)90025-3

Dick, W.A., and M.A. Tabatabai. 1983. Activation of soil pyrophosphate by metal ions. Soil Biol. Biochem. 15:359–363. doi:10.1016/0038-0717(83)90084-6

Dick, W.A., and M.A. Tabatabai. 1984. Kinetic parameters of phosphatases in soils and organic waste materials. Soil Sci. 137:7–15. doi:10.1097/00010694-198401000-00002

Dick, W.A., and M.A. Tabatabai. 1987. Kinetics and activities of phosphates-clay complexes. Soil Sci. 143:5–15. doi:10.1097/00010694-198701000-00002

Dick, W.A., and M.A. Tabatabai. 1993. Significance and potential uses of soil enzymes. p. 95–127. In F.B. Metting, Jr (ed.) Soil microbial ecology: Applications in agricultural and environmental management. Marcel Dekker, New York.

Dighton, J. 1983. Phosphatase production by mycorrhizal fungi. Plant Soil 71:455–462. doi:10.1007/BF02182686

Dormaar, J.F. 1963. Humic acid associated phosphorus in some soils of Alberta. Cancer Sci. 43:235–241.

Eivazi, F., and M.R. Bayan. 2001. Effect of long-term fertilization and cropping systems on selected soil enzyme activities. Dev. Plant Soil Sci. 92:686–687.

Eivazi, F., and M.A. Tabatabai. 1977. Phosphatases in soils. Soil Biol. Biochem. 9:167–172. doi:10.1016/0038-0717(77)90070-0

Ekenler, M., and M.A. Tabatabai. 2003. Responses of phosphatases and arylsulfatase in soils to liming and tillage systems. J. Plant Nutr. Soil Sci. 166:281–290. doi:10.1002/jpln.200390045

Emmerling, C., M. Schloter, A. Hartmann, and E. Kandeler. 2002. Functional diversity of soil organisms—a review of recent research activities in Germany. J. Plant Nutr. Soil Sci. 165:408–420. doi:10.1002/1522-2624(200208)165:4<408::AID-JPLN408>3.0.CO;2-3

Estermann, E.F., and A.D. McLaren. 1961. Contribution of rhizoplane organisms to the total capacity of plants to utilize organic nutrients. Plant Soil. 15:243–260.

Fares, F., J.C. Fardeau, and F. Jaquin. 1974. Etude quantitative du phosphore organique dans different types de sols. Phosphore Agric. 63:25–41.

Feder, J. 1973. The phosphatases. p. 475–508. In E.J. Griffith, A. Beeton, J.M. Spencer, and D.T. Mitchell (ed.) Environmental phosphorus handbook. John Wiley & Sons, New York.

Florkin, M., and E.H. Stotz. 1964. Comprehensive biochemistry. Vol. 13:126–134. Elsevier North-Holland, New York.

Frankenberger, W.T. Jr., and J.B. Johanson. 1986. Use of plasmolytic agents and antiseptics in soil enzyme assays. Soil Biol. Biochem. 18:209–213.

Ghonsikar, C.P., and R.H. Miller. 1973. Soil inorganic polyphosphates of microbial origin. Plant Soil 38:651–655. doi:10.1007/BF00010703

Gilliam, J.W., and E.C. Sample. 1968. Hydrolysis of pyrophosphate in soils: PH and biological effects. Soil Sci. 106:352–357. doi:10.1097/00010694-196811000-00004

Grindel, N.M., and N.G. Zyrin. 1965. Method of determination and the dynamics of organic P compounds in the plow horizon of slightly cultivated Sod-Podsolic soils. Sov. Soil Sci. 17:1391–1410.

Hance, R.J., and G. Anderson. 1963. Identification of hydrolysis products of soil phospholipids. Soil Sci. 96:157–161. doi:10.1097/00010694-196309000-00002

Harter, R.D. 1969. Phosphorus adsorption sites in soils. Soil Sci. 33:630–632. doi:10.2136/sssaj1969.03615995003300040039x

Hashimoto, I., J.D. Hughes, and O.D. Philen, Jr. 1969. Reaction of triammonium pyrophosphate with soils and soil minerals. Soil Sci. Soc. Am. Proc. 33:401–405. doi:10.2136/sssaj1969.03615995003300030020x

Hossner, L.R., and D.P. Phillips. 1971. Pyrophosphate hydrolysis in flooded soil. Soil Sci. Soc. Am. Proc. 35:379–383. doi:10.2136/sssaj1971.03615995003500030018x

Ivey, F.J., and K. Shaver. 1977. Enzymatic hydrolysis of polyphosphate in the gastrointestinal tract. J. Agric. Food Chem. 25:128–130. doi:10.1021/jf60209a021

Iyyemperumal, K., and W. Shi. 2008. Soil enzyme activities in two forage systems following application of different rates of swine lagoon effluent or ammonium nitrate. Appl. Soil Ecol. 38:128–136. doi:10.1016/j.apsoil.2007.10.001

Juma, N.G., and M.A. Tabatabai. 1977. Effects of trace elements on phosphatase activity in soils. Soil Sci. Soc. Am. J. 41:343–346. doi:10.2136/sssaj1977.03615995004100020034x

Juma, N.G., and M.A. Tabatabai. 1978. Distribution of phosphomonoesterases in soils. Soil Sci. 126:101–108. doi:10.1097/00010694-197808000-00006

Juma, N.G., and M.A. Tabatabai. 1988a. Hydrolysis of organic phosphates by corn and soybean roots. Plant Soil 107:31–38. doi:10.1007/BF02371541

Juma, N.G., and M.A. Tabatabai. 1988b. Phosphatase activity in corn and soybean roots: Conditions for assay effects of metals. Plant Soil 107:39–47. doi:10.1007/BF02371542

Juma, N.G., and M.A. Tabatabai. 1988c. Comparison of kinetic and thermodynamic parameters of phosphomonoesterases of soils and of corn and soybean roots. Soil Biol. Biochem. 20:533–539. doi:10.1016/0038-0717(88)90069-7

Klose, S., and M.A. Tabatabai. 2002. Response of phosphomonoesterases in soils to chloroform fumigation. J. Plant Nutr. Soil Sci. 165:429–434. doi:10.1002/1522-2624(200208)165:4<429::AID-JPLN429>3.0.CO;2-S

Kornberg, S.R. 1956. Tripolyphosphate and trimetaphosphate in yeast extracts. J. Biol. Chem. 218:23–31.

Kowalenko, C.G., and R.B. McKercher. 1970. An examination of methods for extraction of soil phospholipids. Soil Biol. Biochem. 2:269–273. doi:10.1016/0038-0717(70)90033-7

Kroll, L., and M. Kramer. 1955. Der Einfluss der Tonmineralien auf die Enzym-Aktivitat der Bodenphosphatase. Naturwissenschaften 42:157–158. doi:10.1007/BF00600010

Kuo, S. 1996. Phosphorus. p. 869–919. In D.L. Sparks (ed.) Methods of soil analysis. Chemical methods. Part 3. SSSA Book Ser. 5. SSSA and ASA, Madison, WI.

Lee, Y.B., N. Lorenz, L.K. Dick, and R.P. Dick. 2007. Cold storage and pretreatment incubation effects on soil microbial properties. Soil Sci. Soc. Am. J. 71:1299–1305. doi:10.2136/sssaj2006.0245

Lindsay, W.L. 1979. Chemical equilibria in soils. John Wiley & Sons, New York.

Margesin, R., and F. Schinner. 1994. Phosphomonoesterase, phosphodiesterase, phosphotriesterase, and inorganic pyrophosphatase activities in forest soils in an alpine area: Effect of pH on enzyme activity and extractability. Biol. Fertil. Soils 18:320–326. doi:10.1007/BF00570635

Mattenheimer, H. 1956. Die Substratspezifität "anorganischer" Poly- und Metaphosphatasen: III. Papierchromatographische Untersuchungen beim enzymatischen Abbau von anorganischen Poly- und Metaphosphaten. Hoppe Seylers Z. Physiol. Chem. 303:125–138.

McKercher, R.B., and G. Anderson. 1968. Content of inositol penta-and hexaphosphates in some Canadian soils. J. Soil Sci. 19:47–55. doi:10.1111/j.1365-2389.1968.tb01519.x

McLaren, A.D., R.A. Luse, and J.J. Skujins. 1962. Sterilization of soil by irradiation and some further observations on soil enzyme activity. Soil Sci. Soc. Am. Proc. 26:371–377. doi:10.2136/sssaj1962.03615995002600040019x

Meyerhof, O., R. Shatas, and A. Kaplan. 1953. Heat of hydrolysis of trimetaphosphate. Biochim. Biophys. Acta 12:121–127. doi:10.1016/0006-3002(53)90130-9

Nannipieri, P., S. Grego, and B. Ceccanti. 1990. Ecological significance of the biological activity in soil. p. 293–355. In J.M. Bollag and G. Stotzky (ed.) Soil biochemistry. Vol. 6. Marcel Dekker, New York.

Newman, R.H., and K.R. Tate. 1980. Soil phosphorus characterization by ^{31}P nuclear magnetic resonance. Commun. Soil Sci. Plant Anal. 11:835–842. doi:10.1080/00103628009367083

Ogner, G. 1983. ^{31}P-NMR spectra of humic acids: A comparison of four different raw humus types in Norway. Geoderma 29:215–219. doi:10.1016/0016-7061(83)90088-5

Parham, J.A., S.P. Deng, W.R. Raun, and G.V. Johnson. 2002. Long-term cattle manure application in soil. I. Effect on soil phosphorus levels, microbial biomass C, and dehydrogenase and phosphatases activities. Biol. Fertil. Soils 35:328–337. doi:10.1007/s00374-002-0476-2

Pascual, J.A., C. Garcia, and T. Hernandez. 1999. Lasting microbiological and biochemical effects of the addition of municipal solid waste to an arid soil. Biol. Fertil. Soils 30:1–6. doi:10.1007/s003740050579

Pearson, R.W., A.G. Norman, and C. Ho. 1941. The mineralization of the organic phosphorus of various compounds in soils. Soil Sci. Soc. Am. Proc. 6:168–175. doi:10.2136/sssaj1942.036159950006000C0029x

Pepper, I.L., R.H. Miller, and C.P. Ghonsikar. 1976. Microbial inorganic polyphosphates: Factors influencing their accumulation in soil. Soil Sci. 40:872–875. doi:10.2136/sssaj1976.03615995004000060022x

Peshakov, G., V. Varov, V. Rankov, G. Ampova, N. Toskov, and T. Tzirkov. 1984. An attempt at optimizing fertilization with a view to maximize capacity of soil enzymes. p. 37–44. In J. Szegi (ed.) Soil biology and conservation of the biosphere. Vol. 1. Akademiai Kiado, Budapest.

Pierpoint, W.S. 1957. The phosphatase and metaphosphatase activities of pea extracts. Biochem. J. 65:67–76.

Privat de Garilhe, M. 1967. Enzymes in nucleic acid research. p. 259–278. Holden-Day, San Francisco.

Ramirez-Martinez, J.R., and A.D. McLaren. 1966. Determination of phosphatase activity by a fluorimetric technique. Enzymologia 30:243–253.

Razzell, W.E., and H.G. Khorana. 1959. Studies on polynucleotides: III. Enzymatic degradation. Substrate specificity and properties of snake venom phosphodiesterase. J. Biol. Chem. 234:2105–2113.

Rogers, H.T. 1942. Dephosphorylation of organic phosphorus compounds by soil catalysts. Soil Sci. 54:439–446. doi:10.1097/00010694-194212000-00005

Rotini, O.T., and L. Carloni. 1953. La transformazione die metafosfati in ortofasfati promossa dal terreno Agrario. Ann. Sper. Agrar. 7:1789–1799.

Schmidt, G., and M. Laskowski, Sr. 1961. Preparation and some properties of a phosphate ester cleavage (Survey). p. 3–35. *In* P.D. Boyer (ed.) The enzymes. 2nd ed. Academic Press, New York.

Skujiņš, J. 1967. Enzymes in soil. p. 371–414. *In* A.D. McLaren and G.H. Peterson (ed.) Soil biochemistry. Vol. 1. Marcel Dekker, New York.

Skujiņš, J.J., L. Braal, and A.D. McLaren. 1962. Characterization of phosphatase in a terrestrial soil sterilized with an electron beam. Enzymologia 25:125–133.

Speir, T.W., and D.J. Ross. 1978. Soil phosphatase and sulphatase. p. 197–250. *In* R.G. Burns (ed.) Soil enzymes. Academic Press, New York.

Stewart, J.W.B., and H. Tiessen. 1987. Dynamics of soil organic phosphorus. Biogeochemistry 4:41–60. doi:10.1007/BF02187361

Stossel, P., G. Lazarovits, and E.W.B. Ward. 1981. Cytochemical staining and in vitro activity of acid trimetaphosphatase in etiolated soybean hypocotyls. Can. J. Bot. 59:1501–1508. doi:10.1139/b81-204

Stott, D.E., W.A. Dick, and M.A. Tabatabai. 1985. Inhibition of pyrophosphatase activity in soils by trace elements. Soil Sci. 139:112–117. doi:10.1097/00010694-198502000-00003

Stroo, H.F., and E.M. Jencks. 1982. Enzyme activity and respiration in mine soil. Soil Sci. Soc. Am. J. 46:548–553. doi:10.2136/sssaj1982.03615995004600030021x

Sun, H.Y., S.P. Deng, and W.R. Raun. 2004. Bacterial community structure and diversity in a century-old manure-treated agroecosystem. Appl. Environ. Microbiol. 70:5868–5874. doi:10.1128/AEM.70.10.5868-5874.2004

Tabatabai, M.A. 1994. Soil Enzymes. p. 801–834. *In* R.W. Weaver, J.S. Angle, and P.S. Bottomley (ed.) Methods of soil analysis. Part 2. Microbiological and biochemical properties. SSSA, Madison, WI.

Tabatabai, M.A., and J.M. Bremner. 1969. Use of *p*-nitrophenyl phosphate for assay of soil phosphatase activity. Soil Biol. Biochem. 1:301–307. doi:10.1016/0038-0717(69)90012-1

Tabatabai, M.A., and J.M. Bremner. 1971. Michaelis constants of soil enzymes. Soil Biol. Biochem. 3:317–323. doi:10.1016/0038-0717(71)90041-1

Tabatabai, M.A., and W.A. Dick. 1979. Distribution and stability of pyrophosphatase in soils. Soil Biol. Biochem. 11:655–659. doi:10.1016/0038-0717(79)90035-X

Tate, K.R., and R.H. Newman. 1982. Phosphorus fractions of a climosequence of soils in New Zealand tussock grassland. Soil Biol. Biochem. 14:191–196. doi:10.1016/0038-0717(82)90022-0

Thornton, J.I., and A.D. McLaren. 1975. Enzymatic characterization of soil evidence. J. Forensic Sci. 20:674–692.

Torres-Dorante, L.O., N. Claassen, B. Steingrobe, and H.W. Olfs. 2006. Fertilizer-use efficiency of different inorganic polyphosphate sources: Effects on soil P availability and plant P acquisition during early growth of corn. J. Plant Nutr. Soil Sci. 169:509–515. doi:10.1002/jpln.200520584

Torsvik, V.L., and J. Goksoyr. 1978. Determination of bacterial DNA soil. Soil Biol. Biochem. 10:7–12. doi:10.1016/0038-0717(78)90003-2

Turner, B.L., and P.M. Haygarth. 2005. Phosphatase activity in temperate pasture soils: Potential regulation of labile organic phosphorus turnover by phosphodiesterase activity. Sci. Total Environ. 344:27–36. doi:10.1016/j.scitotenv.2005.02.003

Turner, B., and S. Newman. 2005. Phosphorus cycling in wetland soils: The importance of phosphate diesters. J. Environ. Qual. 34:1921–1929. doi:10.2134/jeq2005.0060

Tyler, G. 1974. Heavy metal pollution and soil enzymatic activity. Plant Soil 41:303–311. doi:10.1007/BF00017258

Tyler, G. 1976a. Influence of vanadium on soil phosphatase activity. J. Environ. Qual. 5:216–217. doi:10.2134/jeq1976.00472425000500020023x

Tyler, G. 1976b. Heavy metal pollution, phosphatase activity and mineralization of organic phosphorus in forest soils. Soil Biol. Biochem. 8:327–332. doi:10.1016/0038-0717(76)90065-1

Van Wazer, J.R. 1958. Phosphorus and its compounds. Vol. 1. Chemistry. Interscience Publ., New York.

Veinot, R.L., and R.L. Thomas. 1972. High molecular weight organic P complexes in soil organic matter: Inositol and metal content of various soil fractions. Soil Sci. Soc. Am. Proc. 36:71–73. doi:10.2136/sssaj1972.03615995003600010016x

Wodzinski, R.J., and A.H.J. Ullah. 1996. Phytase. Adv. Appl. Microbiol. 42:263–302. doi:10.1016/S0065-2164(08)70375-7

Chapter 9

Carbohydrate Hydrolases

Shiping Deng* and Inna Popova

9–1 INTRODUCTION

Carbohydrates are the most abundant of the four main classes of biomolecules, exceeding proteins, lipids, and nucleic acids. In biological systems, carbohydrates function as the storage of energy (starch) and structural components (cellulose and hemicellulose in plants, chitin in fungi and insects). In nature, carbohydrates include mono-, di-, oligo- and polysaccharides. Most carbohydrates in the soil environment originate from plant biomass, which is composed of 15 to 60% cellulose, 10 to 30% hemicelluloses, 5 to 30% lignin, and 2 to 15% protein (Sylvia et al., 2005). In addition to cellulose and hemicelluloses, important carbohydrates also include starch, chitin, xylan, sucrose, glucose, glucosamine, galactose, fructose, and xylose. In soil, carbohydrates and their derivatives play major roles in supporting microbial life as carbon and energy sources. Enzymes that hydrolyze carbohydrates are key players in mineralization and degradation of organic compounds as well as formation and development of soil organic matter and structural components.

Numerous enzymes are involved in hydrolyzing carbohydrates. Noteworthy enzymes include those that break down large organic complexes, such as cellulase, amylase, chitinase and xylanase, and those that convert disaccharides and soluble oligosaccharides to simple sugars such as invertase, α- and β-glucosidases, α- and β-galactosidases, and N-acetyl-β-glucosaminidase. Many assay methods for detecting activities of these enzymes in soil been developed based on colorimetric quantifications. More recently, there is a growing interest to develop enzyme assay methods using fluorimetric substrates for increased sensitivity. Perhaps the most challenging are methods for assaying activities involved in degradation of complex carbohydrates. These hydrolytic transformation processes often involve multiple enzymes and produce multiple products ranging from mono- and disaccharides, to soluble and insoluble oligosaccharides. It is in this area that new and improved assay methods and/or protocols have been established. Particularly noteworthy are those for cellulase evaluation due to their central role in the global carbon cycle and energy flow. In this chapter, improvements and challenges in performing colorimetric-based enzyme assays and precautions in data interpretations will

Shiping Deng, Oklahoma State University, Department of Plant and Soil Sciences, 368 Agricultural Hall, Stillwater, OK 74078 (shiping.deng@okstate.edu), *corresponding author;
Inna Popova, Oklahoma State University, Department of Plant and Soil Sciences, 368 Agricultural Hall, Stillwater, OK 74078 (inna.popova@okstate.edu).

doi:10.2136/sssabookser9.c9

be discussed. It is hoped that this information will help soil and environmental scientists choose the method that best suits their needs in evaluating carbohydrate hydrolases in complex soil ecosystems.

9–1.1 Carbohydrates and Derivatives

Assays of carbohydrate hydrolases are often based on the determination of monosaccharides, which are end products of their hydrolytic activities. Monosaccharides are the simplest carbohydrates, which include aldehydes or ketones that have two or more hydroxyl groups. Six-carbon carbohydrates, such as D-glucose and D-fructose, are the building blocks of many important complex biopolymers in nature. The predominant form of glucose is β-D-glucopyranose, while the predominant form of fructose is β-D-fructofuranose (Fig. 9–1). When the substituent at C-2 of β-D-glucopyranose is an acetylated amino group instead of a hydroxyl group, the resulting sugar is N-acetyl-β-D-glucosamine (NAG) (Fig. 9–2). The designation α and β indicates that the hydroxyl group attached to the anomeric C atom (C-1 for glucopyranose and C-2 for fructofuranose) is below or above the plane of the ring, respectively. Sugars containing a free aldehyde or keto group attached to the anomeric C have reducing power and are termed reducing sugars. All monosaccharides are reducing sugars. Therefore, determination of reducing sugars is often employed in assaying activities of these enzymes. Sucrose, however, is not a reducing sugar because its anomeric C atom of the glucose is bound to the anomeric C atom of the fructose residue in the disaccharide.

Monosaccharides are joined by various bonds to form complex sugars such as starch, cellulose, xylan, and chitin. Starch is a polysaccharide composed of D-glucose residues linked by α-1,4 or α-1,6-O-glycosidic bonds, which are the bonds between C-1 of glucose and the oxygen atom at C-4 of another sugar molecule. As a major food reserve, starch comprises up to 75% of the mass in seeds of cereals and 20 to 30% in plant bulbs. There are two types of starch: amylose, the unbranched glucose chains joined by α-1,4 linkages, and amylopectin, the branched glucose chains containing about one α-1,6 linkage per 30 α-1,4 linkages. Each amylose molecule is composed of 500 to 20,000 D-glucose residues depending on the source (Hoover, 2001), while each amylopectin molecule contains up to

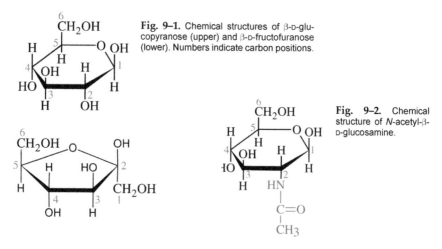

Fig. 9–1. Chemical structures of β-D-glucopyranose (upper) and β-D-fructofuranose (lower). Numbers indicate carbon positions.

Fig. 9–2. Chemical structure of N-acetyl-β-D-glucosamine.

two million glucose residues in a compact structure with a hydrodynamic radius from 21 to 75 nm (Parker and Ring, 2001).

Similar to starch, celluloses are also polysaccharides of D-glucose residues, but linked by β-1,4-O-glycosidic bonds. Each cellulose molecule is a linear homopolymer of glucose with approximately 14,000 glucose units in grasses and 2000 to 5,000 glucose units in woods (Paul and Clark, 1996). In plant biomass, cellulose molecules are bundled by hydrogen bonding in crystalline, paracrystalline, or amorphous forms, and exist within a matrix of primarily hemicellulose and lignin polymers. Cellulose in soils is derived mainly from plant residues incorporated into the soil with a limited amount derived from fungi and bacteria (Richmond, 1991).

The backbone chain of xylan is a homopolymeric β-1,4-linked D-xylose. Xylan occurs as heteropolysaccharides with homopolymeric backbone chains of 1,4-linked β-D-xylopyranose units, which consist of O-acetyl, α-L-arabinofuranosyl, α-1,2-linked glucuronic, or 4-O-methylglucuronic acid substituent. Xylan presents as a plant cell wall polysaccharide in the hemicellulose fraction of the wall matrix. As the most abundant hemicellulosic polysaccharide, xylan comprises up to 30% of the cell wall material of annual plants, 15 to 30% of hardwoods, and 7 to 10% of softwoods (Viikari et al., 1993).

Chitin, a structural component of the cell walls of fungi and the exoskeletons of insects and arthropods, is composed of long, straight chains of NAG residues in β-1,4 linkage. Chitin has a chemical structure that is similar to cellulose, but with a different substitution at C-2 of D-glucose. N-Acetylglucosamine is also a major component of bacterial cell wall polysaccharides.

These complex sugars are ubiquitous in the environment; cellulose and chitin are the first and second most abundant biopolymers on earth. Degradation of these sugars leads to the release of simple sugars, and thus plays a crucial role in supporting microbial life in soil and in global C cycling. Pure cellulose, chitin, hemicellulose, and xylan are easily decomposed. In nature, they form complexes with other substances such as lignin, making breakdown difficult. Moreover, these complexes have limited solubility and are much too large to enter cells. They must first be hydrolyzed to small and soluble oligosaccharides by extracellular enzymes.

9–1.2 Carbohydrate Hydrolases

Although many carbohydrates in nature have similar chemical formulas, their degradations require different enzymes or enzyme systems because of their different configurations and glycosidic linkages. All enzymes that hydrolyze O- and S-glycosidic bonds are termed glycosidases or glycoside hydrolases with an enzyme classification code of EC 3.2.1 (Webb, 1992; Moss, 2010). In enzyme nomenclature, the nature of the bond hydrolyzed is indicated by the second figure, while the nature of the substrate is indicated by the third figure in the code number of hydrolases. There are 165 enzymes currently listed under glycosidases.

The 10 carbohydrate hydrolases discussed in this chapter belong to O-glycosidases and are summarized in Table 9–1. Of these enzymes, activities of α- and β-glucosidases, α- and β-galactosidases, N-acetyl-β-glucosaminidase, and invertase result in the release of monosaccharides from disaccharides and soluble oligosaccharides, while activities of amylase, cellulase, xylanase, and chitinase undertake endohydrolysis of complex carbohydrates and release mixtures of di- and oligosaccharides. Enzymes involved in the hydrolysis of complex

Table 9–1. Carbohydrate hydrolases discussed in this chapter (Webb, 1992; Moss, 2010).

Common/accepted name	Systematic name	Enzyme classification number	Reactions
α-Amylase	4-α-D-glucan glucanohydrolase	3.2.1.1	endohydrolysis of 1,4-α-D-glucosidic linkages in polysaccharides containing three or more 1,4-α-linked D-glucose units
Cellulase	4-(1,3;1,4)-β-D-glucan 4-glucanohydrolase	3.2.1.4	endohydrolysis of 1,4-β-D-glucosidic linkages in cellulose, lichenin, and cereal β-D-glucans
Xylanase/*endo*-1,4-β-xylanase	4-β-D-xylan xylanohydrolase	3.2.1.8	random hydrolysis of 1,4-β-D-xylosidic linkages in 1,4-β-D-xylans
Chitinase	(1-4)-2-acetamido-2-deoxy-β-D-glucan glycanohydrolase	3.2.1.14	random hydrolysis of N-acetyl-β-D-glucosaminide 1,4-β linkages in chitin and chitodextrins
α-Glucosidases	α-D-glucoside glucohydrolase	3.2.1.20	hydrolysis of terminal, nonreducing 1,4-linked α-glucose residues with release of α-D-glucose
β-Glucosidases	β-D-glucoside-glucohydrolase	3.2.1.21	hydrolysis of terminal, nonreducing 1,4-linked β-glucose residues with release of β-D-glucose
α-Galactosidase	α-D-galactoside-galactosidase	3.2.1.22	hydrolysis of terminal, nonreducing 1,4-linked α-galactose residues in α-D-galactosides, including galactose oligosaccharides, galactomannans and galactolipids
β-Galactosidase/lactase	β-D-galactoside-galactosidase	3.2.1.23	hydrolysis of terminal, nonreducing 1,4-linked β-galactose residues in β-D-galactosides
Invertases/saccharase/β-fructofuranosidase	β-D-fructofuranoside fructohydrolase	3.2.1.26	hydrolysis of terminal nonreducing β-D-fructofuranoside residues in β-D-fructofuranosides
β-N-Acetylhexosaminidase/β-N-acetylglucosaminidase	β-N-acetyl-D-hexosaminide N-acetylhexosaminohydrolase	3.2.1.52	hydrolysis of terminal nonreducing N-acetyl-D-hexosamine residues in N-acetyl-β-D-hexosaminides

carbohydrates are as complex as the substrates undertaken. Understanding the complexity of these enzyme systems is important in data interpretation.

9–1.2.1 Amylases

According to the Nomenclature Committee of the International Union of Biochemistry and Molecular Biology (IUBMB) (Webb, 1992; Moss, 2010), degradation of starch involves three different amylases. Alpha-amylases (EC 3.2.1.1), which are also termed endoamylases, hydrolyze α-1,4-glycosidic linkages, randomly yielding dextrins, oligosaccharides, and monosaccharides. Exoamylases hydrolyze the α-1,4-glycosidic linkage from only the nonreducing ends of the polysaccharide chains. Exoamylases include β-amylases (EC 3.2.1.2) and γ-amylases (amyloglucosidases) (EC 3.2.1.3). β-Amylases successively remove β-maltose units successively from the nonreducing ends of the chain, while γ-amylases successively hydrolyze the terminal α-glucose residues from the nonreducing ends of the chain and release β-D-glucose. α-Glucosidase (EC 3.2.1.20) can also catalyze the reaction of γ-amylases, but with the release of α-D-glucose. By quantifying glucose or reducing sugars produced following addition of starch into a soil solution, activities of all amylases are accounted for. Often, α-amylase activity is the rate-limiting step in the synergistic hydrolysis of the large and complex starch molecules. Thus, rates of glucose released from complex starch molecules are used to indicate activities of α-amylase.

9–1.2.2 Cellulases

Similar to amylases, cellulases are an enzyme system consisting of a mixture of different glycosyl hydrolases, each with distinctly different functions that act synergistically to degrade cellulose. A cellulase system can degrade crystalline cellulose, which is often called a "true cellulase" or a "cellulase complex." Currently, it is generally accepted that the cellulase system comprises three categories of cellulases (Fig. 9–3): (i) endocellulase or *endo*-1,4-β-glucanase (EC 3.2.1.4), which attacks the cellulose polymers randomly; (ii) cellobiosidase or *exo*-1,4-β-glucanase (EC 3.2.1.91), which removes cellobiose (two glucose units) from the nonreducing end of the cellulose chains; and (iii) β-D-glucosidase (EC 3.2.1.21), also an exocellulase, which hydrolyzes cellobiose and other

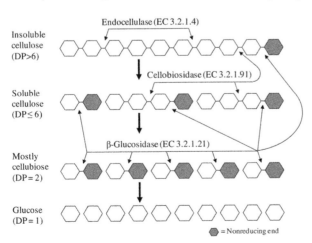

Fig. 9–3. Enzymatic hydrolysis of cellulose (modified from Saddler, 1986). Endocellulase (*endo*-1,4-β-glucanase) attacks the cellulose polymers randomly. Cellobiosidase (exocellulase or *exo*-1,4-β-glucanase) removes cellobiose (two glucose units) from the nonreducing end of the cellulose chains. β-D-Glucosidase (also exocellulase or *exo*-1,4-β-glucanase) hydrolyzes cellobiose and other water-soluble cellodextrins from the nonreducing end and releases simple glucose.

water-soluble cellodextrins from the nonreducing end and releases glucose (Webb, 1992; Moss, 2010). Both cellobiosidase and β-glucosidase are exocellulases. Cellulase enzymes can either be free, particularly in aerobic microorganisms, or grouped in a multicomponent enzyme complex (cellulosome) as in anaerobic cellulolytic bacteria (Bayer et al., 1998). However, these are simplified general categories of enzymes involved in degrading cellulose biopolymers. In reality, the cellulase system is very complex and poorly understood. For example, exo-1,4-β-glucosidase (EC 3.2.1.74) hydrolyzes 1,4-β linkages in 1,4-β-glucans to remove successive glucose units and can also hydrolyze cellobiose (Webb, 1992; Moss, 2010). Revealing functions of these enzymes will require a much more in-depth understanding of their regulation, production, activity, and synergism as a tightly controlled, highly organized system.

9–1.2.3 Xylanases

Xylanases are also an enzyme system, which is usually composed of β-xylanase, β-xylosidase, and enzymes such as α-L-arabinofuranosidase, α-glucuronidase, acetylxylan esterase and hydroxycinnamic acid esterases that cleave side chain residues from the xylan backbone. These enzymes act cooperatively to convert xylan to its constituents (Anand et al., 1990; Sunna and Antranikian, 1997). Therefore, xylanase (1,4-β-D-xylan xylanohydrolase, EC 3.2.1.8) is defined as a class of enzymes that catalyze the endohydrolysis of β-1,4-xylosidic linkages in xylans to produce short-chain oligomers, xylobiose, and xylose (Poutanen and Puls, 1988; Anand et al., 1990). The hydrolysis end product, xylose, is sometimes referred to as wood sugar which is a five-carbon reducing sugar containing an aldehyde functional group (an aldopentose).

9–1.2.4 Chitinases

Two classes of chitin hydrolases are recognized by IUBMB (Webb, 1992; Moss, 2010). These include (i) chitinase, β-1,4-poly-N-acetylglucosaminidase (EC 3.2.1.14), which randomly hydrolyzes N-acetyl-β-D-glucosaminide 1,4-β linkages in chitin and chitodextrins; and (ii) N-acetyl-β-D-glucosaminidase (EC 3.2.1.52), which hydrolyzes chitin chains from the terminal nonreducing end and releases simple N-acetylglucosamine (NAG) units. Although not recognized by IUBMB, a third class of chitinase was reported by Tronsmo and Harman (1993), which releases dimeric NAG units. This enzyme activity showed similarity in function to cellobiosidase (EC 3.2.1.91) in the cellulase system. By analogy, Tronsmo and Harman (1993) suggested that these enzymes be termed chitobiosidase. The enzyme systems hydrolyzing chitin and cellulose bear close resemblance. Similar to the cellulase system, only chitinase (EC 3.2.1.14) has the ability to hydrolyze natural forms of chitin. Studies on chitinase from *Trichoderma atroviride* P. Karst. 1984 showed that chitinase requires at least a tetramer of NAG residues for activity, chitobiosidase requires a trimer, and N-acetyl-β-glucosaminidase requires at least a dimer (Tronsmo and Harman, 1993).

9–1.3 Detection and Quantification of Carbohydrate Hydrolases in Soil

In the complex soil system, enzyme activities are often determined by quantifying an end product released upon enzymatic reactions. Theoretically, hydrolysis of all carbohydrates ultimately leads to the production of monosaccharides, which can be

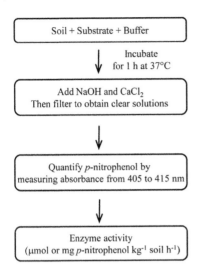

Fig. 9–4. General scheme for assaying activities of α- and β-glucosidases, α- and β-galactosidases, and N-acetyl-β-glucosaminidase in soil using chromogenic p-nitrophenyl substrates and by quantifying p-nitrophenol released.

quantified by the determination of reducing sugars. However, such methodological approaches are not as exclusive and are used only in the absence of methods for the quantification of a specific end product. Enzyme assay methods discussed in this chapter employ two general categories of approaches based on the nature of the substrate used (Fig. 9–4 and 9–5). The first approach employs the use of synthetic chromogenic p-nitrophenyl substrates followed by the quantification of p-nitrophenol released (Fig. 9–4). These assay methods are not only specific for enzyme activities tested, but also are simple and reproducible, making them most popular in soil enzymology studies. Enzyme activities that directly lead to the release of monosaccharides from disaccharides and soluble oligosaccharides, such as α- and β-glucosidases, α- and β-galactosidases, and N-acetyl-β-glucosaminidase, employ this approach (Fig. 9–4, Table 9–2). The second approach employs the use of native, purified, or colloidal substrates followed by the quantification of a specific reaction end product or reducing sugar released (Fig. 9–5, Table 9–3). The activity of invertase can be determined using the second approach described here. Invertase cleaves sucrose and releases glucose and fructose, and is often measured by quantifying rates of reducing sugars produced following addition of sucrose (Fig. 9–5).

The activities of cellulase, amylase, xylanase, and chitinase are also quantified using the second approach. It is challenging to quantify the activities of these enzymes because they undertake endohydrolysis of complex carbohydrates and release mixtures of di- and oligosaccharides. Subsequent activities of exoenzymes are required to ultimately release monosaccharides that are quantified to estimate activities of these enzymes. Detection of monosaccharides following the addition of respective complex substrates provides measurements of the synergistic actions of multiple enzymes, with the rate-limiting reaction dictating the rate of the hydrolysis. Often, rates of monosaccharide release from complex carbohydrates are used to indicate endohydrolysis activities. In quantifying cellulase activities, for example, rates of glucose release are indicative of synergistic reactions of the cellulase system including endo- and exocellulases as well as β-glucosidase (see section 9–1.2). In this complex enzyme system, endocellulase activity is often the rate-limiting step. Therefore, the generation rate of glucose or reducing sugar

Table 9–2. Substrates and buffers used for assaying activities of α- and β-glucosidases, α- and β-galactosidases, and N-acetyl-β-glucosaminidase in soil.

Enzyme	Substrate	Catalog number (Sigma-Aldrich, St. Louis, MO, USA)	Buffer and pH	Reference
α-Glucosidases	p-nitrophenyl-α-D-glucopyanoside (50 mM, 1.506 g/100 mL buffer)	N-1377	MUB†, pH 6.0	Eivazi and Tabatabai, 1988
β-Glucosidases	p-nitrophenyl-β-D-glucopyanoside (50 mM, 1.506 g/100 mL buffer)	N-7006	MUB, pH 6.0	Eivazi and Tabatabai, 1988
α-Galactosidases	p-nitrophenyl-α-D-galactopyranoside (50 mM, 1.506 g/100 mL buffer)	N-0877	MUB, pH 6.0	Eivazi and Tabatabai, 1988
β-Galactosidases	p-nitrophenyl-β-D-galactopyranoside (50 mM, 1.506 g/100 mL buffer)	N-1252	MUB, pH 6.0	Eivazi and Tabatabai, 1988
N-Acetyl-β-glucosaminidase	p-nitrophenyl-N-acetyl-β-D-glucopyanoside (10 mM, 0.342 g/100 mL buffer)	N-9376	acetate buffer, 100 mM, pH 5.5	Parham and Deng, 2000

† MUB, modified universal buffer prepared as described for Reagent 3, section 9–2.3.1.2.

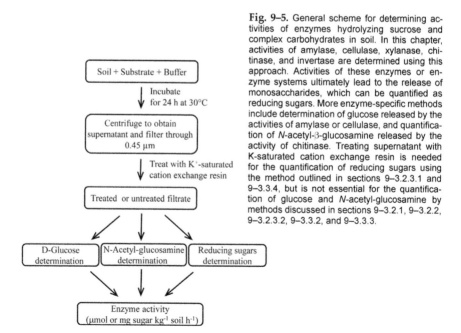

Fig. 9–5. General scheme for determining activities of enzymes hydrolyzing sucrose and complex carbohydrates in soil. In this chapter, activities of amylase, cellulase, xylanase, chitinase, and invertase are determined using this approach. Activities of these enzymes or enzyme systems ultimately lead to the release of monosaccharides, which can be quantified as reducing sugars. More enzyme-specific methods include determination of glucose released by the activities of amylase or cellulase, and quantification of N-acetyl-β-glucosamine released by the activity of chitinase. Treating supernatant with K-saturated cation exchange resin is needed for the quantification of reducing sugars using the method outlined in sections 9–3.2.3.1 and 9–3.3.4, but is not essential for the quantification of glucose and N-acetyl-glucosamine by methods discussed in sections 9–3.2.1, 9–3.2.2, 9–3.2.3.2, 9–3.3.2, and 9–3.3.3.

Soil + Substrate + Buffer

Incubate for 24 h at 30°C

Centrifuge to obtain supernatant and filter through 0.45 μm

Treat with K⁺-saturated cation exchange resin

Treated or untreated filtrate

D-Glucose determination | N-Acetyl-glucosamine determination | Reducing sugars determination

Enzyme activity (μmol or mg sugar kg⁻¹ soil h⁻¹)

Table 9–3. Methods for detecting activities of amylase, cellulase, xylanase, chitinase, and invertase in soil.[†]

Enzyme	Substrate and concentration	Buffer and pH	End product released	Detection	Reference
Amylase	soluble starch solution 2%, w/v	acetate buffer, 100 mM, pH 5.5	glucose	glucose or reducing sugar	Al-Turki and Dick, 2003; Ross, 1965
Cellulase	carboxymethyl cellulose (CMC) 2%, w/v	acetate buffer, 100 mM, pH 5.5	glucose	glucose or reducing sugar	Al-Turki and Dick, 2003; Deng and Tabatabai, 1994a
Xylanase	xylan 1.2%, w/v	acetate buffer, 100 mM, pH 5.5	xylose	reducing sugar	Schinner and von Mersi, 1990; Deng and Tabatabai, 1994a
Chitinase	purified chitin 1.5%, w/v	phosphate buffer, 120 mM, pH 6.0	N-acetyl-β-glucosamine	N-acetyl-β-glucosamine or reducing sugar	Aminoff et al., 1952; Reissig et al., 1955; Rodriguez-Kabana et al., 1983
Invertase	sucrose 10%, w/v	acetate buffer, 100 mM, pH 5.0	glucose and fructose	reducing sugar	Al-Turki and Dick, 2003; Frankenberger and Johanson, 1983a,b

[†] Reducing sugar in soil extracts can be quantified by Somogyi–Nelson or Prussian blue method (Deng and Tabatabai, 1994a; sections 9–3.4 and 9–3.5). The latter is more sensitive (see section 9–1.4).

following addition of cellulose indicates activities of endocellulase even though activities of exocellulases also are involved in the release of glucose from cellulose. In this chapter, methods described for quantifying activities of amylase, cellulase, xylanase, and chitinase employ this approach (Fig. 9–5).

In the above-discussed enzyme assays, purified or colloidal substrates often are used because of limited solubility of the native substrates. Activities of amylase and cellulase can be quantified by determining glucose released, while chitinase activity can be determined by rates of NAG released. These assay methods are relatively more selective than the quantification of reducing sugars, which is employed in the assay of xylanase activity. Currently, there is no method that is specifically for the quantification of xylose in soil. However, activities of all enzymes listed in Table 9–3 can be determined by quantifying reducing sugars released. Although the detected reducing sugars in the solution may not be specific for the enzymatic reaction of interest, colorimetric quantification of reducing sugars are more sensitive than the enzymatic quantification of glucose in soil solutions. When reducing sugars are quantified in an enzyme assay, the specificity of the method lies at the addition of a specific substrate and by calculating enzyme activity with the subtraction of reducing sugars in controls without the addition of substrate. The presence of substantial native substrates for various enzymes in soil could lead to high concentrations of reducing sugars in the control. This was evident in a previous study showing that reducing sugar concentrations increased following incubation of soils with acetate buffer (50 mM, pH 5.5) at 30°C for 24 h, suggesting hydrolysis of native substrates (Deng and Tabatabai 1994a).

9–1.4 Detection and Quantification of Reducing Sugars in Soil

As discussed above, quantification of reducing sugar contents is often employed in assays of soil carbohydrate hydrolases. Colorimetric methods are most common and adaptable in determining reducing sugar concentration in soil, including dinitrosalicylic acid (DNS), Prussian blue (reaction with potassium ferric hexacyanide reagent), and Somogyi–Nelson (molybdenum blue) methods. Following evaluations of these methods for the determination of reducing sugars extracted from soils, Deng and Tabatabai (1994a) concluded that the DNS method was not sensitive enough to be used in the determination of acetate buffer-extractable reducing sugars in agricultural soils; Prussian blue and Somogyi–Nelson methods were the most sensitive. In a 15-mL reaction mixture, Prussian blue can detect as little as 1 µg, while Somogyi–Nelson detects 5 µg of reducing sugars in soil extracts.

However, the Prussian blue method has a narrow detection range (Deng and Tabatabai, 1994a) and is subject to considerable errors due to its high sensitivity. Soil extracts require many-fold of dilution before analysis for reducing sugars by this method. Schinner and von Mersi (1990) reported a 30- to 50-fold dilution for agricultural soils and a 40- to 60-fold dilution for forestry soils. The Somogyi–Nelson method has a wider detection range and soil extracts are seldom diluted before analysis (Deng and Tabatabai, 1994a). Moreover, the Prussian blue complex has only 1 h color stability, while the molybdenum blue color is stable for at least 24 h. On the other hand, metals extracted from soil interfere with molybdenum blue color development but not Prussian blue (Woods and Mellon, 1941; Deng and Tabatabai, 1994a). Studies showed that the interfering metals in soil extracts could be removed by treating the extracts with K-saturated exchange resin prior to analysis of reducing sugars (Deng and Tabatabai, 1994a).

In the literature, the Shaffer–Hartmann method for determination of reducing sugars is also used in soil enzyme assays (Speir et al., 1984). This method is based on the use of iodometric titration of cuprous oxide formed on reduction of copper(II) by glucose (Shaffer and Hartman, 1921). However, the accuracy and reproducibility of this method depends strongly on the conditions of the analysis and the sample (Shaffer and Somogyi, 1933; Somogyi, 1945). Autoreduction of the copper reagent leads to overestimation of enzyme activities. The reaction rate is also sensitive to slight changes in basicity of the solution.

It is also important to understand that soil enzyme assays provide measures of potential activities of multiple enzymes that carry similar functions in nature. Some of the enzymes could be isozymes existing within the same living system or originating from different organisms with different amino acid sequences. Some soil enzymes, such as cellulase, amylase, xylase, and chitinase, are enzyme systems comprised of multiple enzymes that catalyze different reactions that ultimately lead to the release of the measured monosaccharides. Sodium azide and toluene are commonly used in the assay of soil carbohydrate hydrolases to inhibit microbial growth during incubation. Sodium azide is bacteriostatic andcan inhibit cytochrome oxidase in Gram-negative bacteria, but not in Gram-positive bacteria (Lichstein and Soule, 1944). Therefore, toluene has been the most widely used microbial inhibitor in soil enzyme assays. Toluene is believed to stop enzyme synthesis by living cells and prevent assimilation of the reaction products during the assay procedure. Studies by Al-Turki and Dick (2003) showed that toluene prevented assimilation of solution glucose by soil microbes during incubation, allowing quantitative recovery of glucose in soil mixtures. Toluene, however, enhances the intracellular contribution to the measured activities in soils (Frankenberger and Johanson, 1986; Tabatabai, 1994).

As discussed earlier, the 10 enzyme assay methods discussed in this chapter can be divided into two groups. Five enzymes can be quantified using p-nitrophenyl substrates followed by determination of p-nitrophenol released; the remaining five enzymes are quantified by the detection of reducing sugars or a specific sugar released following addition of respective substrates.

9–2 GLYCOSIDASES THAT RELEASE MONOSACCHARIDES (α- and β-Glucosidases, α- and β-Galactosidases, and N-Acetyl-β-Glucosaminidase)

9–2.1 Introduction

Glycosidases catalyze hydrolysis of water-soluble di- and oligosaccharides and release monosaccharides. Activities of these enzymes are important in providing labile C and energy sources to support microbial life in soil. These enzymes also are responsible for the ultimate release of the measured end products in quantifying synergistic action of enzyme systems that hydrolyze complex carbohydrates. In this section, five glycosidases with variations in the fourth EC numbers are discussed (EC 3.2.1.20, 21, 22, 23, and 52, respectively). α-Glucosidases can hydrolyze degradation products of α-amylase, β-glucosidases hydrolyze degradation products of cellulase, while the activities of N-acetyl-β-glucosaminidase lead to the ultimate release of NAG residues from chitin.

9–2.2 Principles

The assay of α- and β-glucosidases, α- and β-galactosidases, and N-acetyl-β-glucosaminidase activities are based on colorimetric determination of p-nitrophenol released when soil is incubated with respective p-nitrophenyl substrates at the pH optimal for the specific enzymatic reaction. The incubation is performed at 37°C for 1 h. The p-nitrophenol released is quantified after addition of 0.5M CaCl$_2$ and 0.1M THAM buffer pH 12.

The reaction involving β-glucosidase hydrolyzing p-nitrophenyl substrates follows:

$$[1]$$

The substrate is p-nitrophenyl-β-D-glucoside when R =

The substrate is p-nitrophenyl-α-D-glucoside when the –OR group of the substrate is below the ring (the sugar unit would be α-D-glucose instead of β-D-glucose). When the –OH at C-4 is above the ring, the sugar is galactose. Similarly, galactose can have α- and β-configurations, becoming α- and β-galactosides. When the sugar unit is replaced by a NAG (the –OH at C-2 is replaced by $-NH-\overset{\overset{\displaystyle O}{\|}}{C}-CH_3$, the substrate is p-nitrophenyl-N-acetyl-β-D-glucosaminide.

Assay methods for detecting activities of these monosaccharide releasing glycosidases are similar, but employ the use of different substrates and buffering pH. Based on findings by Eivazi and Tabatabai (1988), the optimum pH for activities of α- and β-glucosidase and α- and β-galactosidase in soil is pH 6.0 in modified universal buffer (MUB). Data reported by Parham and Deng (2000) suggested that the optimum pH for N-acetyl-β-glucosaminidase activity in soil is at pH 5.5 in 100 mM acetate buffer.

9–2.3 Assay Methods

9–2.3.1 β-D-Glucosidase (Eivazi and Tabatabai, 1988)

9–2.3.1.1 Apparatus

- 50-mL Erlenmeyer flasks fitted with No. 2 rubber stoppers
- Incubator (37°C)
- Whatman No. 2V folded filter paper
- Spectrophotometer or colorimeter that can be adjusted to measure absorbance from 405 to 415 nm

9–2.3.1.2 Reagents

1. Toluene
2. Stock solution of modified universal buffer (MUB): Dissolve 12.1 g of tris(hydroxymethyl)aminomethane (THAM), 11.6 g of maleic acid, 14.0 g of citric acid, and 6.3 g of boric acid (H_3BO_3) in about 800 mL of 0.5 M sodium hydroxide (NaOH), adjust to 1 L with 0.5 M NaOH, and store at under 4°C.
3. Modified universal buffer (pH 6.0): Place 200 mL of MUB stock solution in a 1-L beaker containing a magnetic stirring bar, which is placed on a magnetic stirrer. Titrate the pH of the solution to 6.0 with HCl (0.1–0.5 M), and adjust the volume to 1 L with DI water.
4. p-Nitrophenyl-β-D-glucoside (PNG) (50 mM): Prepare by dissolving 0.753 g of PNG (N-7006, Sigma-Aldrich, St. Louis, MO, USA) in about 40 mL of MUB pH 6.0, and adjusting to 50 mL with the same buffer. Prepare daily or the solution can be stored at 4°C for days and −20°C for weeks.
5. Calcium chloride ($CaCl_2$) (0.5 M): Dissolve 73.5 g of $CaCl_2·2H_2O$ in DI water with the final volume adjusted to 1 L.
6. Tris(hydroxymethyl)aminomethane (THAM) buffer (100 mM, pH 12): Dissolve 12.1 g of THAM in about 800 mL of DI water. Adjust the pH of the solution to 12 by titration with 0.5 M NaOH, and adjust the volume to 1 L with DI water.
7. Standard p-nitrophenol solution (10 mM): Dissolve 1.391 g of p-nitrophenol in about 800 mL of DI water in a 1-L volumetric flask, and adjust to 1 L with DI water. Store the solution in the dark at 4°C.

9–2.3.1.3. Procedure

In a 50-mL Erlenmeyer flask, place 1 g of soil (<2 mm), add 0.2 mL of toluene, mix and let it sit for about 15 min in a fume hood. Then add 4 mL of MUB pH 6.0 and 1 mL of PNG solution, stopper the flask, mix thoroughly, and incubate the soil suspension at 37°C. After 1 h, remove the stopper, add 1 mL of 0.5 M $CaCl_2$ and 4 mL of 0.1 M THAM buffer (pH 12), mix the contents, and filter the soil suspension through a Whatman No. 2V folded filter paper. Measure the yellow color intensity of the filtrate with a spectrophotometer at 405 nm, and calculate the amount of p-nitrophenol released by reference to a calibration curve developed with standards containing 0, 100, 200, 300, 400, or 500 nmol of p-nitrophenol in each flask. To prepare this calibration curve, dilute 1 mL of the standard solution (10 mM) to 100 mL in a volumetric flask with water and mix the solution thoroughly. Then pipette 0, 1, 2, 3, 4, or 5 mL of this diluted standard solution (0.1 mM) into 50-mL Erlenmeyer flasks, adjust to 50 mL by adding water, and follow with the addition of $CaCl_2$ and THAM as described for enzyme assays of soil samples. When filtrates have color intensity exceeding that of the highest p-nitrophenol standard solution, dilute the filtrate with a 1:1 mixture of MUB pH 6.0 and 0.1 M THAM pH 12 until the absorbance readings fall within the limits of the calibration curve. Express enzyme activity as μmol p-nitrophenol kg^{-1} soil h^{-1}.

Controls should be included for each assay by following procedure described above, but adding the substrate PNG solution after termination of the reaction using THAM buffer (pH 12).

9–2.3.1.4. Comments

Calcium chloride (CaCl$_2$) is added to prevent dispersion of clay and extraction of soil organic matter. Tabatabai (1994) suggested that the substrates are hydrolyzed with time in the presence of excess NaOH. Therefore, treating the incubated soil samples with THAM buffer pH 12 was recommended. The rate of such chemical hydrolysis of substrates is highly variable, depending on the substrates. Absorbance of *p*-nitrophenol can be measured at wavelengths from 405 to 415 nm. Detection is more sensitive at 405 nm, but is subject to reduced analytical errors at 415 nm. For each assay, it is recommended to include another control for each specific substrate to determine rates of chemical hydrolysis following the procedure described for the assay, but without adding soil. Any detectable absorbance values should be subtracted from those obtained in sample assays before calculating enzyme activities. All enzyme activities should be expressed on soil dry weight basis.

9–2.3.2. Other Enzymes (α-Glucosidase, α- and β-Galactosidases, and N-Acetyl-β-glucosaminidase (Eivazi and Tabatabai, 1988; Parham and Deng, 2000)

Apparatus, reagents, and procedures are the same as described above for the detection of β-glucosidase activity, but substrates used are different. For the assay of N-acetyl-β-glucosaminidase, acetate buffer (100 mM, pH 5.5) is used in the place of MUB (pH 6.0). Substrates and buffers used in these assays are summarized in Table 9–2.

Acetate buffer (100 mM, pH 5.5) is prepared by dissolving 13.6 g of sodium acetate trihydrate (CH$_3$COONa·3H$_2$O) in about 800 mL of DI water, titrating to pH 5.5 with 99% glacial acetic acid (CH$_3$COOH), and adjusting the volume to 1 L with DI water.

9–3 AMYLASE, CELLULASE, XYLANASE, CHITINASE, AND INVERTASE

9–3.1 Introduction

Substrates for the activities of amylase, cellulase, xylanase, chitinase, and invertase often have limited water solubility. Thus, low concentrations, suspensions, or treated derivatives of the complexes are commonly used as the substrates in enzyme assays. Cellulose, for example, is not soluble when the degree of polymerization is >6. Carboxymethyl cellulose (CMC) is often used as the substrate in cellulase assays. Carboxymethyl cellulose molecules are shorter than native cellulose and are rod-like at low concentrations. Its solution is viscous. Activities of these enzymes all lead to the release of reducing sugars, which are glucose for amylase and cellulase, glucose and fructose for invertase, xylose for xylanase, and N-acetyl-β-glucosamine (NAG) for chitinase. Therefore, their activities are indicated by the releasing rates of sugars from these complexes. The released sugars can be quantified colorimetrically.

9–3.2 Principles

The general principles for assaying activities of these enzymes are similar. Upon incubation of soil with respective substrates in a buffered soil suspension at 30°C, the enzymatically released end product is quantified and expressed per unit of time and mass of soil to indicate enzyme activities under the assay conditions. Comparison of these enzyme assay methods is shown in Table 9–3. These enzyme activities can be determined by the quantification of reducing sugars or more specifically by glucose for amylase and cellulase, and NAG residues for chitinase. Glucose or NAG standard solutions are used to develop the calibration curves. Cleavage of sucrose by invertase releases glucose and fructose at a 1:1 ratio. Therefore, the calibration curve for quantifying invertase activity is often made of glucose and fructose at a 1:1 ratio. As discussed in section 9–1.4, the assay method would be more specific if a particular product is quantified because reducing sugars can be produced from several different enzymatic reactions. Unfortunately, glucose quantification by the currently available method is interfered with by the presence of sucrose, while a method for the specific quantification of xylose in soil extracts is not available yet. Principles for the quantifications of glucose, NAG, and reducing sugars are discussed below.

9–3.2.1 Glucose

Glucose is water-soluble and can be readily extracted from soil (Frey et al., 1999). The concentration of glucose can be determined by an enzymatic approach, which has been tested for the determination of glucose concentration in soils (Raba and Mottola, 1995; Frey et al., 1999; Al-Turki and Dick, 2003). Glucose oxidase or hexokinase-glucose-6-phosphate dehydrogenase have been used as analytic reagents in the measurement of glucose in soils. Such approaches of glucose quantification are not only accurate, but also fast and simple. The reagents for the measurements are commercially available as assay kits.

In the enzymatic approach for the quantification of glucose using a commercial glucose determination kit (Trinder), two reactions are employed. In the first reaction, glucose is oxidized to gluconic acid and hydrogen peroxide by glucose oxidase. Subsequently, hydrogen peroxide reacts with 4-aminoantipyrine and p-hydroxybenzene sulfonate in the presence of peroxidase to produce a quinoneimine complex. These reactions are described as:

$$\text{Glucose} + 2H_2O + O_2 \xrightarrow{\text{Glucose Oxidase}} \text{Gluconic Acid} + H_2O_2$$

$$H_2O_2 + \text{4-Aminoantipyrine} + \text{p-Hydroxybenzene Sulfonate} \xrightarrow{\text{Peroxidase}} \text{Quinoneimine Dye} + 4H_2O$$

[2]

The formed complex has a pinkish color and can be detected at 505 nm. This glucose assay kit has been thoroughly evaluated by Al-Turki and Dick, (2003). Glucose recovery from soil extracts was about 100% in soils tested.

9–3.2.2 N-Acetyl-β-glucosamine (NAG)

Hydrolysis of chitin by chitinase leads to the ultimate release of NAG units, which can be quantified by the coloring reaction of NAG with p-(dimethylamino)-benzaldehyde (DMAB). The color development reactions are referred to as Morgan–Elson reactions (Kuhn and Kruger, 1956; 1957). The proposed reaction schemes are shown in Fig. 9–6. Under alkaline conditions, N-acetylglucosamine reacts in its pyranose form to yield two chromogenic compounds, α and β forms, which are often referred to as chromogen I and II (Kuhn and Kruger, 1956; 1957; Beau et al., 1977; Muckenschnabel et al., 1998). With the addition of acid, chromogen III is

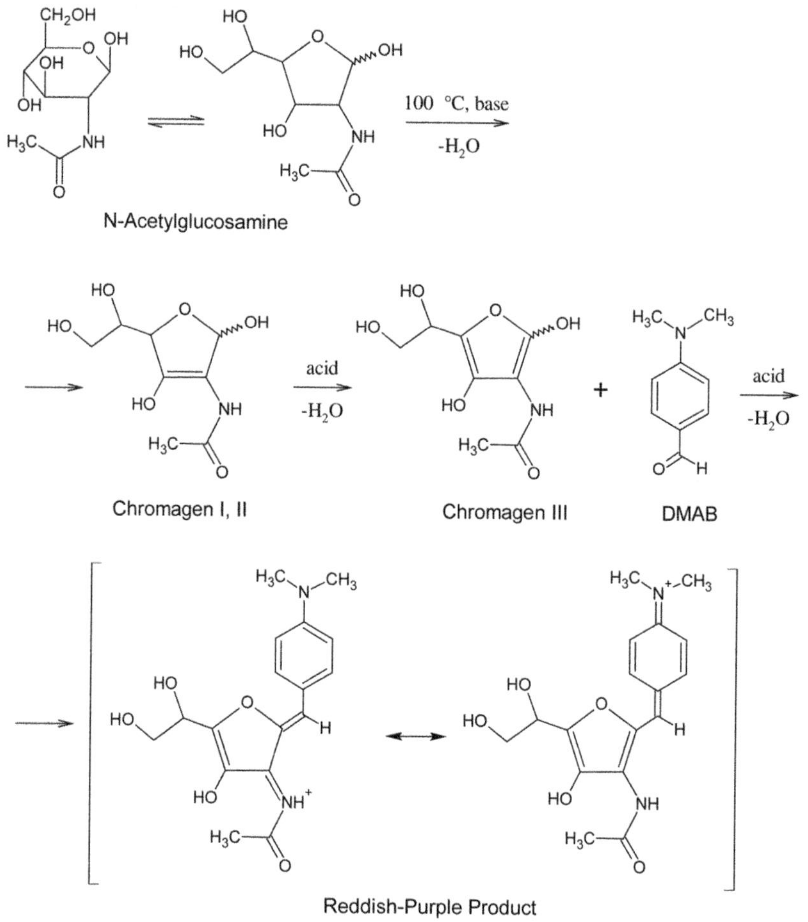

N-Acetylglucosamine

Chromagen I, II   Chromagen III   DMAB

Reddish-Purple Product

Fig. 9–6. Proposed structures of the colored product and chromogens I, II, and III in the Morgan–Elson reaction (Kuhn and Kruger, 1956; Kuhn and Kruger, 1957; Beau et al., 1977; Muckenschnabel et al., 1998; Takahashi et al., 2003)

formed, which further reacts with DMAB reagent to yield a reddish-purple product that can be measured by absorbance at 590 nm (Muckenschnabel et al., 1998; Takahashi et al., 2003).

9–3.2.3 Reducing Sugars

Methods for the determination of reducing sugars have been discussed in section 9–1.4 and the Somogyi–Nelson method is recommended for use in most agricultural soils. The Prussian blue method can be used when a more sensitive detection method is needed. Other methods may be used depending on specific objectives according to discussions above and referenced studies. In this chapter, Somogyi–Nelson and Prussian blue methods will be discussed.

9–3.2.3.1 Somogyi–Nelson Method

In the quantification of reducing sugars by the Somogyi–Nelson method, glucose reduces Cu(II) to Cu(I) ions on addition of Somogyi reagents (Nelson, 1944; Somogyi, 1945). Subsequent addition of Nelson reagent leads Cu(I) to react with polymolybdate ion and form a blue complex. The reactions involved can be summarized as follows.

$$\text{Glucose}_{\text{Red}} + \text{Cu(II)} \rightarrow \text{Glucose}_{\text{Ox}} + \text{Cu(I)} \qquad\qquad [3]$$
$$\text{Cu(I)} + (\text{NH}_4)_2\text{MoO}_4 \rightarrow [\text{Blue Complex}]$$

Although the specific structure of the blue complex is not clear, the intensity of the blue color is proportional to glucose concentration in the solution.

9–3.2.3.2 Prussian Blue Method

The Prussian blue method applied to determine sugar is a variation of the method of Park and Johnson (1949). Schinner and von Mersi (1990) adapted the method for soil enzyme assays. The Prussian blue method involves the reduction of ferricyanide ions in alkaline solution followed by formation of Prussian blue (ferric ferrocyanide), which is measured quantitatively with a spectrophotometer at 690 nm (Schinner and von Mersi, 1990). The key colored complex in Prussian blue pigment is an insoluble inorganic compound composed of iron and cyanide ions, with an approximate formula of $\text{Fe}_7(\text{CN})_{18}$ hydrated with 14 to 16 H_2O molecules. The color of the pigment depends partly on the size of precipitated particles formed from addition of iron(III) to soluble ferrocyanide. Although Prussian blue is insoluble, it tends to form small crystalline colloids that behave like solutions. According to Dunbar and Heintz (1997), these "soluble" forms of Prussian blue have compositions with the approximate formula $\text{KFe}_2\text{Fe(CN)}_6$. It is not surprising that the Prussian blue color developed from soil extracts was stable for only 1 h (Deng and Tabatabai, 1994a).

9–3.3 Assay Methods

9–3.3.1 Incubation of Soil with Respective Substrates (Deng and Tabatabai, 1994b; Frankenberger and Johanson, 1983a,b; Rodriguez-Kabana et al., 1983; Schinner and von Mersi, 1990)

9–3.3.1.1. Apparatus

- Erlenmeyer flasks, 50 mL
- Centrifuge tubes with caps, 50 mL, polypropylene
- Volumetric flasks, 100 mL
- 0.45-μm MF-Millipore cellulose acetate membrane filter (Millipore Corp., Bedford, MA)
- Incubator (30°C)
- Centrifuge, high-speed

9–3.3.1.2. Reagents

1. Toluene ($\text{C}_6\text{H}_5\text{CH}_3$)
2. Buffers:

   For amylase, cellulase, xylanase, and invertase: Acetate buffer (100 mM) with pH 5.5 for amylase, cellulase, and xylanase and pH 5.0 for invertase. Prepare by dissolving 13.6 g of sodium acetate trihydrate ($CH_3COONa·3H_2O$) in 700 mL of DI water. Titrate the solution to pH 5.5 or 5.0 with 99% glacial acetic acid (CH_3COOH) and adjust the volume to 1 L with DI water.

For chitinase: Phosphate buffer (0.12 M, pH 6.0). Prepare by dissolving 16.13 g of KH_2PO_4 and 0.2 g of $Na_2HPO_4·2H_2O$ in 700 mL of DI water. Adjust the solution to pH 6.0 with NaOH or HCl, and adjust the volume to 1 L with DI water. Store this solution at 4°C.

3. Substrates: Prepare all in respective buffer (w/v). Can be stored at 4°C for days, −20°C for weeks.

   Amylase: 2% starch solution. Prepare by dissolving 20 g of soluble starch in 1 L of acetate buffer (100 mM, pH 5.5).

   Cellulase: 2% carboxymethyl cellulose (CMC) solution. Prepare by dissolving 20 g of CMC in 1 L of acetate buffer (100 mM, pH 5.5). Note: CMC dissolves slowly in acetate buffer. It may take several hours to prepare the 2% CMC-acetate buffer solution.

   Xylanase: 1.2% xylan solution. Prepare by suspending 12 g xylan from oat spelts (Sigma-Aldrich 95590) in 1 L of acetate buffer (100 mM, pH 5.5) and stirring at 45°C for 2 h.

   Chitinase: 1.5% chitin solution. Prepare by dissolving 1.5 g of purified chitin (Sigma-Aldrich 9752) in 100 mL of freshly prepared phosphate buffer.

   Invertase: 10% sucrose solution. Prepare by adding 10 g of sucrose in 100 mL acetate buffer (100 mM, pH 5.0).

9–3.3.1.3. Procedure

Place 1 to 5 g of soil (<2 mm) in a 50-mL Erlenmeyer flask, add 0.2 to 0.5 mL of toluene; mix the contents, and place under a fume hood for 15 min. Then add 20 mL of respective buffered substrate solution. Stopper the flask, mix thoroughly, and incubate the soil suspension for 24 h at 30°C. Following incubation, mix the suspension well, transfer to a 50-mL centrifuge tube, and centrifuge at 17,000 g for 10 min at 4°C. Filter the supernatant through a 0.45-μm MF-Millipore membrane filter. The filtrates are used for the quantification of glucose, NAG, or reducing sugars as discussed below.

9–3.3.2 Quantification of Glucose in Soil Extracts (Al-Turki and Dick, 2003)

9–3.3.2.1 Apparatus

- Test tube, 15 mL
- Spectrophotometer or colorimeter that can be adjusted to measure absorbance at 505 nm

9–3.3.2.2 Reagents

1. Glucose diagnostic kit (Trinder, Catalog No. 315-100, Sigma-Aldrich, St. Louis, MO, USA).

2. Silver nitrite (1 M): Dissolve 1.7 g of silver nitrite ($AgNO_3$) in 7 mL of DI water. Adjust the volume to 10 mL with DI water.

3. Glucose standard solution (5.0 mM or 0.900 g L^{-1}): Dissolve 0.900 g of glucose in 1 L of DI water. Store solution at 4°C.

9–3.3.2.3 *Procedure*

Add 1 mL of the filtered supernatant to a test tube containing 2 mL of warm (room temperature) glucose reagent from the diagnostic kit. Mix the content gently by inverting the tube, incubate at room temperature (21–23°C) for 20 min, and then add 25 µL of $AgNO_3$ (1 M) to stop the enzymatic reaction. The pink color intensity of the supernatant is measured at 505 nm relative to DI water within 20 min.

Two types of controls are included, one without addition of soil to correct for impurities in the starch solution as well as nonenzymatic hydrolysis of the substrate, and another without starch to correct for background glucose present in the soil and that generated from native substrates of various enzymatic reactions. The former is included for each set of analysis and the latter is included for each soil sample. The two controls are subtracted from values obtained from the starch-treated soils.

To prepare a glucose standard curve, pipette 0, 60, 120, 180, 300, or 600 µL of glucose stock solution (5.0 mM or 0.900 g L^{-1}) into test tubes and adjust to 3 mL with DI water. Add 1 mL of each standard solution to 2 mL of glucose reagent from the diagnostic kit and perform the same procedure as described above for the determination of glucose in soil supernatants. This will result in glucose standard calibration solutions of 0, 0.1, 0.2, 0.3, 0.5, and 1.0 µmol glucose. If the developed color intensity exceeds that of the 1.0-µmol glucose standard, an aliquot of the supernatant should be diluted with acetate buffer until the absorbance reading falls within the limits of the calibration graph. Calculate the glucose concentration of the analyzed aliquot by reference to a calibration curve. The enzyme activity can be expressed as µmoles glucose kg^{-1} soil h^{-1}.

9–3.3.2.4 *Comments*

Cooling of the solution during centrifugation would slow down the enzymatic reactions significantly. If a temperature-controlled centrifuge is not available, the solutions can be chilled by placing the tubes on ice immediately following incubation. Repeated high-speed centrifugation (about 17,000 *g*) can be used in the place of filtration. It is critical to obtain clear supernatants of the soil suspension for the subsequent determination of glucose. Gentle shaking between centrifugations may be needed to obtain clear supernatants. As discussed earlier, other commercial glucose assay kits can also be used as long as quantitative recovery of glucose standard added to soils can be achieved.

9–3.3.3 Quantification of *N*-Acetyl-β-Glucosamine in Soil Extracts (Aminoff et al., 1952; Reissig et al., 1955; Rodriguez-Kabana et al., 1983)

9–3.3.3.1 *Apparatus*

- Volumetric flasks, 50 and 100 mL
- Test tubes, 15 mL
- Spectrophotometer or colorimeter
- Boiling water bath
- Incubator (37°C)

9–3.3.3.2 Reagents

1. Borate buffer (0.8 M, pH 9.1): Dissolve 49.5 g of H_3BO_3 and 12.8 g of NaOH in 700 mL of DI water. Adjust the solution to pH 9.1 with NaOH, and adjust the volume to 1 L with DI water. Store the solution at 4°C.

2. Potassium chloride solution (2 M): Dissolve 149 g of KCl in 700 mL of DI water. Adjust the volume to 1 L with DI water. Store this solution at 4°C.

3. 4-(dimethylamino)benzaldehyde (DMAB) reagent: Dissolve 10 g of DMAB in 87.5 mL of glacial acetic acid (CH_3COOH) and 12.5 mL of concentrated HCl. This stock solution can be stored at 4°C for a month without significant deterioration. Prepare the working DMAB solution daily by diluting 1 part of the stock solution with four parts of glacial acetic acid.

4. N-acetylglucosamine solution (10 g L^{-1}): Dissolve 1.000 g of N-acetylglucosamine in 70 mL of DI water. Adjust the volume to 100 mL with DI water. Store the solution at 4°C.

9–3.3.3.3 Procedure

To analyze for N-acetylglucosamine concentration, transfer 0.5 mL of filtrate to a 15-mL test tube, and adjust to 2 mL with DI water. Then, add 0.4 mL of borate buffer (0.8 M, pH 9.1), mix the tube thoroughly, and place in a boiling water bath to the level of the liquid for 3 min. Cool the tube in a pan with cold water until the solution reaches room temperature (21–23°C). Add 5 mL of working solution of DMAB reagent to the test tube, shake it thoroughly, and incubate at 37°C for exactly 30 min. Chill the tubes in ice-cold water immediately after incubation. Take an aliquot of supernatant and measure the absorbance at 590 nm relative to DI water within 20 min. The color fades slowly (<1% per min at 20°C). It is, therefore, important to keep the solutions cool and take the absorbance reading as soon as possible on completion of the color reaction.

Two types of controls should be included. One control is prepared by incubating the buffered chitin solution without addition of soil to correct for background absorbance reading caused by impurities in the chitin solution. This type of control is included for each set of analysis. Another control is prepared by treating 1 g of soil with toluene, and then adding 10 mL of phosphate buffer (0.12 M, pH 6.0) without chitin to correct for background N-acetylglucosamine present in the soil. This control should be included for each soil sample. Values of these controls should be subtracted from values obtained for the tested unknown samples.

Calculate NAG content of the aliquot analyzed by reference to a calibration curve plotted from results obtained with standards containing 0, 12.5, 25, 50, 100, or 200 μg of NAG. To prepare this curve, pipette 0, 0.125, 0.25, 0.5, 1.0, or 2.0 mL of the standard stock solution into six 50-mL volumetric flasks. Add 12.5 mL of phosphate buffer and 25 mL of potassium chloride solution, and adjust the volume to 50 mL with DI water. Then pipette 0.5 mL of each of the diluted NAG standard solutions into 15-mL test tubes, and perform the color reaction as described for NAG in the aliquots of the incubated samples. If the absorbance for the aliquot is too high, dilute an aliquot with DI water in two-fold steps until the signal falls into the calibration range. Enzyme activity can be expressed as μmole N-acetylglucosamine kg^{-1} soil h^{-1}.

9–3.3.3.4 *Comments*

If the commercial colloid chitin solution is not available, the colloidal chitin can be prepared from crab shells as described by Hirano and Nagao (1988). The chitin concentration used for the assay is about 50% of the K_m values for chitinases in soils (Rodriguez-Kabana et al., 1983). To maintain substrate saturation during the incubation, about 10% chitin solution is desired. However, concentrations of chitin above 1.5% result in little increase of the detected activity due to considerable viscosity of the substrate solution.

The colorimetric reaction is sensitive to the reaction conditions. The pH of the borate buffer affects the required heating time. Studies by Aminoff et al. (1952) showed that a decrease by 1 pH unit resulted in a twofold reduction of the heating time. Meanwhile, the rate of chromogen decomposition also will decrease. Moreover, the chromogen formed by heating N-acetylglucosamine in alkaline solution is not stable and decomposes rapidly with prolonged heating. Thus, it is important to take precautions to maintain exact pH and heating time for all analyses. At 37°C, full color develops in 30 min and begins to fade at a rate of 0.3% per minute (Reissig et al., 1955). The fading rate was decreased to 0.1% per minute by cooling the solution to 22 to 25°C (Reissiget al., 1955).

NAG released from chitinase activity can also be quantified by methods that quantify reducing sugars as discussed below.

9–3.3.4 Quantification of Reducing Sugar in Soil Extracts by the Somogyi–Nelson Method (Deng and Tabatabai, 1994a)

9–3.3.4.1 *Apparatus*

- Test tubes with caps, 15 mL
- Volumetric flasks, 100 mL
- Spectrophotometer or colorimeter
- Boiling water bath
- End-to-end shaker

9–3.3.4.2. *Reagents*

1. K-saturated cation-exchange resin: Add KCl (1 M) to analytical grade cation-exchange resin, AG 50W-X8, 20 to 50 mesh, hydrogen form (Bio-Rad Lab., Richmond, CA, USA) and shake for 15 min. Decant the KCl solution and repeat the procedure two more times before washing the resin thoroughly with DI water. To regenerate the used resin, wash it three times with HCl (1 M), then three times with KCl (1 M), and then three times with DI water to remove the excess KCl.

2. Somogyi reagent 1: Dissolve 30 g of anhydrous Na_2CO_3, 15 g of sodium potassium tartrate ($C_4H_4KNaO_6$), 20 g of $NaHCO_3$, and 180 g of anhydrous Na_2SO_4 in 0.8 L of boiled DI water. Mix thoroughly until all salts are dissolved and adjust the volume to 1 L with DI water. This solution should be stored at room temperature above 20°C. Filter solution if precipitation is formed.

3. Somogyi reagent 2: Dissolve 45 g of anhydrous Na_2SO_4 and 5 g of $CuSO_4 \cdot 5H_2O$ in about 200 mL of boiled DI water. Mix the solution thoroughly until all salts are dissolved and adjust the volume to 250 mL with DI water.

This solution should be stored at room temperature above 20°C.

4. Nelson reagent: Dissolve 25 g of ammonium molybdate [$(NH_4)_2MoO_4$] in 450 mL of DI water; then add 21 mL of H_2SO_4 (concentrated). Mix the solution thoroughly and incubate at 37°C for 24 to 48 h. Do not use the solution if bright yellow precipitation is formed. Store this reagent in a glass-stoppered brown bottle.

5. Glucose standard solution (5.0 mM or 0.900 g L^{-1}): Dissolve 0.900 g of glucose in 1 L of DI water. Store solution at 4°C.

9–3.3.4.3 Procedure

Transfer an aliquot (~10 mL) of the filtered solution to a centrifuge tube, add about 2 g of K-saturated cation-exchange resin, and shake the tube in an end-to-end shaker for 30 min. Keep the treated solutions on ice until analysis for reducing sugars released.

To analyze the reducing sugar concentration, transfer an aliquot of clear filtrate (1–6 mL) containing 0 to 60 μg of reducing sugars to a 15-mL test tube, and adjust to 6 mL with acetate buffer. Add a 2-mL mixture of freshly prepared Somogyi 1 and 2 (4:1), mix thoroughly, and heat in a boiling water bath for exactly 20 min. Cool the test tube to room temperature in cold water. Subsequently, add 2 mL of Nelson reagent and mix thoroughly. The absorbance of the developed green color is measured at 710 nm after 30 min using DI water as the reference.

Two types of controls are included. One is performed as for the analysis but without addition of soil samples to correct for the background caused by impurities as well as nonenzymatic hydrolysis of the substrate. This type of control is included for each set of analyses. The readings from this type of control are low and often can be ignored. Another control is performed as for the analysis but without the addition of starch to correct for background reducing sugars present in the soil and those generated from native substrates of various enzymatic reactions. This control should be included for each soil sample. The values of these controls are subtracted from values obtained for the starch-treated soils.

The calibration curve is constructed by diluting 1 mL of the glucose standard solution (5.0 mM) with acetate buffer to 100 mL in a volumetric flask, which yields 50 nmol glucose mL^{-1}. Then pipette 0-, 1-, 2-, 3-, 4-, 5-, or 6-mL aliquots of this diluted standard solution into 15-mL test tubes, adjust the volume to 6 mL by addition of acetate buffer, and proceed as described for reducing sugar analysis of the incubated soil sample. If the developed color intensity exceeds that of the 300 nmol glucose standard, an aliquot of the supernatant should be diluted with acetate buffer until the absorbance reading falls within the limits of the calibration graph. Express enzyme activity as μmol glucose kg^{-1} soil h^{-1}.

9–3.3.5 Quantification of Reducing Sugar in Soil Extracts by the Prussian Blue Method (Park and Johnson, 1949; Schinner and von Mersi, 1990; Deng and Tabatabai, 1994a)

9–3.3.5.1 Apparatus

- Test tubes with caps, 15 mL
- Volumetric flasks, 100 mL and 10 mL

- Spectrophotometer or colorimeter
- Boiling water bath

9–3.3.5.2 Reagents

1. Reagent A: Dissolve 16.0 g of anhydrous sodium carbonate (Na_2CO_3) and 0.9 g potassium cyanide (KCN) in 1 L of DI water.
2. Reagent B: Dissolve 0.5 g of potassium ferric hexacyanide [$K_3[Fe(CN)_6]$] in 1 L DI water. Store the solution in the dark, protected from light.
3. Reagent C: Dissolve 1.5 g of ferric ammonium sulfate [$NH_4Fe(SO_4)_2 \cdot 12 H_2O$], 1.0 g of sodium dodecyl sulfate ($C_{12}H_{25}SO_4Na$), and 4.2 mL of concentrated H_2SO_4 in about 700 mL of DI water at 50°C. Adjust the volume to 1 L with DI water.

9–3.3.5.3. Procedure

Transfer 1 mL of the soil filtrate to a 15-mL test tube, add 1 mL Reagent A and 1 mL Reagent B, mix, and set in a boiling water bath to the level of the liquid for 15 min. After cooling in a room temperature water bath (approximately 20°C) for 5 min, add 5 mL of Reagent C, mix, and stand at room temperature for 60 min for color development. Measure the absorbance of the developed blue color at 690 nm against DI water within 30 min.

Similar to the Somogyi–Nelson method, two types of controls are included and a glucose standard calibration curve is prepared. Express enzyme activity as µmole glucose kg^{-1} soil h^{-1}.

9–3.3.5.4 Comments

It is important to mention that the pH value in the solution for color development must be >10.5 after adding Reagents A and B, but must be <2 following the addition of Reagent C. The soil extract may be diluted if these values are not achieved. The linear relationship between color and glucose concentration is in the range of 2.8 to 28 µg mL^{-1} according to Schinner and von Mersi (1990). Based on studies of Deng and Tabatabai (1994a), the Prussian blue color is stable for 1 h and is a sensitive method for determination of low concentrations (<80 pg glucose g^{-1} soil) of reducing sugars in soil extracts.

REFERENCES

Al-Turki, A.I., and A.W. Dick. 2003. Myrosinase activity in soil. Soil Sci. Soc. Am. J. 67:139–145. doi:10.2136/sssaj2003.0139

Aminoff, D., W.T.J. Morgan, and W.M. Watkins. 1952. Studies in immunochemistry. 2. The action of dilute alkali on the N-acetylhexosamines and the specific blood-group mucoids. Biochem. J. 51:379–389.

Anand, L., S. Krishnamurthy, and P.J. Vithayathil. 1990. Purification and properties of xylanase from the thermophilic fungus, Humicola lanuginosa (Griffon and Maublanc) Bunce. Arch. Biochem. Biophys. 276:546–553. doi:10.1016/0003-9861(90)90757-P

Bayer, E.A., L.J.W. Shimon, Y. Shoham, and R. Lamed. 1998. Cellulosomes—structure and ultrastructure. J. Struct. Biol. 124:221–234. doi:10.1006/jsbi.1998.4065

Beau, J.M., P. Rollin, and P. Sinaÿ. 1977. Structure du chromogène i de la réaction de Morgan-Elson. (In French, with English abstract.) Carbohydr. Res. 53:187–195. doi:10.1016/S0008-6215(00)88086-6

Deng, S.P., and M.A. Tabatabai. 1994a. Colorimetric determination of reducing sugars in soils. Soil Biol. Biochem. 26:473–477. doi:10.1016/0038-0717(94)90179-1

Deng, S.P., and M.A. Tabatabai. 1994b. Cellulase activity of soils. Soil Biol. Biochem. 26:1347–1354. doi:10.1016/0038-0717(94)90216-X

Dunbar, K.R., and R.A. Heintz. 1997. Chemistry of transition metal cyanide compounds: Modern perspectives. Prog. Inorg. Chem. 45:283–391. doi:10.1002/9780470166468.ch4

Eivazi, F., and M.A. Tabatabai. 1988. Glucosidases and galactosidases in soils. Soil Biol. Biochem. 20:601–606. doi:10.1016/0038-0717(88)90141-1

Frankenberger, W.T., and J.B. Johanson. 1983a. Method of measuring invertase activity in soils. Plant Soil 74:301–311. doi:10.1007/BF02181348

Frankenberger, W.T., and J.B. Johanson. 1983b. Factors affecting invertase activity in soils. Plant Soil 74:313–323. doi:10.1007/BF02181349

Frankenberger, W.T., and J.B. Johanson. 1986. Use of plasmolytic agents and antiseptics in soil enzyme assays. Soil Biol. Biochem. 18:209–213. doi:10.1016/0038-0717(86)90029-5

Frey, S.D., E.T. Elliott, and K. Paustian. 1999. Application of the hexokinase-glucose-6-phosphate dehydrogenase enzymatic assay for measurement of glucose in amended soil. Soil Biol. Biochem. 31:933–935. doi:10.1016/S0038-0717(98)00176-X

Hirano, S., and N. Nagao. 1988. An improved method for the preparation of colloidal chitin by using methanesulfonic acid. Agric. Biol. Chem. 52:2111–2112.

Hoover, R. 2001. Composition, molecular structure, and physicochemical properties of tuber and root starches: A review. Carbohydr. Polym. 45:253–267. doi:10.1016/S0144-8617(00)00260-5

Kuhn, R., and G. Kruger. 1956. 3-Acetamino-furan aus N-Acetyl-D-glucosamin; ein Beitrag zur Theorie der Morgan-Elson-Reaktion. Chem. Ber. 89:1473–1486. doi:10.1002/cber.19560890615

Kuhn, R., and G. Kruger. 1957. Das Chromogen III der Morgan-Elson-Reaktion. Chem. Ber. 90:264–277. doi:10.1002/cber.19570900222

Lichstein, H.C., and M.H. Soule. 1944. Studies of the effect of sodium azide on microbic growth and respiration. I. The action of sodium azide on microbic growth. J. Bacteriol. 47:221–230.

Moss, G.P. 2010. Enzyme nomenclature: Recommendations of the nomenclature committee of the International Union of Biochemistry and Molecular Biology on the nomenclature and classification of enzymes by the reactions they catalyse. Available at http://www.chem.qmul.ac.uk/iubmb/enzyme/ (verified 27 May 2011).

Muckenschnabel, I., C. Bernhardt, T. Spruss, B. Dietl, and A. Buschauer. 1998. Quantitation of hyaluronidases by the Morgan-Elson reaction: Comparison of the enzyme activities in the plasma of tumor patients and healthy volunteers. Cancer Lett. 131:13–20. doi:10.1016/S0304-3835(98)00196-7

Nelson, N. 1944. A photometric adaption of the Somogyi method for the determination of glucose. J. Biol. Chem. 153:375–380.

Parham, J.A., and S.P. Deng. 2000. Detection, quantification and characterization of β-glucosaminidase activity in soil. Soil Biol. Biochem. 32:1183–1190. doi:10.1016/S0038-0717(00)00034-1

Park, J.T., and M.J. Johnson. 1949. A submicrodetermination of glucose. J. Biol. Chem. 181:149–151.

Parker, R., and S.G. Ring. 2001. Aspects of the physical chemistry of starch. J. Cereal Sci. 34:1–17. doi:10.1006/jcrs.2000.0402

Paul, E.A., and F.E. Clark. 1996. Soil microbiology and biochemistry. 2nd ed. Academic Press, New York.

Poutanen, K., and J. Puls. 1988. Characteristics of Trichoderma reesei β-xylosidase and its use in the hydrolysis of solubilized xylans. Appl. Microbiol. Biotechnol. 28:425–432. doi:10.1007/BF00268208

Raba, J., and H.A. Mottola. 1995. Glucose oxidase as an analytic reagent. Crit. Rev. Anal. Chem. 25:1–42. doi:10.1080/10408349508050556

Reissig, J.L., J.L. Strominger, and L.F. Leloir. 1955. A modified colorimetric method for the estimation of N-acetylamino sugars. J. Biol. Chem. 217:959–966.

Richmond, P.A. 1991. Occurrence and functions of native cellulose. p. 5–23. In C.H. Haigler, and P.J. Weimer (ed.) Biosynthesis and biodegradation of cellulose. Marcel-Dekker, New York.

Rodriguez-Kabana, R., G. Godoy, G. Morganjones, and R.A. Shelby. 1983. The determination of soil chitinase activity: Conditions for assay and ecological studies. Plant Soil 75:95–106. doi:10.1007/BF02178617

Ross, D.J. 1965. A seasonal study of oxygen uptake of some pasture soils and activities of enzymes hydrolysing sucrose and starch. J. Soil Sci. 16:73–85.

Saddler, J.N. 1986. Factors limiting the efficiency of cellulase enzymes. Microbiol. Sci. 3:84–87.

Schinner, F., and W. von Mersi. 1990. Xylanase-, CM-cellulase-, and invertase activity in soil: An improved method. Soil Biol. Biochem. 22:511–515. doi:10.1016/0038-0717(90)90187-5

Shaffer, P.A., and A.F. Hartman. 1921. The iodometric determination of copper and its use in sugar analysis. II. Methods for the determination of reducing sugars in blood, urine, milk, and other solutions. J. Biol. Chem. 45:365–390.

Shaffer, P.A., and M. Somogyi. 1933. Copper-iodometric reagents for sugar determination. J. Biol. Chem. 100:695–713.

Somogyi, M. 1945. A new reagent for the determination of sugars. J. Biol. Chem. 160:61–68.

Speir, T.W., D.J. Ross, and V.A. Orchard. 1984. Spatial variability of biochemical properties in a taxonomically-uniform soil under grazed pasture. Soil Biol. Biochem. 16:153–160. doi:10.1016/0038-0717(84)90106-8

Sunna, A., and G. Antranikian. 1997. Xylanolytic enzymes from fungi and bacteria. Crit. Rev. Biotechnol. 17:39–67. doi:10.3109/07388559709146606

Sylvia, D.M., J.J. Fuhrmann, P.G. Hartel, and D.A. Zuberer. 2005. Principles and applications of soil microbiology. 2nd ed. Prentice Hall, New Jersey.

Tabatabai, M.A. 1994. Soil enzymes. p. 775–833. In R.W. Weaver, S. Angle, P. Bottomley, D. Bezdicek, S. Smith, M.A. Tabatabai, and A. Wollum (ed.) Methods of soil analysis. Part 2. Microbiological and biochemical properties. SSSA, Madison, WI.

Takahashi, T., M. Ikegami-Kawai, R. Okuda, and K. Suzuki. 2003. A fluorimetric Morgan–Elson assay method for hyaluronidase activity. Anal. Biochem. 322:257–263. doi:10.1016/j.ab.2003.08.005

Tronsmo, A., and G.E. Harman. 1993. Detection and quantification of N-acetyl-β-D-glucosaminidase, chitobiosidase and endochitinase in solutions and on gels. Anal. Biochem. 208:74–79. doi:10.1006/abio.1993.1010

Viikari, L., M. Tenkanen, J. Buchert, M. Ratto, M. Bailey, M. Siika-Aho, and M. Linko. 1993. Hemicellulases for industrial applications. p. 131–182. In J. N. Saddler (ed.) Bioconversion of forest and agricultural plant residues. CAB International, Wallingford.

Webb, E.C. 1992. Enzyme nomenclature 1992: Recommendations of the Nomenclature Committee of the International Union of Biochemistry and Molecular Biology on the nomenclature and classification of enzymes. Academic Press, San Diego, CA.

Woods, J.T., and M.G. Mellon. 1941. The molybdenum blue reaction. Ind. Eng. Chem. Anal. Ed. 13:760–764. doi:10.1021/i560099a003

Chapter 10

Nitrogen Cycle Enzymes

Ellen Kandeler,* Christian Poll, William T. Frankenberger Jr., and M. Ali Tabatabai

10-1 INTRODUCTION

The nature of organic and inorganic nitrogen compounds in soils, or added to soils, has a crucial influence on the nitrogen (N) cycle. This chapter reviews methods of enzyme activities involved in the transformation of different organic and inorganic N compounds. Detailed descriptions and comments provide guidelines of the theory, procedures, and performance of these methods. Whereas N fixation (transformation of N_2 into organic compounds), proteases as well as peptidases, are described in Chapter 11 (Landi et al., 2011, this volume), this chapter covers most of the subsequent transformations: *N mineralization* converting native or added organic nitrogen compounds into ammonium (amidohydrolases: urease, L-asparaginase, L-glutaminase, and amidase), *ammonium oxidation* converting ammonia (NH_3) to nitrite (NO_2^-), and *denitrification* converting nitrate (NO_3^-) to nitrous oxide (N_2O) and then dinitrogen gas (N_2).

Each enzyme assay should be performed under conditions where substrate concentration, pH, temperature, and time of incubation have been optimized and where the enzyme reaction is terminated either by adding an enzyme inhibitor or by cooling the reaction mixture to prevent further reaction. Most methods described have been developed for agricultural soils, but can also be applied to other soils (grassland soils, forest soils, sediments).

The enzymes presented in this chapter contribute toward many ecosystem functions (waste decomposition, soils acting as a water filtration system, degradation of environmental pollutants). Aminohydrolases have been used to understand changes of management practices of agricultural soils (tillage, crop rotation, fertilization) (Deng and Tabatabai, 1996; Marschner et al., 2003; Sessitsch et al., 2005; Bastida et al., 2008; Hallin et al., 2009; Lagomarsino et al., 2009). The determination of urease activity is a frequently used method because this enzyme is a very sensitive indicator of eutrophication of agricultural and

Ellen Kandeler, Institute of Soil Science and Land Evaluation, Soil Biology Section (310b), University of Hohenheim, Emil-Wolff-Str. 27, D-70593 Stuttgart, Germany (Ellen.Kandeler@uni-hohenheim.de), *corresponding author;
Christian Poll, Institute of Soil Science and Land Evaluation, Soil Biology Section (310b), University of Hohenheim, Emil-Wolff-Str. 27, D-70593 Stuttgart, Germany (Christian.Poll@uni-hohenheim.de);
William T. Frankenberger, 2326 Geology, Department of Environmental Sciences, University of California, Riverside, CA 92521 (william.frankenberger@ucr.edu);
M. Ali Tabatabai, Dep. of Agronomy, Iowa State Univ., 2403 Agronomy Hall, Ames, IA 50011 (malit@iastate.edu).

doi:10.2136/sssabookser9.c10

grassland sites (Kandeler and Eder, 1993; Kandeler et al., 1994; Drissner et al., 2007) and a good indicator for heavy metal pollution of agricultural and anthropogenic soils (Lorenz and Kandeler, 2006; Lorenz et al., 2006; Kim et al., 2008; Chaer et al., 2009).

Nitrogen is a key element in agricultural production as it is normally the most limiting nutrient for plants. A long-standing objective has been to develop soil measurements that predict soil N mineralization. Limited research has shown some hydrolytic enzyme assays to hold potential as N mineralization indictors. Burket and Dick (1998) found the activities of amidase, asparaginase, urease, dipeptidase (Chapter 9), and fluorescein diacetate [3′,6′-diacetylfluorescein (FDA)] hydrolysis (Chapter 6) have correlation coefficients ranging from 0.65 to 0.84 with N mineralization (as measured by N uptake in the greenhouse). Similarly, Tabatabai et al. (2010) showed β-glucosaminidase activity (Chapter 9) to be highly correlated with the cumulative amounts of N mineralized in a 20-wk incubation of 32 soils at 20°C ($r = 0.87^{***}$) or at 30°C ($r = 95^{***}$).

The cycling and fate of N in terrestrial and aquatic ecosystems is critical for many aspects of environmental quality. Consequently, enzymes involved in N reactions are important in controlling N in ecosystems and for potential applications to assess soil quality or degradation. One example is ammonium oxidation, which is performed by only a selected group of soil bacteria and archaea. Therefore, inhibition of ammonia oxidizers might be expected to affect ecosystem functions. Guidelines for long-term monitoring of soil quality and ecotoxicology (e.g., degradation of pesticides and test of new pesticides) have been developed that utilize the ammonium oxidation assay (ISO 15685, 2004; Tscherko et al., 2007). The activity of nitrate reductase as well as denitrification can be monitored to better understand greenhouse gas fluxes from terrestrial ecosystems (Deiglmayr et al., 2004, 2006; Kandeler et al., 2009; Chroňáková et al., 2009). Specific groups of soil microorganisms, as well as different enzymes involved in N cycling, differ in their response to environmental change like N deposition or N addition (Gundersen, 1998; Butterbach-Bahl et al., 2002; Hungate et al., 2007; Enowashu et al., 2009). It is recommended that several indicators should be tested to monitor the most important steps within N cycling (e.g., protease, urease, ammonium oxidation and denitrification enzyme activity). The understanding of agroecosystem functioning could be further improved by linking enzyme activities involved in N cycling with measures of community composition and size of selected functional guilds (Hallin et al., 2009).

10–2 AMIDOHYDROLASES (UREASE, L-ASPARAGINASE, L-GLUTAMINASE, AND AMIDASE)

Several amidohydrolases are present in soils. All are involved in hydrolysis of native and added organic N to soils. Among these, L-asparaginase, L-glutaminase, amidase, and urease are important enzymes releasing ammonium into the environment.

10–2.1 Urease

10–2.1.1 Introduction

Urease (urea amidohydrolase, EC 3.5.1.5) is the enzyme that catalyzes the hydrolysis of urea to CO_2 and NH_3:

$$CH_4N_2O + H_2O \xrightarrow{\text{Urease}} CO_2 + 2\,NH_3$$

[1]

$$\underset{H_2N}{\overset{O}{\underset{\|}{C}}}_{NH_2} + H_2O \xrightarrow{\text{Urease}} CO_2 + 2\,NH_3$$

It acts on C–N bonds other than peptide bonds in linear amides and thus belongs to a group of enzymes that include L-asparaginase, L-glutaminase, and amidase (acylamide amidohydrolases). Urease cleaves two C–N bonds during hydrolysis.

Urease is very widely distributed in nature. It has been detected in microorganisms, plants, and animals. Urease was the first enzyme protein to be crystallized in 1926 by Sumner (1951) and its presence in soils was first reported by Rotini (1935). Early studies by Conrad (1940a,b,c; 1942a,b; 1944) provided the basic information about this enzyme in soil systems. Several reasons could account for the large amount of literature on soil urease activities during the past eighty years. These include (i) its reaction products (CO_2 and NH_3) are relatively easy to determine, (ii) it can be purified from various sources, (iii) it provides a model for a simple enzyme-catalyzed reaction, and (iv) urea is an important N fertilizer (Tabatabai, 1994).

At the molecular level, urease is associated within humus and clay complexes stabilizing the enzyme for long periods (Borghetti et al., 2003; Renella et al., 2007; Bastida et al., 2008). Physical fractionation procedures revealed that this enzyme is bound mainly to the clay fractions (Kandeler et al., 1999a), giving evidence that bacteria are the dominant urease producers in soils. At larger scales, spatial autocorrelation of urease activity of a pasture soil was found for sample spacings of <1 m, while autocorrelation extended up to 15 m for two other sites (von Steiger et al., 1996). The expected range of urease activity in many soils is between 0.14 and 14.3 μmol NH_4–N g^{-1} h^{-1}, corresponding to 3.92 to 400 μg NH_4–N g^{-1} $2h^{-1}$ (Nannipieri et al., 2002; Kandeler and Dick, 2006). Tscherko et al. (2007) revealed 2.6 μg NH_4–N g^{-1} $2h^{-1}$ as a minimum value of urease activity, 192 μg NH_4–N g^{-1} $2h^{-1}$ as a median, and 3228 μg NH_4–N g^{-1} $2h^{-1}$ as a maximum value for 900 grassland samples originating from the surface horizon of eight soil types across the European continent.

In agricultural studies, urease assays often are used to predict N mineralization in surface soil and soil profiles when organic amendments (e.g., slurry, farmyard manure, or sewage sludge) are added (Kandeler and Eder, 1993; Zaman et al., 2002; Fernandes et al., 2005). Urease activity is less affected by pesticide treatment (Ingram et al., 2005; Scelza et al., 2008) than by other soil management treatments like tillage or cropping systems (Kandeler et al., 1999b; Klose and Tabatabai, 2000). The importance of ureolytic microorganisms in soil fertility has been recently reviewed by Hasan (2000). This review discusses not only ureolytic production and activity in agricultural soils, but also problems encountered in the use of urea as a fertilizer resulting from its rapid hydrolysis to ammonium carbonate and the concomitant rise in pH and accumulation of ammonium. In a range of environmental studies, the response of urease activity as well as other enzyme activities to climate change has been tested (Ebersberger et al., 2003; Tscherko et al., 2003, 2005; Sardans et al., 2008). In ecotoxicology, urease assays are used to quantify the toxic effect of heavy metals in soils (Kandeler et al., 2000; Lorenz and

Kandeler, 2005; Chaperon and Sauveì, 2007; Chen et al., 2008; Cloutier-Hurteau et al., 2008). Sodium chloride derived from irrigation water in arid ecosystems also might reduce urease activity to a high extent (Cookson, 1999).

A variety of methods have been used for assays of urease activity in soils (Tabatabai and Bremner, 1972; Bremner and Mulvaney, 1978; Kandeler and Gerber, 1988; Sinsabaugh et al., 2000). Most of these involve determination of the ammonium released after incubation of soil with buffered or nonbuffered urea solution. In addition, assays also have been performed estimating the CO_2 released (Conrad, 1940c; Porter, 1965; Skujiņš and McLaren, 1969; Guettes et al., 2002). The assay of urease based on the quantification of carbon dioxide evolvement into the head space of gas-tight vessels might be advantageous when the assay is used for ecotoxicological evaluation of contaminated soils. Transformation of urea also can be quantified in the laboratory at in situ temperature ($12°C$) with the use of the ^{14}C-tracer method, which measures the release of $^{14}C\text{-}CO_2$ as a direct result of urease activity (Thoren, 2007). This method can be applied to wetland soils where high organic matter content might interfere with the colorimetric measurement of ammonium released. The microplate technology also can be used for the rapid assay of urease activity in environmental samples (Sinsabaugh et al., 2000).

The rapid and sensitive assay proposed by Kandeler and Gerber (1988) involves the estimation of the released ammonium after a 2-hour incubation of soils with a buffered or nonbuffered urea solution. This method is described in this chapter. Alternatively, a method can be used that is based on the estimation of the rate of urea hydrolysis in soils by determination of the urea remaining after incubation of soil with a urea solution (Douglas and Bremner, 1970a). The difference between the amount of urea added and that recovered after incubation for a specific time is taken as an estimate of urease activity. Urea may be estimated by a colorimetric method involving the reaction of urea with p-dimethylaminobenzaldehyde or diacetyl monoxime (DAM) and thiosemicarbazide (TSC) (see review by Lambert et al., 2004). Colorimetric determination of urea by DAM and TSC, however, is more sensitive than by p-dimethylaminobenzaldehyde and is described.

10–2.1.2 Principles

Methods, designed for the assay of urease under optimum conditions, are based on the determination of the NH_4 released or urea remaining when soil is incubated with a buffered urea solution. For the first method, released ammonium is extracted with a potassium chloride solution after an incubation of the samples for 2 h at $37°C$ (Kandeler and Gerber, 1988) followed by a colorimetric determination of ammonium or by steam distillation. The colorimetric method is based on the modified Berthelot reaction, which involves the reaction of sodium salicylate with NH_3 in the presence of dichloroisocyanurate, which forms a green-colored complex under alkaline pH conditions. Results are calculated as $\mu g\ NH_4\text{-}N\ g^{-1}$ soil $2\ h^{-1}$ from the difference between the produced ammonium and the initial ammonium content of the soil. Alternatively, the steam distillation method can be used after extracting ammonium with KCl. This gives quantitative recovery of $NH_4\text{-}N$ added to soil in the presence of THAM (tris) buffer (Tabatabai and Bremner, 1972). The NH_4 released is determined by treatment of the incubated soil sample with 2 M of KC1 containing Ag_2SO_4 (to stop enzyme activity). The method for determining NH_4 by steam distillation with MgO is based on the finding that $NH_4\text{-}N$ in solutions containing glutamine and other alkali-labile organic-N compounds can be determined quantitatively from the NH_4-N liberated by steam-distilling these solutions with a small amount of MgO for 3 to 4 min (Bremner and Keeney, 1965).

For the second method, the rate of urea hydrolysis in soils is estimated colorimetrically after incubation at 37°C for 5 h (Douglas and Bremner, 1970a,b; Zantua and Bremner, 1975a,b). The amount of urea hydrolyzed g^{-1} of soil 5 h^{-1} is estimated from the difference between the initial amount of urea added and that recovered after incubation.

10–2.1.3 Urease by Colorimetric Determination of Ammonium Method (Kandeler and Gerber, 1988)

10–2.1.3.1 Apparatus

1. Incubation flasks, 50 mL
2. Volumetric flasks, 100, 500, 1000, and 2000 mL
3. Ordinary incubator or temperature-controlled water bath
4. Shaker
5. N-free filters (folded)
6. Spectrophotometer

10–2.1.3.2 Reagents

1. Borate buffer 0.1 M, pH 10: Dissolve 57.2 g of disodium tetraborate in about 1500 mL warm water. After cooling to room temperature, titrate the pH of the solution to 10 by addition of 3 M NaOH, and bring up with water to 2-L volume.

2. Urea solution, 720 mM: Dissolve 21.6 g of urea in water and bring up the volume to 500 mL with water. Prepare this reagent daily for each set of samples, because this solution cannot be stored for longer than 1 d.

3. Potassium chloride (2 M)–hydrochloric acid (0.01 M) solution (KCl–HCl): Dissolve 149.2 g of KCl in water, add 10 mL of 1 M HCl, and bring up the volume to 1 L with water. Store this solution no longer than 3 wk at 4°C.

4. Sodium hydroxide solution, 0.3 M: Dissolve 12 g of NaOH in about 500 mL water and bring up the solution to 1 L with water.

5. Sodium salicylate solution, 1.06 M: Dissolve 17 g of sodium salicylate and 120 mg of sodium nitroprusside in water and bring up the solution to 100 mL with water. Prepare the solution shortly before use.

6. Sodium salicylate–sodium hydroxide solution: Mix 100 mL of Reagent 4 and 100 mL of Reagent 5 with 100 mL of water shortly before use.

7. Sodium dichloroisocyanurate solution, 3.91 mM: Dissolve 0.1 g of dichloroisocyanurate in water and bring up the solution to 100 mL with water. Prepare the solution shortly before use.

8. Ammonium chloride stock solution for standard curve, 1000 µg NH_4-N mL^{-1}: Dissolve 3.8207 g of ammonium chloride in water and bring up the volume to 1 L with water. Store the solution at 4°C.

9. Ammonium chloride standards: Pipette 0 (reagent blank), 1.0, 1.5, 2.0 and 2.5 mL of the ammonium chloride stock solution into 100-mL volumetric

flasks, and bring up the volume to 100 mL with KCl (2 M)–HCl (0.01 M) solution. The standards contain 0, 10, 15, 20, and 25 μg NH_4–N mL^{-1}.

10–2.1.3.3 Procedure

Place 5 g of soil into each of three 50-mL incubation flasks, and treat two of them with 2.5 mL of substrate solution and 20 mL of borate buffer; pipette only 20 mL of borate solution into the third flask (control). Stopper the flasks and incubate for 2 h at 37°C. After incubation, treat the control samples with 2.5 mL of substrate solution and all samples with 30 mL of potassium chloride (2 M)–hydrochloric acid (0.01 M) solution. Shake samples for 30 min on a rotatory shaker. Filter the soil suspensions using folded filters.

To determine the ammonium released, pipette 1 mL of filtrate and add 9 mL of water into a test tube. To prepare the standard curve, mix 1 mL of the ammonium chloride standards with 9 mL of water. Concentrations of the standards are 0, 1.0, 1.5, 2.0 and 2.5 μg N mL^{-1}. For the color reaction, add 5 mL of sodium salicylate–sodium hydroxide solution and 2 mL of sodium dichloroisocyanurate solution to the extracts and standards, swirl the tubes for a few seconds to mix the contents, and allow color development for 30 min at room temperature. Measure the absorbance of the color at 660 nm. Calculate the ammonium content of the extracts (μg NH_4–N mL^{-1}) from the calibration curve. To calculate urease activity (μg NH_4–N g^{-1} 2 h^{-1}), subtract the ammonium content of the control (μg NH_4–N mL^{-1}) from the sample average (μg mL^{-1}), multiply by 10 (dilution factor), multiply by 2.5 (volume of extract), divide by 5 (initial soil weight), and correct the values for moisture content.

10–2.1.3.4 Comments

For successful application of this method, the following advice should be considered: (i) The calibration curve is linear up to 3.5 μg N mL^{-1}. Extracts containing more than 3.5 μg N mL^{-1} have to be diluted with water before performing the color reaction. (ii) The color reaction is stable for at least 8 h. If it is not possible to process ammonium determination directly after the incubation period, filtered extracts can be stored overnight at 4°C or for longer time periods at −20°C. (iii) Nitrogen contamination of solutions, filters, and flasks must be avoided. Therefore, frequently checking the quality of the assay by following the procedure omitting the addition of soil is recommended. (iv) The use of field-moist soils is preferred, because air-drying induces an unexpected increase or decrease of urease activity depending on the chemical soil properties (Kandeler and Gerber, 1988). (v) The use of a detergent can be avoided with the short-term incubation of 2 h because microbial growth does not occur.

If the aim of the study is to simulate urea fertilizer treatment of agricultural soils and to follow the direct response of urease activity without any microbial growth, a nonbuffered method can be used. For the nonbuffered method, 5 g of soil with 2.5 mL of urea solution (79.9 mM) are incubated for 2 h at 37°C. Released ammonium is determined as described above. A direct comparison between urease activities derived from buffered or nonbuffered methods is not recommended (Taylor et al., 2002).

The method is also applicable for small-scale studies using rhizosphere samples (Kandeler et al., 2002; Sessitsch et al., 2005) or different soil fractions (Kandeler et al., 1999a,b, 2000). For the determination of urease at the small scale, 0.2 to 0.5 g of soil or soil fraction are incubated with 1.5 mL of 79.9 mM urea solution for 2 h at 37°C (Kandeler et al., 1999a,b). This can be adapted easily to a microtiter assay (Sinsabaugh et al., 2000) or to a robot system as described for phosphatases (Sadowsky et al., 2006). This high-throughput method is applied, for example, for long-term monitoring programs or for enzyme kinetics and temperature-sensitivity studies.

10–2.1.4 Urease by Determination of Ammonium Using Steam Distillation Method (Tabatabai, 1994; Mulvaney, 1996)

10–2.1.4.1 Apparatus

- Volumetric flasks, 50 mL.
- Ordinary incubator or temperature-controlled water bath.
- Steam distillation apparatus (O'Brien Scientific Glass Blowing, Monticello, IL). The unit is a modified version of the apparatus described by Bremner (1965) and Keeney and Nelson (1982). The major modifications are: (i) addition of a stopcock to the steam-bypass assembly to allow complete termination of the steam supply to the distillation head in distillation of Kjeldahl digests (this stopcock is left open at all times in distillation of soil extracts) and (ii) addition of a Teflon sleeve (3 cm long, 10-mm o.d., 8-mm i.d.) to permit use of interchangeable steam-inlet tubes with different types of distillation flasks. The apparatus is equipped with an O-ring ball joint (standard taper joint 28/15) for attachment of Pyrex boiling flasks (via a pinch clamp) used as distillation chambers. Steam for distillation is generated by boiling deionized water in a 5-L flask containing a few Teflon boiling chips and equipped with an electric heating mantle controlled by a variable transformer and a needle valve stopcock to supply deionized water. Before use, the apparatus should be steamed out for about 10 min to remove traces of NH_3, during which the transformer is adjusted so that distillate is collected at a rate of 7 to 8 mL min^{-1}. A trap is installed below the water jacket of the condenser to prevent condensate that collects on the outside surface of the condenser from contaminating the distillate.
- Distillation flasks. The flasks employed are 100-mL Pyrex round bottom boiling flasks with standard taper joint (24/40) that have been fitted with a socket joint (standard taper joint 28/15) for attachment to the distillation apparatus. Their dimensions should be such that when the flasks are connected to the steam-distillation apparatus, the distance between the tip of the steam-inlet tube and the bottom of the flask does not exceed 4 mm.
- Microburette (5 mL, graduated at 0.01-mL intervals) or automatic titrator.

10–2.1.4.2 Reagents

1. Toluene (Fisher certified reagent, Fisher Scientific Co., Chicago).
2. Tris(hydroxymethyl)aminomethane (THAM) buffer (0.01 M, pH 9.0): Dissolve 12.2 g of THAM in about 700 mL of water, titrate the pH of the solution to 9.0 by addition of approximately 0.2 M H_2SO_4, and bring up with water to 1 L.
3. Urea solution, 0.2 M: Dissolve 1.2 g of urea (Fisher certified reagent, Fisher Scientific Co., Chicago) in about 80 mL of THAM buffer and dilute the solution to 100 mL with THAM buffer. Store the solution in a refrigerator.
4. Potassium chloride (2.5 M)–silver sulfate (0.32 mM) solution: Dissolve 100 mg

of reagent-grade Ag_2SO_4 in about 700 mL of water, dissolve 188 g of reagent-grade KCl in this solution, and bring up the solution to 1 L with water.

5. Reagents for determination of NH_4–N.

    a. Magnesium oxide (MgO): Heat heavy MgO in an electric muffle furnace at 600 to 700°C for 2 h. Cool the product in a desiccator containing potassium hydroxide (KOH) pellets, and store it in a tightly stoppered bottle.

    b. Boric acid-indicator solution: Add 400 g of reagent-grade boric acid (H_3BO_3) to 18 L of deionized water in a 20-L Pyrex bottle marked to indicate a volume of 20 L, and stir vigorously with a motorized stirrer to dissolve the H_3BO_3. Then add 400 mL of indicator solution (prepared by dissolving 0.495 g of bromocresol green and 0.33 g of methyl red in 500 mL of ethanol), and bring the volume to 20 L with deionized water. With continuous stirring, adjust the pH to 4.8 to 5.0, or until the solution assumes a reddish purple tint, by cautiously adding 1 M sodium hydroxide (NaOH) or single NaOH pellets. If excess NaOH is added, the pH can be reduced by adding dilute HCl. Dispense the solution through Teflon tubing.

    c. Sulfuric acid (H_2SO_4), 0.0025 M standard.

    d. Standard NH_4–N solution: Dissolve 0.2358 g of ammonium sulfate [$(NH4)_2SO_4$] (available from Fisher Scientific as a primary-standard grade) in deionized water, dilute the solution to a volume of 1 L in a volumetric flask, and mix thoroughly. If pure, dry reagents are used, this solution contains 50 µg of NH_4–N mL^{-1}. Store the solution in a refrigerator.

10–2.1.4.3 Procedure

UREASE ASSAY Place 5 g of soil (<2 mm) in a 50-mL volumetric flask, add 0.2 mL of toluene and 9 mL of THAM buffer, swirl the flask for a few seconds to mix the contents, add 1 mL of 0.2 M urea solution, and swirl the flask again for a few seconds. Then stopper the flask and place it in an incubator at 37°C. After 2 h, remove the stopper, add approximately 35 mL of KCl–Ag_2SO_4 solution, swirl the flask for a few seconds, and allow the flask to stand until the contents have cooled to room temperature (about 5 min). Bring the contents to 50 mL by addition of KCl–Ag_2SO_4 solution, stopper the flask, and invert it several times to mix the contents. To determine NH_4–N in the resulting soil suspension, pipette a 20-mL aliquot of the suspension into a 100-mL distillation flask, and determine the released NH_4–N by steam distillation of this aliquot with 0.2 g of MgO for 4 min (Mulvaney, 1996) as described below. Controls should be performed in each series of analyses to allow for NH_4–N not derived from urea through urease activity. To perform controls, follow the procedure described for the urease assay activity, but make the addition of 1 mL of 0.2 M urea solution after the addition of 35 mL of KCl–Ag_2SO_4 solution.

DETERMINATION OF AMMONIUM BY STEAM DISTILLATION To determine the NH_4 in the post-incubation extracts, start by putting 5 mL of H_3BO_3–indicator solution in a 50-mL beaker marked to indicate a volume of 35 mL. Position the beaker under the condenser of the distillation apparatus so that the tip of the condenser is 1 cm below the top of the beaker that is in contact with the sidewall. Pipette an aliquot (normally 10–20 mL) of the soil suspension into a distillation flask, and add

0.2 g of MgO through a dry powder funnel with a long stem that reaches down into the bulb of the flask. Immediately attach the flask to the steam-distillation apparatus and commence steam distillation by closing the lower stopcock on the steam-bypass assembly. When the volume of distillate reaches 35 mL, rinse the tip of the condenser, and stop the distillation by opening the lower stopcock on the steam-bypass assembly. Titrate the distillate with 0.0025 M H_2SO_4. At the endpoint, the color changes from green to a permanent, faint pink.

CALCULATIONS Calculate the amount of N liberated by steam distillation from the expression, $(S - C) \times T$, where S is the volume of H_2SO_4 used in titration of the sample, C is the volume used in titration of a control (obtained by steam-distilling an aliquot of soil suspension), and T is the titer of the titrant (for 0.0025 M H_2SO_4, $T = 70\ \mu g\ N\ mL^{-1}$).

10–2.1.4.4 Comments

UREASE ASSAY In the assay methods based on determination of the NH_4 released, the buffer described should be used, because other buffers (e.g., phosphate, acetate, and universal buffers) give lower results for L-asparaginase, L-glutaminase, amidase, and urease activities in soils. Of the buffers tested by Tabatabai and Bremner (1972), the buffer described prevents NH_4 fixation by soils (this has been confirmed by quantitative recovery of NH_4 added to Montana vermiculite). The THAM buffer should be prepared by using H_2SO_4 because work by Wall and Laidler (1953) showed that buffers prepared by treatment of THAM solutions with HCl have an activation effect on hydrolysis of urea by jackbean (*Canavalia ensiformis* L.) urease.

The $KCl–Ag_2SO_4$ solution must be prepared by the addition of KCl to Ag_2SO_4 solution as specified (Ag_2SO_4 will not dissolve in KCl solution), and the soil suspension analyzed for NH_4 must be mixed thoroughly immediately before sampling. The $KCl–Ag_2SO_4$ treatment terminates the enzyme activity, and no additional NH_4 release is expected if the soil suspension is allowed to stand for 2 h before NH_4 analysis. If the soil suspension cannot be analyzed within 4 h after the addition of the $KCl–Ag_2SO_4$ reagent, the flask should be stored in a refrigerator. The reagent should be stirred continuously before use; if this solution is allowed to stand without stirring, KCl will precipitate slowly out of the solution with time.

For a valid assay of enzyme activity, it is necessary to ensure that the enzyme substrate concentration is not a limiting factor in the assay procedure. The substrate concentration recommended (0.05 M L-asparagine, L-glutamine, and formamide, and 0.02 M of urea) have been shown to be satisfactory for assay of amidohydrolase activities in a variety of soils. The D-isomers of asparagine and glutamine also are hydrolyzed in soils, but at about 16 and 7% of the L-isomers, respectively, at saturating concentration of the substrates (Frankenberger and Tabatabai, 1991a,c). The sensitivity of the assay procedure that involves determination of the NH_4 released is such that precise results can be obtained with most soils even if the 2-h incubation time recommended is reduced to 1 h. No hydrolysis of L-asparagine, formamide, or urea is found when autoclaved soils are incubated with solutions of the substrates at 37°C for 2 h under the conditions of the assay procedures described. L-glutamine is hydrolyzed slightly during incubation, and appropriate controls should be included to account for the NH_4 produced through chemical hydrolysis. The results obtained by these methods are not affected when the amount of toluene is increased from 0.2 to 2.0 mL.

DETERMINATION OF NH$_4$–N BY STEAM DISTILLATION   The steam distillation apparatus as described, with a sidearm, can be used for distillation of NH$_4$–N for ^{15}N analysis (see Hauck, 1982) and for nitrate and nitrite analysis. However, a simpler apparatus can be constructed for enzyme assays that only measures NH$_4$, with parts connected by rubber or plastic tubing instead of ground-glass joints. For more detailed guidance on the construction of distillation equipment see Mulvaney (1996).

The MgO must be ignited before use to remove any MgCO$_3$ present and the ignited material must be stored in an airtight container to avoid the formation of this impurity. Impurities can interfere with titrimetric determination of NH$_4$–N in the distillate. The use of heavy MgO is recommended, as the light variety tends to creep up the walls of the distillation flask and migrate through the distillation apparatus. The MgO does not have to be accurately weighed, as it just needs to be accessible to drive the NH$_3$ volatilization reaction.

The sharpness of the titration endpoint using 0.0025 M H$_2$SO$_4$ to determine NH$_4$ collected in H$_3$BO$_3$–indicator solution depends on the source of the indicators. If necessary, the recommended proportions of bromocresol green and methyl red may be altered to obtain a more satisfactory endpoint. The sharpness of the end-point also depends on the purity of the H$_3$BO$_3$ and the concentration of the H$_3$BO$_3$ solution (Yuen and Pollard, 1953). A sharper endpoint is obtained with a more dilute solution of H$_3$BO$_3$ (Yuen and Pollard, 1953), but the concentration must be high enough to ensure complete retention of NH$_4$. The H$_3$BO$_3$–indicator solution used in this chapter contains 20 g of H$_3$BO$_3$ L^{-1} and 5 mL of this solution can adsorb about 5 mg of NH$_4$–N (Yuen and Pollard, 1953).

The H$_2$SO$_4$ used for titration can be standardized using accurately known amounts of tris(hydroxymethyl)aminomethane, which is available from several commercial sources as a highly purified acidimetric standard under the trade name THAM. The THAM can be weighed directly (after drying for 2 h at 105°C), but it is usually more convenient to pipette aliquots from an aqueous solution. Such a solution is stable for several months if stored in a refrigerator, but the container should be stoppered tightly to exclude atmospheric CO$_2$, which is slowly absorbed.

Titrations with standard H$_2$SO$_4$ are conveniently performed using an automatic titrator with potentiometric (pH) endpoint detection, but titrations also can be done manually using a microburette. In the latter case, the burette should be equipped with a three-way stopcock for rapid refilling from a reservoir, and titrations should be performed with continuous stirring, using a variable-speed magnetic stirrer and a Teflon- or glass-coated stirring bar.

Extensive foaming can occur during steam distillation of 2 M KCl soil extracts with MgO, and this will cause interference if any of the foam contaminates the distillate. One solution is to use larger flasks (150- to 300-mL capacity) when excessive foaming occurs. However, this will require a longer steam-inlet tube.

Controls should be performed in each series of analyses to allow for NH$_4$ derived from reagents employed.

10–2.1.5 Urease by Determination of Urea Remaining Method (Modified from Douglas and Bremner 1970a,b and Zantua and Bremner 1975a,b)

10–2.1.5.1 Apparatus

1. Volumetric flasks, 50 mL with glass stoppers
2. Ordinary incubator or temperature-controlled water bath

3. Water bath (boiling water)

4. Suction funnels (polyethylene) and filtering funnel stand (Soil Moisture Equipment Co., Santa Barbara, CA)

5. Klett–Summerson photoelectric colorimeter or spectrophotometer

10–2.1.5.2 Reagents

1. Urea (substrate) solution: Dissolve 2.0 g of urea in about 700 mL of water and adjust the volume to 1 L with water. This solution contains 2 mg of urea mL^{-1}. Store this solution in a refrigerator.

2. Phenylmercuric acetate (PMA) solution: Dissolve 50 mg of PMA in 1 L of water.

3. Potassium chloride–phenylmercuric acetate (2 M KCl–PMA) solution: Dissolve 1500 g of reagent-grade KCl in 9 L of water and add 1 L of PMA solution.

4. Diacetyl monoxime (DAM) solution: Dissolve 2.5 g of DAM in 100 mL of water.

5. Thiosemicarbazide (TSC) solution: Dissolve 0.25 g of TSC in 100 mL of water.

6. Acid reagent: Add 300 mL of 85% phosphoric acid (H$_3$PO$_4$) and 10 mL of concentrated sulfuric acid (H$_2$SO$_4$) to 100 mL of water, and bring up the volume to 500 mL with water.

7. Color reagent: Prepare this reagent immediately before use by adding 25 mL DAM solution and 10 mL of TSC solution to 500 mL of acid reagent.

8. Standard urea stock solution for standard curve: Dissolve 0.500 g of urea in about 1500 mL of 2 M potassium chloride–phenylmercuric acetate (KCl–PMA) solution, and bring up to 2 L with the same solution. If pure dry urea is used, this solution will contain 250 μg of urea mL^{-1}. Store this solution in a refrigerator.

10–2.1.5.3 Procedure

Place 5 g of soil (<2 mm, on oven-dry basis) in a 65-mL glass bottle, and treat it with 5 mL of urea solution (10 mg of urea). Stopper the bottle and incubate at 37°C. After 5 h, remove the stopper and add 50 mL of 2 M KCl–PMA solution (Reagent 3). Stopper the bottle again, and shake it for 1 h. Filter the soil suspension, under suction, through Whatman No. 42 filter paper by using a suction funnel and filtering funnel stand.

To determine the urea remaining, pipette an aliquot (1–2 mL) of the extract containing up to 200 μg of urea into a 50-mL volumetric flask, bring the volume to 10 mL with 2 M KDL–PMA solution, and add 30 mL of the color reagent. Swirl the flask for a few seconds to mix the contents and place it in a bath of boiling water. After 30 min, remove the flask from the water bath and cool it immediately in running water for about 15 min (this can be accomplished by placing the flask in a deep tray containing cold water, 12–20°C). Then bring the volume to 50 mL with water and mix thoroughly. Measure the intensity of the red color produced with a Klett–Summerson photoelectric colorimeter fitted with a green (No. 54) filter.

Calculate the urea content of the extract analyzed by reference to a calibration graph plotted from results obtained with standards containing 0, 25, 50, 100, 150, and 200 μg mL^{-1} of urea. To prepare this graph, dilute 10 mL of the standard urea stock solution to 100 mL with 2 M KCl–PMA solution in a volumetric flask, and mix thoroughly. Pipette 0, 1, 2, 4, 6, and 8 mL of aliquots of this working standard solution into 50-mL volumetric flasks, adjust volumes to 10 mL by adding 2 M KDL–PMA solution, and proceed as described for urea analysis of the soil extract.

To calculate the amount of urea hydrolyzed in soil during incubation, divide the amount of total urea recovered by 5 (μg of urea recovered per g of soil), multiply by 55 or 22.5 (an aliquot of 1 or 2 mL out of 55 mL extraction volume), and subtract this value from 2000 (μg of urea initially added per g of soil).

10–2.1.5.4 Comments

The steps involved in the colorimetric method described for urea analysis should be strictly adhered to, but any colorimeter or spectrophotometer that permits color intensity measurement at 500 to 550 nm can be used for the procedure described. The maximum absorption of the color produced from the reaction of urea, DAM, and TSC as described is 527 nm (Douglas and Bremner, 1970a).

10–2.2 L-Asparaginase and L-Glutaminase

10–2.2.1 Introduction

L-Asparaginase and L-glutaminase, as well as amidase and urease, are amido-hydrolases acting on a C–N bond, releasing ammonia from linear amides other than peptide bonds. These enzymes play an important role in N mineralization of soils. The enzyme L-asparaginase (L-asparagine amidohydrolase EC 3.5.1.1) was first detected in soils by Drobník (1956). This enzyme catalyzes the hydrolysis of L-asparagine to produce L-aspartic acid and NH_3 as shown below:

$$C_4H_8N_2O_3 + H_2O \xrightarrow{\text{L-Asparaginase}} C_4H_7NO_4 + NH_3$$

[2]

Asparaginases have been shown to vary widely in different strains of microorganisms. Campbell et al. (1967) found two asparaginases in *Escherichia coli* B and designated them EC-1 and EC-2. These two enzymes differed in solubility, chromatographic behavior, antilymphoma activity, and optimum pH. Also, *E. coli* K-12 contains two asparaginases but only one appears in cells grown under anaerobic conditions (Cedar and Schwartz, 1967). Cells of *Pseudomonas fluorescens* contain two inducible isoenzymes of asparaginase (Eremenko et al., 1975): one that hydrolyzes only L-asparagine (asparaginase A) and one that hydrolyzes L-asparagine, L-glutamine, and D-asparagine (asparaginase AG).

Soils have been tested for L-asparaginase activity by Beck and Poschenrieder (1963), Frankenberger and Tabatabai (1991a), Senwo and Tabatabai (1999), Acosta-Martínez and Tabatabai (2001a) and Blank (2002). The L-asparaginase assay is used mainly to understand the impact of soil management on N cycling in agricultural ecosystems (Deng and Tabatabai, 1996; Dodor and Tabatabai, 2003). Alternatively, this method can be used to test the response of L- asparaginase to invading plant species (Blank, 2002), different heavy metals (Acosta-Martínez and Tabatabai, 2001a), or liming (Acosta-Martinez and Tabatabai, 2000b). The proposed method by Frankenberger and Tabatabai (1991a) involves determination of the NH_4–N released by L-asparaginase activity when soil is incubated with buffered (0.1 M of THAM, pH 10) L-asparagine solution and toluene at 37°C.

L-Glutaminase (L-glutamine amidohydrolase, EC 3.5.1.2) activity in soils was first detected by Galstyan and Saakyan (1973). The reaction catalyzed by this enzyme involves the hydrolysis of L-glutamine yielding L-glutamic acid and NH_3 as shown below:

$$C_5H_{10}N_2O_3 + H_2O \xrightarrow{\text{L-Glutaminase}} C_5H_9NO_4 + NH_3$$

[3]

L-Glutaminase is widely distributed in nature. It has been detected in several animals (Sayre and Roberts, 1958), plants (Bidwell, 1974), and microorganisms (Imada et al., 1973). Microorganisms that have been shown to contain L-glutaminase activity include bacteria, yeasts, and fungi. They are believed to be the main source in nature. Among the bacteria, very high levels of L-glutaminase activity have been reported in Achromobacteraceae soil isolates (Roberts et al., 1972). Fungal species that are known to produce L-glutaminase include *Tilachlidium humicola*, *Verticillium malthousei*, and *Penicillium urticae* (Imada et al., 1973). Frankenberger and Tabatabai (1991c) proposed a method that involves determination of the NH_4–N released by L-glutaminase activity when soil is incubated with buffered (0.1 M of THAM, pH 10) L-glutamine solution and toluene at 37°C.

10–2.2.2 Principles

The assays of L-asparaginase and L-glutaminase under optimum conditions are based on determination of the NH_4 released when soil is incubated with THAM buffer, substrate (L-asparagine or L-glutamine) and toluene at 37°C for 2 h. The NH_4 released is determined by treatment of the incubated soil sample with 2 M KCl containing Ag_2SO_4 (to stop enzyme activity) and steam distillation of an aliquot of the resulting soil suspension with MgO for 4 min. In these methods the enzymes are assayed at optimum buffer pH and substrate concentration.

10–2.2.3 Assay Methods (Frankenberger and Tabatabai 1991a,b)

10–2.2.3.1 Special Apparatus

As described in section 10–2.1.4.1

10–2.2.3.2 Reagents

1. Toluene (Fisher certified reagent, Fisher Scientific Co., Chicago).

2. Tris(hydroxymethyl)aminomethane (THAM) buffer, 0.1 M, pH 10: Dissolve 12.2 g of THAM in about 700 mL of water, titrate the pH of the solution to 10

by addition of approximately 0.1 M NaOH, and bring up with water to 1 L.

3. Potassium chloride (2.5 M)–silver sulfate (100 mg L^{-1}) ($KCl-Ag_2SO_4$) solution: Prepare as described in section 10–2.1.4.2.

4. L-Asparagine solution, 0.5 M: Dissolve 1.65 g of L-asparagine in about 20 mL of tris(hydroxymethyl)aminomethane (THAM) buffer and dilute the solution to 25 mL with THAM buffer. Mix the contents while running hot tap water over the flasks.

5. Magnesium oxide (MgO): Prepare as described in section 10–2.1.4.2.

6. Boric acid-indicator solution: Prepare as described in section 10–2.1.4.2.

7. Sulfuric acid (H_2SO_4), 0.0025 M standard.

8. Standard NH_4-N solution: Prepare as described in section 10–2.1.4.2.

10–2.2.3.3 Procedure

Follow the procedure described in section 10–2.1.5 for urease activity but use 1 mL of 0.5 M L-asparagine solution instead of 1 mL of 0.2 M urea solution. Controls should be performed in each series of analyses to allow for NH_4-N not derived from L-asparagine through L-asparagine activity. To perform controls, follow the procedure described in section 10–2.1.5 for performing controls in assay of urease activity but use 1 mL of 0.5 M of L-asparagine solution instead of 0.2 M urea solution.

10–2.2.3.4 Comments

A similar procedure can be used for the determination of L-glutaminase using 1 mL of 0.5 M L-glutamine solution instead of 1 mL of 0.5 M L-asparagine solution. Controls should be performed in each series of analyses to account for ammonium not derived from enzyme activity. To perform controls, follow the procedure described for performing controls in assay of L-asparaginase activity but use *steam-sterilized* soil and 1 mL of 0.5 M L-glutamine instead of 1 mL of 0.5 M L-asparagine.

   A related enzyme that has been detected in soils is L-histidine ammonia-lyase (Frankenberger and Johanson, 1982, 1983).

   See section 10–2.1.4.4 for comments that apply to L-asparaginase and L-glutaminase methods, as well as the amidase method described in section 10–2.3.3.

10–2.3 Amidase

10–2.3.1 Introduction

Amidase (acylamide amidohydrolase, EC 3.5.1.4) is the enzyme that catalyzes the hydrolysis of amides and produces NH_3 and the corresponding carboxylic acid. Amidase acts on C–N bonds other than peptide bonds in linear amides as shown below:

$$R\text{-}CONH_2 + H_2O \xrightarrow{\text{Amidase}} R\text{-}CO_2H + NH_3$$

[4]

It is specific for aliphatic amides and aryl amides cannot act as substrates (Kelly and Clarke, 1962; Florkin and Stotz, 1964). This enzyme is widely distributed in nature. It has been detected in animals and microorganisms (Clarke, 1970). Amidase is present in leaves of corn (*Zea mays* L.), sorghum [*Sorghum bicolor* (L.) Moench], alfalfa (*Medicago sativa* L.), and soybean [*Glycine max* (L.) Merr.] (Frankenberger and Tabatabai, 1982). Microorganisms shown to possess amidase activity include bacteria (Clarke, 1970; Frankenberger and Tabatabai, 1985), yeast (Joshi and Handler, 1962), and fungi (Hynes, 1970, 1975). The substrates of this enzyme are sources of N for plants (Cantarella and Tabatabai, 1983).

Soils have been tested for L-amidase activity by Frankenberger and Tabatabai (1980a), Dick (1984), Kay-Shoemake et al. (2000), Dodor and Tabatabai (2003), Ekenler and Tabatabai (2004), and Deng et al. (2006). Research has been conducted to understand the response of amidase activity in agricultural ecosystems to amide-containing pesticides (Kay-Shoemake et al., 2000), animal manure (Deng et al., 2006), tillage (Ekenler and Tabatabai, 2004), cropping (Dick, 1984; Dodor and Tabatabai, 2003), and liming (Acosta-Martinez and Tabatabai, 2000b).

10–2.3.2 Principle
The method proposed involves determination by steam distillation of the NH_4–N released by amidase activity when soil is incubated with buffered (0.1 M of THAM, pH 8.5) amide solution and toluene at 37°C (Frankenberger and Tabatabai, 1980a,b, 1981a,b).

10–2.3.3. Assay Methods (Frankenberger and Tabatabai, 1980a,b, 1981a,b)
10–2.3.3.1 Apparatus
As described in section 10–2.1.4.1

10–2.3.3.2 Reagents
1. Toluene (Fisher certified reagent, Fisher Scientific Co., Chicago).
2. THAM buffer (0.1 M, pH 8.5): Prepare as described in section 10–2.2.3.2 except adjust the pH to 8.5 by titration with about 0.2 M H_2SO_4 and dilute the solution with water to 1 L.
3. Formamide solution (0.50 M): Add 2.0 mL of formamide (Aldrich certified) into a 100-mL volumetric flask. Make up the volume by adding THAM buffer, and mix the contents. Store the solution in a refrigerator.
4. Potassium chloride (2.5 M)–uranyl acetate (0.005 M) solution: Dissolve 2.12 g of reagent-grade $UO_2(C_2H_3O_2)_2$•$2H_2O$ in about 700 mL of water, dissolve 188 g of reagent-grade KCl in this solution, and bring up the solution to 1 L with water.
5. Reagents for determination of NH_4–N (MgO, H_3BO_3–indicator solution, 0.0025 M H_2SO_4): Prepare as described in section 10–2.2.3.3.

10–2.3.3.3 Procedure
Follow the procedure described in section 10–2.2.3.4 for assay of L-asparaginase activity but use 1 mL of 0.5 M formamide solution instead of 1 mL of 0.5 M L-asparagine solution. Controls should be performed in each series of analyses to allow for NH_4–N not derived from formamide through amidase activity. To perform controls, follow the procedure described in section 10–2.2.3.4 but use 1mL of 0.5 M formamide instead of 1mL of 0.5 M L-asparagine. To determine NH_4–N in the

resulting soil suspension, pipette a 20-mL aliquot of the suspension into a 100-mL distillation flask, and determine the released NH_4–N by steam distillation of this aliquot with 0.2 g of MgO for 4 min (see Mulvaney, 1996).

10–2.3.4 Comments

In addition to formamide, acetamide and propionamide may be used as substrates, but with lower specificity. Assay of amidase in the absence of toluene indicated that the substrates may induce the synthesis of this enzyme by soil microorganisms (Frankenberger and Tabatabai, 1980a). Air-drying of field-moist samples decreased amidase activity by about 21%, whereas storage of field-moist-soil samples resulted only in a decrease of about 4% (Frankenberger and Tabatabai, 1981a). Amidase activity is inactivated at temperatures above 50°C.

The $KCl–UO_2(C_2H_3O_2)_2 \cdot 2H_2O$ reagent should be stirred continuously before use; if this solution is allowed to stand without stirring, KCl will precipitate slowly out of the solution with time. The purpose of adding the KCl–uranyl acetate solution to the soil sample after incubation in assay of amidase activity is to inactivate amidase and allow quantitative determination of the NH_4 released (Jakoby and Fredericks, 1964). Potassium chloride–Ag_2SO_4 can be substituted for KCl–uranyl acetate provided the NH_4 release is determined immediately, because $KCl–Ag_2SO_4$ does not completely inactivate amidase activity in soils.

Also see comments in section 10–2.1.4.4 for preparation of solutions and possible pitfalls of the method.

10–3 AMMONIUM OXIDATION

10–3.1 Introduction

Nitrification is the microbial oxidation of reduced forms of nitrogen to less reduced forms, principally NO_2^- and NO_3^- (Robertson and Groffmann, 2007), yielding energy for autotrophic bacteria. Heterotrophic microbes also can nitrify, but autotrophic nitrification appears to be the dominant process in many soils. Autotrophic nitrification is a two-step process performed by separate groups of bacteria: the ammonia and nitrite oxidizers. The first step—oxidation of ammonia to hydroxylamine—is mediated by the membrane-bound enzyme ammonia monooxygenase (EC 1.7.1.10), which also can oxidize a wide variety of nonpolar low molecular weight compounds (Tourna et al., 2008). Hydroxylamine is further oxidized to nitrite by the hydroxylamine oxidoreductase (EC 1.7.3.4). In many soils, the nitrite produced by ammonia oxidizers is quickly oxidized to nitrate by the nitrite-oxidizing bacteria with the enzyme nitrite oxidoreductase (EC 1.7.1.1) (Robertson and Groffmann, 2007).

$$NH_3 + 2H^+ + O_2 + 2e^- \xrightarrow{\text{ammonia monooxidase}} NH_2OH + H_2O$$

$$NH_2OH + H_2O \xrightarrow{\text{hydroxylamine oxidoreductase}} NO_2^- + 4e^- + 5H^+ \qquad [5]$$

$$NO_2^- + H_2O \xrightarrow{\text{nitrite oxidoreductase}} NO_3^- + 2H^+ + 2e^-$$

As revealed by 16S rRNA gene sequence analysis, nitrite-oxidizing bacteria appear in a broader array of phylogenetic groupings, whereas most of the ammonia oxidizers are placed in the β-subclass of the proteobacteria. Only the single species, *Nitrosococcus* is placed in the γ-proteobacteria (Purkhold et al., 2000; Koops et al., 2003). After ammonia-oxidizing archaea first were discovered in marine ecosystems (Francis et al., 2005; Könneke et al., 2005), it was suggested that archaea may be more numerous than bacterial ammonia-oxidizers in soil (Leininger et al., 2006). Cultivation-independent molecular surveys showed that members of the kingdom Crenarchaeota within the domain Archaea represent a substantial component of microbial communities in aquatic and terrestrial environments (Nicol and Schleper, 2006; Weidler et al., 2007). Ammonia-oxidizing archaea (AOA) were more abundant than were ammonia-oxidizing bacteria (AOB) in a Chinese upland red soil under long-term fertilization practices and a significant positive correlation were observed among the population sizes of AOB and AOA, soil pH, and potential nitrification rates (He et al., 2007).

In this chapter, a method is proposed to determine the potential ammonium oxidation. This method was first described by Berg and Rosswall (1985) as a procedure to estimate potential and actual oxidation rates of ammonium oxidizers. Since the activity of ammonia monooxygenase is the rate-limiting step in nitrification, several authors refer to this method as "potential nitrification or ammonium oxidation." A detailed protocol (Kandeler, 1996) has been validated as the standard method by the International Organization for Standardization (ISO 15685, 2004). The following range of activities was detected for different soil habitats: Ammonium oxidation of aggregates derived from arable, grassland, and forest soils showed an activity between 100 and 2500 ng NO_2–N g^{-1} 5 h^{-1}. Long-term monitoring of ammonium oxidation in bulk soil of an agricultural field revealed a range of activity between 300 and 700 ng NO_2–N g^{-1} 5 h^{-1} over a period of 9 yr (Kandeler et al., 1999b). Similar results were obtained by Berg and Rosswall (1985) for two Swedish arable soils. Much higher activities (up to 500 µg NO_2–N g^{-1} 5 h^{-1}) were described for grassland soils (Niklaus et al., 2001).

10–3.2 Principle

The method designed for assay of ammonium oxidation under optimum conditions is based on the determination of NO_2^- released when soil is incubated for 5 h at 25°C with a buffered ammonium sulfate solution and sodium chlorate as an inhibitor of nitrite oxidation. Since generation time of ammonia-oxidizing bacteria is longer than 10 h, the method provides the potential ammonia oxidation of the nitrifying population at the time of sampling and does not measure growth of the nitrifying population. Nitrite released during incubation is extracted and colorimetrically determined at 520 nm. Potential ammonium oxidation is calculated as ng N g^{-1} soil 5 h^{-1} from the difference between the produced nitrite and the initial nitrite content of the soil.

10–3.3 Assay Methods (Berg and Rosswall, 1985)

10–3.3.1 Apparatus

1. 100-mL Erlenmeyer flasks with caps
2. Ordinary incubator or temperature-controlled water bath
3. N-free filters (folded)

10–3.3.2 Reagents

1. Substrate stock solution 10 mM: Dissolve 1.3214 g of ammonium sulfate in water and bring up the volume to 1 L with water.

2. Substrate working solution 1 mM: Bring up 100 mL of the substrate stock solution to 1 L with water.

3. Sodium chlorate solution 1.5 M: Dissolve 15.97 g of sodium chlorate ($NaClO_3$) in water and bring up the volume to 100 mL with water.

4. Potassium chloride solution 2 M: Dissolve 149.12 g of potassium chloride (KCl) in water and bring up the volume to 1 L with water.

5. Ammonium chloride buffer 0.19 M, pH 8.5: Dissolve 10 g of ammonium chloride (NH_4Cl) in water, titrate the pH of the solution to 8.5 by addition of NH_4OH_{conc} and bring up with water to 1 L.

6. Color reagent: Dissolve 2 g of sulphanilamide and 0.1 g of N-(1-naphthyl)-ethylenediamine hydrochloride in 150 mL of water and add 20 mL of concentrated phosphoric acid. After cooling to room temperature, bring up the volume to 200 mL with water. This solution has to be colorless (to guarantee N-free volumetric flask, clean it with boiling water before use). Prepare this solution fresh each day.

7. Sodium nitrite stock solution, 1000 µg NO_2–N mL^{-1}: Dissolve 4.926 g of sodium nitrite ($NaNO_2$) in water and bring up with water to 1 L. Store the solution at 4°C for no more than 3 wk.

8. Sodium nitrite standard, 10 µg NO_2–N mL^{-1}; bring up 1 mL of the $NaNO_2$ stock solution to 100 mL with water.

9. Calibration standards: Pipette 0 (reagent blank), 2, 4, 8, and 10 mL of $NaNO_2$ standard solution into 100-mL volumetric flasks, add 20 mL of KCl solution (2 M), and bring up the volume to 100 mL with water. Calibration standards contain 0, 0.2, 0.4, 0.8 and 1 µg NO_2–N mL^{-1}.

10–3.3.3 Procedure

Place 5 g of soil into three 100-mL flasks and add 20 mL of substrate working solution (1 mM) and 0.1 mL of sodium chlorate solution. Close the flasks with caps and incubate two flasks for 5 h on a rotatory shaker (samples); store the third flask for 5 h at −20°C (control). After incubation, thaw the control sample at room temperature. Add 5 mL of potassium chloride solution (2 M) to samples and controls, mix briefly, and filter suspensions immediately.

To determine the nitrite released, pipette 5 mL of filtrate, 3 mL of ammonium chloride buffer, and 2 mL of color reagent into test tubes. After mixing, allow color development for 15 min at room temperature. To prepare the calibration curve, treat 5 mL of calibration standards like soil filtrates. Measure absorbance of the samples and controls at 520 nm against the reagent blank and calculate the nitrite content of the extracts (µg NO_2–N mL^{-1}) from the calibration curve. To express results of ammonium oxidation activity as µg NO_2–N g^{-1} 5 h^{-1}, multiply the nitrite concentration (µg NO_2–N mL^{-1}) by 25.1 (extraction volume) and divide by 5 (soil weight).

10–3.3.4 Comments

This method was developed and applied for agricultural soils. Ammonium oxidation of acid forest soils is expected to be very low. In many cases, sensitivity of the method is too low to detect ammonium oxidation in soils with pH lower than 4.0. It is recommended to adjust the optimum substrate concentration for each soil, because ammonium oxidation is inhibited at high concentrations of ammonium. In addition, the inhibitor concentration should be tested, because nonspecific adsorption of the ClO_3^- onto organic matter might reduce the efficiency of the inhibitor. Soil storage at +4°C is possible for up to 3 wk before analyses. This method is used as an international standard procedure for testing soils, influence of chemicals, polluted soils, and water extracts of biosolids (ISO 15685, 2004).

10–4 DENITRIFICATION ENZYME ACTIVITY AND NITRATE REDUCTASE

10–4.1 Introduction

Denitrification consists of four reaction steps in which nitrate is reduced to dinitrogen gas. The respiratory nitrate reduction and the dissimilatory reduction of nitrate to ammonia (DNRA) are catalyzed by a membrane-bound or periplasmatic nitrate reductase (Philippot and Hojberg, 1999). The reduction of nitrite (NO_2^-) to nitric oxide distinguishes denitrifiers from other nitrate-respiring bacteria (Philippot et al., 2007). This reaction is catalyzed by two different types of nitrite reductases (Nir), either a cytochrome cd_1 or a Cu-containing enzyme. Nitric oxide reductase further reduces nitric oxide to nitrous oxide, which is then reduced by nitrous oxide reductase during the last step in the denitrification pathway. Whereas some denitrifiers can reduce nitrate into dinitrogen due to the presence of genes encoding all denitrification reductases, others have a truncated pathway (Zumft, 1997). Nitrate-reducing bacteria are widespread in the environment and belong to most of the prokaryotic families (Philippot, 2005). Between 10 and 50% of total bacteria have the capability to reduce nitrate. Among the phyllogenetic diverse group of denitrifiers, several bacteria also are involved in other processes of N cycling (nitrification or N-fixation). The fungal contribution to denitrification has been shown by selective inhibition technique (Laughlin and Stevens, 2002), but the importance of fungi for the overall process is still controversial (Ma et al., 2008).

$$2NO_3^- \xrightarrow{\text{nitrate reductase}} 2NO_2^- \xrightarrow{\text{nitrite reductase}} 2NO \xrightarrow{\text{nitric oxide reductase}} N_2O \xrightarrow{\text{nitrous oxide reductase}} N_2 \qquad [6]$$

A common way to characterize denitrification in soil is to determine the potential denitrifying activity (PDA) or denitrification enzyme activity (DEA) (Smith and Tiedje, 1979; Zechmeister-Boltenstern, 1996). Soil slurries amended with an optimal amount of a terminal electron acceptor (nitrate), an easily available C and energy source (glucose alone or glucose and succinate), as well as an inhibitor of enzyme synthesis (chloramphenicol), are incubated under anaerobic conditions. Acetylene is used to block N_2O reductase so that the total N gas emitted (N_2O + N_2) can be measured easily as N_2O (Yoshinari and Knowles, 1976). This acetylene-based method, originally proposed by Smith and Tiedje (1979), is frequently applied (Rich et al., 2003; Boyle et al., 2006; Davis et al., 2008). Nevertheless, several modifications were suggested that can be used for specific purposes: Pell et al. (1996) recommended a procedure omitting the use of chloramphenicol (CAP) and considering mathematically the growth-associated product formation. Patra et al. (2006), Barnard et al. (2006) and Davis et al. (2008) also recommend not to use chloramphenicol: They made sure that no de novo synthesis of denitrifying enzyme took place by checking the linearity of the reaction (e.g., after 0, 4, and 7 h). Since acetylene concentrations high enough to block the last enzymatic step in the denitrification sequence might cause artifacts, Müller et al. (2002) suggested the quantification of nitrous oxide and N_2 fluxes by a ^{15}N-flux method with no additional enzymatic inhibitors, thus overcoming problems associated with the use of chloramphenicol and acetylene. The ^{15}N-labeled methods usually add one or several ^{15}N-labeled nitrogen compounds, such as nitrate, ammonium, fertilizers, or plant litter to soils and quantify the subsequent production of $^{15}N_2$ and $^{15}N_2O$ by mass spectrometry (Philippot et al., 2007). Application of this method is limited by the high cost of ^{15}N. Median denitrification enzyme activity (DEA) rates ranged from 29.5 to 44.6 mg N_2O–N kg^{-1} d^{-1} for surface soils (0–15 cm) and 0.7 to 1.7 µg N_2O–N kg^{-1} d^{-1} in the subsoil (135–150 cm) (Davis et al., 2008).

Whereas DEA yields information about the N_2O production as a final product of nitrate-, nitrite- as well as NO reductases, the methods proposed by Abdelmagid and Tabatabai (1987) and Kandeler (1996) focus on nitrate reduction as the first step of denitrification and dissimilatory reduction of nitrate to nitrite. The method is based on the determination of the NO_2–N production after adding nitrate as a substrate and 2,4-dinitrophenol as an inhibitor of nitrite reductase. The 2,4-dinitrophenol reacts as an uncoupler of oxidative phosphorylation interfering with electron transport. The measurement of nitrate reductase allows one to study the relationships between activity and abundance of nitrate-reducing prokaryotes (Deiglmayr et al., 2004, 2006; Kandeler et al., 2006). Nitrate-reducing prokaryotes constitute a wide taxonomic group sharing the ability to produce energy from dissimilatory reduction of nitrate into nitrite (Philippot, 2005). Using the proposed method, the potential nitrate reductase activity ranged from 0.1 to 1.2 µg NO_2–N g^{-1} d^{-1} in an acid spruce forest and was modified by N deposition, soil depth, and sampling time (Kandeler et al., 2009). Deiglmayr et al. (2004) applied this method to rhizospheric samples of *Trifolium repens* L. and *Lolium perenne* L. and revealed values in the range of 2 to 45 µg NO_2–N g^{-1} d^{-1}.

10–4.2 Principles

Soils are incubated under anaerobic atmosphere with added substrate (glucose and nitrate) to determine the denitrification potential of the soil under optimum denitrifying conditions (denitrification enzyme activity, DEA). The method is based on the principle that the rate of the process is proportional to enzyme concentration when no other factors are limiting. Nitrous oxide production is obtained from soils incubated with both glucose and nitrate during phase I, reflecting DEA of the existing bacterial enzymes, rather than phase II, which reflects the period of bacterial growth (Smith and Tiedje, 1979; Davis et al., 2008). The addition of chloramphenicol is recommended to ensure the prevention of enzyme protein synthesis and microbial growth during incubation times up to 6 h.

The method for determination of nitrate reductase activity is based on the determination of the NO_2–N production after adding nitrate as a substrate as well as 2,4-dinitrophenol as an inhibitor of nitrite reductase and incubating soil slurries under water-logged conditions for 24 h at 25°C. Nitrate reductase releases NO_2^-, which is extracted after the incubation period by potassium chloride solution and is determined colorimetrically at 520 nm. The method is designed for nitrate reductase under optimum conditions (Abdelmagid and Tabatabai, 1987; Kandeler, 1996). Substrate as well as inhibitor concentrations have to be optimized in pre-experiments. Results are calculated as $\mu g\ NO_2$–N g^{-1} soil 24 h^{-1} from the difference between the produced nitrite and the initial nitrite content of the soil.

10–4.3 Assay Methods (Smith and Tiedje, 1979)

10–4.3.1 Denitrification Enzyme Activity

10–4.3.1.1 Apparatus

- Gas-tight vessels with septa (e.g.,100 mL)
- Vacuum pump
- Gas-tight syringes (1, 10, 50, 100 mL) and bags with valve and septum
- 250-mL gas-washing bottle
- 5-mL vacutainers
- Gas chromatograph equipped with 63Ni ECD (electron capture detector; detection limit: 0.9 mg N_2O–N L^{-1})
- Ordinary incubator or temperature-controlled water bath

10–4.3.1.2 Reagents

1. Standard gas mixture (1 $\mu L\ N_2O\ L^{-1}$ in N_2 or He) and pure N_2O.
2. Substrate solution (1.07 mM KNO_3, 1 mM glucose and 0.7 mM chloramphenicol): Dissolve 108 mg of potassium nitrate, 180 mg glucose, and 225 mg of chloramphenicol in water and bring up the volume to 1 L with water. Degas the solution by evacuating in a desiccator for 15 min.
3. Acetone-free acetylene: Remove acetone by flushing acetylene through concentrated H_2SO_4 in a gas-wash bottle, and collect gas in a gas-tight bag.
4. Calibration standards (0.2– 500 $\mu L\ N_2O\ L^{-1}$): Bring up the standard gas mixture (for standards up to 1$\mu L\ N_2O\ L^{-1}$) and pure N_2O (for standards above 1 $\mu L\ N_2O\ L^{-1}$) into 100-mL gas-tight vessels evacuated and filled with helium to a final concentration of 0.2, 5, 10, 25, 50, 100, 200, 300, 400, and 500 μL $N_2O\ L^{-1}$. Prepare the calibration curve for each new series of experiments. On a daily basis, use two standards that contain 1 and 100 $\mu L\ N_2O\ L^{-1}$.

10–4.3.1.3 Procedure

Mix soils, remove debris, and place 10 g of soil into four, 100-mL flasks. Treat one flask with 10 mL of water, close and weigh the flask; fill with water. Determine the headspace volume of the flask by differential weighing. Treat the other three flasks with 10 mL of substrate solution, close them, apply a vacuum on the headspace and subsequently overpressurize with He. Repeat this procedure three times to ensure that oxygen is removed from the system. Remove 10 mL of He and inject 10 mL of acetone-free acetylene to block the reduction of N_2O to N_2. Shake the flasks at 225 rpm at 25°C to support diffusion of the substrates and the inhibitor. After incubation (0, 3, and 6 h), withdraw 5 mL of the flask headspace into a gas-tight syringe and place into 5-mL vacutainers until they are analyzed for N_2O on a gas chromatograph equipped with a ^{63}Ni ECD detector. Estimate the amount of N_2O in solution by using the Bunsen coefficient (0.544 mL of N_2O per mL of water at 25°C) (Tiedje, 1982). Calculate the nitrous oxide production for both 0- to 3- and 3- to 6-h intervals.

10–4.3.1.4 Comments

If N_2O production in a certain range is expected, commercially available N_2O standard gases can be used. The added water should submerge the soil completely. Therefore, more water might be needed depending on soil (e.g., organic layers). Compare nitrous oxide production for both 0- to 3- and 3- to 6-h intervals. If no significant difference occurs between the 0- to 3- and the 3- to 6-h interval, report only the 0- to 3-h rates as a result. Prepare samples with and without chloramphenicol in a pre-experiment to clarify the linearity of the reaction during the first 6 h. Duff and Triska (1990) showed that N_2O production is linear up to 4 h in soils incubated without chloramphenicol. Gas samples can be stored in evacuated vials for later analysis. Control the storage conditions by including controls of known concentrations in your analyses of stored samples.

This method has been used successfully for comparative studies to determine environmental effects or for modeling studies where DEA was used as input parameter (Tiedje, 1982).

10–4.3.2 Nitrate Reductase (Abdelmagid and Tabatabai, 1987)
10–4.3.2.1 Apparatus

- Test tubes (180 × 18mm) with screw caps
- Ordinary incubator or temperature-controlled water bath
- N-free filters (folded)

10–4.3.2.2 Reagents

1. Substrate solution, 25 mM: Dissolve 2.53 g of potassium nitrate in water and bring up the volume to 1 L with water.
2. Inhibitor stock solution, 0.9 mM: Dissolve 166 mg of 2,4-dinitrophenol solution in water by heating. After cooling the solution to room temperature, bring up to 1 L with water. Estimate the optimal amount of inhibitor for each soil using a range of 5 to 300 μg 2,4-DNP.
3. Potassium chloride solution, 4 M: Dissolve 298 g of potassium chloride (KCl) in water and bring up the volume to 1 L with water.
4. Ammonium chloride buffer, 0.19 M, pH 8.5: Dissolve 10 g of ammonium chloride (NH_4Cl) in 900 mL of water, titrate the pH of the solution to 8.5 by addition of NH_4OH_{conc} and bring up with water to 1 L.
5. Color reagent: Dissolve 2 g of sulphanilamide and 0.1 g of N-(1-naphthyl)-ethylenediamine hydrochloride in 150 mL of water and add 20 mL of concentrated phosphoric acid. After cooling to room temperature, bring up the volume to 200 mL with water. This solution has to be colorless (to guarantee N-free volumetric flask, clean it with boiling water before use). Prepare this solution each day; it needs to be fresh.
6. Sodium nitrite stock solution, 1000 μg NO_2-N mL^{-1}: Dissolve 4.926 g of sodium nitrite ($NaNO_2$) in water and bring up with water to 1 L. Store the solution at 4°C for only 3 wk.

7. Sodium nitrite standard, 10 μg NO_2–N mL^{-1}: Bring up 1 mL of the sodium nitrite stock solution to 100 mL with water.

8. Calibration standards: Pipette 0 (reagent blank), 2, 4, 8, and 10 mL of sodium nitrite standard solution into 100-mL volumetric flasks, add 20 mL of potassium chloride solution (2 M), and bring up the volume to 100 mL with water. Calibration standards contain 0, 0.2, 0.4, 0.8 and 1 μg NO_2–N mL^{-1}.

10–4.3.2.3 Procedure

Place 5 g of soil into each of three test tubes, and treat them with 4 mL of inhibitor solution, 1 mL of substrate solution, and 5 mL of water. Close the test tubes with caps and mix briefly. Incubate two tubes for 24 h at 25°C (samples) and one tube for 24°C at −20°C (control). After incubation, thaw the control at room temperature. Add 10 mL of potassium chloride solution to all samples; mix briefly and filter samples immediately. To determine the nitrite released, pipette 5 mL of filtrate, 3 mL of ammonium chloride buffer, and 2 mL of color reagent into test tubes. After mixing, allow color development for at least 15 min at room temperature. The color complex is stable for at least 4 h. To prepare the calibration curve, treat 5 mL of calibration standards like soil filtrates. Measure absorbance of the samples at 520 nm against the reagent blank. Calculate the nitrite content of the extracts (μg N mL^{-1}) from the calibration curve and express results of nitrate reductase activity as μg NO_2–N g^{-1} 24 h^{-1}.

10–4.3.2.4 Comments

This method is used as either a buffered or nonbuffered method (Abdelmagid and Tabatabai, 1987; Fu and Tabatabai, 1989; Kandeler, 1996). Since TRIS buffer interferes with the color reaction, the nonbuffered method is recommended. Care should be taken to dissolve the 2,4-dinitrophenol completely. An alternative approach is to dissolve 2,4-DNP in ethanol and to evaporate ethanol by using a stream of air for 2 h. It is possible to reduce the lag-phase of the reaction by incubating soils under a nitrogen-saturated atmosphere or by pre-incubating soil with 2,4-DNP solution for 16 h.

To use this method for small samples sizes (e.g., rhizosphere soils or different soil fractions) according to Deiglmayr et al. (2004) and Kandeler et al. (2009), weigh 0.2 g soil in five replicates into 2.0-mL tubes, add 33.3 μg of 2,4-dinitrophenol per g of soil (fresh weigh) to inhibit nitrite reductase and incubate in 1 mM potassium chloride in a total volume of 1 mL at 25°C in the dark. Extract nitrite with 4 M potassium chloride and centrifuge for 1 min at 1400 g. Determine nitrite in the supernatant as described above.

10–5 ARYLAMIDASE

10–5.1 Introduction

The neutral amino acid- β-naphthylamides or ρ-nitroanilides are readily split by neutral amino acid arylamidase (EC 3.4.11.2). This enzyme is distinct from leucine aminopeptidase (EC 3.4.11.1), which hydrolyzes L-leucinamide or L-leucyl-glycine. Patterson et al. (1963) conducted a series of experiments to determine the specific role of these two enzymes that hydrolyze L-leucinamide and L-leucine β-naphthylamide. They concluded that because it is the arylamides of amino acids that are substrates for these enzymes, it would be more accurate to use the name "amino acid naphthylamidases" or more generally "arylamidases." Consequently, the latter name has been used for the enzyme hydrolyzing neutral amino acid β-napthylamides (Marks et al., 1968; Hiwada et al., 1977, 1980). The reaction involved in hydrolysis of an N-terminal amino acid from arylamides is as follows:

$$[7]$$

L-Leucine β-naphthylamide   β-Naphthylamine   Leucine

This enzyme is widely distributed in nature. It has been detected in tissues and body fluids of all animals (Hiwada et al., 1980), in plants and microorganisms (Appel, 1974), and in soils (Acosta-Martínez and Tabatabai, 2000a, 2001b; Acosta-Martínez et al., 2003;). The chemical nature of organic N in soils is such that a large proportion (15–25%) is believed to be present as amides associated with amino acids (Sowden, 1958). The activity of this enzyme deserves special attention because present knowledge indicates that a variety of arylamides could be present in soils (Stevenson, 1994). In addition, work by Senwo and Tabatabai (1998) showed that at least 14 amino acids are associated with soil organic matter, and that the percentage distribution of amino acids recovered in the hydrolysates of 10 Iowa surface soils contained various proportions of acidic (asparagine plus aspartic acid, glutamine plus glutamic acid), basic (arginine, histidine, lysine), and neutral (phenylalanine, tyrosine, glycine, alanine, valine, leucine, isoleucine, serine, threonine, proline) amino acids. Therefore, arylamidase may play an important role as an initial reaction-limiting step in mineralization of organic N in soils as follows:

$$[8]$$

Evaluation of the specificity of this enzyme, purified from four organs of rats and in the livers of five animal species, for seven chromogenic substrates showed differences in the cleavage of various substrates tested relative to the amino acid moiety linked to β-naphthylamine (Nachlas et al., 1962). Therefore, information on the substrate specificity of this enzyme in soils is important for better understanding the chemistry and biochemistry of the N mineralization process.

10–5.2 Principles

The method involves colorimetric determination of the β-naphthylamine produced by arylamidase activity when soil is incubated with 0.1 M THAM buffer (pH 37°C for 1 h). The intensity of the resulting red azo compound is measured at 540 nm. The reaction involved is as follows:

β-Naphthylamine   Red Dye

p-Dimethylaminocinnamaldehyde

[9]

10–5.3 Assay Methods
(Acosta-Martinez and Tabatabai, 2000a)

10–5.3.1 Apparatus

- Incubation flasks, 25-mL Erlenmeyer
- Ordinary incubator or temperature-controlled water bath
- Shaker
- N-free filters (folded)
- Photometer or spectrophotometer
- Centrifuge
- Test tube vortexer

10–5.3.2 Reagents

1. THAM buffer (0.1 M, pH 8.0): Prepare by dissolving 2.44 g of tris(hydroxymethyl) aminomethane (THAM buffer, Fisher Scientific, Chicago) in about 50 mL of water, adjusting the pH by titration with $\approx$0.05 M H_2SO_4, and diluting the solution to 200 mL with water.

2. L-leucine β-naphthylamide solution (8.0 mM): Prepare by dissolving 0.2342 g of hydrochloride salt of L-leucine β-naphthylamide (Sigma Chemical, St. Louis) in water and adjusting the volume to 100 mL with water.

3. Ethanol (95%).

4. Acidified ethanol (0.26 M HCl): Prepare by adding 4.32 mL of concentrated HCl to ethanol and adjusting the volume to 200 mL with ethanol.

5. p-Dimethylaminocinnamaldehyde solution (0.6 mg mL^{-1}) (Sigma Chemical, St. Louis): Prepare by dissolving 0.12 g of p-dimethylaminocinnamaldehyde in ethanol and adjusting the volume to 200 mL with ethanol.

6. Standard β-naphthylamine stock solution (125 µg mL^{-1}): Prepare by dissolving 12.5 mg of β-naphthylamine (Sigma Chemical Co., St. Louis) in 75 mL deionized water containing 5 mL of ethanol in a 100-mL volumetric flask and adjusting the volume with deionized water.

7. Standard β-naphthylamine working solutions: Prepare by transferring 1, 2, 3, 4, 5, or 6 mL of the standard β-naphthylamine stock solution (125 µg mL^{-1}) into a 25-mL volumetric flask and adjusting the volume with deionized water. These standard solutions contain 5, 10, 15, 20, 25, or 30 µg of β-naphthylamine mL^{-1}, respectively.

10–5.3.3 Procedure

A 1-g soil sample (air-dried, <2mm) in a 25-mL Erlenmeyer flask is treated with 3 mL of 0.1 M THAM buffer (pH 8.0) and 1 mL of 8.0 mM L-leucine β-naphthylamide hydrochloride. The flask is swirled for a few seconds to mix the contents and is stoppered and placed on a shaker in an incubator (37°C) for 1 h. After incubation, the reaction is stopped by adding 6 mL of ethanol (95%). The soil suspension is immediately mixed and transferred into a centrifuge tube and centrifuged for 1 min at $17000 \times g$. The supernatant is transferred to a test tube to prevent any further hydrolysis of the substrate, and a 1-mL aliquot of this supernatant is treated (in a second test tube) with 1 mL of ethanol, 2 mL of acidified ethanol, and 2 mL of the p-dimethylaminocinnamaldehyde reagent. The solution is mixed on a vortex mixer after adding each of the reagents. The intensity of the resulting red azo compound is measured at 540 nm.

The calibration graph is prepared by treating 1 mL of each of the standard working solutions in a test tube with 1 mL of ethanol, 2 mL of the acidified ethanol, and 2 mL of reagent containing p-dimethylaminocinnamaldehyde.

Controls are included as described for the assay, but the 1 mL of the substrate is added after incubation. These are selected to give ranges in the activity values and to avoid overlapping curves.

10–5.3.4 Comments

Investigations on substrate specificity of arylamidase activity in soils have shown that the optimal pH values varied among the eight compounds containing individual amino acid moieties linked to naphthylamine (Tabatabai et al., 2002). The optimal pH of activity with the amino acid moieties ranges from 7.0 to 9.0. However, generally pH 8.0 was optimal. Among the amino acid moieties studied, the means of activities in six surface soils decreased in the following order: alanine > leucine > serine > lysine > glycine > arginine > histidine; proline was not hydrolyzed in soils. The chromophore linked to the amino acids L-asparagine, L-tyrosine, L-glutamic acid, and L-aspartic acid does not dissolve. It should be noted that at the optimal pH values for arylamidase activity, the K_m, V_{max}, Ea, and Δ H values varied widely among the soils and the amino acid moieties of the substrates studied. The mean of the Q_{10} values ranged from 1.04 to 1.66. These results suggest that there are more than two isomers of this enzyme in soils, and that the amino acid moieties of ring compounds in soils significantly affect the rate of N mineralization in soils (Tabatabai et al., 2002).

If the color intensity of the red azo compound exceeds the concentration of the highest β-naphthylamine standard, an aliquot of the red azo compound can be diluted with ethanol until the reading is within the limits of the calibration graph.

An alternative method is available for colorimetric determination of the β-naphthylamine produced (Goldbarg and Rutenburg, 1958). This method involves diazotization of the β-naphthylamine released with $NaNO_2$, decomposition of the excess $NaNO_2$ with ammonium sulfamate, and conversion of β-naphthylamine to a blue azo compound at pH 1.2 with N-(1-naphthyl) ethylenediamine dihydrochloride solution. The absorbance of the blue azo compound is measured at 700 nm. This method, however, is complicated and tedious. The color of the red azo compound produced from the reaction described is stable for at least 24 h.

Treatment of soils with toluene, formaldehyde, dimethylsulfoxide, $HgCl_2$, or iodoacetic acid inhibited arylamidase activity, and autoclaving completely destroyed the enzyme protein in soils (Acosta-Martinez and Tabatabai, 2000a). This indicates that the active sites of this enzyme contain sulfhydryl groups. Although toluene is commonly used to inhibit microbial growth during enzyme assays, this should not be done for this assay. However, since this is short incubation, minimal microbial growth would be expected.

REFERENCES

Abdelmagid, H.M., and M.A. Tabatabai. 1987. Nitrate reductase activity of soils. Soil Biol. Biochem. 19:421–427. doi:10.1016/0038-0717(87)90033-2

Acosta-Martínez, V., S. Klose, and T.M. Zobeck. 2003. Enzyme activities in semiarid soils under conservation reserve program, native rangeland, and cropland. J. Plant Nutr. Soil Sci. 166:699–707. doi:10.1002/jpln.200321215

Acosta-Martínez, V., and M.A. Tabatabai. 2000a. Arylamidase activity of soils. Soil Sci. Soc. Am. J. 64:215–221. doi:10.2136/sssaj2000.641215x

Acosta-Martínez, V., and M.A. Tabatabai. 2000b. Enzyme activities of limed agricultural soils. Biol. Fertil. Soils 31:85–91. doi:10.1007/s003740050628

Acosta-Martínez, V., and M.A. Tabatabai. 2001a. Arylamidase activity in soils: Effect of trace elements and relationships to soil properties and activities of amidohydrolases. Soil Biol. Biochem. 33:17–23. doi:10.1016/S0038-0717(00)00109-7

Acosta-Martínez, V., and M.A. Tabatabai. 2001b. Tillage and residue management effects on arylamidase activity in soils. Biol. Fertil. Soils 34:21–24. doi:10.1007/s003740100349

Appel, W. 1974. Peptidases. p. 949–954. In H.U. Bergmeyer (ed.) Methods of enzymatic analysis. Academic Press, New York.

Barnard, R., X. Le Roux, B.A. Hungate, E.E. Cleland, J.C. Blankinship, L. Barthes, and P.W. Leadley. 2006. Several components of global change alter nitrifying and denitrifying activities in an annual grassland. Funct. Ecol. 20:557–564. doi:10.1111/j.1365-2435.2006.01146.x

Bastida, F., E. Kandeler, T. Hernández, and C. García. 2008. Long-term effect of municipal solid waste amendment on microbial abundance and humus-associated enzyme activities under semiarid conditions. Microb. Ecol. 55:651–661. doi:10.1007/s00248-007-9308-0

Beck, T., and H. Poschenrieder. 1963. Experiments on the effect of toluene on the soil microflora. Plant Soil 18:346–357. doi:10.1007/BF01347234

Berg, P., and T. Rosswall. 1985. Ammonium oxidizer numbers, potential and actual oxidation rates in two Swedish arable soils. Biol. Fertil. Soils 1:131–140. doi:10.1007/BF00301780

Bidwell, R.G.S. 1974. Plant physiology. MacMillan, New York.

Blank, R.R. 2002. Amidohydrolase activity, soil N status, and the invasive crucifer Lepidium latifolium. Plant Soil 239:155–163. doi:10.1023/A:1014943304721

Borghetti, C., P.G. Ioachini, C. Marzadori, and C. Gessa. 2003. Activity and stability of urease-hydroxyapatite and urease-hydrooxyapatite-humic acid complexes. Biol. Fertil. Soils 38:96–101. doi:10.1007/s00374-003-0628-z

Boyle, S.A., J.J. Rich, P.J. Bottomly, K. Cromack, Jr., and D.D. Myrold. 2006. Reciprocal transfer effects on denitrifying community composition and activity at forest and meadow sites in the Cascade Mountains of Oregon. Soil Biol. Biochem. 38:870–878. doi:10.1016/j.soilbio.2005.08.003

Bremner, J.M. 1965. Inorganic forms of nitrogen. 1179–1137. In C.A. Black, D.D. Evans, J.L. White, L.E. Ensminger, and F.E. Clark (ed.) Methods of soil analysis. Part 2. Agron. Monogr. 9. ASA, Madison, WI.

Bremner, J.M., and D.R. Keeney. 1965. Steam distillation methods for determination of ammonium, nitrate, and nitrite. Anal. Chim. Acta 32:485–495. doi:10.1016/S0003-2670(00)88973-4

Bremner, J.M., and R.L. Mulvaney. 1978. Urease activity in soils. p. 149–196. In R.G. Burns (ed.) Soil enzymes. Academic Press, New York.

Burket, J.Z., and R.P. Dick. 1998. Microbial and soil parameters in relation to N mineralization in soils of diverse genesis under differing management systems. Biol. Fertil. Soils 27:430–438. doi:10.1007/s003740050454

Butterbach-Bahl, K., R. Gasche, G. Willibald, and H. Papen. 2002. Exchange of N-gases at the Höglwald forest– a summary. Plant Soil 240:117–123. doi:10.1023/A:1015825615309

Campbell, H.A., L.T. Mashburn, E.A. Boyce, and L.J. Old. 1967. Two L-asparaginases from *Escherichia coli* B: Their separation, purification, and antitumor activity. Biochemistry 6:721–730. doi:10.1021/bi00855a011

Cantarella, H., and M.A. Tabatabai. 1983. Amides as sources of nitrogen for plants. Soil Sci. Soc. Am. J. 47:599–603. doi:10.2136/sssaj1983.03615995004700030042x

Cedar, H., and J.H. Schwartz. 1967. Localization of the two L-aspariginases in anaerobically grown *Escherichia coli*. J. Biol. Chem. 242:3753–3755.

Chaer, G.M., D.D. Myrold, and P.J. Bottomley. 2009. A soil quality index based on the equilibrium between soil organic matter and biochemical properties of undisturbed coniferous forest soils of the Pacific Northwest. Soil Biol. Biochem. 41:822–830. doi:10.1016/j.soilbio.2009.02.005

Chaperon, S., and S. Sauveì. 2007. Toxicity interaction of metals (Ag, Cu, Hg, Zn) to urease and dehydrogenase activities in soils. Soil Biol. Biochem. 39:2329–2338. doi:10.1016/j.soilbio.2007.04.004

Chen, W., L. Wu, W.T. Frankenberger, Jr., and A.C. Chang. 2008. Soil enzyme activities of long-term reclaimed wastewater-irrigated soils. J. Environ. Qual. 37:S36–S42.

Chroňáková, A., V. Radl, J. Čuhel, M. Šimek, D. Elhottová, M. Engel, and M. Schloter. 2009. Overwintering management on upland pasture causes shifts in an abundance of denitrifying microbial communities, their activity and N_2O-reducing ability. Soil Biol. Biochem. 41:1132–1138. doi:10.1016/j.soilbio.2009.02.019

Clarke, P.H. 1970. The aliphatic amidase of *Pseudomonas aeruginosa*. Adv. Microb. Physiol. 4:179–222. doi:10.1016/S0065-2911(08)60442-7

Cloutier-Hurteau, B., S. Sauveì, and F. Courchesne. 2008. Influence of microorganisms on Cu speciation in the rhizosphere of forest soils. Soil Biol. Biochem. 40:2441–2451. doi:10.1016/j.soilbio.2008.06.006

Conrad, J.P. 1940a. Catalytic activity causing the hydrolysis of urea in soils as influenced by several agronomic factors. Soil Sci. Soc. Am. Proc. 5:238–241. doi:10.2136/sssaj1941.036159950005000C0040x

Conrad, J.P. 1940b. Hydrolysis of urea in soils by thermolabile hydrolysis. Soil Sci. 49:253–263. doi:10.1097/00010694-194004000-00002

Conrad, J.P. 1940c. The nature of the catalyst causing the hydrolysis of urea in soils. Soil Sci. 50:119–134. doi:10.1097/00010694-194008000-00005

Conrad, J.P. 1942a. The occurrence and origin of ureaselike activities in soils. Soil Sci. 54:367–380. doi:10.1097/00010694-194211000-00012

Conrad, J.P. 1942b. Enzymatic vs. microbial concepts of urea hydrolysis in soils. Agron. J. 34:1102–1113. doi:10.2134/agronj1942.00021962003400120005x

Conrad, J.P. 1944. Some effects of developing alkalinities and other factors upon ureaselike activities in soils. Soil Sci. Soc. Am. Proc. 8:171–174. doi:10.2136/sssaj1944.036159950008000C0030x

Cookson, P. 1999. Spatial variation of soil urease activity around irrigated date palms. Arid Soil Res. Rehabil. 13:155–169.

Davis, J.H., W.R. Horwath, J.J. Steiner, and D.D. Myrold. 2008. Denitrification and nitrate consumption in an herbaceous riparian area and perennial ryegrass seed cropping system. Soil Sci. Soc. Am. J. 72:1299–1310. doi:10.2136/sssaj2007.0279

Deiglmayr, K., L. Philippot, U.A. Hartwig, and E. Kandeler. 2004. Structure and activity of the nitrate-reducing community in the rhizosphere of *Lolium perenne* and *Trifolium repens* under long-term elevated atmospheric pCO_2. FEMS Microbiol. Ecol. 49:445–454. doi:10.1016/j.femsec.2004.04.017

Deiglmayr, K., L. Philippot, and E. Kandeler. 2006. Functional stability of the nitrate-reducing community in grassland soils towards high nitrate supply. Soil Biol. Biochem. 38:2980–2984. doi:10.1016/j.soilbio.2006.04.034

Deng, S.P., J.A. Parham, J.A. Hattey, and D. Babu. 2006. Animal manure and anhydrous ammonia amendment, microbial biomass, and activities of dehydrogenase and amidohydrolases in semiarid agroecosystems. Appl. Soil Ecol. 33:258–268. doi:10.1016/j.apsoil.2005.10.004

Deng, S.P., and M.A. Tabatabai. 1996. Effect of tillage and residue management on enzyme activities in soils. 1. Amidohydrolases. Biol. Fertil. Soils 22:202–207.

Dick, W.A. 1984. Influence of long-term tillage and crop rotation combinations on soil enzyme activities. Soil Sci. Soc. Am. J. 48:569–574. doi:10.2136/sssaj1984.03615995004800030020x

Dodor, D.E., and M.A. Tabatabai. 2003. Amidohydrolases in soils as affected by cropping systems. Appl. Soil Ecol. 24:73–90. doi:10.1016/S0929-1393(03)00067-2

Douglas, L.A., and J.M. Bremner. 1970a. Extraction and colorimetric determination of urea in soils. Soil Sci. Soc. Am. Proc. 34:859–862. doi:10.2136/sssaj1970.03615995003400060015x

Douglas, L.A., and J.M. Bremner. 1970b. Colorimetric determination of microgram quantities of urea. Anal. Lett. 3:79–87.

Drissner, D., H. Blum, D. Tscherko, and E. Kandeler. 2007. Nine years of enriched CO_2 changes the function and structural diversity of soil microorganisms in a grassland. Eur. J. Soil Sci. 58:260–269. doi:10.1111/j.1365-2389.2006.00838.x

Drobník, J. 1956. Degradation of asparagines by the soil enzyme complex. Cesk. Mikrobiol. 1:47.

Duff, J.H., and F.J. Triska. 1990. Denitrification in sediments from the hyporheic zone adjacent to a small stream. Can. J. Fish. Aquat. Sci. 47:1140–1147. doi:10.1139/f90-133

Ebersberger, D., P. Niklaus, and E. Kandeler. 2003. Elevated carbon dioxide stimulates N-mineralisation and enzyme activities in a calcareous grassland. Soil Biol. Biochem. 35:965–972. doi:10.1016/S0038-0717(03)00156-1

Ekenler, M., and M.A. Tabatabai. 2004. Arylamidase and amidohydrolases in soils as affected by liming and tillage systems. Soil Tillage Res. 77:157–168. doi:10.1016/j.still.2003.12.007

Enowashu, E., C. Poll, N. Lamersdorf, and E. Kandeler. 2009. Microbial enzyme activities under reduced nitrogen deposition in a spruce forest soil. Appl. Soil Ecol. 43:11–21. doi:10.1016/j.apsoil.2009.05.003

Eremenko, V.V., A.V. Zhukov, and A.Y. Nikolaev. 1975. Asparaginase and glutaminase activity of *Pseudomonas fluorescens* during continuous culturing. Mikrobiologia. 44:550–555.

Fernandes, S.A.P., W. Bettiol, and C.C. Cerri. 2005. Effect of sewage sludge on microbial biomass, basal respiration, metabolic quotient and soil enzymatic activity. Appl. Soil Ecol. 30:65–77. doi:10.1016/j.apsoil.2004.03.008

Florkin, M., and E.H. Stotz. 1964. Comprehensive biochemistry. 13:26–134. Elsevier, North-Holland, New York.

Francis, C.A., K.J. Roberts, J.M. Beman, A.E. Santoro, and B.B. Oakley. 2005. Ubiquity and diversity of ammonia-oxidizing archaea in water columns and sediments of the ocean. Proc. Natl. Acad. Sci. USA 102:14683–14688. doi:10.1073/pnas.0506625102

Frankenberger, W.T., Jr., and J.B. Johanson. 1982. L-Histidine ammonia-lyase activity in soil. Soil Sci. Soc. Am. J. 46:943–948. doi:10.2136/sssaj1982.03615995004600050012x

Frankenberger, W.T., Jr., and J.B. Johanson. 1983. Distribution of L-histidine ammonia-lyase activity in soils. Soil Sci. 136:347–353. doi:10.1097/00010694-198312000-00003

Frankenberger, W.T., Jr., and M.A. Tabatabai. 1980a. Amidase activity in soils. 1. Method of Assay. Soil Sci. Soc. Am. J. 44:282–287.

Frankenberger, W.T., Jr., and M.A. Tabatabai. 1980b. Amidase activity in soils. 2. Kinetic parameters. Soil Sci. Soc. Am. J. 44:532–536. doi:10.2136/sssaj1980.03615995004400030019x

Frankenberger, W.T., Jr., and M.A. Tabatabai. 1981a. Amidase activity in soils. 3. Stability and distribution. Soil Sci. Soc. Am. J. 45:333–338. doi:10.2136/sssaj1981.03615995004500020021x

Frankenberger, W.T., Jr., and M.A. Tabatabai. 1981b. Amidase activity in soils. 4. Effects of trace elements and pesticides. Soil Sci. Soc. Am. J. 45:1120–1124. doi:10.2136/sssaj1981.03615995004500060021x

Frankenberger, W.T., Jr., and M.A. Tabatabai. 1982. Amidase and urease activities in plants. Plant Soil 64:153–166. doi:10.1007/BF02184247

Frankenberger, W.T., Jr., and M.A. Tabatabai. 1985. Characteristics of an amidase isolated from soil bacterium. Soil Biol. Biochem. 17:303–308. doi:10.1016/0038-0717(85)90065-3

Frankenberger, W.T., Jr., and M.A. Tabatabai. 1991a. L-Asparaginase activity of soils. Biol. Fertil. Soils 11:6–12. doi:10.1007/BF00335826

Frankenberger, W.T., Jr., and M.A. Tabatabai. 1991b. Factors affecting L-asparaginase activity in soils. Biol. Fertil. Soils 11:1–5. doi:10.1007/BF00335825

Frankenberger, W.T., Jr., and M.A. Tabatabai. 1991c. L-Glutaminase activity in soils. Soil Biol. Biochem. 23:869–874. doi:10.1016/0038-0717(91)90099-6

Fu, M.H., and M.A. Tabatabai. 1989. Nitrate reductase activity in soils: Effects of trace elements. Soil Biol. Biochem. 21:943–946. doi:10.1016/0038-0717(89)90085-0

Galstyan, A.S., and E.G. Saakyan. 1973. Determination of soil glutaminase activity. Dokl. Akad. Nauk SSSR 209:1201–1202.

Goldbarg, J.A., and A.M. Rutenburg. 1958. The colorimetric determination of leucine aminopeptidase in urine and serum of normal subjects and patients with cancer and other diseases. Cancer 11:283–291. doi:10.1002/1097-0142(195803/04)11:2<283::AID-CNCR2820110209>3.0.CO;2-8

Guettes, R., W. Dott, and A. Eisentraeger. 2002. Determination of urease activity in soils by carbon dioxide release for ecotoxicological evaluation of contaminated soils. Ecotoxicology 11:357–364. doi:10.1023/A:1020509422554

Gundersen, P. 1998. Effects of enhanced nitrogen deposition in a spruce forest at Klosterhede, Denmark, examined by moderate NH_4NO_3 addition. For. Ecol. Manage. 101:251–268. doi:10.1016/S0378-1127(97)00141-2

Hallin, S., C.M. Jones, M. Schloter, and L. Philippot. 2009. Relationship between N-cycling communities and ecosystem functioning in a 50-year-old fertilization experiment. ISME J. 3:597–605. doi:10.1038/ismej.2008.128

Hasan, H.A.H. 2000. Ureolytic microorganisms and soil fertility: A review. Commun. Soil Sci. Plant Anal. 31:2565–2589. doi:10.1080/00103620009370609

Hauck, R.D. 1982. Nitrogen-isotope-ratio analysis. p. 735–779. In A.L. Page, R.H. Miller, and D.R. Keeney (ed.) Methods of soil analysis. Part 2. 2nd ed. Agron. Monogr. 9. ASA and SSSA, Madison, WI.

He, J.-Z., J.-P. Shen, L.-M. Zhang, Y.-G. Zhu, Y.-M. Zheng, M.-G. Xu, and H. Di. 2007. Quantitative analyses of the abundance and composition of ammonia-oxidizing bacteria and ammonia-oxidizing archaea of a Chinese upland red soil under long-term fertilization practices. Environ. Microbiol. 9:2364–2374. doi:10.1111/j.1462-2920.2007.01358.x

Hiwada, K., T. Ito, M. Yokoyama, and T. Kokubu. 1980. Isolation and characterization of membrane-bound arylamidases from human placenta and kidney. Eur. J. Biochem. 104:155–165. doi:10.1111/j.1432-1033.1980.tb04411.x

Hiwada, K., M. Terao, K. Nishimura, and T. Kokubu. 1977. Comparison of human membrane-bound neutral arylamidases from small intestine, lung, kidney, liver, and placenta. Clin. Chim. Acta 76:267–275. doi:10.1016/0009-8981(77)90106-1

Hungate, B.A., S.C. Hart, P.C. Selmants, S.I. Boyle, and C.A. Gehring. 2007. Soil responses to management, increased precipitation, and added nitrogen in ponderosa pine forests. Ecol. Appl. 17:1352–1365. doi:10.1890/06-1187.1

Hynes, M.J. 1970. Induction and repression of amidase enzymes in Aspergillus nidulans. J. Bacteriol. 103:482–487.

Hynes, M.J. 1975. Amide utilization in Aspergillus nidulans: Evidence for a third amidase enzyme. J. Gen. Microbiol. 91:99–109.

Imada, A., S. Igarasi, K. Nakahama, and M. Isono. 1973. Asparaginase and glutaminase activities of microorganisms. J. Gen. Microbiol. 76:85–99.

Ingram, C.W., M.S. Coyne, and D.W. Williams. 2005. Effects of commercial diazinon and imidacloprid on microbial urease activity in soil and sod. J. Environ. Qual. 34:1573–1580. doi:10.2134/jeq2004.0433

ISO. 2004. ISO 15685. Soil quality— Determination of potential nitrification and inhibition of nitrification—Rapid test by ammonium oxidation. International Organization for Standardization, Geneva, Switzerland.

Jakoby, W.B., and J. Fredericks. 1964. Reactions catalyzed by amidases: Acetamidase. J. Biol. Chem. 239:1978–1982.

Joshi, J.G., and P. Handler. 1962. Purification and properties of nicotinamidase from *Torula cremoris*. J. Biol. Chem. 237:929–935.

Kandeler, E. 1996. Nitrate reductase activity. p. 176–179. *In* F. Schinner, R. Öhlinger, E. Kandeler, and R. Margesin (ed.) Methods in Soil Biology. Springer, Berlin New York.

Kandeler, E., T. Brune, E. Enowashu, N. Doerr, G. Guggenberg, N. Lamersdorf, and L. Philippot. 2009. Response of total and nitrate dissimilating bacteria to reduced N deposition in a spruce forest soil profile. FEMS Microbiol. Ecol. 67:444–454. doi:10.1111/j.1574-6941.2008.00632.x

Kandeler, E., K. Deiglmayr, D. Tscherko, D. Bru, and L. Philippot. 2006. Abundance of *narG*, *nirK*, *nirS* and *nosZ* genes of denitrifying bacteria during primary successions of a glacier foreland. Appl. Environ. Microbiol. 72:5957–5962. doi:10.1128/AEM.00439-06

Kandeler, E., and R.P. Dick. 2006. Distribution and function of soil enzymes in agroecosystems. p. 263–285. *In* Biodiversity in agricultural production systems. G. Benckiser and S. Schnell (ed.) Taylor and Francis, New York.

Kandeler, E., and G. Eder. 1993. Effect of cattle slurry in grassland on microbial biomass and on activities of various enzymes. Biol. Fertil. Soils 16:249–254. doi:10.1007/BF00369300

Kandeler, E., G. Eder, and M. Sobotik. 1994. Microbial biomass, N mineralization and the activities of various enzymes in relation to nitrate leaching and root distribution of a slurry-amended grassland. Biol. Fertil. Soils 18:7–12. doi:10.1007/BF00336437

Kandeler, E., and H. Gerber. 1988. Short-term assay of soil urease activity using determination of ammonium. Biol. Fertil. Soils 6:68–72. doi:10.1007/BF00257924

Kandeler, E., P. Marschner, D. Tscherko, T.S. Gahoonia, and N.E. Nielsen. 2002. Structural and functional diversity of soil microbial community in the rhizosphere of maize. Plant Soil 238:301–312. doi:10.1023/A:1014479220689

Kandeler, E., S. Palli, M. Stemmer, and M.H. Gerzabek. 1999b. Tillage changes microbial biomass and enzyme activities in particle-size fractions of a Haplic Chernozem. Soil Biol. Biochem. 31:1253–1264. doi:10.1016/S0038-0717(99)00041-3

Kandeler, E., M. Stemmer, and E.M. Klimanek. 1999a. Response of soil microbial biomass, urease and xylanase within particle size fractions to long-term soil management. Soil Biol. Biochem. 31:261–273. doi:10.1016/S0038-0717(98)00115-1

Kandeler, E., D. Tscherko, K.D. Bruce, M. Stemmer, P.J. Hobbs, R.D. Bardgett, and W. Amelung. 2000. Structure and function of the soil microbial community in microhabitats of a heavy metal polluted soil. Biol. Fertil. Soils 32:390–400. doi:10.1007/s003740000268

Kay-Shoemake, J.L., M.E. Watwood, R.E. Sojka, and R.D. Lentz. 2000. Soil amidase activity in polyacrylamide-treated soils and potential activity toward common amide-containing pesticides. Biol. Fertil. Soils 31:183–186. doi:10.1007/s003740050643

Keeney, D.R., and D.W. Nelson. 1982. Nitrogen—inorganic forms. p. 643–698. *In* A.L. Page, R.H. Miller, and D.R. Keeney (ed.) Methods of soil analysis. Part 2. 2nd ed. Agron. Monogr. 9. ASA and SSSA, Madison, WI.

Kelly, M., and P.H. Clarke. 1962. An inducible amidase produced by a strain of *Pseudomonas aeruginosa*. J. Gen. Microbiol. 27:305–316.

Kim, B., M.B. McBride, and A.G. Hay. 2008. Urease activity in aged copper and zinc-spiked soils: Relationship to CaCl$_2$-extractable metals and Cu^{2+} activity. Environ. Toxicol. Chem. 27:2469–2475. doi:10.1897/08-023.1

Klose, S., and M.A. Tabatabai. 2000. Urease activity of microbial biomass in soils as affected by cropping systems. Biol. Fertil. Soils 31:191–199. doi:10.1007/s003740050645

Könneke, M., A.E. Bernhard, J.R. De La Torre, C.B. Walker, J.B. Waterbury, and D.A. Stahl. 2005. Isolation of an autotrophic ammonia-oxidizing marine archaeon. Nature 437:543–546. doi:10.1038/nature03911

Koops, H.-P., U. Purkhold, A. Pommerening-Röser, G. Timmermann, and M. Wagner. 2003. The lithoautotrophic ammonia-oxidizing bacteria. p. 778–813. In M. Dworkin, S. Falkow, E. Rosenberg, K.-H. Schleifer, and E. Stackebrandt (ed.) The prokaryotes: An evolving electronic resource for the microbiological community, Vol. 5. Springer Verlag, New York.

Lagomarsino, A., S. Grego, S. Marhan, M.C. Moscatelli, and E. Kandeler. 2009. Soil management modifies micro-scale abundance and function of soil microorganisms in a Mediterranean ecosystem. Eur. J. Soil Sci. 60:2–12. doi:10.1111/j.1365-2389.2008.01113.x

Lambert, D.F., J.E. Sherwood, and P.S. Francis. 2004. The determination of urea in soil extracts and related samples—A review. Aust. J. Soil Res. 42:709–717. doi:10.1071/SR04028

Landi, L., G. Renella, L. Giagnoni, and P. Nanniperi. 2011. Activities of proteolytic enzymes. p. 247–260. In R.P. Dick (ed.) Methods of soil enzymology. SSSA Book Ser. 9. SSSA, Madison, WI. (This volume.)

Laughlin, R.J., and R.J. Stevens. 2002. Evidence for fungal dominance of denitrification and codenitrification in a grassland soil. Soil Sci. Soc. Am. J. 66:1540–1548. doi:10.2136/sssaj2002.1540

Leininger, S., T. Urich, M. Schloter, L. Schwark, J. Qi, G.W. Nicol, J.I. Prosser, S.C. Schuster, and C. Schleper. 2006. Archaea predominate among ammonia-oxidizing prokaryotes in soils. Nature 442:806–809. doi:10.1038/nature04983

Lorenz, N., T. Hintemann, T. Kramarewa, A. Katayama, T. Yasuta, P. Marschner, and E. Kandeler. 2006. Response of microbial activity and composition in soils to long-term arsenic and cadmium exposure. Soil Biol. Biochem. 38:1430–1437. doi:10.1016/j.soilbio.2005.10.020

Lorenz, K., and E. Kandeler. 2005. Biochemical characterization of urban soil profiles from Suttgart, Germany. Soil Biol. Biochem. 37:1373–1385. doi:10.1016/j.soilbio.2004.12.009

Lorenz, K., and E. Kandeler. 2006. Microbial biomass and activities in urban soils in two consecutive years. J. Plant Nutr. Soil Sci. 169:799–808. doi:10.1002/jpln.200622001

Ma, W.K., R.E. Farrell, and S.D. Siciliano. 2008. Soil formate regulates the fungal nitrous oxide emission pathway. Appl. Environ. Microbiol. 74:6690–6696. doi:10.1128/AEM.00797-08

Marks, N., K. Datta, and A. Lajtha. 1968. Partial resolution of brain arylamidases and aminopeptidases. J. Biol. Chem. 213:2882–2889.

Marschner, P., E. Kandeler, and B. Marschner. 2003. Structure and function of the soil microbial community in a long-term fertilizer experiment. Soil Biol. Biochem. 35:453–461. doi:10.1016/S0038-0717(02)00297-3

Müller, C., R.J. Stevens, R.J. Laughlin, F. Azam, and J.C.G. Ottow. 2002. The nitrification inhibitor DMPP had no effect on denitrifying enzyme activity. Soil Biol. Biochem. 34:1825–1827. doi:10.1016/S0038-0717(02)00165-7

Mulvaney, R.L. 1996. Nitrogen—Inorganic forms. p. 1123–1184. In D.L. Sparks (ed.) Methods in soil analysis. Part 3. Chemical analysis. SSSA Book Ser. 5. SSSA, Madison, WI.

Nachlas, M.M., T.P. Goldstein, and A.M. Seligman. 1962. An evaluation of aminopeptidase specificity with seven chromogenic substrates. Arch. Biochem. Biophys. 97:223–231. doi:10.1016/0003-9861(62)90073-5

Nannipieri, P., E. Kandeler, and P. Ruggiero. 2002. Enzyme activities and microbiological and biochemical processes in soil. p. 1–34. In R. Burns and R. Dick (ed.) Enzymes in the environment: Activity, ecology, and application. Marcel Dekker, New York.

Nicol, G.W., and C. Schleper. 2006. Ammonia-oxidising Crenarchaeota: Important players in the nitrogen cycle? Trends Microbiol. 14:207–212. doi:10.1016/j.tim.2006.03.004

Niklaus, P.A., E. Kandeler, P.W. Leadley, B. Schmid, D. Tscherko, and C. Körner. 2001. A functional link between plant diversity, elevated CO_2 and soil nitrate. Oecologia 127:540–548. doi:10.1007/s004420000612

Patra, A.K., L. Abbadie, A. Clays-Josserand, V. Degrange, S.J. Grayston, N. Guillaumaud, P. Loiseau, F. Louault, S. Mahmood, S. Nazaret, L. Philippot, F. Poly, J.I. Prosser, and X. Le Roux. 2006. Effects of management regime and plant species on the enzyme activity and genetic structure of N-fixing, denitrifying and nitrifying bacterial communities in grassland soils. Environ. Microbiol. 8:1005–1016. doi:10.1111/j.1462-2920.2006.00992.x

Patterson, E.K., S.H. Hsiao, and A. Keppel. 1963. Studies on dipeptidases and aminopeptidases: I. Distinction between leucine aminopeptidase and enzymes that hydrolyze L-leucyl-β-naphthylamide. J. Biol. Chem. 238:3611–3620.

Pell, M., B. Stenberg, J. Stenström, and L. Torstensson. 1996. Potential dentrification activity assay in soil—with or without chloramphenicol? Soil Biol. Biochem. 28:393–398. doi:10.1016/0038-0717(95)00149-2

Philippot, L. 2005. Tracking nitrate reducers and denitrifiers in the environment. Biochem. Soc. Trans. 33:200–204. doi:10.1042/BST0330200

Philippot, L., S. Hallin, and M. Schloter. 2007. Ecology of denitrifying prokaryotes in agricultural soil. Adv. Agron. 96:249–305. doi:10.1016/S0065-2113(07)96003-4

Philippot, L., and O. Hojberg. 1999. Dissimilatory nitrate reductases in bacteria. Biochim. Biophys. Acta 1577:1–23.

Porter, L.K. 1965. Enzymes. p. 1536–1549. In C.A. Black, D.D. Evans, J.L. White, L.E. Ensminger, and F.E. Clark (ed.) Methods of soil analysis. Part 2. Agron. Monogr. 9. ASA, Madison, WI.

Purkhold, U., A. Pommerening-Roser, S. Juretschko, M.C. Schmid, H.P. Koops, and M. Wagner. 2000. Phylogeny of all recognized species of ammonia oxidizers based on comparative 16S rRNA and amoA sequence analysis: Implications for molecular diversity surveys. Appl. Environ. Microbiol. 66:5368–5382. doi:10.1128/AEM.66.12.5368-5382.2000

Renella, G., U. Szukics, L. Landi, and P. Nannipieri. 2007. Quantitative assessment of hydrolase production and persistence in soil. Biol. Fertil. Soils 44:321–329. doi:10.1007/s00374-007-0208-8

Rich, J.J., R.S. Heichen, P.J. Bottomley, K. Cromack, Jr., and D.D. Myrold. 2003. Community composition and functioning of dentrifying bacteria from adjacent meadow and forest soils. Appl. Environ. Microbiol. 69:5974–5982. doi:10.1128/AEM.69.10.5974-5982.2003

Roberts, J., J.S. Holcenber, and W.C. Dolowy. 1972. Isolation, crystallization, and properties of Achromobacteraceae, glutaminase-asparaginase with antitumor activity. J. Biol. Chem. 247:84–90.

Robertson, G.P., and P.M. Groffmann. 2007. Nitrogen transformations. p. 341–364. In E.A. Paul (ed.) Soil Microbiology, biochemistry and soil ecology. 3rd ed. Elsevier, San Diego.

Rotini, O.T. 1935. La transformazione enzimatica dell'urea nell terreno. Ann. Labor. Ric. Ferm. Spaallanzani 3:143–154.

Sadowsky, M.J., W.C. Koskinen, J. Seebinger, B.L. Barber, and E. Kandeler. 2006. Automated robotic assay of alkaline phosphomonoesterase activity in soils. Soil Sci. Soc. Am. J. 70:378–381. doi:10.2136/sssaj2005.0156N

Sardans, J., J. Peñuelas, and M. Estiarte. 2008. Changes in soil enzymes related to C and N cycle and in soil C and N content under prolonged warming and drought in a Mediterranean shrubland. Appl. Soil Ecol. 39:223–235. doi:10.1016/j.apsoil.2007.12.011

Sayre, F.W., and E. Roberts. 1958. Preparation and some properties of a phosphatase-activated glutaminase from kidneys. J. Biol. Chem. 233:1128–1134.

Scelza, R., M.A. Rao, and L. Gianfreda. 2008. Response of an agricultural soil to pentachlorophenol (PCP) contamination and the addition of compost or dissolved organic matter. Soil Biol. Biochem. 40:2162–2169. doi:10.1016/j.soilbio.2008.05.005

Senwo, Z.N., and M.A. Tabatabai. 1998. Amino acid composition of soil organic matter. Biol. Fertil. Soils 26:235–242. doi:10.1007/s003740050373

Senwo, Z.N., and M.A. Tabatabai. 1999. Aspartase activity in soils: Effects of trace elements and relationships to other amidohydrolases. Soil Biol. Biochem. 31:213–219. doi:10.1016/S0038-0717(98)00091-1

Sessitsch, A., S. Gyamfi, D. Tscherko, M.H. Gerzabek, and E. Kandeler. 2005. Activity of microbes affected by the cultivation of transgenic glufosinate-tolerant oilseed rape (Brassica napus) and the application of the associated herbicide. Plant Soil 266:105–116. doi:10.1007/s11104-005-7077-4

Sinsabaugh, R.L., H. Reynolds, and T.M. Long. 2000. Rapid assay for amidohydrolase (urease) activity in environmental samples. Soil Biol. Biochem. 32:2095–2097. doi:10.1016/S0038-0717(00)00102-4

Skujiņš, J.J., and A.D. McLaren. 1969. Assay of urease activity using ^{14}C-urea in stored, geologically preserved soils. Enzymologia 34:213–225.

Smith, M.S., and J.M. Tiedje. 1979. Phases of denitrification following oxygen depletion in soil. Soil Biol. Biochem. 11:261–267. doi:10.1016/0038-0717(79)90071-3

Sowden, F.J. 1958. The forms of nitrogen in the organic matter of different horizons of soil profiles. Can. J. Soil Sci. 38:149–154. doi:10.4141/cjss58-023

Stevenson, F.J. 1994. Humus chemistry: Genesis, composition, reactions. 2nd ed. John Wiley & Sons, New York.

Sumner, J.B. 1951. Urease. p. 873–892. In J.B. Sumner and K. Myrbäck (ed.) The enzymes. Vol. 1. Part 2. Academic Press, New York.

Tabatabai, M.A. 1994. Soil enzymes. p. 801–834. In R.W. Weaver, J.S. Angle, and P.S. Bottomley (ed.) Methods of soil analysis. Part 2. Microbiological and biochemical properties. SSSA, Madison, WI.

Tabatabai, M.A., and J. Bremner. 1972. Assay of urease activity in soils. Soil Biol. Biochem. 4:479–487. doi:10.1016/0038-0717(72)90064-8

Tabatabai, M.A., M. Ekenler, and Z.N. Senwo. 2010. Significance of enzyme activities in soil nitrogen mineralization. Commun. Soil Sci. Plant Anal. 41:595–605. doi:10.1080/00103620903531177

Tabatabai, M.A., A.M. Garcia-Manzanedo, and V. Acosta-Martinez. 2002. Substrate specificity of arylamidase in soils. Soil Biol. Biochem. 34:103–110. doi:10.1016/S0038-0717(01)00162-6

Taylor, J.P., B. Wilson, M.S. Mills, and R.G. Burns. 2002. Comparison of microbial numbers and enzymatic activities in surface soils and subsoils using various techniques. Soil Biol. Biochem. 34:387–401. doi:10.1016/S0038-0717(01)00199-7

Thoren, A.K. 2007. Urea transformation of wetland microbial communities. Microb. Ecol. 53:221–232. doi:10.1007/s00248-006-9098-9

Tiedje, J.M. 1982. Denitrification. In A.L. Page, R.H. Miller, and D.R. Keeney (ed.) Methods of soil analysis. Agron. Monogr. 9. Part 2. 2nd ed. ASA and SSSA, Madison, WI.

Tourna, M., T.E. Freitag, G.W. Nicol, and J.I. Prosser. 2008. Growth, activity and temperature response of ammonia-oxidizing archaea and bacteria in soil microcosms. Environ. Microbiol. 10:1357–1364. doi:10.1111/j.1462-2920.2007.01563.x

Tscherko, D., U. Hammesfahr, G. Zeltner, E. Kandeler, and R. Böcker. 2005. Soil microflora function and composition in recently deglaciated alpine terrain: Impact of early and late coloniser plants. Basic Appl. Ecol. 6:367–383. doi:10.1016/j.baae.2005.02.004

Tscherko, D., E. Kandeler, and A. Bárdossy. 2007. Fuzzy classification of soil microbial biomass and enzyme activity in grassland soils. Soil Biol. Biochem. 39:1799–1808. doi:10.1016/j.soilbio.2007.02.010

Tscherko, D., J. Rustemeier, A. Richter, W. Wanek, and E. Kandeler. 2003. Functional diversity of the soil microbial community in primary succession along two glacier forelands in the Central Alps. Eur. J. Soil Sci. 54:685–696. doi:10.1046/j.1351-0754.2003.0570.x

von Steiger, B., K. Nowack, and R. Schulin. 1996. Spatial variation of urease activity measured in soil monitoring. J. Environ. Qual. 25:1285–1290. doi:10.2134/jeq1996.00472425002500060017x

Wall, M.C., and K.J. Laidler. 1953. The molecular kinetics of the urea-urease system. IV. The reaction in an inert buffer. Arch. Biochem. Biophys. 43:299–306. doi:10.1016/0003-9861(53)90124-6

Weidler, G.W., M. Dornmayr-Pfaffenhuemer, F.W. Gerbl, W. Heinen, and H. Stan-Lotter. 2007. Communities of Archaea and Bacteria in a subsurface radioactive thermal spring in the Austrian central alps, and evidence of ammonia-oxidizing Crenarchaeota. Appl. Environ. Microbiol. 73:259–270. doi:10.1128/AEM.01570-06

Yoshinari, T., and R. Knowles. 1976. Acetylene inhibition of nitrous oxide reduction by denitrifying bacteria. Biochem. Biophys. Res. Commun. 69:705–710. doi:10.1016/0006-291X(76)90932-3

Yuen, S.H., and A.G. Pollard. 1953. Determination of nitrogen in soil and plant materials: Use of boric acid in the micro-Kjeldahl method. J. Sci. Food Agric. 4:490–496. doi:10.1002/jsfa.2740041006

Zaman, M., K.C. Cameron, H.J. Di, and K. Inubushi. 2002. Changes in mineral N, microbial biomass and enzyme activities in different soil depths after surface applications of dairy shed effluent and chemical fertilizer. Nutr. Cycling Agroecosyst. 63:275–290. doi:10.1023/A:1021167211955

Zantua, M.I., and J.M. Bremner. 1975a. Comparison of methods of assaying urease activity in soils. Soil Biol. Biochem. 7:291–295. doi:10.1016/0038-0717(75)90069-3

Zantua, M.I., and J.M. Bremner. 1975b. Preservation of soil samples for assaying urease activity. Soil Biol. Biochem. 7:297–299. doi:10.1016/0038-0717(75)90070-X

Zechmeister-Boltenstern, S. 1996. Denitrification enzyme activity. p. 179–184. *In* F. Schinner, R. Öhlinger, E. Kandeler, and R. Margesin (ed.) Methods in soil biology. Springer, Berlin New York.

Zumft, W.G. 1997. Cell biology and molecular basis of denitrification. Microbiol. Mol. Biol. Rev. 61:533–616.

Chapter 11

Activities of Proteolytic Enzymes

Loretta Landi,* Giancarlo Renella, Laura Giagnoni, and Paolo Nannipieri

11–1 INTRODUCTION

11–1.1 Classification, Reactions, and Ecological Functions of Proteases

Proteases, also known as proteinases or proteolytic enzymes, are a large group of hydrolases that catalyze the cleavage of peptide bonds in proteins to produce peptides and/or amino acids.

Classification of proteolytic enzymes is based on three major criteria: (i) type of reaction catalyzed, (ii) functional group of the active site, and (iii) type of molecular structure and evolutionary relationship among the various enzymes (Rao et al., 1998). According to the Nomenclature Committee of the International Union of Biochemistry and Molecular Biology (NC-IUBMB), the proteolytic enzymes can be grouped into proteases and peptidases on the basis of their nature of attack.

Proteases can catalyze the hydrolysis of the terminal amino acid of polypeptide chains (exopeptidases) or the internal peptide bond (endopeptidases, synonymous with proteinases). The exopeptidases act at the N or C terminals. Those capable to remove one, two and three amino acids from the N terminus are called aminopeptidases, dipeptidyl-peptidases, and tripeptidyl-peptidases, respectively. Those that liberate one and two amino acids away from the C terminus are carboxypeptidases and peptidyl-dipeptidases, respectively (Table 11–1). The carboxypeptidases are further subdivided based on catalytic mechanism (serine-type carboxypeptidases, metallocarboxypeptidases, and cysteine-type carboxypeptidases). Endopeptidases (proteinases) are recognized on the basis of the chemical nature of the groups responsible for catalytic activity. As shown in Table 11–1, four distinct classes of proteinases have been identified: serine, cysteine, aspartic, and metalloendopeptidases. According to the optimal pH, they are classified as acidic, neutral, or alkaline proteinases. A fifth class (EC 3.4.99) also has been assigned for those of unknown catalytic mechanisms and for those that do not fit clearly into one of the four other groups.

Loretta Landi, Department of Plant, Soil and Environmental Science, Piazzale Cascine 15, 50144, University of Firenze, Firenze, Italy (loretta.landi@unifi.it), *corresponding author;
Giancarlo Renella, Department of Plant, Soil and Environmental Science, Piazzale Cascine 15, 50144, University of Firenze, Firenze, Italy (giancarlo.renella@unifi.it);
Laura Giagnoni, Department of Plant, Soil and Environmental Science, Piazzale Cascine 15, 50144, University of Firenze, Firenze, Italy (laura.giagnoni@unifi.it);
Paolo Nanniperi, Department of Plant, Soil and Environmental Science, Piazzale Cascine 15, 50144, University of Firenze, Firenze, Italy (paolo.nannipieri@unifi.it).

doi:10.2136/sssabookser9.c11

Table 11–1. Classification of proteases according to the EC nomenclature.

Sub-subclass	Proteases (= peptidases) EC 3.4
3.4.11	Aminopeptidases
3.4.13	Dipeptidases
3.4.14	Dipeptidyl-peptidases
3.4.15	Peptidyl-dipeptidases
3.4.16	Serine-type carboxypeptidases
3.4.17	Metallopeptidases
3.4.18	Cysteine-type carboxypeptidases
3.4.19	Omega peptidases
3.4.21	Serine endopeptidases
3.4.22	Cysteine endopeptidases
3.4.23	Aspartic endopeptidases
3.4.24	Metalloendopeptidases
3.4.99	Endopeptidases of unknown type

In soil, proteases and peptidases can have different locations: in living and active cells; in dead cells; as free enzymes; and adsorbed to organic, inorganic, or organomineral particles (Loll and Bollag, 1983). These enzymes play an important role in N mineralization (Ladd and Jackson, 1982), a process regulating the amount of plant-available N (Stevenson, 1982). Extracellular proteases catalyze the hydrolysis of exogenous proteins to smaller molecules before cellular uptake whereas intracellular proteases are involved in the intracellular protein turnover. All living cells maintain a particular rate of protein turnover by continuous degradation and synthesis of proteins and each intracellular protease has its own target function. Extracellular proteases and peptidases can remain active and be protected against proteolysis by their interaction with surface-reactive particles or after entrapment in humic molecules.

These stabilized extracellular enzymes are not subjected to repression or induction, and thus may not be sensitive to environmental conditions affecting the physiological state of the microorganisms (Burns, 1982; Nannipieri, 1994). Studies of soil microbial populations by cultivation techniques indicated that soil proteases and peptidases are mainly of bacterial origin (Hayano, 1993; Bach and Munch, 2000). However, by selective inhibition of different groups of bacterial proteases and peptidases, it has been shown that neutral metalloenzymes and serine enzymes are mainly responsible for protein and peptide degradation in arable soils (Hayano et al., 1987, 1995; Bach and Munch, 2000; Kamimura and Hayano, 2000; Watanabe et al., 2003). Recently, information on the unculturable proteolytic bacterial communities has been obtained by characterizing both DNA and mRNA extracted from soil and amplified by specific primers (Bach et al., 2001; Mrkonjic Fuka et al., 2008). Using mass spectrometry-based proteomics, Schulze et al. (2005) have identified the type and biological origin of soil proteins, including enzymes.

The proteolytic bacterial community comprises 22 to 89% of the total soil microbial biomass (Kumar et al., 2006), and different proteolytic bacteria release different types and amounts of proteases (Sakurai et al., 2007). These authors performed an innovative approach since they measured N-benzyloxycarbonyl L-phenylalanyl L-leucine (ZPL)-hydrolyzing activity and characterized proteolytic bacterial communities by denaturing gradient gel electrophoresis (DGGE) of the alkaline metalloprotease (*apr*) and neutral metalloprotease (*npr*) genes in both rhizosphere and bulk soil treated with organic and inorganic fertilizer.

The rhizosphere is a zone of active interchange between plants and soil bacteria. Root exudates can be a source of easily degradable N-compounds, such as amino acids and small peptides, able to induce protease synthesis (Badalucco et al., 1996; García-Gil et al., 2004). DeAngelis et al. (2008) found higher protease-specific activity (enzyme activity per cell) in the rhizosphere than in the bulk soil. Microbial synthesis of catabolic enzymes, such as proteases, is normally regulated at the cellular level by induction or repression when the cell is exposed to the substrate or the product of the reaction (Gottschalk, 1979).

Amato and Ladd (1988) observed an increase of protease activity during a 10-d incubation period of a chloroform-fumigated soil, indicating that enzyme activity also may be stimulated when high concentrations of substrates are available (soil fumigated) and growth of surviving or inoculated organism occur. An increase of protease activity was shown in soil samples treated with glucose and nutrients (Asmar et al., 1992; Renella et al., 2007). Allison and Vitousek (2005) reported that glycine aminopeptidase activity was stimulated by the addition of a complex source of N (collagen), but repressed with the addition of ammonium. Nevertheless, both the inducer and the repressor may be degraded and/ or adsorbed by soil particles and thus not survive long enough to reach a sufficiently high concentration to induce or repress the enzyme synthesis; for this reason, constitutive enzyme synthesis is probably the dominant microbial source of enzymatic activity in soil (Burton and McGill, 1991).

This hypothesis seems to be supported by the observations of Tateno (1988), who found that protease activity was limited in soil by substrate availability and not by the amount of enzyme. However, Marx et al. (2005) found that leucine aminopeptidase activity was twice as large in the clay-size fraction as in the bulk soil. Considering that these enzymes can be stabilized and remain active when adsorbed on humic molecules and clay particles, and that proteinaceous material (proteins, glycoproteins, peptides, and amino acids) represents 40% of total N in soil (Schulten and Schnitzer, 1998), protease and peptidase activities can highly affect the rate of organic N mineralization (Nannipieri and Badalucco, 2002). Protease activity has shown a close relationship with N mineralization (Nannipieri et al., 1983; Asmar et al., 1994), microbial respiration, ATP content, and bacterial biomass (Asmar et al., 1992).

Protease and peptidase activity of soils is affected by different soil physico-chemical properties, climatic conditions, presence of substrates, and management practices (Ladd and Butler, 1972; Klein and Koths, 1980; Perucci, 1992; Schloter et al., 2003; Watanabe et al., 2003; Marx et al., 2005; Rahman et al., 2005).

It is well established that protease activity decreases with soil depth (Bach and Munch, 2000; Taylor et al., 2002; Mrkonjic Fuka et al., 2008) and with reduction in soil moisture (Sardans and Penuelas, 2005). Soil texture affects protease with the clay fraction having higher activity than in the bulk soil (Marx et al., 2005).

Crop and soil management does affect protease with diverse crop rotations increasing activity over corn monocultures (Ladd and Butler, 1972; Blagoveshchenskaya and Danchenko, 1974; Marx et al., 2005). Undisturbed soils, such as in woodlands (Burket and Dick, 1998; Renella et al., 2002) or in no-tillage systems, significantly increases protease activity over conventional agricultural systems that include tillage (Klein and Koths, 1980; Burket and Dick, 1998, Kandeler et al.,1999). Similarly, removal of crop residues under conventional tillage systems significantly reduced protease activity (Klein and Koths, 1980) whereas organic inputs such as straw increased activity (Loll and Bollag, 1983). Protease and peptidase activities are positively correlated with organic amendments and soil organic matter and clay contents of soil (Ladd and Butler, 1972; Klein and Koths, 1980; Perucci, 1992; Marx et al., 2005).

A number of studies have shown protease to be an indicator of the degree of soil pollution or remediation. This has been shown where protease activity increased by reduced trace element mobility in remediated soils and in mine spoils (Mench et al., 2006; Renella et al., 2008; Kumpiene et al., 2009; Ascher et al., 2009).

11–1.2 Protease Activity Methodology

Pretreatment and handling of soils does affect protease activities. Air-drying or storage at −8, 4, or 22°C causes a decrease in protease activities on range of soil types (Ladd, 1972; Speir et al., 1980). Air-drying of soil can denature thermosensitive enzymes but also facilitate adsorption of extracellular enzymes such as caseinase by soil colloids, making them more resistant to thermal denaturation (Burns, 1982; Bonmati et al., 2003). Adsorption might protect the enzyme but render it inaccessible and therefore inactive toward its high molecular weight substrates.

Protease can be extracted from soils and the activity estimated in soil extracts. Different buffer solutions have been used to extract proteases from soil, such as 0.1 M Tris-borate at pH 8.1, 0.1 M Tris at pH 8.1, 0.1 M Tris-citrate at pH 8.0, and 0.1 M Tris-EDTA at pH 8.0 (Ladd, 1972); 0.2 M phosphate–0.2 M EDTA buffer at pH 8.0 (Mayaudon et al., 1975; Batistic et al., 1980); 0.1 M phosphate at pH 7.0 (Hayano et al., 1987); and 140 mM sodium pyrophosphate at pH 7.1 (Nannipieri et al., 1980, 1982, 1985; Bonmati et al., 1998), the most efficient in extracting hydrolases from soil (Nannipieri et al., 1996). Studies by Bonmati et al. (1998; 2009) on soil extracts showed that N-benzoyl-L-argininamide (BAA)-hydrolyzing proteases are probably mostly associated with highly condensed humus, ZPL-hydrolyzing proteases with less-condensed humic material, and casein-hydrolyzing proteases with fresh organic matter content of soils. Obviously the enzyme activity of the soil extracts does not reflect the activity in bulk soil, because a complete extraction of the enzyme is almost impossible (see Chapter 16, Fornasier et al., 2011, this volume, for further discussion and methods for extracting enzymes from soils).

Methods developed for the assay of protease activity may differ in terms of substrate, incubation conditions (pH, type and concentration of buffer, temperature, incubation period, etc.) and analytical procedures. Generally, soil proteolytic activity is detected following: (i) the decrease of initial substrate, or more often (ii) the increase of peptides or amino acids released during a given incubation period using high and low molecular weight substrates or fluorogenic substrates, and measuring spectrophotometrically or fluorometrically the released chromogenic or fluorogenic compounds.

Various incubation times, shaking rates, and buffer solutions have been used to estimate protease activity of soil toward high molecular weight substrates (casein, azocasein, gelatin, hemoglobin, albumins), and low molecular weight substrates as dipeptides and dipeptide derivatives containing amino acid with either aromatic, acidic, or nonpolar aliphatic side chains (BAA and N-benzyloxycarbonyl (Z) derivatives of the dipeptides, L-phenylalanyl L-leucine, glycyl L-phenylalanine, glycyl L-leucine, glycyl glycine, L-leucyl L-tyrosine, and L-glutamyl L-tyrosine) (Hoffmann and Teicher, 1957; Ladd and Butler, 1972; Beck, 1973; Ross et al., 1975). Tests with different buffer solutions (2 M acetate buffer pH 4.0–5.0, 1 M maleate buffer pH 5.0–7.5 and 0.1 M Tris-borate buffer pH 7.5–10, Tris-borate buffer at pH 8.1) showed that enzymatic hydrolysis is considerably higher in Tris-borate buffer at pH 8.1 for all dipeptide derivatives studied but not for BAA; borate buffer inhibits BAA hydrolysis by about 50% (Ladd and Butler, 1972).

The maximum enzymatic hydrolysis of BAA occurs with 0.1 M phosphate buffer having pH 7.1 (Nannipieri et al., 1980). Ladd and Butler (1972) observed that (Z) L-phenylalanyl L-leucine (ZPL), containing amino acids with hydrophobic side chains, was most rapidly hydrolyzed respect to other dipeptide derivatives. By contrast, the rate of hydrolysis of the amide BAA varied with each soil studied. Probably, due to the polar nature of BAA, different enzymes were involved in the hydrolysis of BAA (Ladd and Butler, 1972).

The maximum protease activity of soil toward low and high molecular weight substrates occurred at 40 to 50°C and 55°C; and for pH, from approximately 6.0 to 8.5 and 8.0, respectively. At temperatures above 60°C the enzyme is denaturated (Ladd, 1972; Ladd and Butler, 1972).

Recently, aminopeptidase activity has been measured with methods based on the use of fluorogenic 7-amino-4-methylcoumarin (AMC) substrates (e.g., L-leucine-7-AMC; L-tyrosine-7-AMC, L-arginine-7-AMC, L-alanine-7-AMC) and microplates (Marx et al., 2001, 2005; Vepsäläinen et al., 2001). The introduction of 7-amino-4-methylcoumarin (7-AMC)-coupled model substrates, which release highly fluorescing products after enzymatic hydrolysis, increased the sensitivity of the enzymatic assays. The principles of this assay have been described by Hoppe (1983), Hoppe et al. (1988), and Somville (1984) for measurements of enzyme activities in aquatic environments, and by Darrah and Harris (1986) and Freeman et al. (1995) for measurements in soil.

Potential advantages of the fluorogenic method in comparison with colorimetric techniques are: (i) higher sensitivity, (ii) short incubation time, (iii) only small quantities of substrate are needed, (iv) the separation of the reaction product from soil is unnecessary, and (v) the product can be continually monitored over time (Vepsäläinen et al., 2001). These features would provide the opportunity for high throughput and automated assays that utilize 96-well plates and readers.

Although protocols for fluorogenic soil protease and other enzyme assays are described by Vepsäläinen et al. (2001) and Marx et al. (2001, 2005), there is no commonly accepted standard procedure. Major limitations from a quantitative and reproducible perspective are substrate adsorption and quenching of the fluorogenic product. There is no simple control to account for quenching due to the presence of soil phenolics in reaction mixtures. This was shown by Freeman et al. (1995), who measured the amount of quenching on a range of samples and found that it varied widely both temporally and spatially. In addition, the International Organization for Standardization (ISO) is trying to set up a standard protocol for determining soil protease activity by using fluorogenic substrates. As a standard procedure has not yet been accepted, we cannot include a fluorogenic method in this chapter for protease or dipeptidase activity (See Chapter 14, Deng et al., 2011, this volume, for microplate methods).

Methods proposed to estimate soil protease activity are based on colorimetric determination of the products (amino acids or ammonium) released by the activities of the protein, peptide, and amide-hydrolyzing enzymes. Ladd et al. (1976) found that activities toward BAA were less affected by seasonal drying of the soils than by activities toward ZPL and casein. Ladd and Butler (1972) observed that stored, moistened soils hydrolyzed Z-L-phenylalanyl L-leucine at rates one or two orders of magnitude greater in respect to casein and hemoglobin, and that activities of different substrates (casein, hemoglobin, N-benzoyl-L-argininamide, Z-L-phenylalanyl L-leucine, Z-glycyl L-phenylalanine) were proportional to the soil concentration and to incubation time, except for casein. Casein showed a lag period before the linear relationship; probably some time is required before the protein is hydrolyzed to products of sufficiently low molecular weight to be trichloroacetic acid (TCA)-soluble.

11–2 PRINCIPLES

Methods described to estimate soil protease activity in this chapter utilize two substrates: casein, essentially a nonspecific substrate, and N-benzoyl-L-argininamide (BAA), a typical substrate for trypsin-like enzyme. The assay is based on colorimetric estimation of products released by the protein and amide-hydrolyzing enzymes when soil is incubated with buffered solutions of casein and BAA, respectively.

The first method is based on determination of the tyrosine released by casein-hydrolyzing proteases when the soil sample is incubated with casein (or sodium caseinate), a high molecular weight substrate, and 0.1 M Tris buffer at pH 8.1. The soil–substrate mixture is incubated for 2 h at 50°C. At the end of incubation, trichloroacetic acid (TCA) solution is added to precipitate residual protein, and the concentration of TCA-soluble tyrosine is spectrophotometrically determined by the Folin–Ciocalteu reagent (Ladd and Butler, 1972).

The second method is based on the spectrophotometric determination of NH_4^+–N released when soil is incubated with BAA, a low molecular weight substrate, and 0.1 M phosphate buffer pH 7.1 for 1 h at 40°C (Nannipieri et al., 1980). The ammonium released is extracted after addition of 2 M KCl and then determined by a modified indophenol reaction (Kandeler and Gerber, 1988).

11–3 ASSAY METHODS

11–3.1 Protease Assay with High Molecular Weight Substrate (Ladd and Butler, 1972; Alef and Nannipieri, 1995)

11–3.1.1 Apparatus

- Volumetric glass flasks, 100, 500, 1000 mL
- Erlenmeyer flasks with stopper, 25 mL
- Microcuvette
- Pipettes
- Vortex
- pH meter

- Shaking water bath (adjustable to 50°C)
- Centrifuge
- Spectrophotometer

11–3.1.2 Reagents

1. Tris buffer, 50 mM, pH 8.1: Dissolve 6.05 g of Tris (hydroxymethyl) amino methane in about 700 mL of distilled water, adjust the pH to 8.1 with HCl and bring up to 1000 mL with distilled water.

2. Sodium caseinate solution, 2%: Dissolve 10 g of sodium caseinate in warm distilled water (50°C) and bring up with distilled water to 500 mL (use a stirrer). If only casein is available, dissolve 10 g of casein in Tris buffer, adjust the pH to 8.1 with NaOH solution and bring up to 500 mL with Tris buffer (50 mM, pH 8.1).

3. Trichloroacetic acid (TCA), 0.92 M: Dissolve 75 g of TCA in about 300 mL of distilled water and bring up to 500 mL with distilled water.

4. Sodium hydroxide (NaOH), 1 M: Dissolve 40 g of NaOH in distilled water and dilute with distilled water to 1 L.

5. Sodium hydroxide (NaOH)–sodium carbonate (Na_2CO_3) solution: Dilute 60 mL of NaOH (1 M) with distilled water before dissolving 50 g Na_2CO_3 (water free) in the solution and bring up to 1000 mL with distilled water.

6. Copper sulfate ($CuSO_4 \cdot 5H_2O$) solution, 0.02 M: Dissolve 0.5 g of $CuSO_4 \cdot 5H_2O$ in distilled water and bring up to 100 mL with distilled water.

7. Potassium sodium tartrate ($C_4H_4KNaO_6 \cdot 4H_2O$) solution, 0.035 M: Dissolve 1 g of $C_4H_4KNaO_6 \cdot 4H_2O$ in distilled water and bring up to 100 mL with distilled water.

8. Alkaline reagent: Mix 1000 mL of NaOH–Na_2CO_3 solution with 20 mL of $CuSO_4 \cdot 5H_2O$ solution and 20 mL of $C_4H_4KNaO_6 \cdot 4H_2O$ solution.

9. Folin–Ciocalteu reagent, 33%: Dilute 167 mL of Folin–Ciocalteu reagent to 500 mL with distilled water. The diluted Folin–Ciocalteu reagent is unstable and it is necessary to prepare it immediately before use. The Folin–Ciocalteu reagent should be stored at 4°C.

10. Standard tyrosine stock solution, 500 µg mL^{-1}: Dissolve 50 mg of tyrosine in about 50 mL of Tris buffer and dilute to 100 mL with Tris buffer. The solution is stable from 2 to 3 d at 4°C.

11–3.1.3 Procedure

Place 1 g of moist soil (<2 mm), in a 25-mL Erlenmeyer flask, add 5 mL of Tris buffer and 5 mL of sodium caseinate solution. Vortex for a few seconds to mix the contents. Stopper the flask and incubate at 50°C on a shaking water bath for 2 h. After incubation, immediately add 5 mL of trichloroacetic acid (TCA) to stop enzyme activity and to precipitate the residual casein, and mix the contents thoroughly. Controls are conducted by adding 5 mL of sodium caseinate solution at the end of the incubation and immediately before adding the TCA solution. Centrifuge the soil suspensions at 10,000 g for 10 min at 4°C. Pipette 5 mL of the clear supernatant into a new flask, add 7.5 mL of the alkaline reagent, and incubate for 15 min at room temperature. Add 5 mL of the Folin–Ciocalteu reagent, swirl the flasks, and filter the mixtures through paper filter into microcuvettes. After 1 h of incubation at room temperature, measure the absorbance at

700 nm by spectrophotometer. It is recommended to measure the absorbance several times until the value becomes constant.

Calculate the concentration of tyrosine by relating measured absorbance at 700 nm to that of a calibration graph and correct the results for the control value. To prepare this graph, pipette 0, 1, 2, 3, 4, and 5-mL aliquots of the standard tyrosine stock solution into 25-mL Erlenmeyer flasks, add 5 mL of sodium caseinate solution and bring up to 10 mL with Tris buffer. Then add 5 mL of TCA solution and perform the measurements as described above.

To calculate protease activity (μg g tyrosine g^{-1} dwt $2h^{-1}$), multiply the amount of tyrosine measured (μg mL^{-1}) by the final volume of the solutions added to the soil in the assay (15 mL) and divide by the dry weight of the soil.

11–3.1.4 Comments

Toluene was not used because with a short-term assay (1–2 h) it is unnecessary to prevent microbial proliferation (Ladd and Butler, 1972).

To perform controls, follow the procedure described for assay of protease activity by adding sodium caseinate solution immediately before TCA at the end of incubation time; the casein precipitation is not complete if the substrate is added after the TCA solution.

Upon completion of the incubation and after adding the Folin–Ciocalteu reagent, various periods of color development (5 to 60 min) have been reported (Ladd and Butler, 1972; Nannipieri et al., 1979; Watanabe and Hayano, 1994; Geisseler and Horwath, 2008). However, 5 to 10 min are sufficient for the colorimetric determination by Folin–Ciocalteu reagent; there is no microbial degradation of aromatic amino acids up to 1 h.

Tyrosine standard curves may differ slightly from day to day. Therefore, preparing a calibration curve in each series of analyses is recommended.

If the filtrate cannot be analyzed immediately, it should be stored a 4°C and the tyrosine content analyzed within 5 h.

The substrate, pH, and temperature conditions used are those reported by Ladd and Butler (1972). Ladd and Butler (1972) tested a wide range of soil types of Australia in terms of pH, organic matter, and texture; it may be worthwhile to check the buffer pH and temperature optimum of soils from other regions.

Chloroform fumigation did not influence the casein-hydrolyzing activity of different soils (Ladd 1978).

11–3.2 Protease Assay with Low Molecular Weight Substrate (Nannipieri et al., 1980)

11–3.2.1 Apparatus

- Volumetric glass flasks, 25, 50, 100, 1000, 2000 mL
- Erlenmeyer flask with stopper, 15 mL
- Pipettes
- Vortex
- Shaking water bath (adjustable to 40°C)
- Refrigerating centrifuge, 4°C
- Spectrophotometer

11–3.2.2 Reagents

1. Reagent A: Dissolve 27.80 g KH_2PO_4 in 1000 mL of distilled water.
2. Reagent B: Dissolve 53.65 g $K_2HPO_4 \cdot 7H_2O$ in 1000 mL of distilled water.
3. Phosphate buffer solution K_2HPO_4–KH_2PO_4, 0.1 M, pH 7.1: Mix 390 mL of Reagent A with 610 mL of Reagent B and dilute the solution to 2000 mL with distilled water. Store at 4°C.
4. N-benzoyl-L-argininamide (BAA) solution, 0.03 M: Dissolve 0.416 g of BAA in about 20 mL of phosphate buffer solution and bring up to 50 mL with phosphate buffer solution.
5. Potassium chloride (KCl), 2.0 M: Dissolve 149 g of KCl in about 800 mL of distilled water and dilute to 1 L.
6. Sodium hydroxide (NaOH), 0.3 M: Dissolve 12 g of NaOH in about 800 mL distilled water, cool the solution, and bring up to 1 L with distilled water.
7. Sodium salicylate solution: Dissolve 12 g of Na-salicylate and 120 mg of sodium nitroprusside in distilled water and bring up to 100 mL with distilled water.
8. Sodium salicylate–NaOH solution: Mix equal volume of the NaOH solution, Na-salicylate solution, and distilled water. Prepare the solution daily.
9. Sodium dichloroisocyanide solution, 0.1%: Dissolve 0.1 g of Na-dichloroisocyanide in 100 mL of distilled water. Prepare the solution shortly before using.
10. Standard ammonium stock solution, 1000 μg NH_4^+–N mL^{-1}: Dissolve 3.821 g of dry ammonium chloride (NH_4Cl) in a 1-L volumetric flask, and dilute to volume with distilled water. Store at 4°C.

11–3.2.3 Procedure

Place 1.0 g of moist soil (<2 mm) in a 15-mL Erlenmeyer flask, add 4.0 mL of phosphate buffer and 1 mL of BAA solution, and swirl the flask for a few seconds to mix the contents. Stopper the flask and place it on a shaker in an incubator at 40°C for 1 h. After incubation, immediately add 4.0 mL of KCl solution to stop the reaction, and mix the contents thoroughly. Prepare controls following the procedure described, but make the addition of 1 mL of BAA solution at the end of the incubation and immediately before adding the KCl solution. Centrifuge the soil slurries at 6000 g for 10 min at 4°C. Pipette 1 mL of the soil extract into a 25-mL volumetric flask, add 9 mL of distilled water, 5 mL of the Na-salicylate–NaOH solution, and 2 mL of the Na-dichloroisocyanide solution, mix well. Incubate for 30 min at room temperature. Measure the absorbance of the colored complex with a spectrophotometer adjusted to a wavelength of 690 nm against a reagent blank solution (Kandeler and Gerber, 1988).

Calculate the ammonium released by relating measured absorbance at 690 nm to that of a calibration graph containing 0, 1.0, 1.5, 2.0, and 2.5 μg NH_4^+–N mL^{-1}, and correct the results for the control value. To prepare this graph, dilute 0, 1.0, 1.5, 2.0, and 2.5 mL of the standard ammonium stock solution into 100-mL volumetric flasks, bring up to 100 mL with KCl solution, and mix the solution thoroughly. The dilute standard solutions contain 0, 10, 15, 20, and 25 μg of NH_4^+–N mL^{-1}. Then pipette 1 mL of these dilute standard ammonium solutions into 25-mL volumetric flasks, add 9 mL of distilled water, and determine the ammonium concentrations as described for ammonium analysis of soil extracts (i.e., add 5 mL of the Na-salicylate–NaOH solution and 2 mL of the Na-dichloroisocyanide solution, mix, and incubate for 30 min at room temperature; measure the absorbance at 690 nm against a reagent blank solution).

To calculate protease activity, multiply the amount of NH_4^+–N ($\mu g\ mL^{-1}$) recovered by the final volume of the solutions added to the soil in the assay (9 mL) and divide by the dry weight of the soil. Protease activity was expressed as μg of ammonium released per gram dry weight of soil per hour ($\mu g\ NH_4^+$–N g^{-1} dwt h^{-1}).

11–3.2.4 Comments

This method permits rapid analysis, requires relatively simple equipment, and is characterized by high sensitivity and stability of the green-colored complex formed under alkaline pH conditions (Kandeler and Gerber, 1988). The color developed in the ammonium analysis is stable for at least 8 h (Kandeler and Gerber, 1988).

Values obtained by the colorimetric method agree with those obtained by steam distillation of the supernatant and titration of the distillate (Tabatabai and Bremner, 1972).

Ammonium concentration also can be measured colorimetrically by Nessler's reagent (Makboul and Ottow, 1979), or by indophenol method (Keeney and Nelson, 1982), with the ammonium electrode or by a flow injection analyzer (Bonmati et al., 1998; Renella et al., 2002). Some of these methods are more time-consuming or require expensive equipment.

Because ammonium released by enzymatic hydrolysis of the substrate during the assay may be adsorbed by the soil colloids, it is necessary to evaluate NH_4^+–fixing capability of soil by measuring the recovery. Briefly incubate the soil with NH_4^+ solutions at concentrations in the range of those released by protease activity, shake for 1 h at 40°C, then extract with 2 M KCl and quantify the recovery of NH_4^+–N.

Chloroform fumigation did not influence the BAA-hydrolyzing activity of soils subjected to different management regimes (Renella et al., 2002).

Although BAA is a typical substrate for the trypsin-like enzyme (Keil, 1971), other proteolytic enzymes with different optimal pH values are able to hydrolyze this substrate (Glazer and Smith, 1971; Liu and Elliott, 1971; Matsubara and Feder, 1971; Mitchell and Harrington, 1971).

Choice of buffer and buffer pH in the method described is based on findings of Ladd and Butler (1972) showing that BAA hydrolysis is inhibited by about 50% with 0.1 Tris-borate buffer at pH 8.1. Maximal enzymatic hydrolysis of BAA occurs with 0.1 M phosphate buffer at pH 7.1 (Nannipieri et al.,1980).

Protease activity estimated by other low molecular weight substrates such as dipeptide derivatives has shown that the amount of product released by incubation of soil with buffered substrate solution is considerably higher in 0.1 Tris-borate buffer at pH 8.1 than in other buffers (Ladd and Butler, 1972).

The method proposed by Ladd and Butler (1972) involves the determination of leucine released from hydrolysis of N-benzyloxycarbonyl (Z) L-phenylalanyl L-leucine (ZPL), a low molecular weight substrate, when soil is incubated with 0.1 M Tris-sodium borate buffer, pH 8.1 for 1 h at 40°C. Leucine released is then treated with a ninhydrin reagent and measured by spectrophotometer. This assay is described by Ladd and Butler (1972).

REFERENCES

Alef, K., and P. Nannipieri. 1995. p. 313–315. *In* Methods in applied soil microbiology and biochemistry. Academic Press, London.

Allison, S.D., and P.M. Vitousek. 2005. Responses of extracellular enzymes to simple and complex nutrient inputs. Soil Biol. Biochem. 37:937–944. doi:10.1016/j.soilbio.2004.09.014

Amato, M., and J.N. Ladd. 1988. Assay for microbial biomass based on ninhydrin-reactive nitrogen in extracts of fumigated soils. Soil Biol. Biochem. 20:107–114. doi:10.1016/0038-0717(88)90134-4

Ascher, J., M.T. Ceccherini, L. Landi, M. Mench, G. Pietramellara, P. Nannipieri, and G. Renella. 2009. Composition, biomass and activity of microflora, and leaf yields and foliar elemental concentrations of lettuce, after in situ stabilization of an arsenic-contaminated soil. Appl. Soil Ecol. 41:351–359. doi:10.1016/j.apsoil.2009.01.001

Asmar, F., F. Eiland, and N.E. Nielsen. 1992. Interrelationship between extracellular-enzyme activities, ATP content, total counts of bacteria and CO_2 evolution. Biol. Fertil. Soils 14:288–292. doi:10.1007/BF00395465

Asmar, F., F. Eiland, and N.E. Nielsen. 1994. Effect of extracellular-enzyme activities on solubilization rate of soil organic nitrogen. Biol. Fertil. Soils 17:32–38. doi:10.1007/BF00418669

Bach, H.J., A. Hartmann, M. Schloter, and J.C. Munch. 2001. PCR primers and functional probes for amplification and detection of bacterial genes for extracellular peptidases in single strains and in soil. J. Microbiol. Methods 44:173–182. doi:10.1016/S0167-7012(00)00239-6

Bach, H.J., and J.C. Munch. 2000. Identification of bacterial sources of soil peptidases. Biol. Fertil. Soils 31:219–224. doi:10.1007/s003740050648

Badalucco, L., P.J. Kuikman, and P. Nannipieri. 1996. Protease and deaminase activities in wheat rhizosphere and their relation to bacterial and protozoan populations. Biol. Fertil. Soils 23:99–104. doi:10.1007/BF00336047

Batistic, L., J.M. Sarkar, and J. Mayaudon. 1980. Extraction, purification and properties of soil hydrolases. Soil Biol. Biochem. 12:59–63. doi:10.1016/0038-0717(80)90103-0

Beck, T. 1973. Über die Eignung von Modellversuchen bei der Messung der biologischen Aktivität von Böden. Bayer. Landwirtsch. Jahrb. 50:270–288.

Blagoveshchenskaya, Z.K., and N.A. Danchenko. 1974. Activity of soil enzymes after prolonged application of fertilizers to a corn monoculture and crops in rotation. Sov. Soil Sci. 5:569–575.

Bonmati, M., B. Ceccanti, and P. Nannipieri. 1998. Protease extraction from soil by sodium pyrophosphate and chemical characterization of the extracts. Soil Biol. Biochem. 30:2113–2125. doi:10.1016/S0038-0717(98)00089-3

Bonmati, M., B. Ceccanti, P. Nannipieri, and J. Valero. 2009. Characterization of humus–protease complexes extracted from soil. Soil Biol. Biochem. 41:1199–1209. doi:10.1016/j.soilbio.2009.02.032

Bonmati, M., P. Jiménez, and M. Julià. 2003. Soil enzymology: Some aspects of its interest and limitations. p. 63–75. *In* M.C.Lobo and J.J. Ibànez (ed.) Preserving soil quality and soil biodiversity: The role of surrogate indicators. Instituto Madrileno de Investigaciòn Agraria y Alimentaria, Madrid.

Burket, J.Z., and R.P. Dick. 1998. Microbial and soil parameters in relation to N mineralization in soils of diverse genesis under differing management systems. Biol. Fertil. Soils 27:430–438. doi:10.1007/s003740050454

Burns, R.G. 1982. Enzyme activity in soil: Location and a possible role in microbial ecology. Soil Biol. Biochem. 14:423–427. doi:10.1016/0038-0717(82)90099-2

Burton, D.L., and W.B. McGill. 1991. Inductive and repressive effects of carbon and nitrogen on L-histidine ammonia-lyase activity in a black Chernozemic soil. Soil Biol. Biochem. 23:939–946. doi:10.1016/0038-0717(91)90174-I

Darrah, P.R., and P.J. Harris. 1986. A fluorimetric method for measuring the activity of soil enzymes. Plant Soil 92:81–88. doi:10.1007/BF02372269

DeAngelis, K.M., S.E. Lindow, and M.K. Firestone. 2008. Bacterial quorum sensing and nitrogen cycling in rhizosphere soil. Microb. Ecol. 66:197–207. doi:10.1111/j.1574-6941.2008.00550.x

Deng, S., H. Kang, and C. Freeman. 2011. Microplate fluorimetric assay of soil enzymes. p. 311–318. *In* R.P. Dick (ed.) Methods of soil enzymology. SSSA Book Ser. 9. SSSA, Madison, WI. (This volume.)

Fornasier, F., Y. Dudal, and H. Quiquampoix. 2011. Enzyme extraction from soil. p. 371–384. *In* R.P. Dick (ed.) Methods of soil enzymology. SSSA Book Ser. 9. SSSA, Madison, WI. (This volume.)

Freeman, C., G. Liska, N.J. Ostle, S.E. Jones, and M.A. Lock. 1995. The use of fluorogenic substrates for measuring enzyme activity in peatlands. Plant Soil 175:147–152. doi:10.1007/BF02413020

García-Gil, J.C., C. Plaza, N. Senesi, G. Brunetti, and A. Polo. 2004. Effects of sewage sludge amendment on humic acids and microbiological properties of a semiarid Mediterranean soil. Biol. Fertil. Soils 39:320–328. doi:10.1007/s00374-003-0709-z

Geisseler, D., and W.R. Horwath. 2008. Regulation of extracellular protease activity in soil in response to different sources and concentrations of nitrogen and carbon. Soil Biol. Biochem. 40:3040–3048. doi:10.1016/j.soilbio.2008.09.001

Glazer, A.N., and E.L. Smith. 1971 Papain and other plant sulphydryl proteolytic enzymes. p. 502–546. *In* P.D. Boyer (ed.) The enzymes. Vol. III. 3rd ed. Academic Press, New York.

Gottschalk, G. 1979. Regulation of bacterial metabolism. p. 178–207. *In* G. Gottschalk (ed.) Bacterial metabolism. 2nd ed. Springer Verlag, New York.

Hayano, K. 1993. Protease activity in a paddy field soil: Origin and some properties. Soil Sci. Plant Nutr. 39:539–546.

Hayano, K., M. Takeuchi, and E. Ichishima. 1987. Characterization of a metalloproteinase component extracted from soil. Biol. Fertil. Soils 4:179–183. doi:10.1007/BF00270938

Hayano, K., K. Watanabe, and S. Asakawa. 1995. Activity of protease extracted from rice-rhizosphere soils under double cropping of rice and wheat. Soil Sci. Plant Nutr. 41:597–603.

Hoffmann, G., and K. Teicher. 1957. Das Enzymsystem unserer Kulturböden VII. Protease II. Z. Pflanzenernaehr. Dueng. Bodenkd. 77:243–251. doi:10.1002/jpln.19570770308

Hoppe, H.G. 1983. Significance of exoenzymatic activities in the ecology of brackish water: Measurements by means of methylumbelliferyl-substrates. Mar. Ecol. Prog. Ser. 11:299–308. doi:10.3354/meps011299

Hoppe, H.G., S.J. Kim, and K. Gocke. 1988. Microbial decomposition in aquatic environments: Combined process of extracellular enzyme activity and substrate uptake. Appl. Environ. Microbiol. 54:784–790.

Kamimura, Y., and K. Hayano. 2000. Properties of protease extracted from tea-field soil. Biol. Fertil. Soils 30:351–355. doi:10.1007/s003740050015

Kandeler, E., and H. Gerber. 1988. Short-term assay of urease activity using colorimetric determination of ammonium. Biol. Fertil. Soils 6:68–72. doi:10.1007/BF00257924

Kandeler, E., D. Tscherko, and H. Spiegel. 1999. Long-term monitoring of microbial biomass, N mineralisation and enzyme activities of a Chernozem under different tillage management. Biol. Fertil. Soils 28:343–351. doi:10.1007/s003740050502

Keeney, D.R., and D.W. Nelson. 1982. Nitrogen-inorganic forms p. 643–698. *In* A.L. Page, R.H. Miller, and D.R. Keeney (ed.) Methods of soil analysis. ASA, Madison, WI.

Keil, B. 1971. Trypsin. p. 250–275. *In* P.D. Boyer (ed.) The enzymes, Vol. III. 3rd ed. Academic Press, New York.

Klein, T.M., and J.S. Koths. 1980. Urease, protease, and acid phosphatase in soil continuously cropped to corn by conventional or no-tillage methods. Soil Biol. Biochem. 12:293–294. doi:10.1016/0038-0717(80)90076-0

Kumar, K., C.J. Rosen, and M.P. Russelle. 2006. Enhanced protease inhibitor expression in plant residues slows nitrogen mineralization. Agron. J. 98:514–521. doi:10.2134/agronj2005.0261

Kumpiene, J., G. Guerri, L. Landi, G. Pietramellara, P. Nannipieri, and G. Renella. 2009. Microbial biomass, respiration and enzyme activities after in situ aided phytostabilization of a Pb- and Cu-contaminated soil. Ecotoxicol. Environ. Saf. 72:115–119. doi:10.1016/j.ecoenv.2008.07.002

Ladd, J.N. 1972. Properties of proteolytic enzymes extracted from soil. Soil Biol. Biochem. 4:227–237. doi:10.1016/0038-0717(72)90015-6

Ladd, J.N. 1978. p. 51–96. Origin and range of enzymes in soil. *In* R.G. Burns (ed.) Soil enzymes. Academic Press, New York.

Ladd, J.N., P.G. Brisbane, J.H.A. Butler, and M. Amato. 1976. Studies on soil fumigation—III: Effects on enzyme activities, bacterial numbers and extractable ninhydrin reactive compounds. Soil Biol. Biochem. 8:255–260. doi:10.1016/0038-0717(76)90053-5

Ladd, J.N., and J.H.A. Butler. 1972. Short-term assays of soil proteolytic enzyme activities using proteins and dipeptide derivatives as substrates. Soil Biol. Biochem. 4:19–30. doi:10.1016/0038-0717(72)90038-7

Ladd, J.N., and R.B. Jackson. 1982. Biochemistry of ammonification. p. 173–228. *In* F.J. Stevenson (ed.) Nitrogen in agricultural soils. Agron. Monogr. 22. ASA, CSSA, SSSA, Madison, WI.

Liu, T.Y., and S.D. Elliott. 1971. Streptococcal proteinase. p. 609–647. *In* P.D. Boyer (ed.) The enzymes. Vol. III. 3rd ed. Academic Press, New York.

Loll, M.J., and J.M. Bollag. 1983. Protein transformation in soil. Adv. Agron. 36:351–382. doi:10.1016/S0065-2113(08)60358-2

Makboul, H.E., and J.C.G. Ottow. 1979. Clay minerals and the Michaelis constant of urease. Soil Biol. Biochem. 11:683–686. doi:10.1016/0038-0717(79)90039-7

Marx, M.C., E. Kandeler, M. Wood, N. Wermbter, and S.C. Jarvis. 2005. Exploring the enzymatic landscape: Distribution and kinetics of hydrolytic enzymes in soil particle-size fractions. Soil Biol. Biochem. 37:35–48. doi:10.1016/j.soilbio.2004.05.024

Marx, M.C., M. Wood, and S.C. Jarvis. 2001. A microplate fluorimetric assay for the study of enzyme diversity in soils. Soil Biol. Biochem. 33:1633–1640. doi:10.1016/S0038-0717(01)00079-7

Matsubara, H., and J. Feder. 1971. Other bacterial mold and yeast proteases. p. 721–795. *In* P.D. Boyer (ed.) The enzymes. Vol. III. 3rd ed. Academic Press, New York.

Mayaudon, J., L. Batistic, and J.M. Sarkar. 1975. Propriétés des activités proteolytiques extraites des sols frais. Soil Biol. Biochem. 7:281–286. doi:10.1016/0038-0717(75)90067-X

Mench, M., G. Renella, A. Gelsomino, L. Landi, and P. Nannipieri. 2006. Biochemical parameters and bacterial species richness in soils contaminated by sludge-borne metals and remediated with inorganic soil amendments. Environ. Pollut. 144:24–31. doi:10.1016/j.envpol.2006.01.014

Mitchell, W.H., and W.F. Harrington. 1971. Clostripain. p. 699–719. *In* P.D. Boyer (ed.) The enzymes. Vol. III. 3rd ed. Academic Press, New York.

Mrkonjic Fuka, M., M. Engel, A. Gattinger, U. Bausenwein, M. Sommer, J.C. Munch, and M. Schloter. 2008. Factors influencing variability of proteolytic genes and activities in arable soils. Soil Biol. Biochem. 40:1646–1653. doi:10.1016/j.soilbio.2008.01.028

Nannipieri, P. 1994. The potential use of soil enzymes as indicators of productivity, sustainability and pollution. p. 238–244. *In* C.E. Pankhurst, B.M. Doube, V.V.S.R. Gupta, and P.R. Grace (ed.) Soil biota: Management in sustainable farming systems. CSIRO, East Melbourne, Victoria, Australia.

Nannipieri, P., and L. Badalucco. 2002. Biological processes. p. 57–82. *In* D.K. Bembi and R. Niedre (ed.) Processes in the soil-plant system: Modelling concepts and applications. The Haworth Press, Binghamton, NY.

Nannipieri, P., B. Ceccanti, D. Bianchi, and M. Bonmati. 1985. Fractionation of hydrolases-humus complexes by gel chromatography. Biol. Fertil. Soils 1:25–29. doi:10.1007/BF00710967

Nannipieri, P., B. Ceccanti, S. Cervelli, and E. Matarese. 1980. Extraction of phosphatase, urease, protease, organic carbon and nitrogen from soil. Soil Sci. Soc. Am. J. 44:1011–1016. doi:10.2136/sssaj1980.03615995004400050028x

Nannipieri, P., B. Ceccanti, C. Conti, and D. Bianchi. 1982. Hydrolases extracted from soil: Their properties and activities. Soil Biol. Biochem. 14:257–263. doi:10.1016/0038-0717(82)90035-9

Nannipieri, P., L. Muccini, and C. Ciardi. 1983. Microbial biomass and enzyme activities: Production and persistence. Soil Biol. Biochem. 15:679–685. doi:10.1016/0038-0717(83)90032-9

Nannipieri, P., F. Pedrazzini, P.G. Arcara, and C. Piovanelli. 1979. Changes in amino acids, enzyme activities, and biomasses during soil microbial growth. Soil Sci. 127:26–34. doi:10.1097/00010694-197901000-00004

Nannipieri, P., P. Sequi, and P. Fusi. 1996. Humus and enzyme activity. p. 293–328. *In* A. Piccolo (ed.) Humic substances in terrestrial ecosystems. Elsevier Science, New York.

Perucci, P. 1992. Enzyme activity and microbial biomass in a field soil amended with municipal refuse. Biol. Fertil. Soils 14:54–60. doi:10.1007/BF00336303

Rahman, R.N.Z.A., L.P. Geok, M. Basri, and A.B. Salleh. 2005. Physical factors affecting the production of organic solvent-tolerant protease by Pseudomonas aeruginosa strain K. Bioresour. Technol. 96:429–436. doi:10.1016/j.biortech.2004.06.012

Rao, M.B., A.M. Tanksale, M.S. Ghatge, and V.V. Deshpande. 1998. Molecular and biotechnological aspects of microbial proteases. Microbiol. Mol. Biol. Rev. 62:597–635.

Renella, G., L. Landi, J. Ascher, M.T. Ceccherini, G. Pietramellara, M. Mench, and P. Nannipieri. 2008. Long-term effects of aided phytostabilization of trace elements on microbial biomass and activity, enzyme activities, and composition of microbial community in the Jales contaminated mine spoils. Environ. Pollut. 152:702–712. doi:10.1016/j.envpol.2007.06.053

Renella, G., L. Landi, and P. Nannipieri. 2002. Hydrolase activities during and after the chloroform fumigation of soil as affected by protease activity. Soil Biol. Biochem. 34:51–60. doi:10.1016/S0038-0717(01)00152-3

Renella, G., U. Szukics, L. Landi, and P. Nannipieri. 2007. Quantitative assessment of hydrolase production and persistence in soil. Biol. Fertil. Soils 44:321–329. doi:10.1007/s00374-007-0208-8

Ross, D.J., T.W. Speir, D.J. Giltrap, B.A. McNelly, and L.F. Molloy. 1975. A principal components analysis of some biochemical activities in climosequence of soils. Soil Biol. Biochem. 7:349–355. doi:10.1016/0038-0717(75)90048-6

Sakurai, M., K. Suzuki, M. Onodera, T. Shinano, and M. Osaki. 2007. Analysis of bacterial communities in soil by PCR-DGGE targeting protease genes. Soil Biol. Biochem. 39:2777–2784. doi:10.1016/j.soilbio.2007.05.026

Sardans, J., and J. Penuelas. 2005. Drought decreases soil enzyme activity in a Mediterranean Quercus ilex L. Forest. Soil Biol. Biochem. 37:455–461. doi:10.1016/j.soilbio.2004.08.004

Schloter, M., H.J. Bach, S. Metz, U. Sehy, and J.C. Munch. 2003. Influence of precision farming on the microbial community structure and functions in nitrogen turnover. Agric. Ecosyst. Environ. 98:295–304. doi:10.1016/S0167-8809(03)00089-6

Schulten, H.R., and M. Schnitzer. 1998. The chemistry of soil nitrogen: A review. Biol. Fertil. Soils 26:1–15. doi:10.1007/s003740050335

Schulze, W.X., G. Gleixner, K. Kaiser, G. Guggenberger, M. Mann, and E.D. Schulze. 2005. A proteomic fingerprint of dissolved organic carbon of soil particles. Oecologia 142:335–343. doi:10.1007/s00442-004-1698-9

Somville, M. 1984. Measurement and study of substrate specificity of exoglucosidase activity in eutrophic water. Appl. Environ. Microbiol. 48:1181–1185.

Speir, R., R. Lee, E.A. Pansier, and A. Cairns. 1980. A comparison of sulphatase, urease and protease activities in planted and in fallow soils. Soil Biol. Biochem. 12:281–291. doi:10.1016/0038-0717(80)90075-9

Stevenson, F.J. 1982. Nitrogen-organic forms. p. 625–641. In A.L. Page, R.H. Miller, and D.R. Keeney (ed.) Methods of soil analysis. Part 2. Chemical and microbiological properties. 2nd ed. ASA, Madison, WI.

Tabatabai, M.A., and J.M. Bremner. 1972. Assay of urease activity in soils. Soil Biol. Biochem. 4:479–487. doi:10.1016/0038-0717(72)90064-8

Tateno, M. 1988. Limitations of available substrates for the expression of cellulase and protease activities in soil. Soil Biol. Biochem. 20:117–118. doi:10.1016/0038-0717(88)90136-8

Taylor, J.P., B. Wilson, M.S. Mills, and R.G. Burns. 2002. Comparison of microbial numbers and enzymatic activities in surface soils and subsoils using various techniques. Soil Biol. Biochem. 34:387–401. doi:10.1016/S0038-0717(01)00199-7

Vepsäläinen, M., S. Kukkonen, M. Vestberg, H. Sirviö, and R.M. Niemi. 2001. Application of soil enzyme activity test kit in a field experiment. Soil Biol. Biochem. 33:1665–1672. doi:10.1016/S0038-0717(01)00087-6

Watanabe, K., and K. Hayano. 1994. Estimate of the source of soil protease in upland fields. Biol. Fertil. Soils 18:341–346. doi:10.1007/BF00570638

Watanabe, K., J. Sakai, and K. Hayano. 2003. Bacterial extracellular protease activities in field soils under different fertilizer managements. Can. J. Microbiol. 49:305–312. doi:10.1139/w03-040

Chapter 12

Biologically Active Compounds in Soil: Plant Hormones and Allelochemicals

Serdar Bilen, Jeong Jin Kim, and Warren A. Dick*

12–1 INTRODUCTION

Soils contain hundreds of compounds derived from plant, animal, or microbial sources, the majority of which never have been isolated and identified. All components in soil may be considered, in a general and indirect sense, to be biologically active because in some way they influence the total biological activity in the soil. Clay minerals, for example, bind many compounds on their surfaces and can thus influence biological activity by affecting substrate concentrations at microenvironments within the soil and subsequent soil enzyme activities. However, the clay mineral itself is, in general, considered biologically inactive.

A biologically active compound is defined as one that has a direct physiological effect on a plant, animal, or microorganism. For activity to actually be expressed, the biologically active compound must be located in the right place in soil to affect its target and be present at sufficient concentration. Many known compounds with biological activity are found in only trace amounts in soil. Other compounds, such as fulvic acids extracted by classical methods, may show biological activity, but their chemical structure may have been severely altered during the extraction process from soil.

The need for further information on the identification, production, and fate of biologically active compounds in soil has become a productive area of research. This reemergence of interest, which was first expressed due to the early discoveries of antibiotics produced by soil microorganisms, is a result of the realization that there exists in soil a large number of natural products of plant, animal, or microbial origin that are biologically active and can be utilized for industrial or agricultural purposes.

The production and activity of biologically active compounds is highly localized in the soil. These compounds may interact in a synergistic or antagonistic

Serdar Bilen, Atatürk University, Faculty of Agriculture, Department of Soil Science, Erzurum, Turkey (sbilen@atauni.edu.tr);
Jeong Jin Kim, Department of Earth and Environmental Sciences, Andong National University, Andong, Korea (jjkim@andong.ac.kr);
Warren A. Dick, Professor, Soil Science, School of Environment and Natural Resources, The Ohio State University, 101A Hayden Hall, 1680 Madison Ave., Wooster, OH 44691(dick.5@osu.edu), *corresponding author.

doi:10.2136/sssabookser9.c12

manner so that it is often difficult to develop a test for a single compound in soil with a specific biological response. The combined effect of location, concentration, and interaction of response makes the study of biologically active compounds in soil extremely difficult. For example, some compounds may be volatile and temperature will greatly affect their concentration in soil. The ability to bind to soil colloids, such as clays, may inhibit the activity of some compounds, while trapping in soil other compounds so that they build up to high levels. Moderate moisture will stimulate microbial and plant growth and thus stimulate the rate of production of a biologically active compound. At the same time, the increased microbial activity may result in rapid degradation so that the presence of the biologically active compound in soil is transient. Lack of oxygen will stimulate the production of some compounds, such as the small organic acids that elicit allelopathic response, while inhibiting the production of others. Anaerobic conditions may also aid the stabilization of compounds in soil.

Plant hormones, such as ethylene, auxins, cytokinins, and gibberellins, can be produced by microorganisms in soil, especially within the plant root rhizosphere (Frankenberger and Arshad, 1995). These hormones influence plant growth and development. The addition of specific substrates that are immediate precursors in the synthetic pathway of a plant hormone can stimulate hormone concentrations above those normally found in soil. In this way, a particular physiological plant response can be created. Arshad and Frankenberger (1988) showed that methionine-dependent ethylene produced by *Acremonium falciforme* caused the classic triple response in etiolated pea seedlings, which includes reduction in elongation, swelling of the hypocotyls, and change in direction of growth (horizontal).

Allelochemicals of a wide variety have been isolated from soils that inhibit plant growth. The source of the phytotoxin or allelochemical may be either microbial or plant. The allelopathic response may also be due to a single compound or a host of rather nonspecific compounds acting in a synergistic manner. For example, microorganisms commonly produce allelochemical compounds such as small organic acids or phenolics when a large amount of plant residues are added to the soil and degradation is occurring under anaerobic conditions. Plants also produce allelopathic compounds and juglone, from black walnut trees, is a good example.

Information is lacking on biochemical reactions that lead to the production of biologically active compounds in soil. The following three sections provide procedures to measure the enzyme activities in soil that lead to the production of the plant hormones, ethylene and indole-3-acetic acid, and the breakdown of a glucosinolate (sinigrin) to yield isothiocyanates, thiocyanates, or nitriles that have allelochemical properties.

12–2 ACC OXIDASE
12–2.1 Introduction

The phytohormone ethylene plays an important role in many aspects of plant growth and development, including fruit ripening and senescence (Abeles et al., 1992). The precursor of ethylene, 1-aminocyclopropane-1-carboxylate (ACC), is produced by the conversion of S-adenosylmethionine by ACC synthase (Adams and Yang, 1979). The ACC is then converted to ethylene, carbon dioxide, and hydrogen cyanide (HCN) by ACC oxidase (EC 1.14.17.4 aminocyclopropane-carboxylate oxidase) in an oxygen-dependent process.

ACC oxidase is well known in different fruit sources including melon (Smith et al., 1992), apple (Fernandez-Maculet and Yang, 1992; Fernandez-Maculet et al.,1993; Pirrung et al., 1993), avocado (McGarvey and Christoffersen, 1992), and banana (Moya-Leon and John, 1994, 1995). However, it also is found in various microorganisms (Nagahama et al., 1992). *Penicillium digitatum* (Chalutz et al., 1977), *Botrytis cinerea* (Chague et al., 2002), *Pseudomonas syringae* (Sato et al., 1997), and *Trichoderma viride* (Tao et al., 2008) have all been reported to contain or express ACC oxidase activity. However, microorganisms produce ethylene via several additional biochemical pathways that are generally distinct from that of higher plants (Tao et al., 2008).

The pathway of methionine (MET) to ethylene (C_2H_2) in plants involves S-adenosylmethionine (SAM) and ACC, serving as intermediates, in a pathway that can be summarized as follows:

$$MET \rightarrow SAM \rightarrow\!\!\!\rightarrow ACC \rightarrow C_2H_4 \text{ (ethylene)} \qquad [1]$$

For the bacterium, *Pseudomonas syringae*, and a few species of fungi such as *Penicillium digitatum*, ethylene is readily produced from 2-oxoglutarate by an ethylene-forming enzyme (EFE) (Nagahama et al., 1992; Völksch and Weingart, 1997). Frankenberger and Phelan (1985) found that the application of ACC to soil resulted in a large increase in ethylene production in a soil that normally produced very little without the added precursor.

12–2.2 Principle

The reaction catalyzed by ACC oxidase is as follows:

$$\text{1-aminocyclopropane-1-carboxylate (ACC)} + \text{ascorbate} + O_2 \rightarrow \text{ethylene}$$
$$+ \text{cyanide} + \text{dehydroascorbate} + CO_2 + H_2O \qquad [2]$$

Frankenberger and Phelan (1985) studied the above reaction and developed an assay to quantify the activity of ACC oxidase in soil. This assay involved adding ACC to soil and measuring ethylene produced, in the headspace above the soil, using a gas chromatograph.

12–2.3 Assay Method (Frankenberger and Phelan, 1985)

12–2.3.1 Apparatus

- Incubation flasks, 250-mL Erlenmeyer flasks fitted with rubber septum
- Ordinary incubator or temperature-controlled water bath
- Gas chromatograph equipped with an activated alumina column, or other appropriate column, and a flame ionization detector
- Gas-tight glass hypodermic syringe

12–2.3.2 Reagents

1. Toluene: Fisher certified reagent (Fisher Scientific Co., Chicago).

2. Modified universal buffer (MUB) stock solution: Dissolve 12.1 g of tris(hydroxymethyl) aminomethane-hydrochloric acid (THAM), 11.6 g of maleic acid, 14.0 g of citric acid, and 6.3 g of boric acid (H_3BO_3) in 488 mL of 1 N sodium hydroxide (NaOH) and dilute the solution to 1 L with deionized water. Store in a refrigerator.

3. Modified universal buffer, pH 9.0 (working buffer): Place 200 mL of MUB stock solution in a 500-mL beaker containing a magnetic stirring bar, and place the beaker on a magnetic stirrer. Titrate the solution to pH 9.0 with 0.1 M NaOH, and adjust the volume to 1 L with water.

4. ACC (1-aminocyclopropane-1-carboxylic acid).

5. Hydrochloric acid (HCl) (0.1 M): Dilute 8.33 mL concentrated HCl (12 M) to 1 L with deionized water.

6. Sodium hydroxide (NaOH) (0.1 M): Dilute 4 g NaOH in deionized water and dilute to 1 L.

7. Ethanol (70%): 70 mL volume of molecular biology grade ethanol in 100 mL of 0.1 M HCl.

12–2.3.3 Procedure

Weigh 10 g of air-dried soil that was sieved through a 2-mm screen into a 250-mL Erlenmeyer flask. Add 1 mL of toluene, wait a few minutes for the toluene to react with the soil, and then add 20 mg (dry reagent) of ACC and 20 mL of modified universal working buffer (pH 9.0). Swirl the flasks to mix the contents and then seal the flask using a rubber septum (sterilized in 70% ethanol with 0.1 M HCl). Incubate in the dark under static conditions.

After 48 h, determine ethylene (C_2H_2) concentrations by withdrawing gas samples from the headspace above the soil with a gas-tight glass hypodermic syringe. Measure ethylene concentrations in the sample using a gas chromatograph equipped with an activated alumina column, or other appropriate column, and a flame ionization detector. Ethylene dissolved in the aqueous phase is ignored. Activity is expressed as ethylene released (mmol/kg/48 h).

To perform controls to account for non-ACC-derived ethylene production, incubate 10 g of toluene-treated soil plus 20 mL of modified universal buffer (pH 9.0), but without the added substrate, ACC. After 48 h of incubation at 37°C, determine indigenous levels of ethylene by gas chromatograph and subtract this from the biological conversion of ACC to ethylene in soils.

12–2.4 Comments

The enzymatic rate of ethylene released from soil is enhanced with increasing pH, up to the pH optimum of 9.0. The ACC will also degrade nonenzymatically to produce ethylene with increasing buffer pH (Frankenberger and Phelan, 1985). However, the nonenzymatic transformation rates are extremely small compared with those caused by the soil enzyme.

Results by Frankenberger and Phelan (1985) revealed reaction rates approached zero-order kinetics when 9.9 mM ACC was added to soils. Under standard conditions, the accumulation of ethylene was linear with time of incubation for at least 7 d. Maximum reaction rates in the conversion of ACC to ethylene were observed at 50°C with enzyme denaturation occurring at 55°C.

The ACC-derived ethylene production in soils was enhanced under highly oxidized conditions and severely inhibited under anaerobiosis. Toluene, a common component in soil enzyme assays, increased ethylene production, as did shaking the assay mixtures during incubation. Dimethyl sulfoxide, formaldehyde, EDTA, 2-mercaptoethanol, 2,4-dinitrophenol, NaF, KCN, KNO_2, and KNO_3 all inhibited the reaction.

12–3 TRYPTOPHAN-DEPENDENT AUXIN PRODUCTION

12–3.1 Introduction

The presence of auxins in soil may have an ecological impact affecting plant growth. The growth of Douglas fir (*Pseudotsuga menziesii*) was considerably increased when soil was both inoculated with the ectomycorrhizal fungus, *Pisolithus tinctorius*, and treated with L-tryptophan at rates as low as 0.34 µg kg^{-1} soil (Frankenberger and Poth, 1987). Similarly, a pronounced physiological effect on radish (*Raphanus sativus*) was noted when grown in soils treated with tryptophan (Frankenberger et al., 1990).

There are numerous reports of soil microorganisms that can produce auxins from precursor molecules. Generally, microorganisms isolated from the rhizosphere and rhizoplane are more active in producing auxin than are those from root-free soil (Frankenberger and Brunner, 1983). Indole-3-acetic acid (IAA) is considered one of the major auxin products of soil, but soils also contain other compounds that exhibit strong auxin-like activity and differ in their IAA- synthesizing capacity depending on their fertility status and organic matter content.

Plant cells mainly synthesize IAA from tryptophan but also can produce it independently of tryptophan. The synthesis is not direct but involves several intermediates such as indol-3-acetonitrile, indole-3-pyruvic acid, and indole-3-acetaldehyde (Koga et al., 1991). Detection of these 3-substituted indoles in soils incubated with tryptophan suggests that IAA is synthesized by at least two major, but distinct, pathways: (i) tryptophan is converted to IAA by deaminating to form indole-3-pyruvic acid followed by decarboxylation to indole-3-acetaldehyde, which is further oxidized to IAA, and (ii) tryptophan is decarboxylated to form indole-3-acetamide, which is then hydrolyzed to IAA. Several other pathways for IAA formation have been proposed and are represented in Fig. 12–1.

12–3.2 Principle

Although, L-tryptophan is known to be a precursor to IAA production, it is not a single enzymatic step that converts tryptophan to IAA as indicated by Fig. 12–1. However, Sarwar et al. (1992) have developed and tested an assay procedure for soils that uses L-tryptophan as substrate and measures IAA as product. The assay procedure described below is taken from their publication.

12–3.3 Assay Method (Sarwar et al., 1992)

12–3.3.1 Apparatus

- Incubation flasks, 50-mL Erlenmeyer flasks
- Whatman No. 2 filter paper
- Ordinary incubator or temperature-controlled water bath

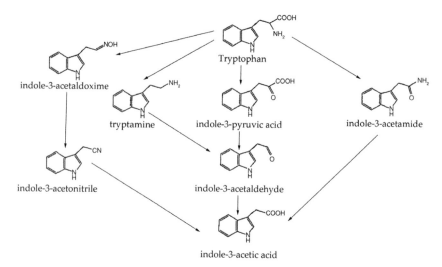

Fig. 12–1. Proposed pathways for the biosynthesis of indole-3-acetic acid (IAA) (Normanly et al., 1995). Reprinted with permission.

- Colorimeter or spectrophotometer, Klett–Summerson photoelectric color-imeter fitted with a blue (No. 42) filter or a spectrophotometer than can be adjusted to a wavelength of 535 nm
- Shaker, end-to-end shaker

12–3.3.2 Reagents

1. Sodium phosphate buffer (0.2 M, pH 7.0): Prepare Reagent A by dissolv-ing 24.0 g of monobasic sodium phosphate (NaH_2PO_4) in 1000 mL wa-ter. Prepare Reagent B by dissolving 28.4 g of dibasic sodium phosphate (Na_2HPO_4) in 1000 mL water. Mix equal volumes of Reagents A and B and adjust to pH 7.0. Store at 4°C in refrigerator.

2. L-Tryptophan (0.02 M), Fisher certified reagent grade (Fisher Scientific Co., Chicago), ≥98%: Dissolve 0.40 g of L-Tryptophan in about 90 mL of water and adjust the volume to 100 mL with water.

3. Trichloroacetic acid (5%), Fisher certified reagent grade (Fisher Scientific Co., Chicago): Dissolve 5.0 g of trichloroacetic acid in about 90 mL of water, and adjust the volume to 100 mL with water.

4. Calcium chloride ($CaCl_2 \cdot 2H_2O$) (0.5 M): Dissolve 7.3 g of $CaCl_2 \cdot 2H_2O$ in about 90 mL of water, and adjust the volume to 100 mL with water.

5. Ferric chloride ($FeCl_3$) (0.5 M): Dissolve 8.10 g of $FeCl_3$ in about 90 mL of wa-ter, and adjust the volume to 100 mL with water.

6. Ferric chloride ($FeCl_3$)–perchloric acid solution (Salkowski reagent): Combine 2 mL of 0.5M $FeCl_3$ and 98 mL of 35% perchloric acid to prepare the color developing reagent.

7. Indole-3-acetic acid (IAA) stock solution for standard curve (100 µg mL⁻¹): Dissolve 0.10 g IAA into 1000 mL water.

12–3.3.3 Procedure

Weigh 3 g of air-dried soil into a 50-mL Erlenmeyer flask and add 6 mL of sodium phosphate buffer (0.2 M, pH 7.0) and 4 mL of L-tryptophan solution (5.3 g kg^{-1} soil). Cover the flasks with parafilm and incubate in the dark at 40°C for 12 h on a shaker (~150 rev min^{-1}). After incubation, add 2 mL of 5% trichloroacetic acid to terminate the reaction and 1 mL of calcium chloride to facilitate filtration. Filter the soil solution through Whatman No. 2 filter paper.

For colorimetric determination of auxins, transfer 3 mL of the soil filtrate to a test tube and treat with 2 mL of the ferric chloride (FeCl$_3$)–perchloric acid solution (color developing reagent). Allow the mixture to stand for 30 min for color development. Measure the intensity of the color developed using a spectrophotometer adjusted to a wavelength 535 nm. If the color intensity of the filtrate exceeds that of the highest IAA standard, take a smaller aliquot of the filtrate and adjust the volume to 3 mL with sodium phosphate buffer (0.2 M, pH 7.0) before adding 2 mL of the ferric chloride (FeCl$_3$)–perchloric acid solution. The amount of L-tryptophan-derived auxin content in soil is reported as indole-3-acetic acid (IAA)-equivalents (mg kg^{-1} soil h^{-1}) using standard IAA solutions.

Conduct controls by using sterilized soil or adding the soil to the reagents at the end of the incubation period and just before addition of trichloroacetic acid.

Prepare a calibration graph to determine IAA content in the soil filtrate by plotting results obtained with standard IAA solutions made in the same reagent matrix as that in the assay procedure. To prepare this graph, dilute the 100 μg IAA mL^{-1} stock solution to make working standards by adding 0, 1.67, 3.33, 6.67, 10.0, 13.3, and 16.7 mL to 100-mL volumetric flasks. Add 40 mL of sodium phosphate buffer (0.2 M, pH 7.0) and then bring the final volume to 100 mL. Mix the solutions thoroughly. Pipette 3-mL aliquots of these working standards into test tubes. This will result in total amounts of IAA being 0, 5, 10, 20, 30, 40, and 50 μg. Add 2 mL of the ferric chloride (FeCl$_3$)–perchloric acid solution (color developing reagent) and proceed as described above for the regular assay procedure. Plot absorbance at 535 nm on the y axis and amount of IAA (in μg) on the x axis. Remember to divide by the amount of soil to obtain the amount of IAA produced g^{-1} soil.

12–3.4 Comments

Auxin biosynthesis in soils essentially achieved zero-order kinetics in regard to L-tryptophan concentrations of about 5.0 mg g^{-1} (Sarwar et al., 1992). Buffer pH had a significant effect on auxin biosynthesis in soils treated with L-tryptophan with optimal activity observed at pH 7.0.

Maximum auxin-producing bioactivity was observed at 40°C.

The addition of glucose to soil caused a rapid increase in L-tryptophan-dependent auxin synthesis in soils while the addition of N (especially NH$_4$NO$_3$) had a strong inhibitory concentration-dependent effect on L-tryptophan-derived auxin formation in soil.

The colorimetric procedure forms chromophoric complexes with other auxin compounds synthesized by soil bacteria in addition to IAA. Thus, Sarwar et al. (1992) proposed the unit of activity as IAA equivalents produced per mass of soil. A sensitive, specific, and precise high- performance liquid chromatography (HPLC) method for detection of IAA in soils had been previously developed by Frankenberger and Brunner (1983). However, the colorimetric procedure to measure of IAA can be conducted in most soil laboratories. The HPLC method was used, however, by Sarwar et al. (1992) to confirm the presence of IAA in filtrates from soil treated with L-tryptophan.

12–4 MYROSINASE

12–4.1 Introduction

Myrosinase (thioglucoside glucohydrolase, EC 3.2.3.1) is an enzyme that breaks the β-thioglucoside bond of glucosinolate molecules, producing glucose, sulfate, and a diverse group of products. This diverse group of products undergoes further nonenzymatic, intramolecular rearrangement to yield isothiocyanates, thiocyanates, or nitriles (Fig. 12–2).

Glucosinolates are sulfur-containing glycosides detected in various amounts in cruciferous vegetables and possibly in some bacteria and fungi (Rask et al., 2000). Myrosinase is normally segregated from glucosinolates in plant tissues, but when plant cells are damaged or decomposed, myrosinase is released and catalyzes the hydrolysis of glucosinolates. Besides the formation of glucose, the products that eventually form are allelochemicals that have been found to inhibit weed seed germination and some pathogens in soil (Angus et al., 1994; Brabban and Edwards, 1995; Brown and Morra, 1997).

Glucosinolates by themselves are not biologically active but must be enzymatically hydrolyzed by myrosinase to produce the allelochemicals. Therefore, myrosinase is the key factor for allelochemical expression derived from glucosinolates. Myrosinase is thought to be released to soils cropped with *Brassica*, or the related *Sinapis*, via root exudation, disruption, and decomposition (Borek et al., 1996). Soils cropped with *Brassica*, therefore, are expected to have enhanced myrosinase activity. Borek et al. (1996) studied this enzyme in soil extracts, but

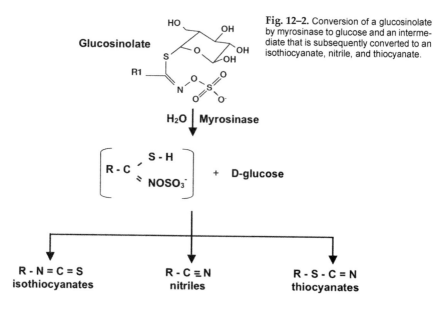

Fig. 12–2. Conversion of a glucosinolate by myrosinase to glucose and an intermediate that is subsequently converted to an isothiocyanate, nitrile, and thiocyanate.

extraction of enzymes from soil is a demanding procedure and is generally incomplete (Tabatabai and Fu, 1992). Moreover, the conditions used by Borek et al. (1996) to extract myrosinase possibly altered its activity (Ludikhuyze et al., 1999). Al-Turki and Dick (2003) developed an assay to measure myrosinase activity in soils based on measuring the concentration of glucose released by the enzymatic hydrolysis of sinigrin (2-propenylglucosinolate) in soils.

12–4.2 Principles

Glucose is a predominant product of glucosinolate hydrolysis and hence a change in its concentration in soil can be used as a direct indication of myrosinase activity. Several studies have measured glucose released from glucosinolates to quantify the activity of myrosinase purified from plant tissues (Wilkinson et al., 1984; Jwanny et al., 1995; Ludikhuyze et al., 1999). Sinigrin, a glucosinolate extracted from horseradish (*Armoracia rusticana*), has been routinely used in these studies as a substrate for myrosinase and produces glucose and allelochemicals.

Glucose is extremely water-soluble and can be readily extracted from soil and analyzed (Frey et al., 1999). Different methods have been used to determine glucose concentrations in soil including nonenzymatic and enzymatic methods. Nonenzymatic methods, however, are time-demanding and some require the addition of concentrated acids that heat the soil. This will likely cause chemical hydrolysis of carbohydrates to produce glucose in soil that will not be distinguished from glucose released from sinigrin hydrolysis. Nonenzymatic methods also may be affected by the presence of metals in soil (Deng and Tabatabai, 1994). An enzymatic approach to glucose measurements, however, is accurate, fast, and simple (Raba and Mottola, 1995). The reagents necessary for enzymatic glucose measurements are commercially available as package kits.

12–4.3 Assay Method (Al-Turki and Dick, 2003)

12–4.3.1 Apparatus

- 0.45-μm MF-Millipore membrane filter (Millipore Corp., Bedford, MA)
- Centrifuge tubes with caps, 50 mL, polypropylene
- Volumetric flasks, 50 mL and 10 mL
- Ordinary incubator or temperature-controlled water bath
- Centrifuge
- Colorimeter fitted with the appropriate filter or a spectrophotometer that can be adjusted to a wavelength of approximately 500 nm

12–4.3.2 Reagents

1. Toluene ($C_6H_5CH_3$) Fisher certified reagent (Fisher Scientific, Chicago, IL).
2. N-Tris(hydroxymethyl)methyl-2-aminoethanesulfonic acid (TES) buffer (0.1 M): Dissolve 22.92 g of TES in about 700 mL of deionized water. Adjust the solution to pH 7 with NaOH (1 M), and adjust the volume to 1 L with deionized water.
3. Silver nitrite (1 M): Dissolve 1.7 g of silver nitrate ($AgNO_3$) in 7 mL of deionized water. Adjust the volume to 10 mL with deionized water.

4. Sinigrin (120 mM): Dissolve 4.76 g of sinigrin in about 70 mL of TES buffer (0.1 M, pH 7). Adjust the volume to 100 mL with water.

5. Glucose solution (1000 μg mL^{-1}): Dissolve 1 g of glucose in 700 μL of deionized water. Adjust the volume to 1.00 mL with water. Store solution in a refrigerator at 4°C.

6. Glucose diagnostic kit (Trinder, Catalog No. 315-100, Sigma-Aldrich, St. Louis, MO): Reconstitute the glucose diagnostic kit according to the manufacturer's specification. Store the reagent at 4°C. It will remain stable for 3 mo.

12–4.3.3 Procedure

Place 1 g of air-dried soil in a 50-mL plastic centrifuge tube, add 0.2 mL of toluene, 2.5 mL of TES buffer (0.1 M, pH 7), and 0.5 mL of sinigrin prepared in 0.1 M TES buffer (pH 7) to obtain a final concentration of 20 mM. Swirl the tube for a few seconds to mix the contents. Stopper the tube and incubate it at 37°C. After 4 h, centrifuge the soil suspension at 8000 × g for 10 min and filter the supernatant through a 0.45 μm MF-Millipore membrane filter into a test tube.

During the centrifugation period, pipette 2 mL of reagents from the diagnostic glucose kit to a labeled test tube and allow it to warm to assay temperature. Add 1 mL of the filtered supernatant to the labeled test tube containing the 2 mL of reagents from the diagnostic kit and mix by gentle inversion. Incubate the tube at room temperature (20–25°C) for exactly 20 min and then immediately add 25 μL of AgNO$_3$ (1 M) to stop activity of all enzymes. Measure the absorbance of the pink color of the quinoneimine complex formed with a spectrophotometer adjusted to a wavelength of 505 nm. Calculate the concentration of glucose by referring values of absorbance to a standard curve. To establish a standard curve, prepare a glucose stock solution of 1000 μg glucose mL^{-1}. Then pipette 0, 50, 100, 200, 400, and 600 μL of glucose stock solution (i.e. 0–600 μg glucose) to a series of labeled test tubes and add doubled deionized water to obtain a total volume of 3 mL in each tube. Pipette 1 mL from each tube to 2 mL of reagents from the diagnostic kit and proceed as described for analysis of glucose in the soil filtrate. Activity is expressed as μg glucose released g^{-1} soil 4 h^{-1}.

Proper controls must be performed in each series of analysis to measure glucose background in soil. To perform controls, add 2.5 mL of TES buffer (pH 7) and 0.2 mL of toluene to 1 g of soil and incubate for 4 h. After incubation, add 0.5 mL of sinigrin to obtain a final concentration of 20 mM and proceed as described above for the standard procedure.

12–4.4 Comments

High-speed centrifugation (about 17,000 g) can be used in the place of the filtration step. It is critical, however, to obtain a clear supernatant of the soil suspension before the colorimetric determination of glucose. Repeated centrifugation may be needed.

Buffer pH, sinigrin concentration, amount of soil, time of incubation, and temperature were all evaluated and optimized in developing this myrosinase assay in soils. The efficiency of glucose recovery from soil in the presence and absence of toluene was tested by adding 3 mL of glucose solution, prepared in TES buffer (pH 7) and containing from 50 to 1000 μg glucose, to 1 g of soil. Toluene (0.2 mL) was added to each sample and controls (without toluene) were also performed. Samples were incubated at 37°C for 4 h. Glucose was extracted and measured as described in the standard procedure. Recovery of glucose from soils treated with toluene was approximately 100% but, without toluene, recovery of glucose was as low as 6% (Al-Turki and Dick, 2003). Thus, there is an absolute requirement for toluene when measuring myrosinase activity in soil based on production of glucose.

Using the standard assay procedure, myrosinase activity was measured in field-moist soil immediately after sampling and in air-dried soil. Myrosinase activity in both soils was calculated based on oven-dry weight and the difference was slight with field-moist soil having approximately 6% lower activity than air-dried soil (Al-Turki and Dick, 2003).

The K_m and V_{max} values of myrosinase in soil were determined by measuring myrosinase activity using a suitable range of sinigrin concentrations (5–50 mM). Enzyme activities were plotted against the corresponding sinigrin concentrations according to the three linear transformations of the Michaelis–Menten equation: Lineweaver–Burk plot ($1/v$ vs. $1/S$), Eadie–Hofstee plot (v vs. v/S), and Hanes–Woolf plot (S/v vs. S) (Hofstee, 1952). The K_m value of myrosinase in four soils averaged 8.1 mM (range 5.9–12.9 mM) and V_{max} values averaged 265 μg glucose g^{-1} soil 4 h^{-1} (range 76–518 μg glucose g^{-1} soil 4 h^{-1}).

REFERENCES

Abeles, F.B., P.W. Morgan, and M.E. Saltveit. 1992. p. 414 Ethylene in plant biology. Academic Press, San Diego.

Adams, D.O., and S.F. Yang. 1979. Ethylene biosynthesis: Identification of 1-aminocyclopropane-1-carboxylic acid as an intermediate in the conversion of methionine to ethylene. Proc. Natl. Acad. Sci. USA 76:170–174. doi:10.1073/pnas.76.1.170

Al-Turki, A.I., and W.A. Dick. 2003. Myrosinase activity in soil. Soil Sci. Soc. Am. J. 67:139–145. doi:10.2136/sssaj2003.0139

Angus, J.F., P.A. Grander, J.A. Kirkegaard, and J.M. Desmarchelier. 1994. Biofumigation: Isothiocyanates released from Brassica roots inhibit growth of the take-all fungus. Plant Soil 162:107–112. doi:10.1007/BF01416095

Arshad, M., and W.T. Frankenberger, Jr. 1988. Influence of ethylene produced by soil microorganisms on etiolated peas seedlings. Appl. Environ. Microbiol. 54:2728–2732.

Borek, V., M.J. Morra, and J.P. McCaffrey. 1996. Myrosinase activity in soil extracts. Soil Sci. Soc. Am. J. 60:1792–1797. doi:10.2136/sssaj1996.03615995006000060026x

Brabban, A.D., and C. Edwards. 1995. The effects of glucosinolates and their hydrolysis products on microbial growth. J. Appl. Bacteriol. 79:171–177.

Brown, P.D., and M.J. Morra. 1997. Control of soil-borne plant pests using glucosinolate-containing plants. Adv. Agron. 61:167–231. doi:10.1016/S0065-2113(08)60664-1

Chague, V., Y. Elad, R. Barakat, P. Tudzynski, and A. Sharon. 2002. Ethylene biosynthesis in Botrytis cinerea. FEMS Microbiol. Ecol. 40:143–149.

Chalutz, E., M. Lieberman, and H.D. Sisler. 1977. Methionine-induced ethylene production by Penicillium digitatum. Plant Physiol. 60:402–406. doi:10.1104/pp.60.3.402

Deng, S., and M. Tabatabai. 1994. Colorimetric determination of reducing sugars in soils. Soil Biol. Biochem. 26:473–477. doi:10.1016/0038-0717(94)90179-1

Fernandez-Maculet, J.C., J.G. Dong, and S.F. Yang. 1993. Activation of 1-aminocyclopropane-1-carboxylate oxidase by carbon dioxide. Biochem. Biophys. Res. Commun. 193:1168–1173. doi:10.1006/bbrc.1993.1748

Fernandez-Maculet, J.C., and S.F. Yang. 1992. Extraction and partial characterization of the ethylene-forming enzyme from apple fruit. Plant Physiol. 99:751–754. doi:10.1104/pp.99.2.751

Frankenberger, W.T., Jr., and M. Arshad. 1995. Phytohormones in soils: Microbial production and function. Marcel Dekker, New York.

Frankenberger, W.T., Jr., and W. Brunner. 1983. Method of detection of auxin-indole-3-acetic acid in soils by high performance liquid chromatography. Soil Sci. Soc. Am. J. 47:237–241. doi:10.2136/sssaj1983.03615995004700020012x

Frankenberger, W.T., Jr., A.C. Chang, and M. Arshad. 1990. Response of *Raphanus sativus* to the auxin precursor, L-tryptophan applied to soil. Plant Soil 129:235–241.

Frankenberger, W.T., Jr., and P.J. Phelan. 1985. Ethylene biosynthesis in soil: I. Method of assay in conversion of 1-aminocyclopropane-1-carboxylic acid to ethylene. Soil Sci. Soc. Am. J. 49:1416–1422. doi:10.2136/sssaj1985.03615995004900060017x

Frankenberger, W.T., Jr., and M. Poth. 1987. Biosynthesis of indole-3-acetic acid by the pine ectomycorrhizal fungus *Pisolithus tinctorius*. Appl. Environ. Microbiol. 53:2908–2913.

Frey, S.D., E.T. Elliott, and K. Paustian. 1999. Application of the hexokinase-glucose-6-phosphate dehydrogenase enzymatic assay for measurement of glucose in amended soil. Soil Biol. Biochem. 31:933–935. doi:10.1016/S0038-0717(98)00176-X

Hofstee, H.J. 1952. On the evaluation of the constants V_m and K_M in enzyme reactions. Science 116:329–331. doi:10.1126/science.116.3013.329

Jwanny, E.W., S.T. El-Sayed, M.M. Rashad, A.E. Mahmoud, and N.M. Abdullah. 1995. Myrosinase from roots of *Raphanus sativus*. Phytochemistry 39:1301–1303. doi:10.1016/0031-9422(95)00099-S

Koga, J., T. Adachi, and H. Hidaka. 1991. IAA biosynthetic pathway from tryptophan via indole-3-pyruvic acid in *Enterobacter cloacae*. Agric. Biol. Chem. 55:701–706.

Ludikhuyze, L., V. Ooms, C. Weemaes, and M. Hendrick. 1999. Kinetic study of the irreversible thermal and pressure inactivation of myrosinase from broccoli (*Brassica oleracea* L. Cv. Italica). J. Agric. Food Chem. 47:1794–1800. doi:10.1021/jf980964y

McGarvey, D.J., and R.E. Christoffersen. 1992. Characterization and kinetic parameters of ethylene-forming enzyme from avocado fruit. J. Biol. Chem. 267:5964–5967.

Moya-Leon, M.A., and P. John. 1994. Activity of 1-aminocyclopropane-1-carboxylate (ACC) oxidase (ethylene-forming enzyme) in the pulp and peel of ripening bananas. J. Hortic. Sci. 69:243–250.

Moya-Leon, M.A., and P. John. 1995. Purification and biochemical characterization of 1-aminocyclopropane-1-carboxylate oxidase from banana fruit. Phytochemistry 39:15–20. doi:10.1016/0031-9422(94)00873-R

Nagahama, K., T. Ogawa, T. Fujii, and H. Fukuda. 1992. Classification of ethylene-producing bacteria in terms of biosynthetic pathways to ethylene. J. Ferment. Bioeng. 73:1–5. doi:10.1016/0922-338X(92)90221-F

Normanly, J., J.P. Slovin, and J.D. Cohen. 1995. Auxin biosynthesis and metabolism. Plant Physiol. 107:323–329.

Pirrung, M.C., L.M. Kaiser, and J. Chen. 1993. Purification and properties of the apple fruit ethylene-forming enzyme. Biochemistry 32:7445–7450. doi:10.1021/bi00080a015

Raba, J., and H.A. Mottola. 1995. Glucose oxidase as an analytic reagent. Crit. Rev. Anal. Chem. 25:1–42. doi:10.1080/10408349508050556

Rask, L., E. Andreasson, B. Ekbom, S. Eriksson, B. Pontoppidan, and J. Meijer. 2000. Myrosinase: Gene family evolution and herbivore defense in Brassicaceae. Plant Mol. Biol. 42:93–113. doi:10.1023/A:1006380021658

Sarwar, M., M. Arshad, D.A. Martens, and W.T. Frankenberger, Jr. 1992. Tryptophan-dependent biosynthesis of auxins in soil. Plant Soil 147:207–215. doi:10.1007/BF00029072

Sato, M., K. Watanabe, M. Yazawa, M. Takikawa, and K. Nishiyama. 1997. Detection of new ethylene-producing bacteria, *Pseudomonas syringae* pvs. cannabina and sesami, by PCR amplification of genes for the ethylene-forming enzyme. Phytopathology 87:1192–1196. doi:10.1094/PHYTO.1997.87.12.1192

Smith, J.J., P. Ververidis, and P. John. 1992. Characterization of the ethylene-forming enzyme partially purified from melon. Phytochemistry 31:1485–1494. doi:10.1016/0031-9422(92)83092-D

Tabatabai, M.A., and M. Fu. 1992. Extraction of enzymes from soils. p. 197–227. *In* G. Stotzky and J.-M. Bollag (ed.) Soil biochemistry. Marcel Dekker, New York.

Tao, L., H.J. Don, X. Chen, S.F. Chen, and T.H. Wang. 2008. Expression of ethylene-forming enzyme (EFE) of *Pseudomonas syringae* pv. glycinea in *Trichoderma viride*. Appl. Microbiol. Biotechnol. 80:573–578. doi:10.1007/s00253-008-1562-7

Völksch, B., and H. Weingart. 1997. Comparison of ethylene-producing *Pseudomonas syringae* strains isolated from kudzu (*Pueraria lobata*) with *Pseudomonas syringae* pv. phaseolicola and *Pseudomonas syringae* pv. glycinea. Eur. J. Plant Pathol. 103:795–802. doi:10.1023/A:1008686609138

Wilkinson, A.P., J.C. Rhodes, and R.G. Fenwik. 1984. Determination of myrosinase (thioglucoside glucohydrolase) activity by a spectrophotometric coupled enzyme assay. Anal. Biochem. 139:284–291. doi:10.1016/0003-2697(84)90004-6

Chapter 13

Enzyme Activities of Root Tips and in situ Profiles of Soils and Rhizospheres

Melanie D. Jones,* Denise D. Brooks, Pierre-Emmanuel Courty, Jean Garbaye, Pauline F. Grierson, and Karin Pritsch

13–1 INTRODUCTION

Almost all methods for estimating activities of soil enzymes, by necessity of the equipment required, are conducted on samples of soil collected in the field and tested in the laboratory. The typical objective of soil enzyme assays is to measure the highest potential activities in the soil. Consequently, assays are generally undertaken at optimal pH and at standard temperatures and saturating concentrations of substrate. Typical sample sizes are large enough to analyze several types of chemical, biochemical, and biological variables, allowing the relationships among them to be quantified. Therefore, results obtained deviate from enzyme activities in the field at the time when the samples were taken. In this chapter, we describe alternative assays that allow investigations of enzyme activities at fine spatial scales (mm scale), and that can be used to test the association between these activities and specific roots or microorganisms.

Greater understanding of how fine-scale soil heterogeneity affects enzyme activities is essential for scaling measurements of individual soil samples up to ecosystem levels (Schimel and Bennett, 2004; Jones et al., 2005; Hobbie and Hobbie, 2008). There are several important sources of fine-scale heterogeneity in soils; one is the presence of plant roots. Within 1 to 5 mm of roots, concentrations of macronutrient cations, pH, types of clay, and both rates and forms of carbon fluxes differ from those in bulk soils (Marschner and Römheld, 1983; Arocena et al., 2004;

Melanie D. Jones, Biology and Physical Geography Unit, Science Building, University of British Columbia, Okanagan campus, 3333 University Way, Kelowna, BC V1V 1V7 Canada (Melanie.Jones@ubc.ca), *corresponding author;
Denise D. Brooks, Forest Sciences Department, University of British Columbia Vancouver campus, 2424 Main Mall, Vancouver, British Columbia, Canada; current address: Okinawa Institute of Science and Technology, 1919-1 Tancha, Onna-son, Kunigami, Okinawa, Japan 904-0412 (brooks.denise@oist.jp);
Pierre-Emmanuel Courty, INRA-Nancy, 54280 Champenoux, France; current address: University of Basel-Botanical Inst., Hebelstrasse 1, CH-4056 Basel, Switzerland (pierre.courty@unibas.ch);
Jean Garbaye, INRA-Nancy, 54280 Champenoux, France (garbaye@nancy.inra.fr);
Pauline F. Grierson, Ecosystems Research Group, School of Plant Biology M090, University of Western Australia, 35 Stirling Highway, Crawley WA 6009 Australia (pauline.grierson@uwa.edu.au);
Karin Pritsch, Inst. of Soil Ecology, Helmholtzzentrum München, German Research Ctr. for Environmental Health, Ingolstaedter Landstraße 1, 85764 Neuherberg, Germany (pritsch@helmholtz-muenchen.de).

doi:10.2136/sssabookser9.c13

Leake et al., 2006). The extent and type of rhizosphere influence depends on the size and type of roots (Dieffenbach and Matzner, 2000).

Typically, the chemical and biological properties of rhizospheres have been studied on soils collected from the field and dried, mixed, sieved, or otherwise manipulated in the lab (e.g., Pijnenborg et al., 1990; Arocena et al., 2004). However, studies of rhizospheres in undisturbed soils have been important in challenging assumptions about ion movements developed using bulk soils (Dieffenbach and Matzner, 2000).

Other, less well-studied sources of fine-scale heterogeneity include soil aggregates and patches of organic matter. Nitrogen and recently fixed C are elevated in the several mm surrounding decaying plant material and feces of soil arthropods (Gaillard et al., 1999; Grieve et al., 2006). Small preferential pathways of water flow can be 40% enriched in nutrients leaching from the soil surface compared with the soil matrix (Bundt et al., 2001). Gradients in oxygen and mineral nutrient cations exist between the interior and exterior of soil aggregates (Hildebrand, 1994).

Such fine-scale heterogeneity can directly affect enzyme activities through, for example, variation in substrate concentrations (Lucas and Casper, 2008), protein binding properties (Fansler et al., 2005; Allison and Jastrow, 2006), and pH (Marschner et al., 1982), but also indirectly through effects on the distribution of the soil microorganisms responsible for excreting many soil enzymes (Aon et al., 2001; Nannipieri et al., 2003). Soil fungi and prokaryotes are more abundant in rhizospheres, in soil particles rich in organic matter, and in the outer shells of macroaggregates (Schack-Kirchner et al., 2000; Vaisanen et al., 2005; Blackwood et al., 2006). Understanding the rates of soil processes in different microsites and applying this information in simulation models is crucial to quantifying overall rates of processes such as mineralization (Schimel and Bennett, 2004); measurements of soil properties taken at fine scales do not always average out to be the same as those measured on bulk samples (Grieve et al., 2006).

It is challenging to evaluate spatial variation of soil enzyme activities and few methods are available to allow such investigations. Soil horizons can be separated and analyzed independently (e.g., Snajdr et al., 2008); however, geostatistical approaches have detected variations in soil chemical and biological properties at mm to cm scales that vary independently of soil horizons (Göttlein and Stanjek, 1996). Researchers have quantified activities associated with different sizes of microaggregates (e.g., Dorodnikov et al., 2009), which allows correlations with physicochemical and general biological characteristics of each aggregate size. Other approaches have been developed that allow the spatial distribution of enzyme activities to be visualized and/or associated with the presence of specific microorganisms (e.g., Grierson and Comerford, 2000; Marschner et al., 2005; Pritsch et al., 2004; Buée et al., 2005; Courty et al., 2005; Dong et al., 2007).

In this chapter, we present two such approaches: (i) assays using imprints made of soil surfaces accessed via root windows and (ii) assays of tips of individual roots. In the future, methods such as fluorescent in situ hybridization (FISH) may become useful for visualizing the spatial distribution of genes of particular enzymes, but currently FISH is limited by requiring high copy numbers of the targeted DNA (Zwirglmaier, 2005). Furthermore, FISH does not quantify enzyme activity.

13–1.1 Principle: Enzyme Imprinting at Root Windows

Transparent plates have been used to make observations of root growth and the interactions among roots, soil, and microorganisms for almost a century (McDougall, 1916). However, there have been relatively few attempts at direct analysis in situ of enzymes or other soil attributes from observational windows except through destructive sampling. A root window, or rhizotron, generally consists of a sheet of transparent and robust plastic or glass inserted against a vertical (occasionally horizontal) soil profile, which permits long-term observation of fine roots, fungal mycelia, and individual mycorrhizae, with limited disturbance (Egli and Kalin, 1990; Smit et al., 2000).

The only disturbance that is caused, after the initial installation of the window, is associated with opening the door of the sampling window, which will cause disruption of hyphae or roots that become attached to the inner surface of the window. The major advantage of root windows over minirhizotrons (clear tubes inserted into holes in the soil created by an auger) is that the soil can be directly accessed, allowing for mm-scale manipulations and sampling. With root windows containing hinged sampling windows, the same patches of soil or roots or fungal hyphae may be repeatedly sampled for enzyme activity. In addition, mm- to cm-sized portions of the soil profile behind the window can be manipulated and the responses to these fine-scale changes studied over time. For example, pH, nutrient concentrations, and organic matter contents can be changed by adding organic residues (e.g., decayed wood, humus, cotton strips) or inorganic minerals to the soil profile behind the window. This allows native microflora to colonize these materials and secrete enzymes in situ. This is a major advantage when studying soils dominated by organisms such as mycorrhizal fungi, which often do not grow in the absence of intact roots of their hosts.

Soil imprinting methods apply cellulose sheets (chromatography or filter paper) or nitrocellulose membranes to the soil and root surfaces exposed behind the window, producing images showing the spatial distribution of enzyme activities over two dimensions (Dinkelaker and Marschner, 1992; Grierson and Comerford, 2000; Dong et al., 2007). Soils are disturbed very little by the imprinting procedure and resolution can be at the mm scale, or even finer if using nitrocellulose membranes. Imprinting can be used to map soil enzyme activities and to correlate the spatial distribution of activities with soil features visible in the window. Correlations can be both qualitative and quantitative, using high-resolution photography and image analysis software (Grierson and Comerford, 2000). By recording activities over an entire soil surface rather than restricting assays to specific regions of soil, unsuspected hotspots of activity may be discovered.

These approaches are especially useful for studies of temporal changes in the location of enzyme activity because the same soil surface can be tested repeatedly. This is important because enzyme activities in the rhizosphere, and likely other parts of the soil, can vary on a diurnal basis (Murray et al., 2004). In addition, soil fungal and prokaryotic communities associated with hotspots of enzyme activity can be described by combining imprint-guided, targeted sampling of the soil profile with molecular genetic methods.

In situ activities of soil enzymes can be detected relatively easily if chromogenic or fluorogenic substrates are embedded in the cellulose sheets before imprinting. The activities detected using this approach will have occurred under ambient temperature and moisture conditions and are more comparable to what happens in the soil in the field than those measured ex situ under optimal lab conditions. Imprints developed in the field can be immediately used to identify areas for direct fine-scale sampling of soil associated with different levels of enzyme activity. Amplification of DNA from roots, fungi, or bacteria associated with hotspots of enzyme activity also can assist in determining the sources of the exoenzymes. However, one of the disadvantages of this approach is that activities may be undetectable when soils are very dry. Furthermore, activities are detected in only parts of the soil in direct contact with the sheet, and soil surfaces against the window may not be representative of the soil as a whole because of greater temperature fluctuations, root growth, or condensation at the soil–window interface.

An alternative approach is to bind enzymes in situ to untreated nitrocellulose membranes and then to develop the imprint under standard conditions in the laboratory. Nitrocellulose membranes are more appropriate than cellulose sheets for quantification and localization at sub-mm scales because there is little diffusion of the colored product within the membrane. Furthermore, running the assay itself under standard laboratory conditions, which detects potential activities, allows for comparisons among studies.

Two-dimensional imprints of enzyme activity associated with roots, fungal hyphae, and localized microbial activity within soil are a useful visualization of the relative patterns within the rhizosphere and bulk soil. However, while these methods elegantly demonstrate enzyme activity in soil, they are not quantitative. The concept of developing a set of standards against which data from imprinting can be calibrated and quantified is similar in principle to methods used in making molecular weight calibrations for analysis of proteins or DNA following gel electrophoresis. By applying enzyme standards of known activities to defined areas on the sheets or membranes, and developing these standard strips under the same environmental conditions as the imprints, spatial, intensity, and/or molecular weight calibrations can be set or loaded from each image to ensure consistent results from one analysis session to another. Therefore, it is desirable to define data, such as color intensity, relative to the surface area sampled, analogous to the calibration of enzyme activity to volume for traditional assays of enzyme activity. In its simplest sense, visual inspection of 2-D images generated from standards is important to detect gross differences in spot patterns or consistent patterns relative to mycelial fans or roots associated with the test conditions under study. For comparative purposes, image analysis of 2-D enzyme prints can use computer software that relies on the detection of hotspots within the image and which can be used to overlay either successive images or to quantify imprints relative to underlying features such as roots or mycorrhizal fungi. The software generally determines the outlines of the separated spots such that densities (or color intensities) can be calculated.

13–1.2 Principle: Microplate Assays of Root Tips

High levels of exoenzyme activities occur in the vicinity of root tips including mycorrhizal fine roots, ectomycorrhizal short roots, and rapidly growing major lateral roots, which are typically nonmycorrhizal (Tarafdar and Jungk, 1987;

Häussling and Marschner, 1989; Grierson and Comerford, 2000; George et al., 2006). Sensitive and rapid assays, developed to quantify the activities associated with individual root tips, have been used primarily on roots of ectomycorrhizal tree species (Buée et al., 2007; Courty et al., 2005, 2006, 2007, 2008; Mosca et al., 2007). Ectomycorrhizal fungi contribute to tree nutrition by mobilizing the nutrients entrapped in litter and soil organic matter through a range of cell-wall-bound and secreted enzymes. Hence, these assays have been useful in determining the influence that different species of ectomycorrhizal fungi have on this process. Nevertheless, there is a much broader potential for application of these assays on individual roots of different plant species, regardless of whether the roots are mycorrhizal (Pritsch et al., 2006).

The principle of these assays is to measure, on the same root tips, several enzymatic activities involved in breaking down macromolecules, and thereby mobilizing nutrients from decaying plant material and soil organisms. In the two methods described here (from Pritsch et al., 2004; Buée et al., 2005; Courty et al., 2005; Pritsch et al., 2011), this aim has been achieved using colorimetric methods or fluorogenic substrates based on 4-methylumbelliferone (MU) or 7-amino-4-methylcoumarin (AMC), and combining tests on the same root tips in 96-well microtitration plates. Because these assays allow the rapid processing of large numbers of individual root tips, they facilitate large-scale ecological studies investigating the functional structure of plant root or mycorrhizal fungal communities and their contribution to plant nutrition.

13–2 METHODOLOGY FOR SOIL IMPRINTING OF ENZYME ACTIVITY

In the methods presented here, root windows are used to access relatively undisturbed soil profiles for imprinting of enzyme activity. Nitrocellulose membranes or cellulose sheets, with or without imbedded substrates, are applied to the soil face to produce imprints of enzyme activity.

13–2.1 Installation of Root Windows (Grierson and Comerford, 2000; Smit et al., 2000)

The root window consists of a transparent and acrylic sheet, e.g., Plexiglas (5 mm thick) inset with a hinged sampling window held in place by a small latch (Fig. 13–1). Windows may vary in size depending on the purpose, but smaller windows tend to maintain better contact with the soil. In the design of Grierson and Comerford (2000), hinges are covered by a flexible rubber seal on the *inside* of the window, which prevents any soil particles falling into the hinge and becoming trapped when the window is open and preventing window re-closure. The hinged sampling window has an internal measurement area of 30 by 30 cm. A small hole (0.5 mm i.d.) is drilled into each corner and in the center of the measurement area to assist in alignment of any images of enzyme activity and photos of the root system or soil surface that may require subsequent analysis.

13–2.1.1 Apparatus

- Plexiglas or other transparent glass or plastic, approx. 80 by 50 by 0.5 cm
- Rubber-sealed hinge, 30 cm in length

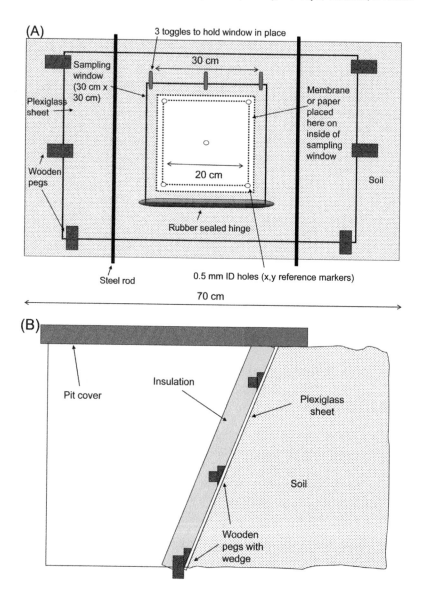

Fig. 13–1. Diagram of root window. A) Front view; B) Side view. The dimensions of the root window and the placement of the inner sampling window can vary, depending on the soil depth under investigation. For example, the sampling window can be located near the top of the window if researchers want to be able to sample the uppermost soil layers. If the root windows are installed on a slope, they should be placed so that only a few cm of the window reaches above the soil surface. Otherwise, runoff or snow pressure can cause the window to lose contact with the soil. Steel rods hold the window in place, but wooden wedges, of various sizes and shapes depending on soil texture, should be pounded into the soil around the edges of the window to assist in maintaining the pressure of the window against the soil.

- Rubber adhesive for hinge seal
- Toggles for opening trap door, approx. 3 by 1.5 by 0.5 cm, three per window
- Three screws, one per toggle, approx. 0.75 cm
- Electric drill to attach toggle openers to window and to drill reference holes
- Steel plate approximately 4 cm wider than the Plexiglas plate, 3 to 5 mm thick
- Two long-handled hand shovels, one with a 30 by 23 cm curved blade and one with an approx. 30 by 18 cm flat blade
- Wooden pegs, 2 cm diameter by 50 cm length, four per window
- Wooden wedges of various dimensions for maintaining contact between window and soil
- Vertical steel rods, 1 m long by 1 cm diameter
- Sledge hammer to pound in steel rods (weight required depends on soil type)
- Insulation block, e.g., Styrodur, Styrofoam (or plywood and fiberglass insulation as an alternative), approximately 4 cm larger than Plexiglas plate, 6 cm thick
- Heavy black landscaping plastic sheeting
- Aluminum screening, chicken wire, or hardware cloth, depending on the size of animal pests present, 100 by 125 cm per window (optional)

13–2.1.2 Procedure
Installation of the window follows the general recommendations of Smit et al. (2000).

1. Hammer the steel plate at a specific angle to the soil surface, usually at an angle of 90°, although the angle may vary depending on the soil depth, the plant species, and the soil type at the site of installation.

2. Excavate all the soil away from the plate using the hand shovels, to the depth of plate installation. Soil from the excavated pit should be separated by depth and stored well away from the window to prevent contamination in case backfilling is required. Construction of a ramp on the side of the pit opposite the plate to facilitate access and observation of the pit may also make working in the pit more comfortable.

3. Remove the steel plate and smooth the surface with the shovel with the flat blade. Where there are large roots or stones, these may be trimmed, the stones removed, and the voids filled with the soil from the appropriate depth.

4. Smit et al. (2000) recommend heating the Plexiglas window in a warm oven before installation to produce slight concavity as this enhances overall soil–window contact on installation; in this instance the hinged sampling portion of the window may become distorted. Consequently, we do not recommend this step. Instead, push the window firmly against the surface, and hammer two vertical steel rods in about 10 cm either side of the hinged sampling window (Fig. 13–1). Wooden pegs and wedges can be used to further improve contact between the window and the soil surface (see Smit et al. (2000) and Fig. 13–1 for details).

5. Cover the window with a layer of insulation and black plastic sheeting to protect against light and temperature fluctuations. This can be comprised of a piece of rigid insulation, such as Styrofoam or Styrodur, or a piece of ply-

wood covered with fiberglass insulation. In either case, the insulation should be protected with aluminum screening or some heavier kind of metal mesh to protect it from attack if the project is in areas where small mammals might use it as nesting material. Alternatively, the root-window pit can be backfilled to the original soil grade and re-excavated as needed, particularly in the period where the window is being left to stabilize across seasons. Backfilling the pit is also more resistant to disturbance by animals and weather.

13–2.1.3 Comments

Installing root windows inevitably results in some initial soil disturbance which influences enzyme activity. Windows should be left undisturbed for a period of time to allow the soil to stabilize and for roots and microbial communities to establish or re-equilibrate against the window. The time required could range from weeks to years depending on the question being asked, the time of year, and the environmental conditions in the area. Consideration also should be given to the degree of backfilling of original soil behind the window that is required. This aspect is of less importance if the objective is experimental and different areas of the soil interface have been manipulated, e.g., backfilling different parts of the window with soils of variable texture, nutrient content, organic matter content, or pH. The integrity of the window should be maintained at all times by limiting disturbance or walking on the ground behind the window.

13–2.2 Principle: Preparation of Imprints (Dinkelaker and Marschner, 1992; Grierson and Comerford, 2000; Dong et al., 2007)

The general principles of imprinting methods are presented here with specific details (e.g., substrates and enzyme reaction chemistry) presented within the Principle Sections associated with each method below.

These in situ enzyme assays use either nitrocellulose or cellulose membranes. The nitrocellulose membrane is exposed (about 10 min) to a soil–rhizosphere profile in the field to allow for adsorption of enzymes. This membrane is returned to the laboratory where it is incubated in the presence of the substrate in solution buffered at an optimal pH for a given enzyme assay. By contrast, the cellulose membrane is first saturated (in the laboratory) with the substrate and then incubated in the field against the soil–rhizosphere profile.

These two approaches can be interpreted to assess different pools of isoenzymes for a given assay. In the case of nitrocellulose, because it adsorbs enzymes at the soil profile or root surface and is then removed back to the laboratory for the assay, measured activities can be interpreted to represent extracellular enzymes that were excreted by viable cells. By contrast, the cellulose membrane detects activities on soil or root surfaces. An interpretation for this approach is that the activity measured is a combination of extracellular enzymes excreted by viable cells plus enzymes stabilized in the soil matrix that are not part of viable cell (Dick, 1997).

Two types of substrates are used: (i) one that when hydrolyzed produces a product that is detected colorimetrically in the laboratory (nitrocellulose membrane) or the field (cellulose membrane) in the presence of color developing reagents, or (ii) a fluorogenic substrate that on hydrolysis is detected on a fluorimeter in the lab. Enzyme activity at the soil surface is indicated by the appearance of spots of colored or fluorescent product on the membranes.

Nitrocellulose membranes must be used to obtain imprints with sufficient resolution to allow correlations of activity with visible soil features at a pixel-by-pixel scale. However, nitrocellulose membranes are more expensive than cellulose sheets (chromatography or filter paper), and consequently the advantages of increased resolution need to be weighed against the higher costs, especially if the experiment involves repeated sampling of many root windows. If resolution at the mm scale is sufficient, then cellulose sheets can be used.

13–2.2.1 Method I Imprinting: Acid Phosphomonoesterase (EC 3.1.3.2) Activity using Imprints of Nitrocellulose Membranes Developed under Standard Lab Conditions (Dinkelaker and Marschner, 1992; Grierson and Comerford, 2000)

13–2.2.1.1 Principle

This is a semiquantitative method that compares the intensity of color development on the membranes with color development of standards applied to known areas of a membrane using a slot blot apparatus. In this method, the nitrocellulose membrane is applied to the soil–rhizosphere profile for 10 min and the enzymes (proteins) are adsorbed to the surface of the membrane. A standard nitrocellulose membrane to estimate concentrations is prepared by exposing the membrane to known phosphatase activities. Both the sample- and standard-imprint membranes are then washed with sterile, deionized water and dried to stabilize the proteins. These membranes are then incubated in the presence of the substrate (α-naphthyl phosphate) in a buffered solution to develop color in locations containing phosphatase.

Following the incubation, the product of α-naphthyl phosphate hydrolysis, naphthol, is measured colorimetrically by reacting it with the diazonium salt Fast Red TR to form a stable red precipitate (Dinkelaker and Marschner, 1992). The buffer pH was selected based on pH optima as determined by Dinkelaker and Marschner (1992) and later confirmed by Grierson and Comerford (unpublished data). The color development is collected by taking a digital image using a high-resolution camera or scanner.

13–2.2.1.2 Apparatus

- Nitrocellulose membranes (0.45μm, 20 cm wide roll, e.g., Hybond C, GE Healthcare; cut to size)
- Latex or nitrile gloves
- Plastic squeeze bottle, 1000 mL, filled with sterile, deionized water
- Plastic squeeze bottle, 500 mL, filled with 70% ethanol
- Glass capillary tubes (<0.5-mm diameter, 75-mm length)
- Plastic or glass tray, 30 by 40 by 4.5 cm
- Flat-tip forceps, 115 mm, two pair, transported in ethanol
- Cling-wrap plastic
- Aluminum foil
- Insulated container, 20 L, and ice
- Large, square, flat-bottomed glass dish (23 by 23 by 8 cm)
- High-resolution digital camera with tripod

- Microfiltration apparatus (Bio-Dot SF Microfiltration Apparatus, Bio-Rad)
- Vacuum pump, of a size recommended by the manufacturer of the microfiltration apparatus
- Digital scanner

13–2.2.1.3 Reagents

Prepare all reagents except buffer on the day of use.

1. Citrate buffer, 50 mM: Dissolve 6.48 g citrate trisodium anhydrous salt (Sigma C3674, $C_6H_5O_7Na_3$) in 500 mL deionized water and adjust pH to 5.6 with 1 M HCl.
2. Substrate solution, 50 mM: Dissolve 0.34 g α-naphthyl phosphate disodium salt (Sigma N7255) in 25 mL of 50 mM citrate buffer (pH 5.6).
3. Colorimetric reagent solution, 10 mM: Dissolve 0.64 g Fast Red TR salt (Sigma F2768) in 250 mL of 50 mM citrate buffer (pH 5.6).

13–2.2.1.4 Standards

Prepare in glass volumetric flasks.

1. Phosphomonoesterase stock solution: Make up stock solution of acid phosphatase from acid phosphatase wheat germ extract (Sigma P3627) by dissolving 0.036 g in 250 mL of 0.05 M citrate buffer, pH 5.6. Prepare on day of use.
2. Working standards: Prepare working standards according to Table 13–1.

13–2.2.1.5 Procedure: In the Field

1. Remove the backing from a 22 by 22 cm precut piece of nitrocellulose membrane and spray with sterile water in the glass dish until the membrane is saturated. Cut one corner of the membrane to assist with subsequent orientation of the image.
2. Open the window and spray the inside with 70% ethanol; wipe dry with clean tissues. Place the membrane on the inside of the open window, using the etched lines on the Plexiglas or Perspex to align the center of the membrane with the center of the window. Close the window; cover with aluminum foil. As good contact between the membrane and the soil surface is paramount, additional pressure can be applied to the outside of the window either by hand or using a wedge. Contact should be maintained for 10 min. Binding of the enzymes to nitrocellulose membranes is theoretically instantaneous and thus longer time periods should not be necessary. However, some assessment should be undertaken to optimize timing for the specific soils being imprinted.
3. Before removing the membrane, pass a fine capillary tube through each of the small holes to mark the corners and center of the membrane. These holes will be used as x, y coordinates for subsequent image analysis. Open the window and remove the membrane (henceforth referred to as the imprint), using sterile forceps, wrap in plastic and then aluminum foil, and transport on ice from the field to the laboratory.
4. Take a digital photograph through the Plexiglas window to record roots, hyphae, and other features on the soil surface. Images should capture the

Table 13–1. Preparation of phosphomonoesterase standard solutions (10 mL each) to create a standard curve using a Bio-Dot SF Microfiltration Apparatus.

Standard solution (10 mL)	Phosphatase activity (katal × 10^{-13})†	Aliquot of phosphomonoesterase standard from column 1 (mL)‡	Citrate buffer (mL)
A	33.33	2	8
B	25.00	1.5	8.5
C	16.67	1	9
D	11.17	2 × A	4
E	8.33	1 × A	3
F	5.50	2 × B	4
G	3.33	1 × A	7
H	2.50	1.5 × C	8.5
I	1.67	1 × C	9
J	0.83	1 × G	4
K	0.55	2 × I	4
L	0.33	1 × G	9
M	0.25	1.5 × I	8.5
N	0.17	1 × I	9
O	0.08	1 × L	3

† One katal will hydrolyze 1.0 mol of *p*-nitrophenyl phosphate per second at 37°C.

‡ The standard solution is made of acid phosphatase wheat germ extract (Sigma P3627) by dissolving 0.036 g in 250 mL of 0.05 M citrate buffer, pH 5.6.

positioning of the corner and center holes marked on the window and be at least 300 dpi in resolution if quantitative analysis is required.

13–2.2.1.6 Procedure: In the Laboratory

1. Prepare the reagents and mix 1 part freshly prepared substrate solution with 10 parts colorimetric reagent solution (1:10 v/v).

2. Standard curve preparation: Following the manufacturer's instructions for the microfiltration apparatus, load three 0.5-mL replicates of each working standard (13–2.2.1.4) to individual slots, under vacuum supplied by the vacuum pump or built-in vacuum system, on to a membrane cut to fit the set area (see Apparatus list 13–2.2.1.2, first item). Cut one corner of the standard membrane to recognize the top and bottom edges of the membrane once it is developed.

3. Unwrap the imprint membrane brought back from the field (13–2.2.1.5), rinse it with sterile, high-purity water and allow to air-dry at room temperature before development. This helps proteins bound to the imprint membrane in the field to bind even more strongly and prevents loss of target material during subsequent washes.

4. After loading the standards membrane, place it with the imprint membrane in a large glass or Pyrex tray and develop in the substrate solution in the dark for 1–2 h at 37°C. The exact development time is not crucial. However, the

standard membranes should be left in the solution until the imprints are clear and, critically, for the same time as the membranes from the root windows.

5. To limit further activity, remove both membranes from the solution and rinse with distilled water to remove excess substrate.

6. Air-dry both imprint and standard membranes on clean tissue before scanning and analysis (see section 13–2.3). The acid phosphomonoesterase-induced color change on the membranes is stable for many months after the imprints have been air-dried.

13–2.2.1.7 Comments

The nitrocellulose membrane used should have a strong ability to absorb water, high strength, and excellent bonding capacity for proteins. Great care should be taken when handling the treated paper to prevent contamination from fingers. Wear gloves and use forceps at all times.

Before development, imprints may be stable for up to 12 h, but this period is dependent on field conditions. It is recommend that imprints be developed as soon as possible after collection. The intensity of the color imprint is dependent on the duration of the developing time. A period of between 1 and 4 h is generally sufficient. Longer time periods usually result in saturation of the membrane and some background discoloration. The total developing time is not crucial if the standard membrane is left to develop for the same period of time as the imprint from the roots. This is advantageous, as it also allows comparison between different sampling events. However, the indicator reaction may be enhanced by increasing the concentration of the dye in the developing solution (this should also decrease the developing time) (Gundlach and Mühlhausen, 1980). All color developing reactions should be kept in the dark.

Small areas of yellow discoloration occasionally appear on imprints, resulting from unknown sources. However, the color is distinct from that developed from acid phosphomonoesterase reactions, and thus does not interfere with the analysis.

13–2.2.2 Method II Imprinting: Acid Phosphomonoesterase Activity using Imprints of Treated Cellulose Sheets Developed under Field Conditions (Dinkelaker and Marschner, 1992; Dong et al., 2007)

13–2.2.2.1 Principle

This semiquantitative method uses the same substrate and color development chemistry of Method I (as described in 13–2.2.1.1). However, in this approach a cellulose membrane is first saturated with the substrate (α-naphthyl phosphate) and the color developing reagents in the laboratory. The membrane is then taken to the field, placed against a soil–rhizosphere profile and incubated under environmental conditions. Those locations on the membrane in contact with soil or roots that have phosphatase present will then hydrolyze the substrate during the incubation period. After the incubation period, the color development is recorded with a digital camera or scanner.

Although Method II is less spatially precise than Method I, calibration strips can again be incubated alongside the treated cellulose sheets. These can then be used to estimate the range of enzyme activities being detected under the conditions at each window and to determine those that are in the linear range of color intensity. Compared with Method I, higher concentrations of enzyme standard must be used because smaller volumes are applied to each spot on the strip. It is generally not practical to use a slot blot apparatus in the field, so the size of the spots of the standards should be constrained by impregnating the perimeter of defined areas with a hydrophobic material, such as lanolin or melted paraffin.

13–2.2.2.2 Apparatus

- Cellulose sheets (Whatman 3030–861 chromatography paper, 20 by 20 cm)
- Large, flat-bottomed glass dish (23 by 23 by 8 cm)
- Acid bath
- Latex or nitrile gloves
- Insulated container, 20 L, and ice
- Aluminum foil
- Flat-tip forceps, 115 mm, two pair, transported in ethanol
- Cling-wrap plastic
- High-resolution digital camera with tripod
- Pipettor and tips
- Small plastic bags with zip closures
- Scanner

13–2.2.2.3 Reagents

As described in 13–2.2.1.3.

13–2.2.2.4 Standards

1. Phosphatase stock solution: Make up stock solution of acid phosphatase from acid phosphatase wheat germ extract (Sigma P3627) by dissolving 0.036 g in 2.5 mL of 50 mM citrate buffer, pH 5.6.
2. Create a series of standards ranging from 0.08×10^{-4} to 33×10^{-4} katal per 5-μL aliquot using the dilutions in Table 13–1, but reducing volumes 100-fold. For example, to create Standard Solution A, add 20 μL of phosphatase stock to 80 μL of citrate buffer.

13–2.2.2.5 Procedure: Before Going to the Field

1. In the laboratory, mix 1 part freshly prepared substrate solution with 10 parts colorimetric reagent solution (1:10 v/v). Immediately soak chromatography paper in the solution, one sheet at a time, for approximately 1 min in an acid-washed, flat-bottomed, glass dish. Air-dry treated papers in the dark on aluminum foil at room temperature.
2. When dry, store in a sealed plastic bag at 4°C for up to 1 wk.
3. Cut strips of the treated cellulose sheets into 2 by 5 cm pieces for use in calibration. Use a stencil to define the perimeters of spots to which the standards will be applied to the strip in the field (circles of 3-mm diameter are appropriate). Apply an inert hydrophobic material, such as lanolin or melted paraffin around the perimeter of the stencils. Prepare sufficient spots for re-

ceiving three replicates of 8 to 10 standards.

13–2.2.2.6 Procedure: In the Field

1. Prepare the calibration strips in the field immediately before making an enzyme imprint by pipetting a 5-µL aliquot of each working standard as a discrete spot along a strip of treated cellulose paper. Once the solutions have been absorbed, seal the calibration strip in a plastic bag.

2. Create soil enzyme imprints by following Step 2 in section 13–2.2.1.5, but with the paper left in place for 60 min. Maintain firm contact between the treated cellulose sheet and the soil surface. Place the calibration strip in its plastic bag against the soil profile next to the treated cellulose sheet and incubate them together for the same length of time.

3. Before removing the cellulose sheet (henceforth referred to as the imprint), take a high-resolution image of the sheet in place. If following Step 5 (below), use a tripod-mounted camera and leave it in place.

4. Remove the imprint and rinse with distilled water to remove any attached soil particles. Wrap washed imprints and calibration strips in aluminum foil and maintain on ice until they can be air-dried in the dark (no later than the end of the day).

5. A photograph can be taken with the window open, using the tripod-mounted camera without having disturbed it from Step 3 if the researcher wishes to record very fine structures, such as individual hyphae.

13–2.2.2.7 Procedure: Upon Return to the Laboratory

Scan the dried imprint and calibration strips at a resolution of at least 300 dpi.

13–2.2.2.8 Comments

For the imprinting to be successful using a cellulose sheet, sufficient moisture must be present in the exposed soil face to allow water to be drawn into the treated cellulose. In cold soils, the time required for imprinting may be longer than 60 min. The concentrations of the standards may also need to be varied to match the activities in specific soils. Some preliminary tests should be done under field conditions to optimize for specific conditions. The acid phosphomonoesterase-induced color change on the cellulose sheets is stable for weeks following air-drying of the imprints.

13–2.2.3 Imprinting of Leucine Aminopeptidase (EC 3.4.11.1) Activity (Humble et al., 1977; Reymond and Wahler, 2002; Dong et al., 2007)

13–2.2.3.1 Principle

This is a semiquantitative method that uses the same basic principle as described in section 13–2.2.2.1, where a cellulose membrane is first saturated with the substrate (L-leucyl 2-naphthylamide) and the color developing reagents in the lab, then incubated in the field. The detection of amino peptidase is based on the hydrolysis of L-leucyl 2-naphthylamide to release naphthol, which reacts with Fast Blue BB Salt hemi (zinc chloride) salt to form an orange-red color (Humble et al., 1977; Reymond and Wahler, 2002). Fast Blue is more effective than Fast Red for this assay (Dong et al., 2007).

Table 13–2. Preparation of protease solutions (0.5 mL) for standard curve used in aminopeptidase imprinting.

Standard solution	Minimum activity (katal × 10⁻¹⁰)†	Aliquot of protease stock‡ (mL)	Deionized water (mL)
A	4.17	0.5	0.0
B	2.50	0.3	0.2
C	1.25	0.25 × B	0.25
D	0.625	0.25 × C	0.25
E	0.313	0.25 × D	0.25

† One katal will hydrolyze 1 mol of l-leucine-*p*-nitroanilide per second.

‡ Make the stock of 0.1 g of fungal protease–peptidase complex of *Aspergillus oryzae* (Sigma P6110) in 1.0 mL of distilled water. There are ≥500 EU per g of the protease stock. Where a letter is shown, use the current standard from the first column as the stock.

13–2.2.3.2 Apparatus

As per 13–2.2.2.2 plus:

- Fine-mist sprayer or atomizer
- Infrared light (150 W)

13–2.2.3.3 Reagents

1. Substrate solution, 20 mM: Dissolve 2.928 g L-leucine 2-naphthylamide hydrochloride (Sigma L0376) into 500 mL 90% ethanol (Fisher A962–4).
2. Colorimetric indicator solution, 2.4 mM: Dissolve 0.8 g Fast Blue BB (Sigma F3378) Salt hemi (zinc chloride) salt in 400 mL deionized water. (This reagent solution should be used within 1 h after preparation.)

13–2.2.3.4 Standards

1. Make up stock solution by mixing 0.1 g of fungal protease–peptidase complex of *Aspergillus oryzae* (Sigma P6110) with 1.0 mL deionized water.
2. Use Table 13–2 to construct a series of standards.

13–2.2.3.5 Procedure: Before going to the Field

Follow Steps 1 to 3 for phosphatase imprinting in section 13–2.2.2.5, except that the cellulose sheets should be soaked in full-strength substrate solution only, not the color reagent.

13–2.2.3.6 Procedure: In the Field

1. Follow Steps 1 to 5 in section 13–2.2.2.6, with the following exception: the treated cellulose sheet needs to be in contact with the soil face for a shorter time than for imprinting of phosphomonoesterase, approximately 30 min.
2. That evening, after sampling, spray each imprint and calibration strip with freshly prepared color reagent using a fine mist. An atomizer is recommended.
3. Expose the imprints and calibration strips to a 150-W heating lamp for 60 s or to a 750-W lamp for 30 s to prevent nonspecific coloring.
4. Dry at room temperature overnight, then store in aluminum foil in a plastic zip bag.

13–2.2.3.7 Procedure: Upon Return to the Laboratory

Scan the imprint and calibration strips at a resolution of at least 300 dpi.

13–2.2.3.8 Comments

This assay results in relatively high levels of background color; however, this method still is useful for visually associating enzyme activity with specific plant structures or for targeted sampling of microbial communities in the soil.

Care must be taken in applying the colorimetric indicator solution as the substrate does not bind strongly to the cellulose sheet and will migrate in the cellulose if it becomes saturated by large droplets of solution. However, once the imprint is dry the induced color change on the imprint is stable for weeks.

See additional comments under phosphomonoesterase methods, above.

13–2.2.4 Imprinting of N-acetyl-β-glucosaminidase (EC 3.2.1.14) or β-glucosidase (EC 3.2.1.74) activity (Hoppe, 1983; Pritsch et al., 2004; Dong et al., 2007)

13–2.2.4.1 Principle

This is a semiquantitative method that uses the same basic approach as described in section 13–2.2.2.1, where a cellulose membrane is first saturated with the substrate, then incubated in the field. However, in this case the substrates fluoresce on hydrolysis, which is recorded on a fluorimeter in the laboratory. Methods to quantify enzyme activities based on the release of fluorescent 4-methylumbelliferone (4-MUB) have been developed for small soil samples and root tips (Hoppe, 1983; Sinsabaugh et al., 1991; Pritsch et al., 2004; Courty et al., 2005).

A method is described here for detecting two activities: N-acetyl- β-glucosaminidase or β-glucosidase because these have been studied on soil imprints (Dong et al., 2007). However, numerous 4-MUB-linked substrate analogs are now available which could be adapted for imprint studies. N-Acetyl-β-glucosaminidase activities are based on the hydrolysis of MU-N-acetyl-β-glucosaminide where as, β-glucosidase is based on the hydrolysis of 4-methylumbelliferyl-β-glucopyranoside.

13–2.2.4.2 Apparatus

As per section 13–2.2.2.2, plus a gel documentation system.

13–2.2.4.3 Reagents

1. N-Acetyl-β-glucosaminidase fluorometric indicator solution, 5 mM: Dissolve 1.04 g 4-methylumbelliferyl N-acetyl-β-D-glucosaminide (Sigma M2133) into 500 mL 2-methoxylethanol (Sigma M5378). This solution should be prepared fresh just before use.

2. β-Glucosidase fluorometric indicator solution, 5 mM: Dissolve 0.85 g 4-methylumbelliferyl β-D-glucoside (Sigma M3633) into 500 mL 2-methoxylethanol (Sigma M5378). This solution should be prepared fresh just before use.

13–2.2.4.4 Standards

1. Make a standard stock of 25 mM by dissolving 44.05 mg 4-methylumbelliferone (Sigma M1381) in 100 mL 2-methoxylethanol (Sigma M5378). This solution should be kept in a dark bottle wrapped in aluminum foil in a −20°C freezer.

2. Place 0, 31.25, 62.5, 125, and 250 µL of standard stock solution into 1.5-mL

Eppendorf microcentrifuge tubes and dilute to 1 mL with 2-methoxylethanol (Sigma M5378) to create standard concentrations of 0, 0.78, 0.156, 0.313, and 0.625 μM.

13–2.2.4.5 Procedure: Before Going to the Field

Follow Steps 1 to 3 in section 13–2.2.2.5, except that the cellulose sheets should be soaked in full-strength substrate solution only.

13–2.2.4.6 Procedure: In the Field

Follow Steps 1–5 in section 13–2.2.2.6, except that the treated cellulose sheet needs to be in contact with the soil face for a shorter time than for imprinting of phosphomonoesterase, approximately 30 min.

13–2.2.4.7 Procedure: Upon Return to the Laboratory

Capture images of air-dried imprints using a gel documentation system. Optimize the F-stop, exposure time, and binning settings under UV transillumination conditions without a filter.

13–2.2.4.8 Comments

MUB-based imprinting results in a detectable fluorescent residue on the soil profile for up to 8 d afterward. Therefore, allow at least a 10-d interval before any 4-MUB-based tests are repeated on the same area of a soil profile. See additional comments under methods for phosphomonoesterase above.

13–2.3 Correlating Activities with Observable Soil Structures

The following section describes how to process the images from the calibration strips to create a standard curve. Then relative enzyme activity detected in specific regions of the soil imprint is determined using a histogram analysis of color intensity and calibrating the color intensity in each pixel with the standard curve.

13–2.3.1 Apparatus

Image processing program such as Adobe Photoshop or GIMP.

13–2.3.2 Procedure

1. Scan the imprints and standards at high resolution (at least 300 dpi) and format as .tiff files.

2. Prepare the imprint for analysis using an image analysis program such as Adobe Photoshop or GIMP. First, sharpen the image (under Filter menu). Second, remove unwanted background color by selecting these pixels with the Magic Wand in Adobe Photoshop and then deleting them. Alternatively, remove any artifacts ("speckles") from the imprint by setting the minimum spot diameter to a level greater than the diameter of the speckles. Then, detect spots using a "spot detection" wizard. To do this, use the normal settings and then select a Gaussian distribution to remove any streaks or discoloration. A 3D-viewer in software can also help to identify artifacts that can be filtered from an image. Finally, use filters to remove any background discoloration.

3. If imprints were marked with capillary tubes, crop the image to remove the edges of the imprints, using the x, y coordinates marked by the capillary tube and adjust the size to 700 by 700 pixels.

4. To align the imprint image with an image of the soil taken through the Plexiglas, the soil image first needs to be sized to exactly match the size of the imprint image. Crop the soil window image using the five holes as a guide for the corner and center points. The window images should then be resized to 700 by 700 pixels, i.e., the same size as the imprint. Go to Step 6.

5. Alternatively, to align an imprint image with an image of the soil taken with the window open (in which case the capillary tubes cannot be used to align the images):

   a. Open the photo of the paper in place in the window in Adobe Photoshop.

   b. Select the polygon lasso tool.

   c. On a new layer, trace the outline of the paper visible in the photo.

   d. Cut the outlined area out of the layer.

   e. Open the photo of the soil taken with the window open.

   f. Copy the modified layer of the photo of the paper in place onto the photo of the soil as a new layer.

   g. Select the polygon lasso tool and trace on the photo layer the outline of the soil visible through the modified layer.

   h. Cut the outlined area out of the photo layer.

   i. Paste this cutout using "Paste as new image." This new image will be of the exact imprint area of the soil profile.

   j. Adjust the dimensions of the new image to match the dimensions of the scanned imprint.

   k. Rotate the image of the scanned imprint as needed to ensure that the imprint is in the original orientation to the soil profile.

   l. Copy the scanned imprint image and paste it into the image of the soil profile as a new layer.

6. Adjust the image to grayscale and 0 to 256 colors.

7. Filters and other tools can then be applied to the soil image to identify key features of interest, such as roots, fungi, or microbial hotspots. To confer an attribute value, such as root type, to the image, it is preferable to recolor the image by selecting desired pixels and changing to the same color to create a root type category. In the case of roots, a number of categories might be used in the classification—for example, long white roots, bifurcated root tips, long brown laterals, mycorrhizal roots, root clusters, "dead" roots, and so on. Categories should be reasonably easy to identify and have some functional significance. Where there are numerous images, resizing, sharpening, and optimizing can be undertaken to a large extent by automated batch actions in a range of software that processes images automatically with parameters you set.

8. To estimate activities associated with visible soil features of interest on the soil profile, first outline the feature with the lasso tool. Use the outlined area(s) to cut the scanned imprint layer. Paste the selected areas of the imprint using "Paste as new image." These can be evaluated as described under Calculations.

13–2.3.3 Calculations

1. To calibrate the imprint, first develop the standard curve by regressing the intensity of color development on slot blot membranes or calibration strips against the enzyme activities of the standards. To do this for the slot blot apparatus, select the core interior of the area developed for each working standard using the marquee tool in Adobe Photoshop set at a standard size (e.g., 60 by 6 pixels). For the calibration strips, select a consistent core area appropriate for the size of spot used. Record the mean color intensity per total number of pixels and then plot color intensity (0–256 in grayscale) against enzyme units or katal/pixel. A range of enzyme activities should be tested to determine the linear range of activity and to span the color intensities found on the imprints (for example, see Grierson and Comerford, 2000). At very low activities, an exponential curve should be fitted.

2. It is relatively simple to develop algorithms to analyze the image files for the correspondence of enzyme activity with observable soil structures or certain root classes as long as the images have been georeferenced. Similarly, the enzyme imprints can be reclassed according to user-defined categories related to known enzyme activities calculated from the standard curves (see above) or to calculate the percentage occurrence of a feature of interest with a specific class of enzyme activity. This is analogous to overlaying the enzyme imprint over the root image. To calibrate the intensity of color development in a region of the imprint with enzyme activities, select an image-defined area for integration, as described in Step 8, and use the histogram function in software such as NIH Image or Adobe Photoshop to obtain a frequency distribution of the number of pixels at each intensity. Then, using the standard curve, sum the product of the number of pixels reporting a particular grayscale value (0–256) by the enzyme concentration correlating with that grayscale value. For example, Grierson and Comerford (2000) classed the phosphomonoesterase imprints into five categories of activity, ranging from no activity (color value 0) to high intensity of activity (range 22.44×10^{-7} EU/pixel to 83.5×10^{-7} EU/pixel), and the roots into lateral roots and root clusters.

13–2.3.4 Comments

There are many software packages that may be suitable for the quantification of 2-D enzyme datasets, including those developed to analyze stained bands of proteins or nucleic acids on gels (e.g., PDQuest, Progenesis). In addition, image analysis software used routinely in GIS applications (IDRISI, ArcGIS, ENVI) is also useful for analyzing the spatial distribution of data captured on membranes and papers. Choice of software must take into consideration the number of images analyzed, the relative cost of the software, and the customer support. Researchers interested in quantifying relative enzyme activities from imprints are advised to try a number of programs. When used in conjunction with software designed for spatial analysis, this procedure should be useful in assessing the relationship of enzyme activity and root class (e.g., white, brown, lateral, mycorrhizal) and to make quantitative estimates of relative enzyme activity associated with root surface area, degree of mycorrhizal infection, or microbial community composition. However, the preparation of root images suitable for spatial analysis can be time-consuming. Likewise, there are some logistic difficulties applying the method to a field situation. For example, both root and phosphomonoesterase imprints must be the exact same dimensions with sufficient coordinates to align the images. The development of software and programs to simplify the process of analyzing the image data is ongoing, and will improve the accuracy and speed of the image analysis.

13–2.4 Use of Imprints for Guiding Soil Sampling for Microbial Community Analysis

13–2.4.1 Principle

Colorimetric soil enzyme imprinting provides a unique opportunity to investigate the microbial community associated with in situ enzyme activity at mm-scale resolution. Fine-scale soil sampling can proceed directly after development of the imprint using a frame to maintain alignment between the imprint and a transparent sheet used as a template. Holes punched in the transparency align with the enzyme activities detected by the imprinting and, hence, can be used to target sampling of roots or soil.

13–2.4.2 Apparatus

- Foam core poster board cut to fit the opening of sampling window, 30 by 30 cm
- Mylar plastic sheets, 30 by 30 cm
- Easy-release paper tape (masking tape)
- Fine-tipped permanent marking pens of several colors
- Hole-punch tool (5-mm, hammer-driven leather punch)
- Toothpicks
- Medium-tipped forceps, 75-mm length
- 70% ethanol for sterilizing forceps and doors of sample windows
- Paper towels
- Sterile microcentrifuge tubes, 1 mL
- Dry ice
- Insulated container to hold dry ice, approximately 40 L
- Materials for imprinting as described in earlier sections

13–2.4.3 Procedure

1. Cut a transparent sheet of Mylar to exactly fit into the opening of the sampling window. Use a piece of foam core poster board cut to fit exactly into the open window to stabilize the transparency with the hinged door open. First cut the foam core poster board precisely to size. Then cut out the center leaving a frame 6 cm wide and tape the transparency to this.

2. Before creating an imprint, fit the frame and transparent sheet into the opening against the soil profile. If relevant to the sampling scheme, trace any visible soil features of interest, such as the interface between organic and mineral soil horizons, fungal mats, or plant roots onto the transparency with a fine permanent marker. Remove the transparency and frame from the window, and remove the transparency from the frame.

Fig. 13–2. Sampling soil microbial community guided by enzyme imprints. (A) Preparing to mark colored imprint areas onto the transparency; (B) Marking sampling areas through holes punched in transparency; (C) Removing transparency; (D) Sampling marked soil profile.

3. Open the hinged door of the sampling window, clean the surface that had been against the soil by spraying with 70% ethanol, dry it, tape the cellulose sheet to the door, and close the door. Follow the instructions for the particular enzyme assay in earlier sections of this chapter.

4. At the end of the assay, place the transparency, soil side up, on top of the cellulose sheet while the sheet is still attached to the trap door aligning the transparency precisely with the edges of the trap door (Fig. 13–2a). Mark selected areas of color development on the transparency with permanent marker. The types of areas marked will depend on the question being asked. For example, some researchers may be interested in only those areas of activity in proximity to roots; others may be interested in activities in certain soil horizons.

5. Remove the transparency from the hinged door and punch holes through the transparency at locations of interest using a sharp, hammer-driven leather punch.

6. Reattach the transparency onto the frame exactly as it was when the soil features were originally marked and replace it in the opening of the root window (Fig. 13–2b). Because the frame is the same size and shape as the hinged door, the punched holes in the transparency are now aligned with the soil features that originally produced the color change on the imprint. Insert toothpicks through the holes in the transparency to mark the locations for targeted soil sampling, and remove the transparency and frame (Fig. 13–2c).

7. Collect soil samples with sterile forceps (Fig. 13–2d) and place them in sterile microcentrifuge tubes. Keep on dry ice during transport to the lab.

13–2.4.4 Comments

Soil fungal and prokaryotic communities associated with hot spots of soil enzyme activity can be investigated and described by combining targeted soil sampling with molecular techniques. DNA can be extracted from small (less than 0.25 g), targeted soil samples using standard DNA extraction kits (e.g., PowerSoil DNA Isolation Kit, MoBio). Genes of interest can be amplified by PCR, and the PCR products can be rapidly and economically sequenced by robotic clone sequencing, or by pyrosequencing. Online databases (e.g., NCBI, RDP, GreenGenes) expedite the process of identifying and categorizing sequences obtained through high-throughput sequencing methods, providing in-depth information about the presence and/or activity of functional genes and microbial species that co-occur with in situ soil enzyme activity.

13–3 EXTRACELLULAR ENZYME ACTIVITY PROFILES OF INDIVIDUAL EXCISED ABSORBING ROOTS
(Pritsch et al., 2004; Buée et al., 2005; Courty et al., 2005)
13–3.1 Principle

This method involves excavation and separation of plant roots or fungal hyphae from soil particles that are then individually incubated in the presence of a substrate. It is assumed these hyphae are not excreting enzymes and the activity is of enzymes on hyphal or root surfaces.

The method is rapid and sensitive, and results in minimal damage to root tips. The method requires only small amounts of root material and, therefore, is an efficient tool for large-scale ecological studies. Tests for activities of eight enzymes have been developed thus far: hydrolases involved in the degradation of hemicelluloses (glucuronidase [EC 3.2.1.31], xylosidase [EC 3.2.1.37]), cellulose (cellobiohydrolase [EC 3.2.1.91], β-glucosidase), chitin (N-acetylglucosaminidase), or in the mobilization of phosphate bound to organic compounds (phosphomonoesterase), or in the degradation of proteins and peptides (leucine aminopeptidase); as well as an oxidative enzyme, laccase (EC 1.10.3.2) (Table 13–3). Laccase is a term grouping all enzymes (ascorbate oxidases, laccases, ferroxidases, etc.) involved in the oxidation of various aromatic (particularly phenolic substrates) from organic compounds.

The activities of the eight enzymes are measured successively on the same root tips. Seven activities are measured using fluorogenic substrates: β-glucuronidase, xylosidase, cellobiohydrolase, chitinase, β-glucosidase, acid phosphatase,

Table 13–3. Enzymes tested and their corresponding substrates used in the microplate assays using Procedures 1 and 2. All fluorogenic substrates are esters of 4-methylumbelliferone (4-MU) or 7-amino-methylcoumarin (7-AMC). Esters of 4-MU or 7-AMC do not fluoresce unless cleaved to release the fluorophore. Fluorometric assays are based on the hydrolysis of 4-MU or 7-AMC containing substrates. The abbreviations for substrates are given in parenthesis following each substrate.

Enzyme	EC No.	Substrates
Leucine aminopeptidase	3.4.11.1	L-leucine 7-AMC (Leu-AMC)
Xylosidase	3.2.1.37	4-MU β-D-xylopyranoside (MU-X)
Glucuronidase	3.2.1.31	4-MU-β-D D-glucuronide hydrate (MU-GU)
Cellobiohydrolase	3.2.1.91	4-MU β-D-cellobioside (MU-C)
N-Acetylglucosaminidase	3.2.1.14	4-MU N-acetyl-β-glucosaminide (MU-NAG)
β -Glucosidase	3.2.1.74	4-MU β-D-glucopyranoside (MU-G)
Phosphomonoesterase	3.1.3.2	4-MU phosphate free acid (MU-P)
Laccase	1.10.3.2	Diammonium 2,2'-azino-bis(3-ethylbenzothiazoline-6-sulfonate) (ABTS)

and leucine aminopeptidase. The substrates are linked with a fluorescent molecule, either methylumbelliferone (MU) or aminomethylcoumarin (AMC). Both MU and AMC have optimal excitation at 364 nm and optimal emission at a wavelength of 445 nm (Hoppe 1983). Excitation and emission wavelengths close to the optimum (±10 nm) also can be used. Esters of MU or AMC do not fluoresce unless cleaved to release the fluorophore.

Fluorimetric enzyme assays are based on the hydrolysis of MU-containing substrates by the specific enzyme (i.e., cleavage of MU-glucopyranoside by β-glucosidase). The yield of the released fluorescent molecule (MU or AMC) is proportional to the substrate turnover rate by the enzyme. Polyphenol oxidase (laccase) activity is the only one measured by colorimetry, using ABTS (diammonium 2,2'-azinobis-3-ethylbenzothiazoline-6-sulfonate) as a substrate. The intensity of the green color of the ABTS radical formed by oxidation is related to the substrate turnover. Of the eight enzymes studied to date, five (laccase, chitinase, xylanase, acid phosphatase, and β-glucosidase) are known to be associated with fungal cell walls (Rast et al., 2003) and at least four enzymes (β-glucosidase, chitinase, cellobiohydrolase, and phosphatase) were found to be stable over the assay time (Pritsch et al., 2004).

All activities are calculated per mm² of projected surface area of individual root tips, determined with image analysis software (Buée et al., 2005). Projected surface areas of ectomycorrhizal root tips are linearly correlated with the surface areas of the ectomycorrhizae considered as cylinders, and with their dry weights (Buée et al., 2005).

Two procedures based on the same measurement principle can be applied, depending on the equipment available. Procedure 1 (section 13–3.4) is based on the transfer of roots from one incubation solution to the next using sieve strips or sieve plates. Procedure 2 (section 13–3.5) transfers the liquids using commercially available filter plates and a vacuum manifold for microplates. All chemicals, concentrations, and measurement conditions are the same for both procedures. They differ primarily in how the samples are handled.

13–3.2 Materials Common to Procedure 1 and 2
13–3.2.1 Apparatus

- Stereomicroscope for selecting roots to be assayed
- Microplate reader for fluorescent assays, with an excitation wavelength in the range of 355 to 375 nm and an emission wavelength in the range of 430 to 460 nm
- Microplate reader for photometric assays ($\lambda = 415–425$ nm)
- WinRhizo software (Regent Instruments, Inc., Quebec, Canada), or any other image analysis software suitable for determining the projection area of the assayed root tips
- Transillumination scanner with sufficiently high resolution (at least 800 dpi)
- Microplate shaker with controlled air temperature (e.g., 21–25°C)
- Multichannel pipette (8 channels)
- Black and clear flat-bottom 96-well microtitration plates
- 50-mL disposable plastic tubes with screw top to prepare and store substrate solutions
- 1-L autoclavable bottles for preparation of sterile buffers
- Aluminum foil for light protection of chemicals

13–3.2.2 Reagents

1. Modified universal buffer stock solution: Weigh the following ingredients in a 1-L glass bottle: Tris (2-Amino-2-(hydroxymethyl)-1,3-propanediol) 12.1 g, maleic acid 11.6 g, citric acid 14.0 g, and boric acid 6.3 g. Add 488 mL of so-

Table 13–4. Preparation of substrate stock solutions used in microplate assays. Abbreviations for substrates: Leu-AMC, L-leucine 7-AMC; MU-X, 4-MU β-D-xylopyranoside; MU-GU, 4-MU-β-D D-glucuronide hydrate; MU-C, 4-MU β-D-cellobioside; MU-NAG, 4-MU N-acetyl-β-glucosaminide; MU-G, 4-MU β-D-glucopyranoside; MU-P, 4-MU phosphate (free acid); ABTS, Diammonium 2,2'-azino-bis(3-ethylbenzothi-azoline-6-sulfonate). Caution: formula weight can change depending on hydration of the chemicals.

Substrate	Formula Weight (g mol⁻¹)	Stock solution (mM)	Substrate needed for 10 mL of stock solution (mg)	Stock solution needed to give 10 mL of working solution (mL)	Working solution (mM)
Leu-AMC	324.8	5	16.2	1.2	1.2
MU-X	308.3	5	15.4	1.5	1.5
MU-GU	352.3	5	17.6	1.5	1.5
MU-C	500.5	5	25.0	1.2	1.2
MU-NAG	379.4	5	19.0	1.5	1.5
MU-G	338.3	5	16.6	1.5	1.5
MU-P	256.2	5	12.8	2.4	2.4
ABTS	548.7	2	11.0	—	—

dium hydroxide (1 M) and adjust with deionized water to 1000 mL. Dissolve the ingredients and store the solution in a refrigerator.

2. Dilution buffer: Place 150 mL of the modified universal buffer stock solution in a 1-L bottle. Add hydrochloric acid (0.1 M) or sodium hydroxide (0.1 M) as needed to reach the desired pH and take to volume with deionized water. Use the dilution buffer to dilute the substrate working solutions. Sterilize the buffer (autoclaving at 121°C for 20 min) and store at room temperature.

3. Rinsing buffer: Place 100 mL of the modified universal buffer stock solution in a 1-L bottle. Add hydrochloric acid (0.1 M) or sodium hydroxide (0.1 M) as needed to reach the desired pH and take to volume with deionized water. Use the dilution buffer to dilute the substrate working solutions. Sterilize the buffer (autoclaving at 121°C for 20 min) and store at room temperature.

4. Substrate stock solutions for all fluorogenic substrates: Prepare 5 mM stock solutions by dissolving each substrate in 10 mL of 2-methoxyethanol (Hoppe, 1983), using the amounts given in the fourth column of Table 13–4. The amounts of the substrates listed in Table 13–4 are valid for only the fresh weights specified. Check the fresh weights given by the manufacturer for each batch and adjust the amount used accordingly. The substrates may not dissolve immediately but may need ultrasonification or mild shaking for several hours.

5. Substrate working solutions of fluorescent substrates: Dilute the 5 mM stock solutions of the fluorogenic substrates with sterile ultrapure water to the desired working concentrations (Table 13–4).

6. Incubation solutions of the fluorogenic substrates: Mix one part of the working solutions with two parts of dilution buffer of the corresponding pH (Table 13–5).

7. Incubation solution of ABTS: Dissolve diammonium 2,2'-azinobis-3-ethyl-benzothiazoline-6-sulfonate in rinsing buffer pH 4.5.

8. Stopping buffer: To prepare the stopping buffer (Tris 1 M pH 10–11) used to stop and alkalinize the fluorescence assays solutions at the end of the incubation, dissolve 121 g Tris in 1 L ultrapure water. Sterilization is not neces-

Table 13–5. Working and incubation concentrations, pH and incubation duration for enzyme assays on roots on microplates. Abbreviations for substrates: Leu-AMC, L-leucine 7-AMC; MU-X, 4-MU β-D-xylopyranoside; MU-GU, 4-MU-β-D D-glucuronide hydrate; MU-C, 4-MU β-D-cellobioside; MU-NAG, 4-MU N-acetyl-β-glucosaminide; MU-G, 4-MU β-D-glucopyranoside; MU-P, 4-MU phosphate (free acid); ABTS, Diammonium 2,2'-azino-bis(3-ethylbenzothiazoline-6-sulfonate).

Order	Substrate	Working solution (μM)	Incubation Solution (mM)	Assay pH	Calibration solution	Incubation time (min)
1	Leu-AMC	1.2	0.4	6.5	AMC	60
2	MU-X	1.5	0.5	4.5	MU	50
3	MU-GU	1.5	0.5	4.5	MU	30
4	MU-C	1.2	0.4	4.5	MU	30
5	MU-NAG	1.5	0.5	4.5	MU	15
6	MU-G	1.5	0.5	4.5	MU	15
7	MU-P	2.4	0.8	4.5	MU	20
8	ABTS	–	0.67	4.5	–	60

sary; store at room temperature.

13–3.2.3 Standards

1. Methylumbelliferone (MU) stock solution: Dissolve 17.62 mg MU in 10 mL of 2-methoxyethanol resulting in a 10 mM stock solution that can be frozen at –20°C. To prepare calibration solutions, dilute the 10 mM stock solution 1000-fold (e.g., by diluting 50 µL of stock in a volume of 50 mL ultrapure water), resulting in a 10 µM solution. To prepare 10 mL of calibration solutions of 1, 2, 3, 4, and 5 µmol L^{-1} concentrations, use 1, 2, 3, 4, or 5 mL and adjust each with ultrapure water to 10 mL. Use 10 mL of ultrapure water for the "0" calibration solution. The resulting calibration solutions contain 0, 100, 200, 300, 400, 500 pmol MU in 100 µL, respectively. Calibration solutions can be used for 1 wk when stored at 4 to 6°C and protected from light. Diluted calibration solutions are not stable when frozen at –20°C.

2. Aminomethylcoumarin stock solution. Dissolve 17.52 mg in 10 mL of 2-methoxyethanol resulting in a 10 mM stock solution and prepare calibration solutions as described for MU.

13–3.3 Collection and Processing of Root Tips

Roots should be collected in a manner that leaves them in contact with the surrounding soil (Courty et al., 2005). For example, if entire tree seedlings or herbaceous plants are sampled, then the root system should be sampled with a spade or shovel, and both the roots and surrounding soil should be placed in a plastic bag. Soil cores, if used, should be transported intact to the laboratory. Roots should be placed on ice or otherwise kept cool during transport to the lab. Conduct the assays as soon as possible, but not more than 4 wk after sampling

13–3.4 Procedure 1

Sieve strips or sieve plates are used to transfer tips simultaneously from one 96-well microplate to another, allowing the same tips to be incubated, rinsed for 3 min, re-incubated, and rinsed until the end of an experimental series.

13–3.4.1 Apparatus

In addition to what is given in 13–3.2.1 for Procedure 1 and 2:

- Polymerase chain reaction (PCR) tubes or alternatively 96-well PCR plate to hold the root tips
- Nylon mesh, 100- to150-µm mesh size
- Scalpel, razor blades
- Sieve strips: Cut 2 to 3 mm from the bottom of each of eight-well PCR reaction tubes (Fig. 13–3). Experiment to determine the exact length to be cut off so that the tops fit tightly to the bottoms, without requiring glue. Cut small discs (approx. 3–4 mm in diameter) of the nylon mesh to exactly fit into the top of a tube. Lay the cut disks into each top and then fit each top to a bottom of a tube (Fig. 13–3). Proceed until all eight tops are fixed to the bottoms of the eight tubes per strip. Cut off the top layer of each of the eight lids so that solutions can be pipetted directly into the tubes. For this purpose, use

A. Cut 2-3 mm from the bottom of each of eight-well polymerase chain reaction (PCR) reaction tubes

Fig. 13–3. Preparation of strip tubes for assays on individual root tips.

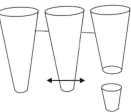

B. Cut small discs (~3-4 mm diameter) of the nylon mesh to exactly fit into the cap of a tube (a). Cut off the top layer of a cap (b) and insert the disc of nylon mesh (c)

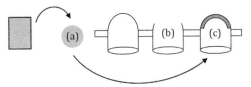

C. Fix each cap to a bottom of a tube by pressure

a stereomicroscope or a strong hand lens. Make sure that the sieve strips loosely fit into a row of eight wells in a flat-bottomed 96-well plate. If not, remove all superfluous parts with a sharp razor blade or a scalpel under a stereomicroscope. Make sure that the hole covered by the mesh is big enough (diameter approx. 2 mm) to let liquids easily pass through. Twelve such sieve strips are required for one entire plate.

- Sieve plates: Alternatively, whole 96-well sieve plates, made from PCR plates and fitting in the microtitration plates, can be used (J.-L. Churin, personal communication, 2008). This makes the transfers faster and easier, but creating an equivalent opening in the bottom of each tube and attaching the nylon mesh is more difficult than for the one-row strips. One approach is to use a hot household iron to seal the mesh onto the bottom of the tubes and then a drill to trim away the excess mesh.

13–3.4.2 Microplate Assays

1. Prepare a separate incubation plate for each assayed enzyme: black plates for the fluorogenic assays and clear plates for the photometric assay for laccase activity. Add 150 µL incubation solution per well, leaving six wells empty for calibration solutions.

2. Prepare the stopping plates for the fluorescence tests, one for each incubation plate, by placing 150 µL of stopping buffer in each well. Add 100 µL calibration solutions (0, 1, 2, 3, 4, 5 µM) to six wells at the position corresponding to the empty wells on the incubation plates. Prepare one plate with AMC for the leucine aminopeptidase assay and six plates with MU for the MU-labeled substrates. The calibration tubes can be sealed with tape to prevent pipetting errors later on in the procedure.

3. Prepare rinsing plates: one plate containing 150 µL of rinsing buffer at pH 6.5, and six plates with 150 µL of rinsing buffer at pH 4.5.

4. Place the sieve strips in a rinsing plate containing 150 µL of 50 mmol L^{-1} rinsing buffer, which should have the same pH as the subsequent enzyme assay (Table 13–5).

5. Carefully wash the fine root samples in tap water to remove soil, and examine in cold water using a stereomicroscope. For each type of root tip to be assayed, excise tips of 2 to 4 mm in length; be sure to remove any soil remaining around each tip using fine forceps. Put each tip into a separate tube. Leave one tube without a tip in each strip or column of a plate to serve as a control.

6. When all root tips are present, remove the sieve strips or sieve plate from the rinsing plate, blot on an absorbing paper to remove the liquid remaining at the bottom and place in the first incubation plate. It is important to note that fluorescent assays should be done first and the colorimetric assay should be performed last. Although the order of the fluorescent assays is not important for the activity, the order given in Table 13–5 is recommended since this is the order of increasing activity. This will reduce the possibility of carryover fluorescence.

7. Place the first incubation plate on the incubation shaker at 21°C. It is important to protect the plate from light because AMC and MU substrates are light sensitive.

8. At the end of the incubation period (Table 13–5), remove the sieve strips, blot on absorbing paper, and place in a rinsing plate for 5 min.

9. Remove 100 µL from each of the eight wells in a row of the incubation plate with the multichannel pipette and add to a row of wells in the stopping plate, leaving out the wells assigned for calibration.

10. Remove the sieve strips from the rinsing plate, blot on absorbing paper, and place in the incubation plate containing the next substrate to be tested. Repeat Steps 7 to 9 for each of the seven fluorogenic substrates.

11. Once a fluorescence assay is finished, the stop plate to which the incubation solutions, substrate controls, and the calibration solutions have been added can now be measured at 364 ± 10 nm excitation and 445 ± 15 nm emission in a fluorescence microplate reader. The fluorescence signal is stable over several hours. Alternatively, all stop plates can be measured at the end of the series. Adjust the other instrument settings, such as gain and sensitivity, for the specific fluorescence spectrometer being used. A slit width of 5 nm, a collection time of 0.5 s, and the sensitivity adjusted according to the strength of the signal has proven optimal.

12. Perform the colorimetric test (for laccase, with ABTS) only after the seven fluorescent assays have been done. Transfer the sieve strips into the clear plate containing the ABTS solution and incubate at 21°C for 1 h. Then re-

move the sieve strips, blot on absorbing paper, and place in a rinsing plate. Transfer 120 µL of the incubation solution into wells on a new clear plate. Measure the intensity of the green color that has developed immediately at 420 ± 5 nm.

13. After performing all eight tests, transfer the root tips carefully with fine, soft tweezers into a clear flat-bottom 96-well microtitration plate containing 100 µL of water per well. Scan this plate and calculate the projected surface area of each root tip using the WinRhizo software.

13–3.4.3 Calculations

Enzyme activities (EA) are calculated from fluorimeter and photometer readings by the following formula:

$$EA = \frac{x - sub}{a \quad pa \quad t} \quad \frac{vol_{tot}}{vol_{meas}} \qquad [1]$$

where x is the measured value of the sample, sub is the measured value with substrate, but without sample, pa is the projection area of the mycorrhizal roots [mm^2], t is the incubation time [min], vol_{tot} is the total volume of the incubation solution per well of an incubation plate, and vol_{meas} is the volume of the incubated solution that was finally measured.

For fluorescence measurements, a is the slope of the regression line of the calibration curve [pmol^{-1}].

For the ABTS test

$$a = \varepsilon_{425} \cdot pl \qquad [2]$$

where ε_{425} is the molar coefficient of extinction for ABTS ($\varepsilon_{425} = 3.6 \cdot 10^4 \, cm^2 \, mol^{-1}$) and pl is the path length within the liquid of the well [cm]. If the photometer does not measure the path length, it can be approximated by dividing the volume (120 µL) added to each well by the circular area ($\pi \cdot r^2$) of the well. EA is expressed as [mol mm^{-2} min^{-1}] released AMC or MU or ABTS, respectively. Unit of "pmol" or "nmol" should be used according to the range of measured values.

13–3.5 Procedure 2

An alternative method for assaying enzyme activities associated with individual roots in a microplate assay is the use of manufactured filter plates in combination with a vacuum manifold. The root tips are placed in the filter plates to which the substrate solutions are added. The liquid is removed after incubation by placing the filter plate on a receiver plate that is positioned in a vacuum chamber. By applying the vacuum to the chamber, all liquid in the upper filter plate is quantitatively transferred to the receiver plate below. The advantage is that this procedure is generally quicker and needs a lower volume of incubation solution; however, it requires additional equipment.

13–3.5.1 Additional Apparatus

- Filter plates (AcroPrep 96 filter plate with 30–40-μm mesh size, Pall, Life Sciences, UK)
- Tape to seal superfluous wells in the filter plate
- Vacuum manifold for microplates
- Laboratory pump to be connected to the vacuum manifold

13–3.5.2 Procedure for Microplate Assays

1. Prepare the stopping plates for the fluorescence tests as described in Step 2 under section 13–3.4.2. Include also calibration solutions and protect the plates from light.

2. Prepare root tips as described under Step 5, section 13–3.4.2.

3. First seal the six calibration wells with laboratory labeling tape. Always support the filter plate on top of an empty plate. Contact between the bottom of each well with a surface will cause leakage from the filter plate. Fill all the remaining wells of the filter plate with 100 μL of rinsing buffer of the pH corresponding to the pH of the enzyme assay that is performed next (Table 13–5).

4. Place a single root tip into each individual well of the filter plate, leaving 12 wells free of roots: the six sealed calibration wells and six other wells are to be used for negative controls. Take care that the calibration wells in the filter plates correspond to the ones in the stopping plates.

5. Place an empty microplate into the chamber of the vacuum manifold, and place the filter plate containing root tips and buffer on top of the chamber. Apply the vacuum to transfer buffer from the filter plate to the empty microplate. Place the filter plate on top of an empty support plate and add 100 μL of the desired incubation solution to each well except for the sealed calibration wells. Place the filled filter plate together with the empty support plate on a microplate shaker and incubate. Protect the filter plate from light.

6. Meanwhile, remove the rinsing plate from the vacuum chamber and discard the contents. The emptied rinsing plate can be reused for the next rinsing step.

7. Place a stopping plate containing the stop buffer and the calibration solution corresponding to the running assay (Table 13–5) into the vacuum chamber. When the incubation time is finished, place the filter plate containing the root tips and the incubation solution on top of the vacuum chamber. Make sure that the stopping plate in the chamber and the filter plate on top have the same orientation. Apply vacuum to transfer the liquids from the filter plate to the stopping plate. The stopping plate now contains 84 samples plus six negative controls and six calibration wells.

8. Remove the stopping plate from the vacuum chamber and protect it from light until measuring it.

9. Place the filter plate containing the root tips on a support plate and add 100 μL of rinsing buffer with the pH appropriate for the next assay to rinse the root tips. Start again at Step 5, section 13–3.5.2, and continue until all fluorescence assays are performed.

10. For the laccase test with ABTS, add 120 μL of incubation solution to the rinsed root tips in the filter plate and incubate on a microplate shaker as for the other substrates. After the incubation time, place a transparent 96-well plate into the vacuum chamber and apply the vacuum as before.

11. Perform the measurements of the photometric and fluorescence tests as described in Step 11, section 13–3.4.2.

12. Transfer root tips to transparent microplates to determine their projection area as described in Step 14, section 13–3.4.2.

13. Use the same calculations to express enzyme activity as given in section 13–3.4.3.

13–3.5.3 Comments

Each lab group should establish the ideal pressure for the pump before starting an experiment. The pressure should be strong enough to transfer the liquid as rapidly as possible into the lower wells, but not so strong that the liquid ends up outside the wells.

The incubation times shown here have been optimized for ectomycorrhizal root tips and therefore may have to be adjusted for other root samples. To date, these multiple enzymatic tests have been applied mainly to ectomycorrhizal communities (Buée et al., 2007; Courty et al., 2005, 2006, 2007, and 2008; Mosca et al., 2007), but they easily can be adapted to studying any other type of roots or mycelia (Pritsch et al., 2006; Courty et al, 2011; Pritsch et al., 2011), e.g., nonmycorrhizal roots, arbuscular mycorrhizae, ericoid or orchid mycorrhizae, bacterial nodules, or cluster roots.

Four types of limitations must be considered when interpreting the results of root-tip assays conducted under laboratory conditions.

First, to compare samples from different origins, the tests are run under standard physicochemical conditions (pH, temperature) close to the optimum, rather than under the ever-changing and often limiting conditions prevailing in situ. Therefore, the values measured are not the actual activities expressed in the soil at the time of sampling, but *potential* activities, i.e., the ability of root tips to quickly resume the assessed function whenever environmental conditions are favorable.

Second, excised root tips theoretically have a metabolism different from that of undisturbed ones still attached to the root system of a plant and receiving photosynthates from its leaves. However, this bias is likely to be negligible for two reasons: (i) the short handling and incubating time of the root tips limit the depletion of their carbon reserves (Ritter et al., 1986) and (ii) at least for ectomycorrhizae, the secreted enzymes tested in the procedures described here are either bound to the fungal cell walls of the ectomycorrhizal mantle or very slowly released, and thus do not undergo rapid, active secretion during the incubation time. However, this could be an issue when applying these tests to other types of roots.

Third, when applying the test to ectomycorrhizae, it has to be kept in mind that the extramatrical mycelium plays a major role in nutrient mobilization and uptake by ectomycorrhizae (Leake et al., 2004) and that there might be specific differentiation of mycelia depending on fungal species and nutrient pools in soil. The assays described here determine enzymatic activities in the close vicinity of ectomycorrhizae deprived of most of their emanating hyphae and strands after being extracted from the soil. Whether these activities correlate with the corresponding activities of whole ectomycorrhizae, including their complete extramatrical mycelium, or whether parts of mycelia remote from the roots display activities other than those close to the roots, is a challenging question for further studies. A way to approach this question would be to compare, in the same experiment, this method with excised root tips with the "root window" method described in section 13–2.

Finally, ectomycorrhizal root tips are considered here as multitrophic associations, for which overall enzyme activity is a result of the combined contributions of the root tissues, the symbiotic fungus, and the ectomycorrhiza-associated bacteria (Frey-Klett et al., 2007). Shorter incubation time would limit bias induced by changes in bacterial populations during the assay.

REFERENCES

Allison, S.D., and J.D. Jastrow. 2006. Activities of extracellular enzymes in physically isolated fractions of restored grassland soils. Soil Biol. Biochem. 38:3245–3256. doi:10.1016/j.soilbio.2006.04.011

Aon, M.A., M.N. Cabello, D.E. Sarena, A.C. Colaneri, M.G. Franco, J.L. Burgos, and S. Cortassa. 2001. I. Spatio-temporal patterns of soil microbial and enzymatic activities in an agricultural soil. Appl. Soil Ecol. 18:239–254. doi:10.1016/S0929-1393(01)00153-6

Arocena, J.M., A. Göttlein, and S. Raidl. 2004. Spatial changes of soil solution and mineral composition in the rhizosphere of Norway-spruce seedlings colonized by *Piloderma croceum*. J. Plant Nutr. Soil Sci. 167:479–486. doi:10.1002/jpln.200320344

Blackwood, C.B., C.J. Dell, A.J.M. Smucker, and E.A. Paul. 2006. Eubacterial communities in different soil macroaggregate environments and cropping systems. Soil Biol. Biochem. 38:720–728. doi:10.1016/j.soilbio.2005.07.006

Buée, M., P.-E. Courty, D. Mignot, and J. Garbaye. 2007. Soil niche effect on species diversity and catabolic activities in an ectomycorrhizal fungal community. Soil Biol. Biochem. 39:1947–1955. doi:10.1016/j.soilbio.2007.02.016

Buée, M., D. Vairelles, and J. Garbaye. 2005. Year-round monitoring of diversity and potential metabolic activity of the ectomycorrhizal community in a beech (*Fagus silvatica*) forest subjected to two thinning regimes. Mycorrhiza 15:235–245. doi:10.1007/s00572-004-0313-6

Bundt, M., S. Zimmerman, P. Blaser, and F. Hagedorn. 2001. Sorption and transport of metals in preferential flow paths and soil matrix after the addition of wood ash. Eur. J. Soil Sci. 52:423–431. doi:10.1046/j.1365-2389.2001.00405.x

Courty, P.-E., N. Breda, and J. Garbaye. 2007. Relation between oak tree phenology and the secretion of organic matter degrading enzymes by *Lactarius quietus* ectomycorrhizas before and during bud break. Soil Biol. Biochem. 39:1655–1663. doi:10.1016/j.soilbio.2007.01.017

Courty, P.-E., A. Franc, J.-C. Pierrat, and J. Garbaye. 2008. Temporal changes in the ectomycorrhizal community in two soil horizons of a temperate oak forest. Appl. Environ. Microbiol. 74:5792–5801. doi:10.1128/AEM.01592-08

Courty, P.E., J. Labbé, A. Kohler, B. Marçais, C. Bastien, J.L. Churin, J. Garbaye, and F. Le Tacon. 2011. Effect of poplar genotypes on mycorrhizal infection and secreted enzyme activities in mycorrhizal and non-mycorrhizal roots. J. Exp. Bot. 62:249–260.

Courty, P.-E., R. Pouységur, M. Buée, and J. Garbaye. 2006. Laccase and phosphatase activities of the dominant ectomycorrhizal types in a lowland oak forest. Soil Biol. Biochem. 38:1219–1222. doi:10.1016/j.soilbio.2005.10.005

Courty, P.-E., K. Pritsch, M. Schloter, A. Hartmann, and J. Garbaye. 2005. Activity profiling of ectomycorrhiza communities in two forest soils using multiple enzymatic tests. New Phytol. 167:309–319. doi:10.1111/j.1469-8137.2005.01401.x

Dick, R.P. 1997. Soil enzyme activities as integrative indicators of soil health. p. 121–156. In C.E. Parkhurst, B. Doube and V. Gupta (ed.) Biological indicators of soil health. CAB International, New York.

Dieffenbach, A., and E. Matzner. 2000. In situ soil solution chemistry in the rhizosphere of mature Norway spruce (Picea abies [L.] Karst.). Plant Soil 222:149–161. doi:10.1023/A:1004755404412

Dinkelaker, B., and H. Marschner. 1992. In vivo demonstration of acid phosphatase activity in the rhizosphere of soil-grown plants. Plant Soil 144:199–205. doi:10.1007/BF00012876

Dong, S., D. Brooks, M.D. Jones, and S.J. Grayston. 2007. A method for linking in situ activities of hydrolytic enzymes to associated organisms in forest soils. Soil Biol. Biochem. 39:2414–2419. doi:10.1016/j.soilbio.2007.03.030

Dorodnikov, M., E. Blagodatskaya, S. Blagodatsky, S. Marhan, A. Fangmeier, and Y. Kuzyakov. 2009. Stimulation of microbial extracellular enzyme activities by elevated CO_2 depends on soil aggregate size. Glob. Change Biol. 15:1603–1614. doi:10.1111/j.1365-2486.2009.01844.x

Egli, S., and I. Kalin. 1990. The root window—A technique for observing mycorrhizae on living trees. Agric. Ecosyst. Environ. 28:107–110. doi:10.1016/0167-8809(90)90023-7

Fansler, S.J., J.L. Smith, H. Bolton, and V.L. Bailey. 2005. Distribution of two C cycle enzymes in soil aggregates of a prairie chronosequence. Biol. Fertil. Soils 42:17–23. doi:10.1007/s00374-005-0867-2

Frey-Klett, P., J. Garbaye, and M. Tarkka. 2007. The mycorrhiza helper bacteria revisited. New Phytol. 176:22–36. doi:10.1111/j.1469-8137.2007.02191.x

Gaillard, V., C. Chenu, S. Recous, and G. Richard. 1999. Carbon, nitrogen and microbial gradients induced by plant residues decomposing in soil. Eur. J. Soil Sci. 50:567–578. doi:10.1046/j.1365-2389.1999.00266.x

George, T.S., B.L. Turner, P.J. Gregory, B.J. Cade-Menun, and A.E. Richardson. 2006. Depletion of organic phosphorus from oxisols in relation to phosphatase activities in the rhizosphere. Eur. J. Soil Sci. 57:47–57. doi:10.1111/j.1365-2389.2006.00767.x

Göttlein, A., and H. Stanjek. 1996. Micro-scale variation of solid-phase properties and soil solution chemistry in a forest podzol and its relation to soil horizons. Eur. J. Soil Sci. 47:627–636. doi:10.1111/j.1365-2389.1996.tb01861.x

Grierson, P.F., and N.B. Comerford. 2000. Non-destructive measurement of acid phosphatase activity in the rhizosphere using nitrocellulose membranes and image analysis. Plant Soil 218:49–57. doi:10.1023/A:1014985327619

Grieve, I.C., D.D. Davidson, N.J. Ostle, P.M.C. Bruneau, and A.E. Fallick. 2006. Spatial heterogeneity in the relocation of added 13C within the structure of an upland grassland soil. Soil Biol. Biochem. 38:229–234.

Gundlach, G., and B. Mühlhausen. 1980. Coupling of 1-naphthol with Fast Red TR. 1. Studies on the optimization of a continuous determination of acid phosphatase. J. Clin. Chem. Clin. Biochem. 18:603–610.

Häussling, M., and H. Marschner. 1989. Organic and inorganic soil phosphates and acid-phosphatase activity in the rhizosphere of 80-year-old Norway spruce [Picea-abies (L.) Karst] trees. Biol. Fertil. Soils 8:128–133. doi:10.1007/BF00257756

Hildebrand, E.E. 1994. The heterogeneous distribution of mobile ions in the rhizosphere of acid forest soils: Facts, causes, and consequences. J. Environ. Sci. Health A 29:1973–1992. doi:10.1080/10934529409376159

Hobbie, E.A., and J.E. Hobbie. 2008. Natural abundance of N-15 in nitrogen-limited forests and tundra can estimate nitrogen cycling through mycorrhizal fungi: A review. Ecosystems 11:815–830. doi:10.1007/s10021-008-9159-7

Hoppe, H.G. 1983. Significance of exoenzymatic activities in the ecology of brackish water: Measurements by means of methylumbelliferyl-substrates. Mar. Ecol. Prog. Ser. 11:299–308. doi:10.3354/meps011299

Humble, M.W., A. King, and I. Phillips. 1977. API ZYM: A simple rapid system for detection of bacterial enzymes. J. Clin. Pathol. 30:275–277. doi:10.1136/jcp.30.3.275

Jones, D.L., J.R. Healy, V.B. Willett, J.F. Farrar, and A. Hodge. 2005. Dissolved organic nitrogen uptake by plants—an important N uptake pathway? Soil Biol. Biochem. 37:413–423. doi:10.1016/j.soilbio.2004.08.008

Leake, J., D. Johnson, D. Donnelly, G. Muckle, L. Boddy, and D. Read. 2004. Networks of power and influence: The role of mycorrhizal mycelium in controlling plant communities and agroecosystem functioning. Can. J. Bot. 82:1016–1045. doi:10.1139/b04-060

Leake, J.R., N.J. Ostle, J.I. Rangel-Castro, and D. Johnson. 2006. Carbon fluxes from plants through soil organisms determined by field $^{13}CO_2$ pulse labelling in an upland grassland. Appl. Soil Ecol. 33:152–175. doi:10.1016/j.apsoil.2006.03.001

Lucas, R.W., and B.B. Casper. 2008. Ectomycorrhizal community and extracellular enzyme activity following simulated atmospheric N deposition. Soil Biol. Biochem. 40:1662–1669. doi:10.1016/j.soilbio.2008.01.025

Marschner, P., P.F. Grierson, and Z. Rengel. 2005. Microbial community composition and functioning in the rhizosphere of three *Banksia* species in native woodland in Western Australia. Appl. Soil Ecol. 28:191–201. doi:10.1016/j.apsoil.2004.09.001

Marschner, H., and V. Römheld. 1983. In vivo measurement of root-induced pH changes at the root-soil interface: Effect of plant species and nitrogen source. Z. Pflanzenphysiol. 111:241–245.

Marschner, H., V. Römheld, and H. Ossenberg-Neuhaus. 1982. Rapid method for measuring changes in pH and reducing processes along roots of intact plants. Z. Pflanzenernaehr. Bodenkd. 105:407–416.

McDougall, W.B. 1916. The growth of forest tree roots. Am. J. Bot. 3:384–392. doi:10.2307/2435018

Mosca, E., L. Montecchio, L. Scattolin, and J. Garbaye. 2007. Enzymatic activities of three ectomycorrhizal types of *Quercus robur* ectomycorrhiza in relation to tree decline and thinning. Soil Biol. Biochem. 39:2897–2904. doi:10.1016/j.soilbio.2007.05.033

Murray, P., N. Ostle, C. Kenny, and H. Grant. 2004. Effect of defoliation on patterns of carbon exudation from *Agrostis capillaris*. J. Plant Nutr. Soil Sci. 167:487–493. doi:10.1002/jpln.200320371

Nannipieri, P., J. Ascher, M.T. Ceccherini, L. Landi, G. Pietramellara, and G. Renella. 2003. Microbial diversity and soil functions. Eur. J. Soil Sci. 54:655–670. doi:10.1046/j.1351-0754.2003.0556.x

Pijnenborg, J.W.M., T.A. Lie, and A.J.B. Zehnder. 1990. Simplified measurement of soil pH using an agar-contact technique. Plant Soil 126:155–160. doi:10.1007/BF00012818

Pritsch, K., P. Courty, J.-L. Churin, B. Cloutier-Hurteau, M. Ali, C. Damon, M. Duchemin, S. Egli, J. Ernst, L. Fraissinet-Tachet, F. Kuhar, E. Legname, R. Marmeisse, A. Müller, P. Nikolova, M. Peter, C. Plassard, F. Richard, M. Schloter, M.-A. Selosse, A. Franc, and J. Garbaye. 2011. Optimized assay and storage conditions for enzyme activity profiling of ectomycorrhizae. Mycorrhiza. doi:10.1007/s00572-011-0364-4

Pritsch, K., M.S. Günthardt-Goerg, J.C. Munch, and M. Schloter. 2006. Influence of heavy metals and acid rain on enzymatic activities in the mycorrhizosphere of model forest ecosystems. For. Snow Landsc. Res. 80:289–304.

Pritsch, K., S. Raidl, E. Marksteiner, H. Blaschke, R. Agerer, M. Schloter, and A. Hartmann. 2004. A rapid and highly sensitive method for measuring enzyme activities in single mycorrhizal tips using 4-methylumbelliferone-labelled fluorogenic substrates in a microplate system. J. Microbiol. Methods 58:233–241. doi:10.1016/j.mimet.2004.04.001

Rast, D.M., D. Baumgartner, C. Mayer, and G.O. Hollenstein. 2003. Cell wall-associated enzymes in fungi. Phytochemistry 64:339–366. doi:10.1016/S0031-9422(03)00350-9

Reymond, J.L., and D. Wahler. 2002. Substrate arrays as enzyme fingerprinting tools. ChemBioChem 3:701–708. doi:10.1002/1439-7633(20020802)3:83.0.CO;2-3

Ritter, T., I. Kottke, and F. Oberwinkler. 1986. Nachweis der Vitalität von Mykorrhizen durch FDA-Vitalfluochromierung. Biol. Unserer Zeit 16:179–185. doi:10.1002/biuz.19860160607

Schack-Kirchner, H., K.V. Wilpert, and E.E. Hildebrand. 2000. The spatial distribution of soil hyphae in structured spruce-forest soils. Plant Soil 224:195–205. doi:10.1023/A:1004806122105

Schimel, J.P., and J. Bennett. 2004. Nitrogen mineralization: Challenges of a changing paradigm. Ecology 85:591–602. doi:10.1890/03-8002

Sinsabaugh, R.L., R.K. Antibus, and A.E. Linkins. 1991. An enzymatic approach to the analysis of microbial activity during plant litter decomposition. Agric. Ecosyst. Environ. 34:43–54. doi:10.1016/0167-8809(91)90092-C

Smit, A.L., E. George, and J. Groenwold. 2000. Root observations and measurements at (transparent) interfaces with soil. p. 235–271. In A.L. Smit, A.G. Bengough, C. Engels, M. Van Noordwijk, S. Pellerin, and S.C. Van de Geijn (ed.) Root methods: A handbook. Springer, Berlin.

Snajdr, J., V. Valaskova, V. Merhautova, J. Herinkova, T. Cajthaml, and P. Baldrian. 2008. Spatial variability of enzyme activities and microbial biomass in the upper layers of *Quercus petraea* forest soil. Soil Biol. Biochem. 40:2068–2075. doi:10.1016/j.soilbio.2008.01.015

Tarafdar, J.C., and A. Jungk. 1987. Phosphatase activity in the rhizosphere and its relation to the depletion of soil organic phosphorus. Biol. Fertil. Soils 3:199–204. doi:10.1007/BF00640630

Vaisanen, R.K., M.S. Roberts, J.L. Garland, S.D. Frey, and L.A. Dawson. 2005. Physiological and molecular characterization of microbial communities associated with different water-stable aggregate size classes. Soil Biol. Biochem. 37:2007–2016. doi:10.1016/j.soilbio.2005.02.037

Zwirglmaier, K. 2005. Fluorescence in situ hybridisation (FISH)– the next generation. FEMS Microbiol. Lett. 246:151–158. doi:10.1016/j.femsle.2005.04.015

Chapter 14

Microplate Fluorimetric Assay of Soil Enzymes

Shiping Deng,* Hojeong Kang, and Chris Freeman

14–1 INTRODUCTION

The enzyme assay methods discussed so far in this book are bench-scale assays, which are generally labor intensive, time consuming, and often constrained to the analysis of one enzyme for a limited number of samples at a time. In bench-scale assays, it is unavoidable that samples need to be stored for an extended period before analysis. Enzyme activities can be impacted by storage, creating challenges in data analysis, and sometimes resulting in misleading data interpretation (Bonin et al., 1999). In many studies, there is a need for simultaneous analysis of multiple soil enzyme activities within a limited time following collection of a relatively small quantity of soil samples.

Use of the microplate format in soil enzyme assays offers the advantages of simultaneous analysis of multiple enzymes using a small quantity of soil, which meets the needs as stated above. A microplate reader can measure absorbance or fluorescence of many samples (e.g., in 96 wells) simultaneously in microliter volumes, which allows researchers to substantially reduce reagent costs and assay time in a manner that would not be possible using conventional bench-scale assays. Several such assays have been developed that utilize 4-methylumbelliferyl (MUF) labeled substrates (Marx et al., 2001; Pritsch et al., 2004; Saiya-Cork et al., 2002; Sinsabaugh et al., 2000). The detection of MUF is very sensitive, making it possible to quantify picomoles of MUF in 200- to 300-μL solutions.

Fluorescence measurements are significantly more immune to the effects of turbidity compared to absorption-based detection. A major concern for MUF-based methods is the significant quenching in the detection of MUF in soil (Freeman et al., 1995). The degree of quenching varies greatly, both temporally and spatially. For example, Freeman et al. (1995) found quenching in one soil led to the reduction of fluorescence by 27 and 61% for May and October samples, respectively, whereas quenching in another soil was consistently less than 20%. Fluorescence is also known to be affected by pH and temperature (Lakowicz, 1983). As temperature increases, fluorescence decreases due to an increase in molecular motion

Shiping Deng, Oklahoma State University, Dep. of Plant and Soil Science, 368 Ag. Hall, Stillwater, OK 74078 (shiping.deng@okstate.edu), *corresponding author;
Hojeong Kang, School of Civil and Environmental Engineering, Yonsei University, Seoul, Korea;
Chris Freeman, Bangor University, School of Biological Sciences, Bangor University Deiniol Road, Bangor, Gwynedd LL57 2DG, UK (c.freeman@bangor.ac.uk).

doi:10.2136/sssabookser9.c14

that results in more frequent molecular collisions and subsequent loss of energy (Guilbault, 1990).

Consequently, quantification of MUF in soil requires a calibration curve be developed for each soil. For quantitative detection, it is also important to treat standards, blanks, and samples in exactly the same manner, prepare all solutions using the same reagents and preparation techniques, and measure at the same temperature after the same amount of time. The additional precautions and calibration steps add considerable labor and expense to the assay, as well as additional sources of error.

More recently, Popova and Deng (2010) reported a micro assay for the simultaneous quantification of multiple soil enzyme activities based on the use of ρ-nitrophenyl substrates. ρ-Nitrophenol released on enzymatic reactions has a distinctive yellow color when pH is above 7.6, which is relatively stable, and can be quantified in soil solutions with little interference (Tabatabai and Bremner, 1969). However, in the microassay reported by Popova and Deng (2010), turbidity resulting from soil suspensions interferes with detection and additional controls are needed for reproducible, accurate results. Additionally, colorimetric detection of ρ-nitrophenol in soil extracts is less sensitive (10–20 fold) than the detection of fluorescence using similar soil solution matrices. Therefore, longer incubation times are required for detectable measures of enzyme activities, making it more challenging in data interpretation due to potential microbial growth and associated consumption and production of compounds during the incubation period (personal communication, S. Deng, 2011).

Limited research also suggests that enzyme kinetics using MUF-based methods may be significantly different from those of colorimetric-based methods that utilize ρ-nitrophenyl substrates (Marx et al., 2001). The observed discrepancies were hypothesized to be related to differences in mobility and adsorption of the substrates. The two assays are conducted at different scales, with MUF-based methods using microplates, and colorimetric-based methods being conducted at bench scale. In the bench assay, 1 g of soil is incubated in 5 mL of buffer/substrate solution. In microplate assays, about 0.83 mg of soil is incubated with substrate in 200 μL of solution, which is equivalent to 0.02 g in 5 mL. Theoretically, some discrepancies in the results are expected. Molecules will interact more effectively in a more dilute solution, leading to substantially lower K_m values in the microplate than the bench-scale assays, as reported by Marx et al. (2001). Further research is needed to fully understand the mechanisms that control the kinetics of enzyme activities evaluated using these assay approaches.

Nevertheless, a large number of MUF and ρ-nitrophenol labeled substrates for different classes of enzymes are commercially available, which makes assays based on the detection of MUF or ρ-nitrophenol widely adaptable. The MUF-based assay methods, however, are already widely adapted and tested. In this chapter, we describe MUF-based microplate format enzyme assays using four enzymes involved in transforming carbon, nitrogen, phosphorus, and sulfur as examples. Buffering pH in these assays is based on reports by Tabatabai (1994) and Parham and Deng (2000). Interferences from soil matrices in the detection of MUF are corrected by employing additional controls and calibration curves that are developed in the matrix for each assayed soil.

14–2 PRINCIPLES

As discussed above, these assays are based on detection of MUF released by enzymatic hydrolysis of specific substrates when incubated with soil at the optimal pH of the assayed enzyme. The reaction involved is shown below:

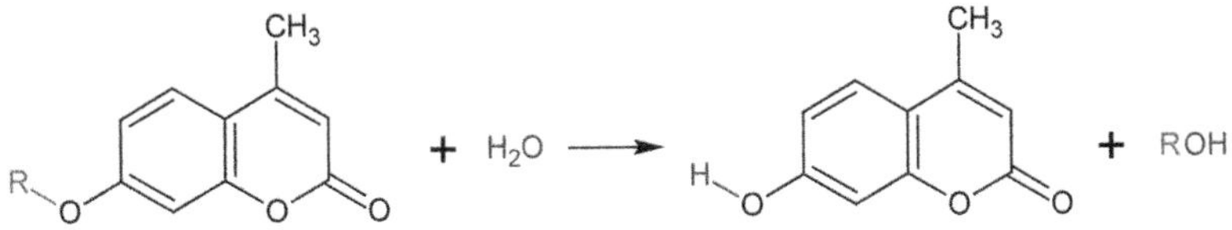

MUF

where R groups are shown below for the indicated enzymes.

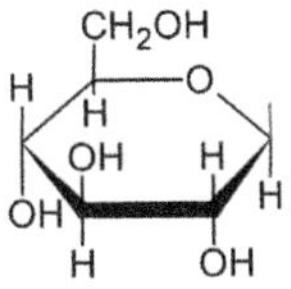

for β-glucosidase

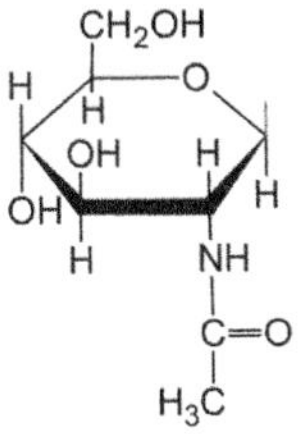

for β-N-acetyl-glucosaminidase

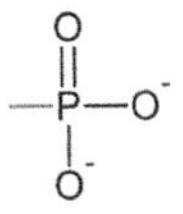

for phosphomonoesterase

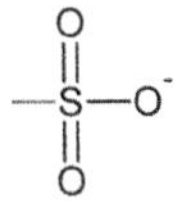

for arylsulfatase

Following incubation for a defined time at the desired temperature, MUF is quantified on termination of the enzymatic reaction by addition of 0.5 M NaOH. Concentrations of MUF are calculated using a calibration curve and enzyme activities are expressed as millimoles or micromoles MUF released kg^{-1} soil h^{-1}.

14–3 ENZYME ASSAYS USING FLUORESCENCE QUANTIFICATION— β-GLUCOSIDASE, β-N-ACETYL-GLUCOSAMINIDASE, ACID PHOSPHOMONOESTERASE, AND ARYLSULFATASE

The following procedure is modified based on a report by Freeman et al. (1995).

14–3.1. Apparatus

- 150-mL beaker
- 100-mL volumetric flasks

- 1-L volumetric flask
- Stir plate and 3.75-cm magnetic stir bar
- Multichannel pipette (0–250 µL)
- Black solid polystyrene microplate, 96-well, 360-µL well capacity
- Incubator (37°C)
- Fluorescence microplate reader

14–3.2. Reagents

1. Modified universal buffer (MUB) stock solution: Dissolve 12.1 g of tris(hydroxymethyl)aminomethane (THAM), 11.6 g of maleic acid, 14.0 g of citric acid, and 6.3 g of boric acid (H_3BO_3) in 488 mL of 1 M sodium hydroxide (NaOH) and adjust the solution to 1 L with deionized (DI) water. Store the stock solution in a refrigerator at approximately 4°C.

2. Modified universal buffer (pH 5.5 or 6.0): Place 200 mL of MUB stock solution in a 500-mL beaker containing a magnetic stir bar, and place the beaker on a magnetic stir plate. Titrate the solution to pH 5.5 or 6.0 with 0.1 M HCl, and then adjust the volume to 1 L with DI water.

3. Methylumbelliferyl (MUF) substrates (1 mM) (catalog numbers for Sigma-Aldrich Chemicals, St. Louis, MO are shown in parentheses): Based on the enzyme being assayed, weigh 0.034 g of methylumbelliferyl-β-D-glucoside (MUF-G, M3633), 0.038 g of methylumbelliferyl N-acetyl-β-D-glucosaminide (MUF-NAG, M2133), 0.026 g of methylumbelliferyl phosphate (MUF-P, M8883), or 0.029 g of methylumbelliferyl sulfate (MUF-S, M7133) into a 100-mL flask, and then adjust the volume to 100 mL with dionized (DI) water. These solutions are best be made daily, but can be stored for days or weeks if kept in the dark at 4 or –20°C.

4. NaOH (0.5 M): Dissolve 20 g of NaOH in about 700 mL of DI water, and adjust the volume to 1 L with DI water. Store the solution in a plastic container at room temperature.

5. Methylumbelliferone stock solution (100 µM): Dissolve 0.0202 g of 4-methylumbelliferone sodium salt (98%, M1508, Sigma-Aldrich Chemical, St. Louis, MO) in about 700 mL of DI water, then adjust the volume to 1000 mL. Store the solution in the dark at 4 or –20°C.

6. Methylumbelliferone working standards: 0, 5, 10, 20, 30, and 50 µM MUF standards are prepared by diluting 0, 5, 10, 20, 30, or 50 mL of the standard MUF stock solutions (100 µM) to 100 mL with DI water, respectively. Store the solution in the dark at 4 or –20°C.

14–3.3. Procedure

1. Two soil suspensions are prepared for each sample. Each soil suspension is prepared by placing 1 g of soil in a 150-mL beaker, adding 120 mL of DI water, and homogenizing for 30 min using a 3.75-cm magnetic stir bar at a speed that is sufficient for complete homogenization of the soil suspension (about 600 rpm on a stir plate).

2. Take the soil suspension during continuous stirring using a multichannel pipette set at 100 µL with four tips loaded and place into microplate wells that

each contains 50 μL of MUB (pH 5.5 for the assay of β-N-acetyl-glucosamini-
dase, and 6.0 for the assay of β-glucosidase, acid phosphomonoesterase, and
arylsulfatase). It is recommended to perform this procedure using the same
stirring plate and speed for all suspensions and samples tested.

3. Add 50 μL of MUF substrate to each well, bringing the total volume of the
reaction mixture to 200 μL. The reaction mixtures are thoroughly mixed by
pipetting up and down several times before incubating at 37°C for 1 h.

4. After incubation, 50 μL of 0.5 M NaOH is added to each microplate well to
terminate the enzymatic reaction.

5. Fluorescence is measured using a fluorescence microplate reader (365-nm
excitation and 450-nm emission) immediately on termination of the reaction
(fluorescence decreases with time).

6. The controls are performed using the same procedure but with the substrate
added after the addition of NaOH.

7. Calibration curves are developed at the same time samples are assayed using the same soil suspension, buffer, and procedure, except MUF standards
are used in the place of MUF substrates. Briefly, 50 μL of each MUF working
standard solution are placed into microplate wells that each contains 50 μL of
MUB, followed by the addition of 100 μL of soil suspension. After incubation,
50 μL of NaOH is added to each well as done for enzyme assays described
above. The total volume in each well is 250 μL and contains MUF standards
of 0, 250, 500, 1000, 1500, or 2500 pmol. Average data obtained from both soil
suspensions are used to develop a calibration curve for the sample.

8. Autohydrolysis during the incubation period should be evaluated by incubating the substrate as done for samples but without soil (adding water in
the place of soil suspension).

9. Calculations:

Corrected fluorescence ($F_{corrected}$) = ($F_{sample} - F_{avg\ control} - F_{avg\ autohydrolysis}$)

Example: For Sample 2A in the lower plate in Fig. 14–1.

($F_{corrected}$) = ($F_{2A} - F_{avg\ of\ 1A\ to\ 1D} - F_{avg\ autohydrolysis}$)

Assuming columns 9 and 10 from A to D are used to test autohydrolysis of
the substrate

$F_{avg\ autohydrolysis} = F_{avg\ of\ 10A\ to\ 10D} - F_{avg\ of\ 9A\ to\ 9D}$

The schematics for the microplate layout are shown in Fig. 14–1. Four replicate
analyses and four controls for each of two replicate soil suspensions are conducted
for each sample. MUF concentrations in reaction mixtures are calculated using the
corrected fluorescence readings against a calibration curve that is constructed for
each soil at the time of sample assay. For the calibration curve, the average reading
from zero MUF standards should be subtracted from all other readings.

14–5 COMMENTS

In each study, it is critical to uniformly mix the soil samples. Therefore, the same
dimensions of glassware and stir bar should be used in the preparation of soil
suspensions for consistent and comparable results. Moreover, it is also best to use

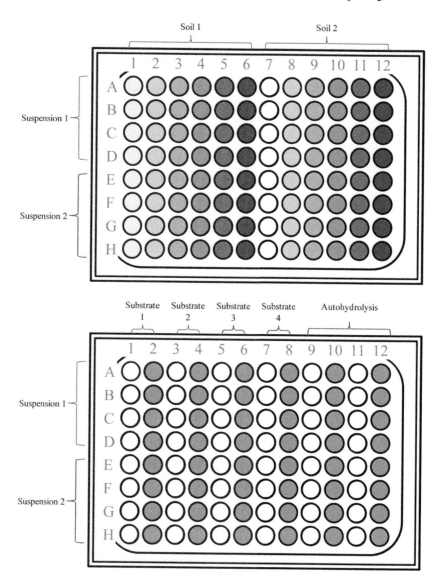

Fig. 14–1. Schematics of typical layouts of 96-well microplates for assaying activities of four different enzymes in soil. The upper layout is for calibration curves, with the filled circles representing wells containing MUF standards with increasing concentrations from columns 1 to 6 or 7 to 12. There are four replications for each concentration, placed A to D or E to H in a column. Calibrations for two different suspensions in each of two soils can be done in one plate. Calibrations should be done simultaneously when assays are conducted using exactly the same procedure. The lower plate is a typical layout for assaying activities of four different enzymes in one soil sample. Open circles represent controls with substrate added on termination of the reaction (by the addition of NaOH), while the filled circles represent assay of samples. Wells from row A to D are used for suspension 1 and E to H for suspension 2. Columns 9 to 12 can be used to test autohydrolysis of enzyme substrates. A third microplate is needed to complete the assay of the four enzyme activities in another soil sample. Tests on autohydrolysis of substrates do not need to be repeated on the third microplate. Therefore, columns 9 to 12 of this plate can be left open or used for repeating assays or additional tests as needed.

the same stir plate and stirring speed during pipetting soil suspensions. Water (instead of a buffer or electrolyte, such as KCl or K_2SO_4) is used in soil suspensions to allow the analyses of multiple enzymes using the same soil suspension. Additional turbidity is generated when soil is incubated in MUB. Therefore, the same amount of MUB is added to each well, and substrate solutions are prepared in water instead of buffer. The same buffer, MUB, is used for the simultaneous assay of multiple enzymes with different pH optima.

All calibrations are done with soil suspensions because of interference in the detection from soil particles and quenching for fluorescence. It is also important to develop calibration curves simultaneously when assays are conducted because fluorescence is sensitive to environmental conditions.

Autohydrolysis may or may not be detectable depending on the nature of the substrates, its storage time and condition, as well as the specific assay conditions.

Pre-testing is needed to check whether the buffer strength is sufficient to hold at the desired pH during incubation and that there is a sufficient shift to an alkaline pH following the addition of NaOH to terminate the enzymatic reaction. For soils having relatively low enzyme activities, longer incubation times may be used for quantitative determination.

The recommended substrate concentrations are about or exceed 10-fold of K_m values based on tests of a limited number of soils (unpublished data, 2011). These concentrations are expected to be sufficient to achieve maximum reaction velocity throughout the incubation period for most agricultural soils under the defined experimental conditions. Higher substrate concentrations may be used for samples with high enzyme activities to ensure substrate saturation in the assay. For convenience and efficiency, substrate solutions can be made in a relatively large quantity, and aliquots can be stored in a freezer at −20°C.

Theoretically, the described assay procedures can be applied to all enzyme assays that employ the use of MUF substrates. Pre-testing may be needed to determine if the products are within the detection range for a specific soil and enzyme assay. When the readings for unknown samples exceed the range of the standards, dilutions should be done using the same solution matrix. Alternatively, one may use less soil suspension for the assay, but add DI water to maintain volume of the reaction mixture. In theory, the activities may also be determined by monitoring changes of fluorescence over time (i.e., Marx et al., 2001), providing that calibration curves are determined on the same plate under the same condition and assay pH at each time point.

REFERENCES

Bonin, H.L., R.P. Griffiths, and B.A. Caldwell. 1999. Effects of storage on measurements of potential microbial activities in stream fine benthic organic matter. J. Microbiol. Methods 38:91–99. doi:10.1016/S0167-7012(99)00081-0

Freeman, C., G. Liska, N.J. Ostle, S.E. Jones, and M.A. Lock. 1995. The use of fluorogenic substrates for measuring enzyme-activity in peatlands. Plant Soil 175:147–152. doi:10.1007/BF02413020

Guilbault, G.G. 1990. Practical fluorescence. 2nd ed. Marcel Dekker, New York.

Lakowicz, J.R. 1983. Principles of fluorescence spectroscopy. Plenum Press, New York.

Marx, M.C., M. Wood, and S.C. Jarvis. 2001. A microplate fluorimetric assay for the study of enzyme diversity in soils. Soil Biol. Biochem. 33:1633–1640. doi:10.1016/S0038-0717(01)00079-7

Parham, J.A., and S.P. Deng. 2000. Detection, quantification and characterization of beta-glucosaminidase activity in soil. Soil Biol. Biochem. 32:1183–1190. doi:10.1016/S0038-0717(00)00034-1

Popova, I., and S.P. Deng. 2010. A high-throughput microplate assay for simultaneous colorimetric quantification of multiple enzyme activities in soil. Appl. Soil Ecol. 45:315–318. doi:10.1016/j.apsoil.2010.04.004

Pritsch, K., S. Raidl, E. Marksteiner, H. Blaschke, R. Agerer, M. Schloter, and A. Hartmann. 2004. A rapid and highly sensitive method for measuring enzyme activities in single mycorrhizal tips using 4-methylumbelliferone-labelled fluorogenic substrates in a microplate system. J. Microbiol. Methods 58:233–241. doi:10.1016/j.mimet.2004.04.001

Saiya-Cork, K.R., R.L. Sinsabaugh, and D.R. Zak. 2002. The effects of long term nitrogen deposition on extracellular enzyme activity in an *Acer saccharum* forest soil. Soil Biol. Biochem. 34:1309–1315. doi:10.1016/S0038-0717(02)00074-3

Sinsabaugh, R.L., H. Reynolds, and T.M. Long. 2000. Rapid assay for amidohydrolase (urease) activity in environmental samples. Soil Biol. Biochem. 32:2095–2097. doi:10.1016/S0038-0717(00)00102-4

Tabatabai, M.A., 1994. Soil enzymes. p. 775–833. *In* A. Weaver et al. (ed.) Methods of soil analysis. Part 2. Microbiological and biochemical properties. SSSA Book Ser. 5. SSSA, Madison, WI.

Tabatabai, M.A., and J.M. Bremner. 1969. Use of p-nitrophenyl phosphate for assay of soil phosphatase activity. Soil Biol. Biochem. 1:301–307. doi:10.1016/0038-0717(69)90012-1

Chapter 15

Stabilizing Enzymes as Synthetic Complexes

Liliana Gianfreda* and Maria A. Rao

Soil biochemical transformations are extremely complex and variable, with a large array of enzymes playing important roles. Since soil is capable of performing reactions in the absence of living cells, it has been assumed that active enzymatic proteins must act in soil as stable complexes with inorganic and organic soil colloids. As discussed in Chapters 1 and 2 (Dick and Burns, 2011, this volume; Wallenstein and Burns, 2011, this volume), overall enzymatic activities in soils are a composite of different intracellular and extracellular enzymatic components.

Research has demonstrated that a large portion of soil enzymatic activity can be attributed to "accumulated" enzymes, i.e., "enzymes present and active in soil in which no microbial proliferation takes place" (Kiss et al., 1975) or "abiontic" enzymes, i.e., "enzymes in soil except those present in proliferating cells, that comprehend accumulated enzymes" (Skujiņš, 1976). Later, Burns (1982 proposed that there are ten distinct categories of enzymes, one of which was called "soil-bound enzymes" and later defined as "naturally immobilized enzymes" by Gianfreda and Bollag (1996) because of their similarity to enzymes in a living cell and operating within a highly organized structure. These soil enzymes include enzymatic molecules not associated with cellular components that are present in the soil as exoenzymes released from living cells or endoenzymes released from disintegrated cells. In turn, these enzymes can be immobilized on innumerable organic and inorganic soil components. Several properties and particular behaviors shown by soil enzymes are consistent with the assumption that a large portion of enzymes in soil are immobilized enzymes.

The ecological role of soil-bound enzymes could be to transform substrate for a cell without the organism having the enzyme or having to expend energy inducing and producing an enzyme. Moreover, immobilized soil enzymes may cleave off small fragments ("inducer" molecules) from large polymeric substrates in soil. The inducer molecule would then be able to enter the nearby cell and trigger the synthesis of inducible extracellular enzymes in large amounts.

Liliana Gianfreda, Università di Napoli Federico II, Via Università 100, Portici, Italy (liliana.gianfreda@unina.it), *corresponding author;
Maria A. Rao, Università di Napoli Federico II, Via Università 100, Portici, Italy (mariarao@unina.it).

doi:10.2136/sssabookser9.c15

15–1 PRINCIPLES

15–1.1 Immobilization of Enzymes

By definition, immobilization is the conversion of enzymes from a water-soluble, mobile state to a water-insoluble, immobile state (Klibanov, 1983) with retention of their catalytic activity, and usable repeatedly and continuously. This implies that an *immobilized enzyme* is a protein physically localized in a certain region of space and that the mobility of protein molecules or their fragments is limited in space, or relative to each other.

There are various methods by which enzymes can be immobilized, ranging from covalent chemical bonding to physical entrapment (Martinek and Mozhaev, 1985). They are broadly classified as follows:

- Adsorption of the enzyme onto a water-insoluble matrix,
- Covalent bonding of the enzyme to a derivatized, water-insoluble matrix,
- Intermolecular cross-linking of enzyme molecules using multifunctional reagents,

- Entrapment of the enzyme inside a water-insoluble polymer lattice or semipermeable membrane.

The literature provides a comparison of enzyme immobilization methods for biotechnological purposes in terms of easy and cost-effective preparation, binding force between the enzyme and the support, enzyme leakage, applicability, pitfalls, matrix effects, microbial protection, and large diffusional barriers. Each method will respond positively or negatively to each of these factors.

Free enzymes (in a given solution) constitute a homogenous and isotropic catalytic system because they exert their catalytic action in a single soluble phase, where all components of the enzymatic process (hydrogen ions, substrates, products, inhibitors, activators, and cofactors) are present. Conversely, characteristics of an immobilized enzyme are determined to a large extent by the nature of the immobilizing support. The soil enzymes operate under a heterogeneous rather than of a homogeneous environment and numerous factors have to be considered for analyzing and interpreting their catalytic behavior.

Several soil components may act as supports for the immobilization of enzymes in soil. They are: clean clays, dirty clays (i.e., clay minerals coated partially or totally by polymeric species or noncrystalline oxides of aluminum and iron), organic matter (including humic acid, tannins, phenols, quinones), and clay–humic associations (Fusi et al., 1989; Gianfreda et al., 1991, 1992, 1993; Huang et al., 1995; Naidja et al., 1995, 1997; Huang et al., 1999; Huang and Shindo, 2000).

Two methodological approaches have been utilized for understanding the relationships between immobilized soil enzymes and their clay, humic, or humic–clay supports. The first approach is based on the isolation, purification, and characterization of active enzymatic fractions from soils. As reported in **Chapter 16** (Fornasier et al., 2011, this volume), few enzymes have been extracted from soils as pure enzymes. Experimental evidence has shown that several extracted enzymatic activities (considered as free enzymes) are usually strongly associated with carbohydrate–enzyme complexes or humic matter (Mayaudon, 1986;

Tabatabai and Fu, 1992; Nannipieri et al., 1996; Nannipieri, 2006; Fornasier et al., 2011, this volume).

In the second approach, model systems in which enzymes are artificially immobilized by their attachment to soil components have been used for studying the properties of soil-bound enzymes. Synthetic enzyme complexes are obtained in vitro by simulating interactions between enzymatic proteins and clay minerals, humic substances, or organo–mineral complexes. These model systems can be used to evaluate the possible role of a given microenvironmental parameter (e.g., support or enzyme nature, type of bond involved between enzyme and support, or type of chemical or physical mechanism implicated in the immobilization process) in determining the activity of an enzyme in its native environment.

Methods for obtaining enzyme–organic matter and enzyme–organo–mineral complexes also are available (Rowell et al., 1973; Ladd and Butler, 1975; Sarkar. and Burns, 1984; Ruggiero and Radogna, 1988; Grego et al., 1990; Gianfreda et al. (1995a,b); Garzillo et al., 1996; Rao and Gianfreda, 2000; Rao et al., 1999; Rao et al., 2000). Numerous synthetic enzymatic complexes have been studied in detail. Humus–enzyme complexes were formed by utilizing natural or synthetic humic acids. Usually, enzymes and humic acid or phenolic monomers are mixed, and the formation of active enzyme complexes studied after changing the nature of the organic constituent (humic acid or phenolic component), the contact time or ratio between the two components, or after adding an additional biotic or abiotic catalyst, such as phenoloxidases or metallic oxides (Rowell et al., 1973; Sarkar. and Burns, 1984; Garzillo et al., 1996; Rao et al., 1999).

Different types of interactions between enzymes and soil constituents are involved in these synthetic complexes (Boyd and Mortland, 1990). Cation-exchange adsorption mechanisms (Harter and Stotzky, 1971), van der Waals–type forces (Hamzehi and Pflug, 1981), and ionic or hydrophobic bonds have been suggested as possible mechanisms to adhere enzymes to clay surfaces (Boyd and Mortland, 1990). Ion exchange, entrapment within organic networks, ionic bonding, or covalent attachment may account for the stable association between enzymes and humic materials (Ladd and Butler, 1975).

Proposed methods to create immobilized enzymes for biotechnological purposes generally utilize naturally occurring chemical and physical mechanisms as follows:

- Adsorption on either charged or neutral soil components. Adsorption process is the simplest immobilization mechanism taking place in soil. It can occur on mineral soil components, organic matter, or organo–mineral complexes.

- Covalent attachment to soil components, either natural or modified by their treatment with activating molecules. In this process, reactive groups cross-link on the support and external amino or carboxyl groups of the enzyme. An association support-enzyme is stabilized by strong chemical bonds between the enzyme and the support.

- Entrapment in polymeric organic materials. The enzyme is usually embedded in a polymeric organic matrix, resembling soil organic matter obtained by polymerization of humic or humic-like precursors through the action of biotic or abiotic catalysts.

15–1.2 Properties of Immobilized Enzymes

Soil enzymes frequently display particular and peculiar properties compared with enzymes from microbial, animal, or plant sources. They usually have high stability when subjected to thermal denaturation, proteolytic attack, or irradiation. Moreover, they can be stored for a long time without a significant loss of activity (Gianfreda and Bollag, 1996; Gianfreda and Ruggiero, 2006, and references therein). The kinetic, pH, and temperature profiles of soil enzymes are usually different from enzymes in their purified form (Nannipieri and Gianfreda, 1998 and references therein).

Enzymes immobilized on solid supports are different from pure enzymes because they operate in a heterogeneous rather than homogeneous system. Enzymes entrapped in a polymeric organic network or adsorbed on a support will have physical and chemical properties that would be expected to have different catalytic properties from those in the bulk solution. As a result, the components participating in the enzymatic reaction are partitioned between the immobilized enzyme phase and the aqueous phase. Moreover, the microenvironment around the immobilized enzyme could be altered by the physical and chemical nature of the support and subtle chemical and/or conformational changes in the enzyme structure due to its immobilization on a surface (Nannipieri and Gianfreda, 1998).

Stabilized enzymes must maintain the correct three-dimensional structure or conformation to remain catalytic. Although moderate changes in conformation may not affect the functioning of the enzyme, as more dramatic changes occur, there will be increasing reduction in enzyme activity.

Ample research in the field of enzyme biotechnology has demonstrated that the stability of enzymes can be enhanced (Gianfreda and Scarfi, 1991 and references therein) and used to preserve the protein structure. This can be done by immobilizing enzymes on solid supports or by adding in the reaction medium sugars, polyols, salts, and various polymers, which have the capability of strongly binding to proteins (Nannipieri and Gianfreda, 1998 and reference therein).

Enzymes also may increase their stability by means of chemical modifications of the enzyme molecule. Mono- or polyfunctional substitutions with different compounds can occur on the protein surface through amino acid groups, which usually increases rigidity and decreases unfolding. Finally, when working in the presence of organic solvents some enzymes quite unexpectedly can have higher stability, as compared with their free-state counterparts.

Several of these situations are encountered for enzymes in soil and could account for most of the stability of enzymes found exclusive of viable cells in the soil matrix. These stabilized matrix enzymes may be (i) immobilized on clay minerals and protected in humus aggregates, (ii) found inside biological membranes or within particular subcellular structures, and (iii) bound to polysaccharides or inorganic and organic compounds.

Immobilization, however, may increase, decrease, or have no effect on the enzyme stability and/or enzymatic activity. The stability of a free enzyme (i.e., a nonimmobilized or modified enzyme) is principally determined by its intrinsic structure whereas the activity, stability, and kinetic behavior of an immobilized enzyme is highly dependent on many external factors including:

- the type of interaction with the carrier,
- the binding position and the number of bonds,

- the freedom of the conformation to change in the matrix,
- the microenvironment within which the enzyme molecule is located,
- the chemical and physical structure of the carrier,
- the properties of the spacer (for example, charged or neutral, hydrophilic or hydrophobic, size, length) linking the enzyme molecules to the carrier,
- partitioning,
- steric and diffusion interactions,
- changes in the conformation of enzymatic proteins,
- molecular orientation,
- water partition (especially in multiphase systems),
- conformation induction,
- binding mode.

Whatever the reason, the enhanced stability resulting from immobilization can often be related to the intrinsic features of immobilization as follows:

- molecular confinement, with reduction of the molecular mobility and consequently the capacity of unfolding as a result of steric hindrances due to the surrounding matrices,
- reduced accessibility to microbial attack or to action of proteases,
- favorable microenvironment by assuming that the microenvironment of a protein's molecule can affect the intramolecular forces that stabilize its native structure,
- chemical modification effect in covalent bonding (such as formation of an extra hydrogen bond as a result of chemical modification in the covalent immobilization process),
- development of a rigid conformation as a result of multipoint attachment.

In other words, the increased stability against denaturation of immobilized enzymes on synthetic supports or within the soil matrix can be due to (i) greater structural support and attraction of the protein (mono- or multipoint) to the insoluble support by its linkages or by "cage effects" inside a polymeric network and (ii) a change of the enzyme microenvironment where there is steric hindrance created by the surrounding matrices that reduces molecular mobility and the possibility of unfolding.

Among them, the most important factors affecting the behavior of an immobilized enzyme are changes in the conformation of enzymatic proteins, steric effects, microenvironmental effects, and diffusional limitations (Nannipieri and Gianfreda, 1998).

CHANGES IN THE CONFORMATION OF ENZYMATIC PROTEINS. The conformation of protein immobilized on a support usually differs from the native one. Consequently, a reduction in specific activity is often measured. Denaturing effects caused by the supports or the several bindings between the enzyme and the support may also produce conformational dissimilarities between native and immobilized enzymes. Moreover, if oligomeric enzymes are involved, the integrity of their oligomeric structure that assures their catalytic behavior can be altered and a negative deformation in the whole structure of the protein could arise. As a consequence, the function of enzymes may be reduced or even completely lost.

STERIC EFFECTS. Steric hindrances to the penetration of the substrate to the enzyme active site may result in a decreased specific activity and an altered Michaelis and Menten constant. These effects are much more evident in the case of high molecular weight substrates, likely because they diffuse very slowly in the matrix.

MICROENVIRONMENTAL EFFECTS. When an enzyme is immobilized in or on a solid support, the protein is exposed to a different local microenvironment strongly dependent on the chemical nature of the support material. A dissimilar distribution of species participating in the enzymatic reaction between the two phases, the immobilized enzyme particles and the bulk solution, may occur and the equilibrium substrate or effective concentration near the immobilizing support may be different from those in the bulk solution. The changed microenvironment may induce perturbations of the catalytic pathway of the reaction and consequently modification of the intrinsic catalytic properties of the immobilized enzyme might occur.

There are two mechanisms that one must consider for interpreting the microenvironmental effects on the catalytic activity of an immobilized enzyme. One is the effect of the chemistry of the support and the other is the direct effect of the support on the enzyme, both of which are further discussed below in detail.

The first microenvironmental effect is the chemistry imposed by the support. In this case the microenvironment surrounding the immobilized enzyme affects activity by increasing or decreasing substrate concentrations, products, and/or cofactors. These factors may be related to the electrostatic, hydrophobic, or other interactions occurring between the support and species present in the medium.

For example, one important factor is the surface charge of the support. In this case, when the species in the medium have the opposite charge of the support, it may result in these species having a higher affinity to the support than the free enzyme. To account for charged solutes, the partition coefficient can be used to characterize the local and bulk concentrations. If hydrogen ions are involved and the support is negatively or positively charged, attractive or repulsive electrostatic interactions may result between hydrogen ions and the support. As a result, the concentration of hydrogen ions surrounding a charged, immobilized enzyme will be higher or lower than in the bulk solution. Hence the pH will be lower or higher around the immobilized enzyme particles due to the particular charge on the support, which in turn affects enzymatic activity.

Similar partitioning effects may arise when a charged substrate is involved in the enzymatic reaction and the enzyme is immobilized on a charged support. Also, in this case different concentrations of the substrate will be present within the domain of the enzyme particle and the outer solution, thereby influencing the final kinetics of the enzyme.

Mathematical expressions may be used to quantify the effects on pH or substrate concentrations between the charged support and those of the outer solution (Nannipieri and Gianfreda, 1998).

The second microenvironmental effect is due to the changed microenvironment surrounding the enzyme molecules. It might change the intrinsic catalytic properties of the immobilized enzyme and consequently the mechanism of the enzymatic reaction. When the kinetics of a given enzyme is controlled by charges on the enzyme, this enzyme charge can be increased or decreased by the presence of the charged matrix, and a modification in the k_{cat} could then occur.

DIFFUSIONAL LIMITATIONS. These may strongly affect the rate and effectiveness of the enzymatic transformation because of their influence on the availability of the reactants and effectors. As in conventional heterogeneous catalysis, several distinct steps in the overall enzymatic process can be identified: (i) diffusion of the substrate from the bulk solution to the immobilized biocatalyst's surface across a boundary layer of water; (ii) diffusion of the substrate inside the immobilized enzyme particle, i.e., transport of the substrate from the carrier surface to the site of the enzyme; (iii) enzymatic conversion of the substrate in product; (iv) transport of the product to the external surface of the immobilized enzyme particle; and (v) diffusion of the product in the bulk solution. Steps (i) and (v) are external or bulk diffusional effects, while steps (ii) and (iv) are internal or pore diffusional effects. Different theoretical approaches and mathematical analyses are required for describing the effects of external and internal diffusional limitations. These effects, usually named mass-transfer effects, will also affect the values of the kinetic parameters.

Usually, if both microenvironmental and mass-transfer effects control the enzymatic reaction catalyzed by an immobilized enzyme, three types of rate parameters may be defined: *intrinsic* rate parameters, which pertain to the kinetic parameters characteristic of the native enzyme in solution; *inherent* rate parameters, which concern the immobilized enzyme in the absence of any diffusional limitations; and *effective* rate and kinetic parameters (observed and determined in the presence of significant diffusional limitations). A curved pattern in the linearized plots (mainly S/V vs. S) used for the determination of kinetic parameters is indicative of the presence of diffusional resistances and many mathematical treatments can be utilized for evaluating the dependence of the kinetic parameters on the substrate concentration.

When considering the kinetic pattern of a reaction catalyzed by an immobilized enzyme, considerable changes in the values of kinetic parameters can be expected and the observed kinetic parameters are only "apparent."

The models utilized for analyzing and describing all previous effects on the behavior of an immobilized enzyme are based on the assumption that the enzyme is uniformly distributed within the immobilized system. When enzyme molecules, and consequently the activity, are not uniformly distributed, the performance of the catalyst can be greatly affected and different theoretical approaches must be used.

15–2 PREPARATION AND CHARACTERIZATION OF SYNTHETIC ENZYMATIC COMPLEXES

Standardized soil enzyme assays are available but are not presented in this chapter for use on immobilized enzymes. Therefore we present methods we have published in numerous papers that are described in detail in this chapter. Moreover, most of the methods reported in the following sections are illustrated for one enzyme (acid phosphatase), although they have been and can be applied to other enzymes. When available, references for other enzymes that have been successfully immobilized by the same method are also reported.

When using immobilized enzymes some general points should be considered:

1. During the reaction, adsorption of substrates and/or products may occur on the support. This will possibly affect the kinetics of the reaction because the substrate concentration will be lower than initially. In addition, the quantitative determination of the product could be underestimated.

2. The presence of the solid phase in the activity assay may affect determination of the product by colorimetric methods. In this case, the solid phase may interfere with the absorbance at a given wavelength of the colored product. To overcome this, high-speed centrifugation (>20,000 rpm) and/or filtering may be needed to remove colloidal particles before colorimetric analysis (see below).

3. The choice of buffer is important because it can affect binding between the support and the enzymes, which can cause the release of enzymes into solution, resulting in a mixture of free and immobilized enzymes. Furthermore, the pH should be as close as possible to the pH where the enzyme displays its highest activity and stability.

15–3 ADSORPTION

15–3.1 Introduction

Adsorption of several enzymes has been performed on supports chosen among soil components. Physical adsorption is the simplest procedure for immobilizing enzymes. Its driving force is hydrophobic interactions and salt bridges and its principal drawback is the easy desorption of the enzymes by changes in pH, temperature, ionic strength, substrate concentration, or buffer used in the activity measurement. Adsorption simply consists of adding an enzyme solution to the support, stirring for a given time, and then removing the unadsorbed enzyme by washing. The amount of the immobilized enzyme is calculated by the difference between the amount of enzyme put in contact with the support and that found in washings.

Natural and dirty clays, oxides, humic acids, organo–mineral complexes, or even soil fractions and whole soils have been used as supports (Harter and Stotzky, 1971; Ruggiero and Radogna, 1988; Fusi et al., 1989; Boyd and Mortland, 1990; Gianfreda et al., 1991, 1992, 1993; Carrasco et al., 1995; Huang et al., 1995; Gianfreda and Bollag, 1996 and references therein; Naidja et al., 1995, 1997; Marzadori et al., 1998a,b; Huang et al., 1999; Huang and Shindo, 2000; Gianfreda et al., 2002 and reference therein; Safari Sinegani et al., 2005; Ahn et al., 2007). The swelling ability of the naturally occurring phyllosilicate minerals provides both unusual properties and appreciable surface area for adsorption of organic molecules. Since clays are aluminosilicates, they possess acidic sites that can interact with amino groups of enzymes leading to ionic binding. This linkage is much stronger than mere physical binding and hence the enzyme will be retained on the support for a longer duration.

Fewer studies have been devoted to the adsorption of enzymes to humic substances probably because of their physicochemical and biochemical complexity. Inhibitory effects of enzyme activity can be caused by several humic compounds and related substances. Tannins, melanins, and synthetic and natural humic acids differentially inhibit enzyme activity (Ladd and Butler, 1975; Pflug, 1980, Pflug and Ziechmann, 1981; Ruggiero and Radogna, 1988; Rao et al., 1998). Research findings have demonstrated that direct interaction of enzyme and humic substances may be strongly inhibited through bonding mechanisms such as adsorption, steric effects, hydrogen, or amidic bonds. A direct inhibitory effect of humic substances on the active sites of enzymes, in addition to a modification of the active site by structural conformational changes of the protein, also can be considered.

Due to their high biochemical heterogeneity, humic substances may also act as analog substrates and interfere with the equilibrium of the enzymatic reaction. Finally, humic substances, due to their cation exchange properties, may interact with cations involved as cofactors in the enzymatic catalysis or in the stabilization of the active enzymatic structure. Experimental evidence is available showing that adsorption of enzymes occurs on preformed mucigel present at the root–soil interface (Ciurli et al., 1996; Marzadori et al., 1998a).

Among the experimental procedures reported in papers dealing with enzymes immobilized on soil components through adsorption mechanisms, the methods presented below have been applied to more than one enzyme and tested for different properties and characteristics of the immobilized and stabilized enzymes.

15–3.2 Adsorption of Invertase, Urease, and Acid Phosphatase on Montmorillonite, Aluminum Hydroxide, and Al(OH)$_x$– Montmorillonite Complexes

15–3.2.1 Introduction

In several acidic soils and sediments, short-range ordered Al or Fe precipitation products, as well as dirty clays, are particularly abundant and play an important role in the adsorption and surface-catalyzed processes involving enzymatic proteins. Three clay minerals, (i) a montmorillonite (M), (ii) a noncrystalline aluminum hydroxide (AL), and (iii) an Al(OH)$_x$–montmorillonite complex (a montmorillonite partially coated by OH-Al ions, AM), and three enzymes—invertase, urease, and acid phosphatase—have been used and the corresponding complexes prepared and studied.

The three enzymes are particularly important because they are key enzymes in the carbon, nitrogen, and phosphorus cycles, respectively. The description and importance of the three enzymes has been reported in **Chapters 7, 9, and 10** (Klose et al., 2011; Deng and Popova, 2011; Kandeler et al, 2011, this volume).

Usually, the adsorption of an enzyme on clay mineral surfaces produces a decrease, sometimes substantial, of enzyme activity. At the same time, resistance of enzyme molecules to proteolytic attack and thermal denaturation is usually observed. The presence on clay mineral surfaces of Fe, Al, and Mn oxides and hydroxides may modify the interactions between enzymes and adsorbent supports and subsequently affect the activity of the adsorbed enzyme.

The methods for the three enzymes are described in detail, although several steps are common to all three methods. This provides additional information on the modifications needed when similar supports are used on different enzymes.

15–3.2.2 Invertase Adsorption Method (Gianfreda et al., 1991)

15–3.2.2.1 Apparatus

- Volumetric flasks, 10, 50, 250, 1000 mL
- Test tubes, 10, 12 mL
- Nalgene centrifuge tubes, 12 mL
- Millipore M.F. filters (<0.025 μm)
- High-speed temperature-controlled centrifuge (>20,000 rpm)
- Atomic absorption spectrophotometer
- UV-Vis spectrophotometer
- Thermostatic bath for use at 100°C or boiling water bath shaker
- Cryostat (temperature interval –20 +100°C)
- Vortex
- Tip sonicator (500 W)
- Freeze-dryer
- Erlenmeyer reaction flasks (25, 50, 100, 250, 1000 mL)
- pH meter
- Andreasen pipette

15–3.2.2.2 Reagents

1. Sodium acetate (NaOAc), 0.2 M: Add 16.40 g of anhydrous NaOAc to 700 mL of bidistilled water, mix, and make up to 1 L final volume with additional water. Store at 4°C for a max. of 3 mo.

2. Acetic acid (HOAc), 0.2 M: Add 11.44 mL of reagent-grade 99.99% acetic acid, specific gravity 1.05, per liter of the final solution desired. Mix, thoroughly cool, and store at 4°C for a max. of 5 mo.

3. Sodium acetate buffer, 0.2 M pH 4.65: Mix suitable volumes of Reagent 1 and of Reagent 2 (usually 490 and 510 mL, respectively) to reach pH 4.65 and 1 L final volume. Check pH. Store at 4°C for a max. of 3 mo.

4. Sodium chloride (NaCl), 0.1 M: Add 5.844 g of NaCl to 700 mL of bidistilled water, mix, and make up to 1 L final volume with additional water. Store at room temperature for up to 1 yr.

5. Sodium hydroxide (NaOH), 1 M: Add 40 g of NaOH to 700 mL of bidistilled water, mix, and make up to 1 L final volume with additional water. Store at room temperature for up to 1 yr.

6. Sodium hydroxide (NaOH), 0.1 M: Dilute 1:10 Reagent 5 with bidistilled water. Store at room temperature for up to 1 yr.

7. Aluminum chloride (AlCl$_3$), 0.1 M: Add 7.80 g of AlCl$_3$ H$_2$O to 700 mL of bidistilled water, mix, and make up to 1 L, final volume with additional water. Store at room temperature for up to 1 yr.

8. Hydrochloric acid (HCl), 0.1 M: Add 8.21 mL of reagent-grade 37% HCl, specific gravity 1.19, per liter of the final solution desired. Mix, thoroughly cool, and store at 4°C for up to 1 yr.

9. Invertase (β-D-fructofuranoside fructohydrolase, EC 3.2.1.26, from yeast, Fluka, I-9274) solution: Prepare by starting from the solid commercial prepa-

rations. Prepare an enzyme solution containing from 0.2 mg mL^{-1} to 5 mg mL^{-1} in acetate buffer, 0.2 M pH 4.65 (Reagent 3). Dispense in small aliquots and store at 4°C. Periodically measure the residual activity of enzyme solutions. When the enzymatic activity is reduced by more than 10 to 20%, prepare fresh enzyme solutions.

10. Clays:

- Na-saturated montmorillonite (M): Ultrasonically disperse a montmorillonite sample from, for instance, Wyoming, Uri, or Crock, in water (usually 10 g of clay in 1000 mL of water, 0.01 w/v ratio) and separate the clay-sized fraction (<2 μm) by sedimentation with an Andreasen pipette. Wash the sedimented clay in centrifuge tubes by adding 1000 mL of 0.1 M HCl, shaking for 1 to 2 min with the vortex, and centrifuging at 15,000 g for 10 min. Then saturate the montmorillonite in sodium by suspending in 1000 mL of 0.1 M NaCl solution, shaking for 1 to 2 min with the vortex and centrifuging at 20,000 g for 10 min. Suspend the centrifuged pellet in 1000 mL of bidistilled water, shake for 1 to 2 min with the vortex, and centrifuge at 20,000 g for 10 min. Repeat this procedure three times. Then suspend the centrifuged pellet in water (usually 1000 mL), dialyze until free of Cl$^-$ ions (equilibrium concentration of ~5 dS m^{-1}), and then freeze-dry. Store the samples at 4°C for no more than 1 yr.

- Aluminum hydroxide (AL): Add slowly (2 mL min^{-1}), drop by drop by using a burette, 0.1 M NaOH under vigorous stirring to a 1-L 0.1 M AlCl$_3$ solution until to pH 8. After 24 h, centrifuge the suspension at 10,000 g for 10 min and wash the precipitate with bidistilled water at a 0.01 w/v ratio. Repeat this step until no Cl$^-$ ions are detected (equilibrium concentration of ~5 dS m^{-1}) in the supernatant. Suspend the precipitate in water at 0.01 w/v ratio) and then freeze-dry.

- Al(OH)$_x$–montmorillonite complexes (AM3, AM9, AM18): Prepare mixtures of AlCl$_3$ and montmorillonite to have 3, 10, or 18 cmol (+) kg^{-1} Al per gram of clay. To each mixture, add 0.1 M NaOH up to pH 8. Wash the final suspension free of Cl$^-$ ions and then freeze-dry.

11. Saccharose (substrate), 0.2 M (saturated): Weigh 68.46 g of reagent-grade saccharose in 700 mL of Reagent 3, measure pH and, if necessary, adjust to exactly pH 4.65 with few drops of either 1 M NaOH or concentrated acetic acid. Make up to 1 L with additional Reagent 3. Mix, thoroughly cool, and store at 4°C. The solution is stable for more than 1 mo at 4°C.

12. Bio-Rad protein assay kit (Bradford, 1976): Use the commercial solutions without any further modification.

13. Reagent for the determination of reducing sugars (Nelson–Somoji Method, Nelson, 1944):

- Reagent A: Add 25 g of anhydrous Na$_2$CO$_3$, 25 g of potassium sodium tartrate [COOK(CHOH)$_2$COONa·4H$_2$O], 20 g of NaHCO$_3$, and 200 g of anhydrous Na$_2$SO$_4$ to 800 mL of bidistilled water, mix, and make up to 1 L final volume with additional water. Store at room temperature (not less than 20°C) to avoid the precipitation of low soluble salts. Stable for 1 mo.

- Reagent B: Add 15 g of CuSO$_4$·15H$_2$O to 100 mL of bidistilled water, mix, add a few drops of reagent-grade 99.5% sulfuric acid, and store at room temperature for up to 1 mo.

- Reagent C: Add 25 g of $(NH_4)_6Mo_7O_{24}\cdot4H_2O$ to 450 mL of bidistilled water, add 21 mL of reagent-grade 99.5% sulfuric acid, and mix. Then add 3 g of $Na_2HAsO_4\cdot7H_2O$, previously dissolved in 25 mL of bidistilled water, mix, and make up to 500 mL final volume with additional water. Incubate the solution for 24 to 48 h at 37°C in a dark bottle to stabilize the solution. Store at room temperature for up to 2 wk.

14. Nelson color reagent (Reagent A + Reagent B): Prepare this reagent immediately before use by mixing 25 mL of Nelson Reagent A with 1 mL of Nelson Reagent B.

15–3.2.2.3 Standard Solutions and Calibration Curves

1. Stock solution of bovine serum albumin (BSA): Prepare a BSA solution containing 1.5 mg mL^{-1} in bidistilled water. Store the solution at –20°C.

2. Prepare different samples in test tubes with three to five dilutions of Reagent 1 (0.2 M NaOAc) and containing from 0.2 to about 1.5 mg mL^{-1} protein. Pipette 100 µL of standards into clean, dry test tubes and determine the concentration of the protein by the Bio-Rad reagent, used without any further modification, following the instructions given by the supplier.

3. Plot a standard calibration curve using from 2.5 to 20 µg BSA.

4. Stock solution (0.2 mg mL^{-1}) of glucose: Dissolve 4 g of glucose in 200 mL of bidistilled water. Mix and thoroughly cool. Store the solution in small aliquots (5 mL) at –20°C.

5. Prepare different samples in test tubes with different volumes of glucose Stock solution 4 and containing up to 200 µg of glucose. Add complementary volumes of bidistilled water to the final volume of 1 mL. Add 1 mL of Nelson color reagent (Reagent 14). Incubate at 100°C for 30 min in a thermostatic bath set at 100°C.

6. Cool at room temperature, then add 1 mL of Nelson Reagent C. Mix on vortex until no CO_2 is evolved from the solution. Add water up to 10 mL final volume. Mix accurately and read at 520 nm within 10 h.

7. Plot a standard calibration curve using the different glucose concentrations and determine the Nelson factor (NF) as the slope of the curve. Sensitivity of the method is 5 to 200 µg of glucose.

15–3.2.2.4 Preparation of Complexes

1. Weigh 40 mg of the selected adsorbant (M, AL, or AM3, AM9, AM18) into 10-mL Nalgene centrifuge tubes. Add 1 mL of 0.2 M pH 4.65 sodium acetate buffer (Reagent 3) containing a suitable amount of enzyme (usually from 0.2 to 5 mg mL^{-1} of invertase) and place in a thermostatically controlled room at 10°C. A cryostat, a thermostatic bath, or an incubator set at 10°C also could be used.

2. Incubate for 1 h at 10°C, swirling the suspension every 10 min.

3. Centrifuge at 10,000 g for 30 min at 4°C.

4. Collect the supernatant in a test tube, wash the precipitate (enzyme–clay complexes) by adding 1 mL of sodium acetate buffer and centrifuging at

10,000 g for 30 min at 4°C. Collect the washing. Repeat the washing procedure two more times; collect each washing separately in a test tube.

5. Determine the protein content in each supernatant and washing using the Bio-Rad protein assay kit.

6. Suspend the enzyme–clay complex in 0.2 M pH 4.65 sodium acetate buffer and store in small aliquots at 4°C. Periodically measure the residual activity of the enzyme suspensions. When the enzymatic activity is reduced by more than 10 to 20%, prepare fresh enzyme complexes.

7. For all determinations, use aliquots of the suspended complexes after rapid equilibration at room temperature. Do not reuse the remaining amount.

8. Also prepare a control sample with the free enzyme solution (Reagent 9), processing as the complexes but omitting the adsorbant.

15–3.2.2.5 Activity Assays

Measure the activity of invertase in the control, supernatants, and washings (free enzyme) and in the suspension of the immobilized enzyme as follows:

For free invertase:

1. Incubate a suitable volume of free enzyme solution (generally 0.05 mL of 0.01 mg mL^{-1} invertase) with 1 mL of 0.2 M saccharose buffered solution (Reagent 11) at 10°C. Usually 20 min is a suitable incubation time. It can be reduced or increased depending on the initial activity of the enzymatic solution.

2. Determine the reducing sugars, expressed as glucose, by the Nelson–Somoji method as described in No. 5 and 6 of 15–3.2.2.3.

For adsorbed invertase:

3. Incubate at 10°C a suitable volume of adsorbed enzyme suspension (generally 0.030 or 0.050 mL) with 1 mL of 0.2 M saccharose buffered solution (Reagent 11). Usually 20 min is a suitable incubation time. It can be reduced or increased depending on the initial activity of the enzymatic solution.

4. Filter the suspension.

5. Determine the concentration of the reducing sugars in the filtrate by the Nelson reagent as described in No. 5 and 6 of 15–3.2.2.3.

The addition of the Nelson color reagent (Reagent 14) stops the enzymatic reaction by a sudden change of the pH.

15–3.2.2.6 Calculations

Protein content:

1. Determine the amount of protein in the supernatants and washings by the calibration curve with BSA.

2. Determine the adsorbed protein (mg g^{-1}clay) by the difference between the amount initially added and that determined by the sum of amounts determined in the supernatant and washings.

Activity of the free enzyme:

The activity of free invertase is calculated by Eq. [1] and is expressed in Units mL^{-1}, i.e., mmol of reducing sugar min^{-1} mL^{-1} of free enzyme solution:

$$U\ mL^{-1} = (A_{520} \cdot V_r \cdot F_d)/(NF \cdot t \cdot V_s) \qquad [1]$$

Where A_{520} is the absorbance at 520 nm, V_r is the reaction volume in mL, F_d is the dilution factor by Nelson reagent addition of 10, NF is the Nelson factor as calculated by the calibration curve of the Nelson reagent, V_s is the volume of enzyme solution in mL, and t is the reaction time in minutes.

Activity of the adsorbed enzyme:

The activity of adsorbed invertase is calculated by Eq. [1], where V_s is the volume of the adsorbed enzyme suspension and expressed in Units mL^{-1}, i.e., mmol of reducing sugar min^{-1} mL^{-1} of immobilized enzyme suspensions.

Specific activity of the free and adsorbed enzyme:

The specific activity is calculated by dividing U mL^{-1} as from Eq. [1] per milligram of free or immobilized enzyme contained in the volume (mg mL^{-1}):

$$Specific\ activity = U\ mL^{-1}/mg\ mL^{-1} \qquad [1a]$$

15–3.2.3 Urease Adsorption Method (Gianfreda et al., 1992)

15–3.2.3.1 Apparatus
As described in section 15–3.2.2.1.

15–3.2.3.2 Reagents

1. Sodium phosphate monobasic (NaH$_2$PO$_4$), 0.1 M: Add 11.99 g of anhydrous NaH$_2$PO$_4$ to 700 mL of bidistilled water, mix, and make up to 1 L final volume with additional water. Store at 4°C for up to 3 mo.
2. Sodium phosphate bibasic (Na$_2$HPO$_4$), 0.1 M: Add 14.20 g of anhydrous Na$_2$HPO$_4$ to 700 mL of bidistilled water, mix, and make up to 1 L final volume with additional water. Store at 4°C for up to 3 mo.
3. Sodium phosphate buffer, 0.1 M pH 7.0: Mix suitable volumes of Reagent 1 and of Reagent 2 (usually 195 and 305 mL, respectively) to reach pH 7.0 and 1 L final volume. Check pH. If necessary, correct using 1 M NaOH or concentrated phosphoric acid. Store at 4°C for up to 3 mo.
4. Sodium phosphate buffer, 0.1 M pH 7.0–1 mM ethylenediaminetetraacetic acid (EDTA): Prepare the sodium phosphate buffer as in No. 3. Add 372.24 mg of EDTA disodium salt dihydrate (C$_{10}$H$_4$N$_2$Na$_2$O$_8$·2H$_2$O) to 800 mL of buffer, stir, check pH, and add additional buffer to reach 1 L final volume. The EDTA is added as a complexing agent for ions possibly interfering with the SH group present in the urease active site and involved in the catalytic mechanism. Store at 4°C for up to 3 mo.
5. Sodium chloride (NaCl), 1 M: Prepare as described in section 15–3.2.2.2.
6. Sodium hydroxide (NaOH), 1 M: Prepare as described in section 15–3.2.2.2.
7. Sodium hydroxide (NaOH), 0.1 M: Prepare as described in section 15–3.2.2.2.

8. Aluminum chloride ($AlCl_3$), 0.1 M: Prepare as described in section 15–3.2.2.2.

9. Hydrochloric acid (HCl), 0.1 M: Prepare as described in section 15–3.2.2.2.

10. Urease (amidohydrolase E.C. 3.5.1.5 from jack bean, Sigma U-0376) solution: Prepare enzyme samples by starting from available commercial preparations. Depending on the formulation of these latter, prepare an enzyme solution containing from 0.1 to 1 mg mL^{-1} of protein in phosphate buffer, 0.1 M pH 7 (Reagent 3). The enzyme solutions are dispensed in small aliquots and store at 4°C. Periodically measure the residual activity of the enzyme solutions. When the enzymatic activity is reduced by more than 10 to 20%, prepare fresh enzyme solutions.

11. Sulfuric acid, 98%, reagent grade.

12. Clays: Prepare as described in section 15–3.2.2.2.

13. Urea (substrate) solution, 0.150 mM: Weigh 9.0 g of reagent-grade urea in sodium phosphate buffer, 0.1 M pH 7.0–1 mM EDTA (Reagent 4). Store at –20°C for up to 2 mo.

14. Bio-Rad protein reagent: Use the commercial solutions without any further modification.

15. Reagents for determination of NH_4^+ (hypochlorite–alkaline phenol method, Charney and Myrback, 1962):

    • Reagent A (phenolic reagent): Add 50 g of phenol and 0.25 g of sodium nitroprussiate ($Na_2Fe(CN)_5NO\cdot2H_2O$) to 800 mL of bidistilled water, mix, and make up to 1 L final volume with additional water. Store at 4°C in a dark bottle (stable for at least 2 mo).

    • Reagent B (sodium hydroxide–sodium hypochlorite, NaOH–NaOCl): Add 25 g of NaOH and 2.1 g of NaOCl to 800 mL of bidistilled water, mix, and make up to 1 L final volume with additional water (21 mL of sodium hypochlorite solution, active Cl$^-$ 10%). Store at 4°C in a dark bottle (stable for almost 2 mo).

15–3.2.3.3 Standard Solutions and Calibration Curves

Prepare a calibration curve with BSA as described in section 15–3.2.2.3.

1. Prepare a calibration curve with BSA as described in No. 1 to 3 of section 15–3.2.2.3.

2. Stock solution of NH_4^+: Dissolve 0.472 g of ammonium sulfate ((NH_4)$_2$SO$_4$) in 700 mL of bidistilled water. Then add, drop by drop, reagent-grade concentrated sulfuric acid until pH 2.0 is reached, make up to 1 L with additional water, and mix. Store the solution at 4°C up to 1 mo.

3. Prepare calibration samples in test tubes with 0, 0.05, 0.1, 0.15, and 0.2 mL of Stock solution 2. Prepare the same calibration samples using Stock solution 2 diluted 1:10 with distilled water. Add complementary volumes of water to the final volume of 0.2 mL. Rapidly add 1 mL of phenolic Reagent A, then 1 mL of sodium hydroxide–sodium hypochlorite Reagent B. Incubate in water bath at 37°C for 20 min.

4. Cool at room temperature, and then add 4 mL of water. Mix accurately and read at 630 nm within 24 h. The samples should be diluted 1:10 before reading at 630 nm if color intensity of the filtrate falls outside the calibration curve.

5. Plot a standard calibration curve using the different NH_4^+ concentrations and determine a proportionality factor (F_{NH4}) as the slope of the curve.

15–3.2.3.4 Preparation of Complexes

Prepare enzymatic complexes by following the procedure described in detail at 15–3.2.2.4. The only differences are:

- the use of urease (from 0.1 to 1 mg mL^{-1} of enzyme) instead of invertase,
- the use of 0.1 M sodium phosphate buffer pH 7.0 (Reagent 3) for the preparation and washing,
- the use of 0.1 M sodium phosphate buffer pH 7.0–1 mM EDTA for suspending the washed enzyme–clay complexes.

Store the enzyme–clay complex suspensions in small aliquots at 4°C. Periodically measure the residual activity of the enzyme suspensions. When the enzymatic activity is reduced by more than 10 to 20%, prepare fresh enzyme complexes.

For all determinations, use aliquots of the suspended complexes after rapid equilibration at room temperature.

Also prepare a control sample with the free enzyme solution following the same procedure reported above but without the adsorbant and processed as the complexes.

15–3.2.3.5 Activity Assays

Measure the activity of urease in the control, supernatants, and washings (free enzyme) and in the suspension of the immobilized enzyme as follows:

For free urease:

1. Incubate a suitable volume of free enzyme solution (generally 0.015 mL of 0.085 mg mL^{-1} urease solution) with 1 mL of 0.150 M urea solution at 10°C. Usually 10 min is a suitable incubation time. It can be reduced or increased depending on the initial activity of the enzymatic solution.
2. Determine the concentration of NH_4^+ by the hypochlorite–alkaline phenol method as described in No. 3 and 4 of 15–3.2.3.3

For adsorbed urease:

3. Incubate at 10°C a suitable volume of urease complex suspension (generally 0.030 mL) with 1 mL of 0.150 M urea solution. Filter the suspension. Usually 10 min is a suitable incubation time. It can be reduced or increased depending on the initial activity of the enzymatic solution.
4. Determine the concentration of NH_4^+ by the hypochlorite–alkaline phenol method as described in No. 3 and 4 of 15–3.2.3.3.

The rapid and successive addition of the two hypochlorite–alkaline phenol reagents will allow the enzymatic reaction to be stopped by the sudden change of the pH.

15–3.2.3.6 Calculations

Protein content:

As described in section 15–3.2.2.6.

Activity of the free enzyme:

The activity of free urease is calculated by Eq. [2] and expressed in Units mL^{-1}, i.e., μmol of NH$_4^+$ min^{-1} mL^{-1} of free enzyme solution:

$$U \text{ mL}^{-1} = (A_{630} \cdot V_r \cdot F_d)/((F_{NH4}) \cdot 0.2 \cdot t \cdot V_s) \qquad [2]$$

Where A_{630} is the absorbance at 630 nm, V_r is the reaction volume in mL, F_d is the dilution factor, F_{NH4} is the slope of the calibration curve in mM^{-1} cm^{-1}, V_s is the volume of the enzyme solution in mL, 0.2 is the volume (in mL) assayed by the hypochlorite–alkaline phenol method in mL, and t is the reaction time in minutes.

Activity of the adsorbed enzyme:

The activity of adsorbed urease is calculated by Eq. [2], where V_s is the volume of the immobilized enzyme suspension and expressed in Units mL^{-1}, i.e., μmol of NH$_4^+$ min^{-1} mL^{-1} of immobilized enzyme suspension.

Specific activity of the free and adsorbed enzyme:

The specific activity is calculated by dividing U mL^{-1} as from Eq. [2] per milligram of free or immobilized enzyme contained in the volume (mg mL^{-1}) by Eq. [1a] in section 15–3.2.2.6.

15–3.2.4 Acid Phosphatase Adsorption Method (Rao et al., 1996, 2000)

15–3.2.4.1 Apparatus

As described in section 15–3.2.2.1.

15–3.2.4.2 Reagents

1. Sodium acetate (NaOAc), 0.1 M: Add 8.20 g of anhydrous NaOAc to 700 mL of bidistilled water, mix, and make up to 1 L final volume with additional water. Store at 4°C for a max. of 3 mo.

2. Acetic acid (HOAc), 0.1 M: Add 5.72 mL of reagent-grade 99.99% acetic acid, specific gravity 1.05, per liter of the final solution desired. Mix, thoroughly cool, and store at 4°C for a max. of 5 mo.

3. Sodium acetate buffer, 0.1 M pH 5.0: Mix suitable volumes of Reagent 1 and of Reagent 2 (usually 700 and 300 mL, respectively) to reach pH 5.0 and 1 L final volume. Check pH. Store at 4°C for a max. of 3 mo.

4. Sodium chloride (NaCl), 0.1 M: Prepare as described in section 15–3.2.2.2.

5. Sodium hydroxide (NaOH), 1 M: Prepare as described in section 15–3.2.2.2.

6. Sodium hydroxide (NaOH), 0.1 M: Prepare as described in section 15–3.2.2.2.

7. Aluminum chloride (AlCl$_3$), 0.1 M: Prepare as described in section 15–3.2.2.2.

8. Hydrochloric acid (HCl), 0.1 M: Prepare as described in section 15–3.2.2.2.

9. Acid phosphatase (E.C 3.1.3.2) solution: Prepare enzyme samples by starting from available commercial preparations. Different procedures may be adopted depending on their formulation. A commercial preparation from sweet potato (type XA, Sigma P-1435) consisting of 57 mg protein mL^{-1} suspension in 1.8 M (NH$_4$)SO$_4$, 10 mM MgCl$_2$, pH 5.3, can be used without any further dilution. Stock the enzyme solutions in small aliquots and stored at –20°C. Periodically measure the residual activity of the enzyme solutions. When the enzymatic activity is reduced by more than 10 to 20%, prepare fresh enzyme solutions.

10. Clays: Prepare as described in section 15–3.2.2.2.

11. *p*-Nitrophenylphosphate (*p*NPP) (substrate, Sigma 73737), 6 mM: Weigh 223 mg of reagent-grade *p*-nitrophenylphosphate in 80 mL of Reagent 3, measure pH and if necessary, adjust to exactly pH 5.0 with a few drops of either 1 M NaOH or concentrated acetic acid. Make up to 100 mL final volume with additional Reagent 3. Mix, thoroughly cool, and store at 4°C. The solution is stable for 2 wk at 4°C. The solution tends to achieve a yellow color indicative of the spontaneous hydrolysis of the product and therefore of the decrease of its concentration.

12. Bio-Rad protein reagent: Use the commercial solutions without any further modification.

15–3.2.4.3 Standard Solutions and Calibration Curves

1. Prepare a calibration curve with BSA as described in No. 1 to 3 in section 15–3.2.2.3.

2. Prepare a calibration curve with different volumes of standard *p*-nitrophenol solution (10 μmol mL^{-1}, Sigma N7660) and containing 0.0 to 0.12 mM *p*-nitrophenol. Add complementary volumes of water to 1 mL final volume. Add 1 mL of 1 M NaOH and read the absorbance at 405 nm with the spectrophotometer. The theoretical extinction molar coefficient of *p*-nitrophenol at 405 nm should be 18.5 cm^{-1} mM^{-1}.

15–3.2.4.4 Preparation of Complexes

Enzymatic complexes are prepared by following the procedure described in detail in 15–3.2.2.4. The only differences are:

- the use of phosphatase (from 0.005 to 0.2 mg mL^{-1} of enzyme) instead of invertase,
- the use of 0.1 M sodium acetate buffer pH 5.0 (Reagent 3) for the preparation and washing,
- the use of 0.1 M sodium acetate buffer pH 5.0 (Reagent 3) for suspending the washed enzyme–clay complexes.

Store the enzyme–clay complex suspensions in small aliquots at 4°C. Periodically measure the residual activity of the enzyme suspensions. When the enzymatic activity is reduced by more than 10 to 20%, prepare fresh enzyme complexes.

For all determinations, use aliquots of the suspended complexes after rapid equilibration at room temperature.

Prepare a control sample with the free enzyme solution following the same procedure reported above but without the adsorbant and processed as the complexes.

15–3.2.4.5 Activity Assays

Measure the activity of phosphatase in the control, supernatants, and washings (free enzyme) and in the suspension of the immobilized enzyme.

For free phosphatase:

1. Incubate at 10°C a suitable volume of free enzyme solution (generally 0.030 or 0.050 mL) with 1 mL of 6 mM *p*NPP buffered solution. Add 1 mL of 1 M

NaOH to stop the enzymatic reaction. Usually 20 min is a suitable incubation time. It can be reduced or increased depending on the initial activity of the enzymatic solution.

2. Determine the concentration of p-nitrophenol, product of the hydrolytic reaction, by its absorbance at 405 nm with the spectrophotometer.

For adsorbed phosphatase:

3. Incubate at 10°C a suitable volume of adsorbed enzyme suspension (generally 0.03 or 0.05 mL) with 1 mL of 6 mM pNPP buffered solution. Usually 20 min is a suitable incubation time. It can be reduced or increased depending on the initial activity of the enzymatic solution.

4. Add 1 mL of 1 M NaOH.

5. Filter the suspension.

6. Determine the concentration of p-nitrophenol in the filtrate by its absorbance at 405 nm with the spectrophotometer.

15–3.2.4.6 Calculations

Protein content:

As described in section 15–3.2.2.6.

Activity of the free enzyme:

The activity of free phosphatase is calculated by Eq. [3] and expressed in Units mL^{-1}, i.e., μmol of p-nitrophenol min^{-1} mL^{-1} of free enzyme solution:

$$U\ mL^{-1} = (A_{405} \cdot V_r \cdot F_d)/(\varepsilon_{405} \cdot t \cdot V_s) \qquad [3]$$

Where A_{405} is the absorbance at 405 nm, V_r is the reaction volume in mL, F_d is the dilution factor by NaOH addition, ε_{405} is the extinction molar coefficient of p-nitrophenol at 405 nm in cm^{-1} mM^{-1} (usually ε_{405} is 18.5), V_s is the volume of enzyme solution in mL, and t is the reaction time in minutes.

Activity of the adsorbed enzyme:

The activity of adsorbed phosphatase is calculated by Eq. [3], where V_s is the volume of the adsorbed enzyme suspension and expressed in Units mL^{-1}, i.e., μmol of p-nitrophenol min^{-1} mL^{-1} of immobilized enzyme suspension.

Specific activity of the free and adsorbed enzyme:

The specific activity is calculated by dividing U mL^{-1} as from Eq. [3] per milligram of free or immobilized enzyme contained in the volume (mg mL^{-1}) by Eq. [1a] in section 15–3.2.2.6.

15–3.3 Comments

The procedures reported in sections 15–3.2.2, 15–3.2.3, and 15–3.2.4 can be applied to other clays such as illite, kaolinite, bayerite, sepiolite, montmorillonite, clays saturated with different cations (Mg, Ca, and others), synthetic beidellite clays (having only Al in the tetrahedral sheets and with variable Si/Al ratio in the tetrahedral layers), aluminum-pillared interlayered clay oxides (goethite, Al-substituted hematite), montmorillonite coated with different metal hydroxides (Fe oxyhydroxides), and hydroxyapatite. At the same time, other enzymes such as alkaline phosphatase, cellulase, glucoamylase, laccase, lipase, peroxidase, and tyrosinase can be used for immobilization (Fusi et al., 1989; Carrasco et al., 1995; Huang et al., 1995; Naidja et al., 1995, 1997; Marzadori et al., 1998b; Huang et al., 1999; Huang, and Shindo 2000; Safari Sinegani et al., 2005; Gopinath and Sugunan, 2005; Cheng et al., 2006; Ahn et al., 2007; Secundo et al., 2008). The preparation of these supports is performed by following the procedure reported above and possibly by changing the solutions used for saturating the clay or for preparing the coated clay (refer to the reported published papers). Substrates, buffers, pHs, and procedures for activity assays are chosen according to the enzyme under investigation.

Special attention, however, must be paid to the following points:

1. When different clay types are used, the procedure for preparing the complexes can be modified for adapting it to the different solid phase. Usually, solid phase/buffer volume or solid phase/enzyme ratios are changed to choose the best conditions to obtain the maximum adsorption of the enzyme on the selected matrix.

2. Activity assays of the complexes should be performed at the same temperature utilized in the preparation experiments to assure that the interaction equilibrium reached by enzyme molecules and support particles in the complexes were not disturbed.

3. The residual activity of the immobilized enzyme, i.e., that retained after its immobilization on the support, is often very low when compared with the activity of the enzyme in the control. This could render difficult the measurement of the kinetics and stability of the immobilized enzyme. Therefore, when the residual activity is less than 10% of that of the control, preparing complexes with higher initial amounts of enzyme to increase the retained activity is suggested. Since this aspect is strongly influenced by both the specific activity of the commercial product used and other properties, another commercial preparation loading more enzymatic units for mg of protein could be used. Preparation of the complexes should be performed at the pH at which the enzyme shows its maximum activity and stability. Usually these two factors coincide; if not, the choice of the pH will depend on the objective of the research.

4. Fresh preparations of adsorbed enzyme complexes are usually made when the activity of the complexes is reduced by 10 to 20% of that initially determined.

5. The activity of the adsorbed enzyme should be expressed as residual activity, i.e., by referring to the activity of the free enzymes in the control. Lower activity usually is measured as the result of the immobilization process. In some cases, however, no reduction or even an increase of the enzymatic activity can be measured for the immobilized enzyme. In this last case, the immobilization procedure might have favored a better conformation of the enzyme or caused its activation by means of residues or groups existing on the support.

The methodologies reported in sections 15–3.2.2, 15–3.2.3, and 15–3.2.4 refer to one amount of support and a range of protein concentrations. Adsorption isotherms can be determined by plotting the amount of protein adsorbed vs. equilibrium protein concentrations. Similarly, to investigate the effect of pH or temperature on the adsorption of the protein, experiments can be performed by fixing the amount of both the clay and the protein and by changing the pH or the temperature of the buffer used in the preparation of the enzymatic complexes.

Several analyses can be performed for characterizing the structural characteristics of the enzymatic complexes. X-ray diffraction and Fourier transform infrared (FTIR) analyses may give information on the allocation of the protein molecules and/or their binding to the inorganic support. Further information can be derived by determining the surface areas of the clays before and after their interaction with the proteins.

Moreover, the kinetics, the stability, and the activity–pH profiles of the adsorbed enzymes as compared with those of the free enzyme also can be determined.

Research results reported in the literature show that the amount of enzymes adsorbed, as well as the resulting properties of the formed complexes, are strictly dependent on the nature and properties of both enzymes and clays. For instance, studies devoted to investigating the influence of the coating of a pure montmorillonite with different amounts of OH-Al species on the adsorption process indicated that the partial coating strongly influenced the affinity of enzymes for the phyllosilicates, as well as the activity of the adsorbed enzymes. Both invertase and urease were adsorbed with a higher binding energy and adsorption intensity on pure (bare) montmorillonite as compared with AM and AL (Gianfreda et al., 1991, 1992). A negative effect on the amount of enzyme adsorbed by OH-Al species coated on M surfaces was also observed for urease. By increasing the Al species loadings on montmorillonite surfaces from 3 to 18 cmol $_{(+)}$ kg^{-1} clay, the percentage of adsorbed urease decreased from 82 to 69% (Gianfreda et al., 1992). Similar results were obtained by Naidja et al. (1995) when tyrosinase was adsorbed on a montmorillonite with different levels of hydroxyaluminum coatings. Conversely, the presence of polymeric oxyhydroxides of Fe(III) on montmorillonite, illite, and kaolinite did not strongly influence the extent of β-lactoglobulin adsorption, whereas lower levels of catalase were usually loaded by the dirty clays as compared with the corresponding Ca-saturated phyllosilicates (Fusi et al., 1989).

The presence of OH-Al species on montmorillonite surfaces resulted in lower activity levels for invertase and urease, whereas a higher activity was measured for phosphatase (Gianfreda et al., 1991, 1992; Rao et al., 1996). This effect was much more evident when urease and phosphatase were adsorbed on Al–montmorillonite complexes. Higher amounts of Al species on M surfaces resulted in lower activity of urease and the higher phosphatase activity.

Also, the nature of the saturating cation may influence the amount of adsorbed enzyme and consequently its activity level. Among three homoionic sepiolites (H$^+$, Na$^+$, and Ca^{2+}), Na-sepiolite presented the highest level of alkaline phosphatase activity, retaining nearly 50% of the added enzyme (Carrasco et al., 1995). No substantial changes of activity, or even of kinetic parameters, were shown by laccase from *Trametes villosa* when adsorbed on short-range ordered aluminum hydroxide. Moreover, attenuated total reflectance infrared (ATR-IR) spectroscopy did not show significant changes in the secondary structure of laccase due to adsorption on aluminum hydroxide (Ahn et al., 2007).

In the study of Secundo et al. (2008), it was demonstrated that the amount of lipase immobilized on the clay surface increased with decreasing levels of aluminum substitution in the tetrahedral layer. The specific activity of the immobilized lipases depends on both the clay composition (the samples with low content of aluminum exhibited the highest specific activity) and the amount of adsorbed enzyme it increases as the amount of enzyme decreases. Also, the catalytic activity of lipase deposited on H^+-beidellite is higher than that of enzyme deposited on the Na^+ form (Secundo et al., 2008).

15–4 ENZYME COMPLEXES WITH ORGANIC AND ORGANO–MINERAL SUPPORTS

15–4.1 Introduction

In soil, associations between enzymes and organic matter, and much more so with organo–mineral complexes, are expected to be even more common than clay–enzyme complexes. Indeed, several soil constituents (e.g., pure clays, phenolic compounds, inorganic components such as Fe, Mn, and Al oxides and hydroxides, primary minerals, and OH-Al polymers) may interact with each other, giving rise to stable organo–mineral complexes. Enzymatic proteins, present in the environment, may be involved in the formation of one or more of these complexes or may interact with them after their formation. Interactions between enzymes and polysaccharidic materials also may occur, mainly in the rhizosphere.

In these processes, monomers, polymers, and noncrystalline products of aluminum and iron, as well as ions and manganese compounds, play an important role. They may facilitate the interaction between organic molecules and clay interfaces and produce organo–mineral complexes, showing peculiar physicochemical properties and reactivity toward nutrients, organics, and pollutants (Violante and Gianfreda, 2000). Furthermore, they may favor the interaction between the enzyme and the organic matrix or accelerate the abiotic polymerization of simple and complex aromatic compounds (e.g., mono- and polyphenols, starting humic-like precursors), favoring the formation of polymeric aggregates (Burns, 1986; Huang, 1990) (see below).

15–4.2 Complexes of Acid Phosphatase with Tannic Acid, Hydroxyl–Aluminum (OH-Al) Species, and Montmorillonite (Rao et al., 1996, 2000)

15–4.2.1 Complex Synthesis Method

15–4.2.1.1 Apparatus

As described in section 15–3.2.2.1.

15–4.2.1.2 Reagents

1. Sodium acetate (NaOAc), 0.1 M: Prepare as described in section 15–3.2.4.2.
2. Acetic acid (HOAc), 0.1 M: Prepare as described in section 15–3.2.4.2.
3. Sodium acetate buffer, 0.1 M pH 5.0: Prepare as described in section 15–3.2.4.2.
4. Sodium chloride (NaCl), 0.1 M: Prepare as described in section 15–3.2.2.2.
5. Sodium hydroxide (NaOH), 1 M: Prepare as described in section 15–3.2.2.2.
6. Sodium hydroxide (NaOH), 0.2 M: Prepare by diluting 1:5 Reagent 5.
7. Aluminum chloride ($AlCl_3$), 0.2 M: Prepare as described in section 15–3.2.2.2 but doubling the amount of $AlCl_3$.
8. Hydrochloric acid (HCl), 0.1 M: Prepare as described in section 15–3.2.2.2.

9. Acid phosphatase (E.C 3.1.3.2) solution: Prepare as described in section 15–3.2.4.2.

10. Tannic acid: Use reagent grade as powder (Fluka 96311).

11. Na-saturated montmorillonite (M): Prepare as described in section 15–3.2.2.2.

12. Soluble OH-Al species [Al(OH)$_x$]: Add 0.2 M NaOH solution slowly (2 mL min^{-1}) under vigorous stirring to 0.2 M AlCl$_3$ solution until pH 4.5 is reached. After aging for 2 wk at room temperature, filter the solution through a <0.025-μm filter to remove solid precipitation products. Determine aluminum concentration in the filtrate by atomic absorption spectrometry. Store at 4°C for a max. of 1 yr.

13. p-Nitrophenylphosphate (pNPP) (substrate, Sigma 73737), 6 mM: Prepare as described in section 15–3.2.4.2.

14. Protein reagent: Use a commercial protein assay kit such as Protein Assay ESL (Roche, Germany). This method does not present interference with phenolic molecules. Another possibility is the use of the Lowry assay as modified by Winters and Minchin (2005).

15–4.2.1.3 Standard Solutions and Calibration Curves

1. Prepare a calibration curve with BSA as described in No. 1 to 3 in section 15–3.2.2.3 and using the appropriate protein reagent.

2. Standard p-nitrophenol solution: Prepare as described in section 15–3.2.4.3.

3. Prepare a calibration curve with different volumes of standard p-nitrophenol solution as described in section 15–3.2.4.3.

15–4.2.1.4 P reparation of Complexes

Tannic–enzyme complexes:

1. Weigh 0.2 mg of tannic acid into a 12-mL Nalgene centrifuge tube. Add 1 mL of 0.1 M sodium acetate buffer pH 5.0 (Reagent 3) containing 0.02 mg of acid phosphatase (tannic acid/phosphatase ratio w/w of 10) and place in a thermostatic bath at 30°C.

2. Incubate at 30°C for 1 h. Swirl the suspension every 10 min.

3. Centrifuge at 10,000 g at 4°C for 30 min and collect the supernatants in test tubes and the precipitates in centrifuge tubes.

4. Wash the solid precipitates (enzyme complexes) with 1 mL of Reagent 3. Centrifuge for 30 min at 10,000 g at 4°C and collect the washing separately.

5. Repeat three times as described in No. 3 and collect the washings separately.

6. Determine the protein content of supernatants and washings using the Protein Assay ESL kit (Roche, Germany) or another appropriate protein assay.

7. Suspend the enzyme complex in 1 mL of 0.1 M sodium acetate buffer pH 5.0 (Reagent 3) and use immediately.

8. When necessary, store the enzyme complex suspensions at 10°C and pH 5.0 and measure the residual activity periodically. When the enzymatic activity is reduced by more than 10 to 20% compared with that of the initial preparation, prepare fresh enzymatic complexes.

9. Prepare a control sample with the free enzyme solution following the same procedure reported above but without the matrices and processed as the complexes.

Tannic–OH-Al–enzyme complexes:

Weigh 0.2 mg of tannic acid into a 12-mL Nalgene centrifuge tube. Add 1 mL of 0.1 M sodium acetate buffer pH 5.0 (Reagent 3) containing 0.02 mg of acid phosphatase (tannic acid/phosphatase ratio w/w of 10) and a suitable volume (usually 0.06 mL) of soluble OH-Al species ($Al(OH)_x$) (Reagent 12) to have $Al(OH)_x$/tannic acid molar ratios of 50, place in a thermostatic bath at 30°C. Continue from No. 2 to 9 previously described.

Tannic–OH-Al–montmorillonite–enzyme complexes:

Weigh 0.2 mg of tannic acid into a 12-mL Nalgene centrifuge tube. Add 1 mL of 0.1 M sodium acetate buffer pH 5.0 (Reagent 3) containing 0.02 mg of acid phosphatase (tannic acid/phosphatase ratio w/w of 10), a suitable volume (usually 0.06 mL) of soluble OH–Al species ($Al(OH)_x$) (Reagent 12) to have $Al(OH)_x$/tannic acid molar ratios of 50, and suitable amounts of montmorillonite to have $Al(OH)_x$/montmorillonite ratios of 10 or 15 cmol of Al kg^{-1} clay. Place in a thermostatic bath at 30°C and continue from No. 2 to 9 previously described.

15–4.2.1.5 Activity Assays

Measure the activity of phosphatase in the control, supernatants, and washings (free enzyme) and in the suspensions of the immobilized enzyme.

For free phosphatase, measure as described in section 15–3.2.4.5 but at 30°C.

For immobilized phosphatase:

1. Incubate a suitable volume of adsorbed enzyme suspension (generally 0.030 or 0.050 mL) with 1 mL of 6 mM p-nitrophenylphosphate (pNPP) buffered solution at 30°C. Usually 10 min is a suitable incubation time. It can be reduced or increased depending on the initial activity of the enzymatic solution. Add 1 mL of 1 M NaOH to stop the hydrolysis.

2. Filter the suspension.

3. Determine the concentration of p-nitrophenol in the filtrate by its absorbance at 405 nm with the spectrophotometer.

15–4.2.1.6 Calculations

Protein content:

As described in section 15–3.2.2.6 but using the appropriate protein reagent.

Activity of the free enzyme:

The activity of the free phosphatase is calculated by Eq. [3] reported in section 15–3.2.4.6 and is expressed in Units per mL. The specific activity is calculated by dividing U mL^{-1} per mg of free enzyme contained in the volume (mg mL^{-1}) by Eq. [1a] in section 15–3.2.2.6.

Activity of the immobilized enzyme:

The activity of immobilized phosphatase is calculated as described in section 15–3.2.4.6. The specific activity is calculated as described in 15–3.2.4.6 if it is possible to measure the protein content.

15–4.2.2 Comments

The same experimental procedure reported for phosphatase can be applied to other enzymes, such as invertase (Gianfreda et al., 1993) and urease (Gianfreda et al. (1995a,b). In these cases, different experimental conditions in terms of pH, temperature, tannic acid/enzyme, $Al(OH)_x$/tannic acid molar ratios, and $Al(OH)_x$/montmorillonite ratios were used. The reagents, standards, activity assays and analytical procedures, and calculations are those described in sections 15–3.2.2 and 15–3.2.3.

15–4.3 Complexes of Tyrosinase with Natural Humic Acids

15–4.3.1 Complex Synthesis Method (Ruggiero and Radogna, 1988)

15–4.3.1.1 Apparatus

As described in section 15–3.2.2.1, plus:

- Spectra/Por 1 Dialysis membranes (MW cutoff 6000–8000 Da)
- Biological oxygen monitor equipped with a Clark oxygen electrode and a linear recorder
- Muffle furnace

15–4.3.1.2. Reagents

1. Sodium phosphate monobasic (NaH_2PO_4), 0.05 M: Add 5.99 g of anhydrous NaH_2PO_4 to 700 mL of bidistilled water, mix, and make up to 1 L final volume with additional water. Store at 4°C for a max. of 2 to 3 mo.

2. Sodium phosphate bibasic (Na_2HPO_4), 0.05 M: Add 7.10 g of anhydrous Na_2HPO_4 to 700 mL of didistilled water, mix, and make up to 1 L final volume with additional water. Store at 4°C for a max. of 2 to 3 mo.

3. Sodium phosphate buffer, 0.05 M pH 7.0: Mix suitable volumes of Reagent 1 and of Reagent 2 (usually 390 and 610 mL, respectively) to reach pH 7.0 and 1 L final volume. Check pH. Store at 4°C for a max. of 3 mo.

4. Calcium acetate [$Ca(OAc)_2$], 0.1 M: Add 15.82 g of anhydrous $Ca(OAc)_2$ to 700 mL of bidistilled water, mix, and make up to 1 L final volume with additional water. Store at 4°C for a max. of 2 to 3 mo.

5. Acetic acid (HOAc), 0.2 M: Prepare as described in section 15–3.2.2.2.

6. Calcium acetate buffer, 0.1 M pH 4.5: Mix suitable volumes of Reagent 4 and of Reagent 5 (usually 120 and 880 mL, respectively) to obtain pH 4.5 and 1 L final volume. Store at 4°C for a max. of 3 mo.

7. Catechol (substrate, Sigma-Aldrich C9510) solution, 36 mM: Dissolve 396.60 mg of reagent-grade catechol in 80 mL of Reagent 3, check pH and if necessary, adjust to exactly pH 7.0 with few drops of either 1 M NaOH or concentrated phosphoric acid. Make up to 100 mL final volume with additional Reagent 3. Mix, thoroughly cool, and store at 4°C. The solution is stable for up to 2 wk a 4°C. The solution tends to develop a yellow color

indicative of the spontaneous oxidation of the product and therefore of the decrease of its concentration.

8. Sodium hydroxide (NaOH), 0.5 M: Add 20 g of NaOH to 700 mL of didistilled water, mix and make up to 1 L final volume with additional water. Stable at room temperature for more than 1 yr.

9. Hydrochloric–hydrofluoric (HCl-HF) solution, 0.5% (v/v): To reduce the ash content of the lyophilized humic acid (HA), perform an additional purification by shaking HA in a large excess of HCl–HF solution, centrifuging at 10,000 g for 30 min, and then washing the residue with distilled water until free of Cl^- by following the protocol in 15–3.2.2.2. Lyophilize the HA and then dry at room temperature in a dessicator over P_2O_5.

10. Calcium chloride ($CaCl_2$), 0.5 M: Add 55.45 g of anhydrous $CaCl_2$ to 700 mL of bidistilled water mix and make up to 1 L final volume with additional water. Stable at room temperature for more than 1 yr.

11. Polyethylene glycol: Use the commercial preparation without purification.

12. Humic acid (HA): Prepare HA from a soil according to classical methods. In this case, extract the HA from the A1 horizon of an Andosol with 0.5 M NaOH under N_2. Precipitate HA by adding concentrated HCl to pH 2.0. Purify the HA by dialyzing against water acidified at pH 2.0 and then against water using dialysis tubes with cutoff 8000 Da. Further purify by passing over OH- and H-saturated exchange resins and by treating with HCl–HF solution. Then determine the ash content of the product gravimetrically by weighing and combusting approximately 10 mg of HA in a platinum crucible at a temperature of 600°C until the mass remains constant.

13. Tyrosinase (EC 1.14.18.1 monophenol monooxygenase from mushrooms, Sigma T3824) solution: Prepare enzyme samples by starting from available commercial preparations. Depending on the formulation of the latter, prepare an enzyme (used without any further purification) solution in the buffer. Stock the enzyme solutions in small aliquots and store at 4°C. Periodically measure the residual activity of the enzyme solutions. When the enzymatic activity is reduced by more than 10 to 20%, prepare fresh enzyme solutions.

14. Bio-Rad protein assay kit: Use the commercial solutions without any further modification.

15. Sodium hydroxide (NaOH), 1 M: Prepare as described in section 15–3.2.2.2.

15–4.3.1.3 Preparation of Complexes

1. Mix different amounts of HA and tyrosinase (HA/Tyr ratios from 0.25 to 5.0 [w/w]) in 0.05 M sodium phosphate buffer pH 7.0 and place in a thermostatically controlled room or a simple incubator at 20°C.

2. Add 2 mL of $CaCl_2$, 0.5 M to flocculate the HA–tyrosinase complex. A decrease of pH to 4.5 is usually reached; to be sure, check pH.

3. Shake the HA–enzyme mixture for 6 h at 20°C on an orbital shaker.

4. Centrifuge at 20,000 g at 4°C for 30 min and collect the supernatant in test tubes and the precipitate in centrifuge tubes.

5. Wash the solid precipitates (enzyme–HA complexes) with 1 mL of 0.1 M calcium acetate buffer pH 4.5. Centrifuge at 20,000 g at 4°C for 30 min and collect the supernatant in test tubes and the precipitate in centrifuge tubes.

6. Repeat No. 5 until the washings do not show tyrosinase activity. Usually fives times is sufficient.

7. Dialyze the supernatant against 0.05 M sodium phosphate buffer pH 7.0 and pooled washings.

8. Concentrate the dialyzed samples using polyethylene glycol as the dehydrating agent.

9. Suspend the enzyme–HA complex in 1 mL of 0.1 M calcium acetate buffer pH 4.5 and store in a small aliquot at 4°C. Periodically measure the residual activity of the enzyme suspensions. When the enzymatic activity is reduced by more than 10 to 20%, prepare fresh enzyme suspensions.

10. Also prepare control samples with the free enzyme solution and HA alone following the same procedure reported above but without the matrix and processed as the complexes.

15–4.3.1.4 Activity Assays

Measure the activity of tyrosinase in the control, supernatants, and washings (free enzyme) and in the suspension of the immobilized enzyme. Tyrosinase activity is determined polarographically by measuring O_2 consumption with a Clark oxygen electrode:

1. Incubate at 37°C 2.5 mL of 0.05 M sodium phosphate buffer pH 7.0 containing a suitable volume of free or immobilized tyrosinase and saturated with O_2 in the sample chamber under agitation.

2. Start the reaction by adding 0.5 mL of 36 mM catechol buffered solution through a side port of the electrode.

3. Record the consumption of O_2 with the time.

15–4.3.1.5 Calculations

Activity of the free and immobilized enzyme:

The activity of free or immobilized tyrosinase is calculated by the consumption of O_2 with time. One unit of activity is the amount of catalyst causing the consumption of 1 mol of O_2 min^{-1}.

Specific activity of the free and adsorbed enzyme:

The specific activity is calculated by dividing the units measured in 1 mL of free or immobilized enzyme solution per milligram of free or immobilized enzyme contained in the volume. The specific activity of HA–enzyme complexes also can be calculated per milligram of organic C.

15–4.4 Complexes of Acid Phosphatase with Ca-polygalacturonate

15–4.4.1 Complex Synthesis Method (Marzadori et al., 1998a)

15–4.4.1.1 Apparatus

- Volumetric flasks, 10, 50, 250 mL
- Test tubes, 10, 12 mL
- Erlenmeyer reaction flasks, 10, 50, 250, 1000 mL

- High-speed temperature-controlled centrifuge (>20,000 rpm)
- Atomic absorption spectrophotometer
- UV-Vis spectrophotometer
- Thermostatic bath or water bath shaker
- Nalgene centrifuge tubes, 38 mL
- Microburette
- Horizontal shaker
- "Fast" filters
- pH meter

15–4.4.1.2. Reagents

1. Sodium acetate (NaOAc), 0.1 M: Prepare as described in section 15–3.2.4.2.
2. Acetic acid (HOAc), 0.1 M: Prepare as described in section 15–3.2.4.2.
3. Sodium acetate buffer, 0.1 M pH 6.0: Mix suitable volumes of Reagent 1 and of Reagent 2 (usually 960 and 40 mL, respectively) to reach pH 6 and 1 L final volume. Check pH. Store at 4°C for a max. of 3 mo.
4. Sodium hydrogen maleate ($NaHCH_4H_2O_4 \cdot 3H_2O$), 2 mM: Dissolve 0.232 g of maleic acid in water and mix with 2 mL of 1 M NaOH and dilute to 1 L final volume. Store at 4°C for a max. of 3 mo.
5. Sodium maleate buffer, 2 mM pH 6.0: Mix 250 mL of 2 mM sodium maleate (Reagent 4) with 26.9 mL of 0.01 M NaOH. Dilute to 1 L final volume and measure the pH. Store at 4°C for a max. of 3 mo.
6. Calcium chloride ($CaCl_2$), 0.05 M: Add 5.545 g of anhydrous $CaCl_2$ to 700 mL of didistilled water, mix, and make up to 1 L final volume with additional water. Store at room temperature for a max. of 1 yr.
7. Sodium hydroxide (NaOH), 1 M: Prepare as described in section 15–3.2.2.2.
8. Sodium hydroxide (NaOH), 0.1 M: Prepare as described in section 15–3.2.2.2.
9. Polygalacturonate (Na-PG, Fluka 48280) stock solution: Suspend 1 g of reagent-grade polygalacturonic acid (PGA) in 150 mL of deionized water and adjust the pH to 6.0 with 0.1 M NaOH. Dissolve the PGA and add water to a final volume of 200 mL. The final concentration of the monomer is 26 mM. Store at 4°C for a max. of 3 mo.
10. Ca-polygalacturonate (Ca-PG) gel suspension: Using a microburette, add 5 mL of 0.05 M $CaCl_2$ solution continuously and slowly (1 mL min^{-1}) while stirring (500 rpm) to 5 mL of Na-PG stock solution (0.13 mmol) in a 38-mL Nalgene centrifuge tube. Shake the gelatinous suspension for 15 min on a horizontal shaker (50 cycles min^{-1}). Centrifuge at 11,000 g for 10 min. Discard the supernatant, wash the gel by shaking with 10 mL of deionized water (15 min, 50 cycles min^{-1}), and centrifuge at 11,000 g for 10 min. Discard the supernatant. Wash the gel twice with 10 mL of 2 mM sodium maleate buffer pH 6.0 (15 min, 50 cycles min^{-1}). After each washing, centrifuge at 11,000 g for 10 min and discard the supernatant. Store the resulting Ca-PG gel in 2 mM sodium maleate buffer pH 6.0 at 4°C for a max. of 2 to 3 mo.
11. Acid phosphatase (E.C 3.1.3.2) solution: Prepare enzyme samples by starting from available commercial preparations. In this cited method, a purified

lyophilized acid phosphatase from potato was used. It was purchased from Boeringher (Mannheim). Currently this commercial preparation is not available and other preparations of enzyme must be used.

12. *p*-Nitrophenylphosphate (*p*NPP) (substrate, Sigma 73737) 10 mM: Weigh 371.67 mg of reagent-grade *p*-nitrophenylphosphate in 80 mL of sodium acetate buffer pH 6.0. Check pH and, if necessary, adjust to exactly pH 6.0 with few drops of either 1 M NaOH or concentrated acetic acid; make up to 100 mL final volume with additional Reagent 3. Mix, thoroughly cool, and store at 4°C. The solution is stable for 2 wk a 4°C. The solution tends to develop a yellow color indicative of the spontaneous hydrolysis of the product and therefore of the decrease of it concentration.

15–4.4.1.3. Standard Solutions and Calibration Curves

Prepare a calibration curve with *p*-nitrophenol as described in No. 2 and 3 in section 15–3.2.4.3 but reading the absorbance at 398 nm and determining ε at that wavelength.

15–4.4.1.4 Preparation of Complexes

1. Dissolve 0.165 mg of lyophilized acid phosphatase in 1 mL of 2 mM sodium maleate buffer pH 6.0.

2. Place 4 mL of buffer containing 1 g of Ca-PG slowly (1 mL min^{-1}) in a centrifuge tube. Add Solution 1 with a microburette while stirring at 500 rpm.

3. Shake the mixture on a horizontal shaker (50 cycles min^{-1}) for 2 h at 25°C.

4. Centrifuge at 11,000 g for 10 min at 4°C and collect the supernatant in a test tube and the precipitate in a centrifuge tube.

5. Wash the gel (enzyme–Ca-PG complexes) twice by shaking (50 cycles min^{-1}) with 10 mL of buffer for 15 min and centrifuging at 11,000 g for 10 min at 4°C. Collect the washings separately.

6. Suspend the enzyme–Ca-PG complex in 3 mL of maleate buffer and store in small aliquots at 4°C. Measure the activity periodically. Prepare fresh complexes when it is reduced by more than 10 to 20%.

7. Also prepare a control sample with the free enzyme solution following the same procedure reported above but without the matrix and processed as the complexes.

15–4.4.1.5 Activity Assays

Measure the activity of phosphatase in the control, supernatants, and washings (free enzyme) and in the suspension of the immobilized enzyme.

For free phosphatase:

1. Incubate 3 mL of free enzyme solution with 1 mL of 10 mM *p*NPP buffered solution at 25°C for 10 min. Then add 4 mL of 1 M NaOH to stop the enzymatic reaction.

2. Determine the concentration of *p*-nitrophenol, product of the hydrolytic reaction, by its absorbance at 398 nm with the spectrophotometer.

For adsorbed phosphatase:

3. Incubate 3 mL of immobilized enzyme suspension with 1 mL of 10 mM
   pNPP buffered solution at 25°C for 10 min. Then add 4 mL of 1 M NaOH to
   stop the enzymatic reaction.
4. Filter the suspension with a fast filter.
5. Determine the concentration of p-nitrophenol in the filtrate by its absorbance
   at 398 nm with the spectrophotometer.

15–4.4.1.6 Calculations

Activity of the free enzyme:

The activity of the free or immobilized phosphatase is expressed in units
defined as the amount of enzyme capable of producing 1 mmol of p-nitrophenol
min^{-1} at 25°C:

$$U\ mL^{-1} = (A_{398} \cdot V_r \cdot F_d)/(\varepsilon_{398} \cdot t \cdot V_s) \qquad [4]$$

Where A_{398} is the absorbance at 398 nm, V_r is the reaction volume in mL, F_d is the
dilution factor by NaOH addition, ε_{398} is the extinction molar coefficient of p-nitro-
phenol at 398 nm in cm^{-1} mM^{-1} (usually ε_{398} is 18.9), V_s is the volume of enzyme
solution in mL, and t is the reaction time in minutes.

15–4.4.2 Comments

The same experimental procedure reported for phosphatase was applied to urease,
but different experimental conditions in terms of pH, temperature, and Ca-PG/
enzyme ratios were used (Ciurli et al., 1996). For the reagents, standards, activity assays and analytical procedures, and calculations, refer to Ciurli et al. (1996).

15–4.5 General Comments

Most of the comments reported in Section 15–3.3 are appropriate for the preparation of organo- and organo-mineral complexes. As for the adsorption procedure, it
should also be taken into account for the organic and organo–mineral matrices that:

1. Activity assays of the complexes should be performed at the same temperature utilized in the preparation experiments.
2. The residual activity of the immobilized enzyme, i.e., activity retained after
   its immobilization on the support, is usually very low when compared with
   the activity of the enzyme in the control. This could compromise the measurement of the kinetics and stability of the immobilized enzyme. Therefore,
   when the residual activity is less than 10% of that of the control, preparing
   complexes with higher initial amounts of enzyme to increase the retained
   activity is suggested.
3. Preparation of the complexes should be performed at the pH where the enzyme shows its maximum activity and stability.
4. Immobilized enzyme complexes usually are discarded when the activity of
   the complexes is reduced by 10 to 20% of the initial one.
5. The activity of the adsorbed enzyme should be expressed as residual activity,
   i.e., by referring to the activity of the free enzymes in the control.

Some problems may arise in the preparation and characterization of the enzymatic complexes in the presence of organic compounds.

1. The presence of residual tannic acid molecules in the supernatants and washings (i.e., tannic acid molecules not involved in the interaction with the enzyme) could interfere with the determination of the protein content by the Bio-Rad method. There can be overestimation of the protein amount. In this case, other methods, such as Protein Assay ESL (Roche, Germany) or the modified Lowry method (Winters and Minchin, 2005), capable of quantitatively detecting protein molecules in the presence of phenolic compounds, could be adopted.

2. When it is not possible to measure the amount of protein in mg present in the supernatants and washings, the activity of the immobilized enzyme is expressed in enzymatic units per mL of immobilized enzyme suspension.

3. Organic molecules such as humic acid, tannic acid, and also other phenolic compounds may inhibit the activity of enzymes (Rao et al., 1998; Ruggiero and Radogna, 1988). Therefore, preliminary investigations are needed to evaluate and quantify, if any, the degree of inhibition, and consequently to limit the concentration of soluble organic molecules in the activity assay. The possible direct inhibition of the compound on the enzyme activity is thus avoided. For instance, in the case of phosphatase, previously described, classical inhibition tests performed in the presence of different concentrations of tannic acid indicated that 20 μM tannic acid gave 50% inhibition of the phosphatase activity when tested in the conditions previously reported. Therefore, residual tannic acid concentration in the activity assay must be significantly lower than 20 μM.

The association with organic matrices may influence the binding of enzymes and consequently their residual activity. For instance, depending on the humic acid concentration, from 32 to 76% of initial tyrosinase was bound to humic acid and retained its catalytic activity with substrate specificity similar to that of the free enzyme (Ruggiero and Radogna, 1988). Similarly, phosphatase retained only 40% of its initial activity after its immobilization on Ca-polygalacturonate (Marzadori et al., 1998a).

When enzymes interact with tannic acid, the formation of active tannate–enzyme complexes influences both the contact time and the ratio between tannic acid and enzyme. For example, tannic acid suddenly and tightly interacted with urease molecules to form soluble urease–tannate complexes that were less active at higher tannic acid/urease ratios (Gianfreda et al. (1995a,b). A similar behavior, both in terms of residual activity of enzymatic complexes and formation of insoluble materials, was observed with phosphatase (Rao et al., 1996, 2000). Interaction between invertase and tannic acid occurred at a lower rate than in the case of phosphatase and urease (Gianfreda et al., 1993). However, stable complexes with the three enzymes were formed. When the complexation phenomenon between enzymes and tannic acid occurred in the presence of aluminum species, a higher number of enzymatic molecules were involved in the complexes and showed greater activity levels.

Spectroscopic investigations may provide information on the interaction between the enzyme molecule and the organic support. For instance, IR spectroscopy suggested that carboxyl and phenolic groups of the humic acid were involved in ionic and hydrogen bond formations with the tyrosinase molecule (Ruggiero and Radogna, 1988).

15–5 CHEMICAL BINDING

15–5.1 Introduction

Enzyme immobilization also can occur through chemical binding between the enzymatic molecule and the immobilizing support. While enzyme immobilization by adsorption (section 15–3) or physical entrapment (see section 15–6) has the benefit of a wide applicability and may provide relatively small perturbation of the enzyme's native structure and function. Chemical immobilization may lead to significant and often irreversible deformations of the protein and thereby alteration of its catalytic behavior.

Methods based on chemical binding usually include: (i) enzyme attachment to the matrix by covalent bonds, (ii) cross-linking between enzyme and matrix, and (iii) enzyme cross-linking by multifunctional reagents (Martinek and Mozhaev, 1985; Durán et al., 2002; Dick and Tabatabai, 1999). Covalent binding implies that at least one atom of the protein and one of the support will share electron bonds. Moreover, mono- or multidentate binding may occur between the two components. More than one bond usually is required for the protein to be firmly immobilized.

The enzyme may be linked directly to the support, or a spacer (or arm) may be put between the carrier and the enzyme. In the first case, the risk that one or more groups present in the enzymatic active site are involved in the binding is high, with evident, negative consequences on the catalytic activity of the enzyme. Indeed, the active site can be completely masked and therefore not accessible to the substrate. The presence of a spacer, instead, strongly reduces this risk and the protein can behave as a free enzyme.

Various groups can be used to link the enzyme and support. The most-used are carboxyl, amino, phenolic, sulfhydryl, and hydroxylimidazole groups. The covalent binding of these groups to the support often implies a preventive surface modification or an activation step of the immobilizing support. Silanization, i.e., the coating of the surface with organic functional groups using an organofunctional silane reagent such as 3-aminopropyltriethoxysilane, is a widely used strategy for initial surface modification of inorganic supports. The alkylamine-derivatized support (or even native surface amino groups) can be further derivatized to arylamine group using p-nitrobenzoyl chloride (alkyl amino azo method) or to aldehyde groups using glutaraldehyde (amino alkylation method). Carboxyl and amino groups of the enzyme will be, respectively, involved in the first and second case. Obviously, other chemical methods exist to bind enzymes and supports. For instance, carbodiimide activation is used when a carboxylic group in the support is expected to react with an amino group of the enzyme. This activation is possible by epoxide formation in the side chain or on the support surface. Examples of this activation are activated Sepharose and cellulose.

The cross-linking between enzyme and matrix or between enzyme molecules implies the use of multifunctional reagents such as glutaraldehyde, dimethyl adipimidate, dimethyl superimidate, and aliphatic diamines, capable of creating firm, covalent bridges between the involved components. There are a great variety of such methods, detailed descriptions of which are beyond the scope of this chapter (Martinek and Mozhaev, 1985).

Several supports have been used in the covalent immobilization of enzymes. They include different inorganic carriers and natural or synthetic organic polymers, such as porous glass, ceramics, stainless steel, sand, clay minerals, charcoal, metallic oxides, cellulose, ethylene–maleic acid copolymers, aminoethyl cellulose. In the case of soil enzymes, the most widely used supports have been clays, inert beads, and whole soils, and the most-used immobilization method has been activation of the support by silanization and grafting through glutaraldehyde (Ruggiero et al., 1989; Sarkar et al., 1989; Gianfreda and Bollag, 1994; Gopinath and Sugunan 2005, 2007).

Covalent binding probably is a more advantageous method compared with adsorption or entrapment because covalent linkages provide strong, stable enzyme attachment and may, in some cases, reduce enzyme deactivation rates and, usefully, alter enzyme specificity. Some disadvantages, however, may be present. Beside the risk of perturbing the enzyme's native structure, they include: difficulty of preparation, no possibility for regenerating the carrier, high costs, and the possibility of direct involvement of the active site in the binding.

In practical applications, a proper choice should be made between chemical and physical methods. Usually, a long-time applicable immobilized enzyme with a lower initial activity is preferable to that with a high level of initial activity but with short-time activity retention.

Studies have been performed with enzymes covalently immobilized on soil components. The scope of these studies is limited and has been principally the potential use of such immobilized enzymes in the restoration of polluted or degraded soils. The methods subsequently described can be applied to more than one enzyme.

15–5.2 Covalent Immobilization of Acid Phosphatase, Laccase, and Peroxidase on Glass Beads, Montmorillonite, Kaolinite, and Soil

15–5.2.1 Introduction

Methods are presented for immobilization of three enzymes—acid phosphatase, laccase, and peroxidase—on two clay minerals, a montmorillonite and a kaolinite, and an inert support (glass beads), as well as in whole soils having different physicochemical properties.

The covalent immobilization procedure involves previous activation of the support, followed by its derivatization with glutaraldehyde.

All the methods used with the three enzymes are described in detail, although several steps are common to all three methods. This provides additional information on modification needed when similar supports are used on different enzymes.

15–5.2.2 Acid Phosphatase Covalent Immobilization Method (Gianfreda and Bollag, 1994)

15–5.2.2.1 Apparatus

- Volumetric flasks, 10, 50, 250, 1000 mL
- Test tubes, 10, 12 mL
- Erlenmeyer reaction flasks, 10, 50, 250, 1000 mL
- Centrifuge tubes (high-strength aluminosilicate glass), 25 mL

- Millipore M.F. filters (<0.025 μm)
- High-speed temperature-controlled centrifuge (>20,000 rpm)
- Vacuum desiccator
- UV-Vis spectrophotometer
- Temperature-controlled water bath shaker
- Vortex
- pH meter
- Rotary shaker
- Refrigerator

15–5.2.2.2 Reagents

1. Sodium phosphate monobasic (NaH_2PO_4), 0.1 M: Prepare as described in section 15–3.2.3.2.
2. Sodium phosphate bibasic (Na_2HPO_4), 0.1 M: Prepare as described in section 15–3.2.3.2.
3. Sodium phosphate buffer, 0.1 M pH 7.0: Prepare as described in section 15–3.2.3.2.
4. Sodium acetate (NaOAc), 0.1 M: Prepare as described in section 15–3.2.4.2.
5. Acetic acid (HOAc), 0.1 M: Prepare as described in section 15–3.2.4.2.
6. Sodium acetate buffer, 0.1 M pH 5.0: Prepare as described in section 15–3.2.4.2.
7. Nitric acid (HNO_3), 0.5 M: Add 29.57 mL of reagent-grade 70% HNO_3, specific gravity 1.42, per liter of the final solution desired. Mix and thoroughly cool; may be stored at 4°C for more than 1 yr.
8. Aminopropyltriethoxysilane (APTES, Aldrich 440140) solution, 2% (in pure acetone): Add 2 mL of 3-aminopropyltriethoxysilane reagent grade in about 70 mL of pure acetone, and make up to 100 mL final volume with additional acetone. Store at 4°C for a max. of 2 mo.
9. Glutaraldehyde solution 5% (in 0.1 M sodium phosphate buffer pH 7.0): Add 20 mL of reagent-grade 25% glutaraldehyde (Sigma-Aldrich G5882 Grade I) in 75 mL of 0.1 M sodium phosphate buffer (Reagent 3) and make up to 100 mL final volume with additional buffer. Store at 4°C for 2 mo.
10. Sodium hydroxide (NaOH), 1 M: Prepare as described in section 15–3.2.2.2.
11. Acid phosphatase (E.C 3.1.3.2) solution: Prepare enzyme samples by starting from available commercial preparations. A commercial preparation from wheat germ Type I (Fluka 79410 lyophilized powder 0.1–0.6 U/mg or Sigma P3627 solid >0.4 U/mg) was used. Store enzyme samples in small aliquots at 4°C. Periodically measure the residual activity; prepare fresh enzyme solution when it is reduced by more than 10 to 20%.
12. Glass beads: Use controlled-pore glass beads (pore size 75 Å, 200–400 mesh) from Sigma Chemical, as supplied.
13. Clays and soils:
    - Na-saturated montmorillonite (Wyoming, bentonite) from Ward's Natural Science Establishment (Rochester, NY), as supplied.

- Kaolinite (Georgia, kaolin) from Ward's Natural Science Establishment (Rochester, NY), as supplied.
- Locally available soils could be used after air-drying at room temperature. In the referenced paper, a silt loam soil (Hagerstown, Ap horizon) was collected from grassland (Centre County, PA). For the choice of the soil to use, see comments at section 15–5.3.

14. p-Nitrophenylphosphate (pNPP) (substrate, Sigma 73737) 5 mM: Weigh 186 mg of reagent-grade pNPP in 80 mL of Reagent 6, check pH and if necessary, adjust to exactly pH 5.0 with a few drops of either 1 M NaOH or concentrated acetic acid. Make up to 100 mL final volume with additional Reagent 6. Mix, thoroughly cool, and store at 4°C. The solution is stable for 2 wk at 4°C. The solution tends to achieve a yellow color indicative of the spontaneous hydrolysis of the product and therefore of the decrease of its concentration.

15. Lowry or Bradford Bio-Rad protein assay kit: Use the commercial solutions without any further modification.

15–5.2.2.3 Standard Solutions and Calibration Curves

1. Prepare a calibration curve with BSA using the chosen protein assay kit as described in No. 1 to 3 in section 15–3.2.2.3.
2. Standard p-nitrophenol solution: As described in section 15–3.2.4.3.
3. Prepare a calibration curve with different volumes of standard p-nitrophenol solution as described in section 15–3.2.4.3.

15–5.2.2.4 Preparation of Complexes

A. Activation of supports (clay, glass bead, or soil):

1. Weigh 200 mg of the selected support (clay, glass bead, or soil) into a 10-mL Erlenmeyer reaction flask. Add 5 mL of 0.5 M HNO_3.
2. Incubate at 45°C for 2 h with shaking.
3. Centrifuge at 13,000 g for 30 min at 4°C and collect the supernatants in test tubes and the precipitates in centrifuge tubes.
4. Wash the pellets with 5 mL of deionized water, stir a few minutes, and centrifuge at 11,000 g for 10 min at 4°C; collect the supernatants and the pellets.
5. Repeat No. 4 until the pH of the supernatants reaches a neutral pH (wait for the temperature of the supernatant to reach room temperature and measure the pH with a pH meter). Typically four to five washings are sufficient.
6. Add to the washed pellets 5 mL of 2% solution of 3-aminopropyltriethoxysilane (APTES) in pure acetone.
7. Shake the suspension at 45°C overnight on a rotary shaker.
8. Centrifuge at 13,000 g for 10 min in a high-speed temperature-controlled centrifuge and collect the activated pellets.
9. Treat the activated pellets with 5 mL of 5% glutaraldehyde in 0.1 M phosphate buffer pH 7.0.
10. Evacuate the suspensions in a vacuum desiccator to remove trapped air bubbles.
11. Cover the samples with parafilm and incubate for 1 h at room temperature.

12. Wash the pellets with 5 mL of deionized water, stir a few minutes, and centrifuge at 11,000 g for 10 min at 4°C; collect the supernatants and the pellets.

13. Repeat No. 12 several times (typically four to five times) and then one time with 0.1 M phosphate buffer pH 7.0. Collect the activated pellets in a centrifuge tube.

B. Immobilization of the enzyme:

1. To the activated pellets, add 2 mL of 0.1 M phosphate buffer pH 7.0 containing 6.0 mg of acid phosphatase.

2. Incubate the suspension at 4°C for 24 h under stirring.

3. Centrifuge at 10,000 g for 10 min at 4°C and collect the supernatants and the precipitates separately.

4. Wash the solid precipitates (enzyme–support complexes) with 2 mL of 0.1 M phosphate buffer pH 7.0, centrifuge as in No. 3, and collect the washings. Repeat washing and centrifugation until no enzymatic activity is detected in the washings (usually five times).

5. Determine the protein content of the supernatants and washings using the Lowry or Bradford Bio-Rad protein assay kit.

6. Suspend the enzyme complex in 2 mL of 0.1 M phosphate buffer pH 7.0 and store in small aliquots at 4°C. Periodically measure the residual activity of the enzyme complex. When the enzymatic activity is reduced by more than 10 to 20%, prepare fresh enzyme complexes.

7. For all determinations use aliquots of the suspended complexes after rapid equilibration at room temperature. Do not reuse the remaining sample.

8. Also prepare a control sample with the free enzyme solution following the same procedures reported above but without the matrix and processed as the complexes.

15–5.2.2.5 Activity Assays

Measure the activity of phosphatase in the control, supernatants, and washings (free enzyme) and in the suspension (immobilized enzyme).

For free phosphatase:

1. Incubate at 25°C a suitable volume (typically 0.05 mL) of free enzyme solution with 1 mL of 5 mM pNPP buffered solution (Reagent 14). Usually 10 min is a suitable incubation time. It can be reduced or increased depending on the initial activity of the enzymatic solution. Add 1 mL of 1 M NaOH to stop the enzymatic reaction.

2. Determine the concentration of p-nitrophenol at 405 nm with the spectrophotometer.

For adsorbed phosphatase:

1. Incubate at 25°C a suitable volume of adsorbed enzyme suspension with 1 mL of 5 mM pNPP buffered solution (Reagent 14). Usually 10 min is a suitable incubation time. It can be reduced or increased depending on the initial activity of the enzymatic solution. Add 1 mL of 1 M NaOH.

2. Filter the suspension.

3. Determine the concentration of p-nitrophenol in the filtrate by its absorbance at 405 nm with the spectrophotometer.

15–5.2.2.6 Calculations

Protein content:

As described in section 15–3.2.2.6 (by reference to each of the supports used in the immobilization of the enzyme).

Activity of the free enzyme:

As described in section 15–3.2.4.6.

Activity of the immobilized enzyme:

As described in section 15–3.2.4.6.

Specific activity of the free and immobilized enzyme:

As described in section 15–3.2.4.6.

15–5.2.3 Laccase Covalent Immobilization Method (Ruggiero et al., 1989; Sarkar et al., 1989; Gianfreda and Bollag, 1994)

15–5.2.3.1 Apparatus

As described in section 15–5.2.2.1 plus:

- Biological oxygen monitor equipped with a Clark oxygen electrode and a linear recorder

15–5.2.3.2 Reagents

1. Sodium phosphate monobasic (NaH_2PO_4), 0.1 M: Prepare as described in section 15–3.2.3.2.
2. Sodium phosphate bibasic (Na_2HPO_4), 0.1 M: Prepare as described in section 15–3.2.3.2.
3. Sodium phosphate buffer, 0.1 M pH 7.0: Prepare as described in section 15–3.2.3.2.
4. Citric acid monohydrate, 0.1 M: Add 21.01 g of $C_6H_8O_7 \cdot H_2O$ to 700 mL of bidistilled water, mix, and make up to 1 L final volume with additional water. Store at room temperature for a max. of 3 mo.
5. Citric acid–phosphate buffer, 0.1 M pH 3.8: Mix suitable volumes of Reagent 2 and of Reagent 4 (usually 645 mL and 355 mL, respectively) to reach pH 3.8 and 1 L final volume. Check pH with pH meter. Store at 4°C for a max. of 3 mo.
6. Ethanol: Use reagent-grade ethanol (99%).
7. Nitric acid (HNO_3), 0.5 M: Prepare as described in section 15–5.2.2.2.
8. Aminopropyltriethoxysilane (APTES) solution 2% (in pure acetone): Prepare as described in section 15–5.2.2.2.
9. Glutaraldehyde solution 5% (in 0.1 M sodium phosphate buffer pH 7.0): Prepare as described in section 15–5.2.2.2.
10. Laccase (E.C 1.10.3.2) solution: Choose one of the available commercial preparations of laccase from different fungal origins. An enzyme preparation isolated and partially purified from *Trametes versicolor* (Sigma 53739) was used in the cited method.

11. Clays and soil: prepare as described in section 15–5.2.2.2.

12. 2,6-dimethoxyphenol (2,6-DMP) (substrate), 10 mM: Dissolve 154 mg of 2,6-DMP (99% purity, Aldrich D135559) in 1 mL of ethanol (Reagent 6), add 0.1 M citrate–phosphate buffer (Reagent 5) to 100 mL final volume. Mix, thoroughly cool, and store at 4°C. The solution is stable for 2 wk at 4°C. The solution tends to achieve a yellow color indicative of the spontaneous oxidation of the product and therefore of the decrease of its concentration.

13. Lowry or Bradford Bio-Rad protein assay kit. Use the commercial solutions without any further modification.

15–5.2.3.3 Standard Solutions and Calibration Curves

1. Prepare a calibration curve with BSA, using the chosen protein reagent as described in No. 1 and 2 in section 15–3.2.2.3.

15–5.2.3.4. Preparation of Complexes

A. Activation of supports (clay, glass bead, or soil):

As described in section 15–5.2.2.4.

B. Immobilization of the enzyme:

1. Add to the activated pellets 2 mL of 0.1 M phosphate buffer pH 7.0 containing 0.88 mg of laccase.

2. Proceed as described in No. 2 to 7 of subsection B. *Immobilization of the enzyme,* in section 15–5.2.2.4.

3. Also prepare a control sample with the free enzyme solution following the same procedures reported above but without the matrix and processed as the complexes.

15–5.2.3.5 Activity Assays

Measure the activity of laccase in the control, supernatants, and washings (free enzyme) and in the suspension of the immobilized enzyme polarographically using the biological oxygen monitor equipped with a Clark oxygen electrode and a linear recorder:

1. Incubate at 25°C 2.7 mL of 0.1 M citrate–phosphate buffer pH 3.8 containing a suitable amount of free or immobilized laccase and saturated with O_2 in the sample chamber under agitation with a Teflon-coated magnetic bar.

2. Start the reaction by adding 0.1 mL of 10 mM 2,6-DMP buffered solution, preheated at the same temperature, through a side port of the electrode.

3. Record the consumption of O_2 with the time.

15–5.2.3.6 Calculations

Protein content:

As described in section 15–3.2.2.6 (by reference to each of the supports used in the immobilization of the enzyme).

Activity of the free and immobilized enzyme:

The activity of free or immobilized laccase is calculated by the consumption of O_2 with time. One unit of activity is the amount of free or immobilized laccase causing the consumption of 1 μmol of O_2 min^{-1} at 25°C and pH 3.8.

Specific activity of the free and adsorbed enzyme:

The specific activity is calculated by dividing the units measured in 1 mL of free or immobilized enzyme solution per milligram of free or immobilized enzyme contained in the volume.

15–5.2.4 Comments

The same experimental procedure reported for phosphatase and laccase was applied to peroxidase, using the same procedure for the preparation of supports and different experimental conditions in terms of pH, buffers, temperature, and substrate (Ruggiero et al., 1989; Sarkar et al., 1989; Gianfreda and Bollag, 1994). Lai and Tabatabai (1992) applied a similar procedure to urease. Different experimental conditions in terms of pH, temperature, and preparation of supports were used. Refer to Lai and Tabatabai (1992) for reagents, standards, activity assays and analytical procedures, and calculations.

15–5.3 General Comments to Section 15–5

Immobilization by chemical binding as described above has been very effective in immobilizing enzymes, not only in terms of mg of protein immobilized on the supports, but also in preserving relatively high residual activity levels. Moreover, several kinetic properties of the immobilized enzymes as well as their response to inhibitors such as NaN_3 were very similar to those of the free enzymes, thus showing no effect on the active enzymatic sites occurred after immobilization (for details refer to reported published papers). These advantages could overcome the disadvantages of the immobilization procedures, which are relatively laborious and expensive.

It is important to highlight that the treatment of the different supports with HNO_3 (first step of activation) may very likely modify their surface and properties, making the supports quite similar in terms of capacities for immobilizing the enzymes.

The methods involving use of HNO_3 are not applicable to calcareous soils or to soils having components that degrade on acid treatment.

The immobilized enzymes are quite stable and remain active with long-term storage and with repeated use in reaction cycles as well as in continuous reactors. High activity levels and stability also were shown in the presence of complex solid phases such as soils, sand, and soil–sand mixtures. No, or negligible, release of enzyme molecules occurred. These critical points should be considered for practical applications of immobilized enzymes.

Chemical binding of enzymes to activated supports using glutaraldehyde as a final step of activation has been applied to other materials (chitosan–clay beads, montmorillonite saturated with potassium, humic substances) as well as to other enzymes such as acid phosphatase, β-glucosidase, invertase, glucoamylase, β-amylase, and α-amylase, (Chang and Juang, 2004, 2005, 2007; Rosa et al., 2000; Gopinath and Sugunan, 2005, 2007). The preparation of the supports was performed by following a procedure different from that reported in sections 15–5.2.2 and 15–5.2.3, but the final step involved glutaraldehyde providing the activated arm to link the enzymes. Also, the experimental conditions used in the preparation of the immobilized enzymes changed depending on the support used and the chosen enzyme (for details refer to reported published papers). Obviously, substrates, buffers, pHs, and procedures for activity assays were chosen according to the enzyme under investigation.

15–6 ENTRAPMENT OF ENZYMES IN HUMIC AND HUMIC-LIKE COMPLEXES

15–6.1 Introduction

The preparation of synthetic humic–enzyme complexes requires two basic procedures (Gianfreda and Bollag, 1996): (i) natural humic acids are extracted from soil and their interactions with enzymes are studied (this approach has been addressed in sections 15–4.3 and 15–4.3.1) and (ii) synthetic humus–enzyme complexes are prepared by putting phenolic compounds in contact with enzymatic proteins, and subsequent polymerization. The oxidative transformation of phenolic compounds can be catalyzed by biotic catalysts such as extracellular polyphenoloxidases or peroxidases, and by abiotic catalysts such as clay minerals, Fe, Mn, and Al oxides. Both types of catalysts may naturally occur in soil and can be involved in such a process.

Phenols are transformed through oxidative coupling reactions with production of polymeric products by self-coupling or cross-coupling with other molecules and the formation of less soluble, high molecular mass compounds. The catalytic mechanisms undertaken by biotic and abiotic catalysts are different and strongly dependant on the type of phenolic substrate as the complexity of phenolic samples may, in turn, negatively affect the catalyst and decrease its catalytic efficiency. If protein molecules are involved in the polymerization process, relatively stable protein–soil colloid complexes may be formed. A strong retention and accumulation of the protein on the formed complexes as well as its release can occur under some specific environmental conditions. As demonstrated by several studies (Ladd and Butler, 1975; Burns, 1986; Boyd and Mortland, 1990; Gianfreda et al., 2002), profound structural alterations and conformational changes of protein molecules associated to soil colloids may take place (as a consequence of the interactions with phenols or phenolic polymeric species) with variations of the protein functionality.

Usually, precursors of humic substances in soil such as catechol, pyrogallol, tannic acid, resorcinol, etc., have been used for the preparation of organic networks entrapping enzyme molecules. The phenolic compounds have been subjected to polymerization by the addition of a catalyst in the presence of the investigated enzyme and the properties of the resulting humus-like enzymatic complexes have been studied. The catalyst most widely used has been peroxidase. Experimental evidence is also available on the use of Fe_2O_3 and MnO_2.

Peroxidases are important oxidative enzymes, catalyzing the oxidation of numerous compounds, and very effective agents in the oxidation and transformation of phenolic compounds. They are widespread in nature, being produced by plants and microorganisms. In soil, these enzymes may play an important role in the rhizosphere at the plant root–soil interface and they usually increase in the presence of phenolic compounds. Manganese oxides are widely distributed in soils of different origins. They are very reactive components that can catalyze phenol polymerization. They may exist as a coating on the soil and sediment particles, as discrete particles, and/or associated with many other chemical species. They are involved in several soil processes (e.g., adsorption of ions, surface reactions, and many redox reactions such as oxidation of trace elements, formation of iron oxides and oxyhydroxides, oxidative polymerization, polycondensation, and degradation of organics (Huang, 1990).

15–6.2. Complexes by Biotic and Abiotic Catalysis

15–6.2.1 Method: Preparation of β-D-Glucosidase–Phenolic Copolymers as Analogs of Soil Humic–Enzyme Complexes (Sarkar and Burns, 1984)

15–6.2.1.1 Apparatus

- Volumetric flasks, 10, 50, 100, 1000 mL
- Erlenmeyer reaction flasks, 25, 50, 250, 1000 mL
- Test tubes, 10 mL
- Screw-cap tubes
- A gradient mixer, 25 mL
- High-speed temperature-controlled centrifuge (>20,000 rpm)
- Centrifuge tubes (high-strength aluminosilicate glass), 15, 30 mL
- Dialysis tubes, cutoff 8000 Da
- UV-Vis spectrophotometer
- pH meter
- Magnetic stirrer and bars
- Shaker
- Freeze-dryer
- Vacuum pump

15–6.2.1.2 Reagents

1. Sodium acetate (NaOAc), 0.1 M: Prepare as described in section 15–3.2.4.2.
2. Acetic acid (HOAc), 0.1 M: Prepare as described in section 15–3.2.4.2.
3. Sodium acetate buffer, 0.1 M pH 4.5 and 5.4: Mix suitable volumes of Reagent 1 (usually 430 mL for pH 4.5 and 860 mL for pH 5.4) and of Reagent 2 (usually 570 mL for pH 4.5 and 140 mL for pH 5.4) to reach pH 4.5 or 5.4 and 1 L final volume. Check pH. Store at 4°C for 3 mo.
4. Pyrogallol, reagent grade (Sigma 16040): Use as powder.
5. Resorcinol, reagent grade (Aldrich R406): Use as powder.
6. Peroxidase (E.C. 1.11.1.7 from horseradish, type VI-A, Sigma P6782), 1 mg mL^{-1}: Dissolve 10 mg in 10 mL of 0.1 M pH 5.4 acetate buffer. Periodically measure the residual activity of the enzyme solutions. When the enzymatic activity is reduced by more than 10 to 20%, prepare fresh enzyme solutions.
7. Hydrogen peroxide, 1% (w/v): Verify actual concentration of the commercial product (30% H_2O_2) by reading its absorbance at 240 nm and using the extinction coefficient of 39.4 M^{-1} cm^{-1}. Dilute a suitable volume of 30% H_2O_2 with a suitable volume of bidistilled water up to 0.3 L final volume.

8. Bentonite activated with dioxane and cyanuric chloride: In a 250-mL Erlenmeyer reaction flask suspend by shaking 100 g of commercial bentonite in 200 mL dioxane, add 5 g of cyanuric chloride, and mix with a magnetic bar the suspension at 20°C for 2 h. Separate the bentonite–cyanuric chloride complex by centrifugation at 12,000 g for 30 min at 4°C. Wash the bentonite–cyanuric chloride complex with dioxane and then centrifuge at 12,000 g for 30 min in at 4°C. Repeat the washing at least five times. Dry in a vacuum and store at 20°C.

9. β-D-Glucosidase (EC 3.2.1.21 from sweet almond, Sigma 49290): Use without further purification. Prepare enzymatic solution (4 mg mL^{-1}) in 0.1 M sodium acetate buffer pH 5.4. Store the enzymatic solutions in small aliquots at 4°C. Periodically measure the residual activity of the enzyme solutions. When the enzymatic activity is reduced by more than 10 to 20%, prepare fresh enzyme solutions.

10. p-Nitrophenyl β-D-glucopyranoside, 15 mM: Dissolve 0.452 g of p-nitrophenyl β-D-glucopyranoside (Sigma N7006) in 100 mL of 0.1 M acetate buffer pH 5.4. Store at −20°C. The solution is stable for 1 mo.

11. Neutral Pb acetate, 1.2 M: Weigh 45.52 g of lead(II) acetate trihydrate and add 70 mL of bidistilled water, mix, and make up to 100 mL final volume with additional water. Store at 4°C. The solution is stable for 2 mo.

12. Potassium oxalate, 1.2 M: Weigh 22.11 g and add 70 mL of bidistilled water, mix, and make up to 100 mL final volume with additional water. Store at 4°C. The solution is stable for 2 mo.

13. Dioxane, reagent grade.

14. Cyanuric chloride (98%).

15–6.2.1.3 Standard Solutions and Calibration Curves

1. Standard p-nitrophenol solution: Prepare as described in section 15–3.2.4.3.

2. Prepare a calibration curve with different volumes of standard p-nitrophenol solution as described in section 15–3.2.4.3. The theoretical extinction molar coefficient of p-nitrophenol at 400 nm should be 18.5 cm^{-1} mM^{-1}.

15–6.2.1.4 Preparation of β-D-Glucosidase–Resorcinol and Pyrogallol Copolymers

1. Weigh 250 mg of resorcinol and pyrogallol and dissolve each substance separately in 100 mL of 0.1 M sodium acetate buffer pH 5.4.

2. Use each single solution for making homopolymers or a combination (1:1) of pyrogallol and resorcinol for making heteropolymers. Add the chosen solution in one cylinder of a gradient mixer, and 1% (w:v) H_2O_2 in the other cylinder. Mix (1:1) the phenolic solution and hydrogen peroxide, and add a solution cooled at 0°C (10 mL) of 1 mg mL^{-1} peroxidase in 0.1 M sodium acetate buffer pH 5.4.

3. When the mixture appears yellow-brown, indicating quinone formation in solution (approximately 15 min after the start of the reaction), add dropwise 10 mL of 4 mg mL^{-1} β-D-glucosidase dissolved in 0.1 M sodium acetate buffer pH 5.4.

4. Concentrate in vacuo by vacuum pump the soluble β-D-glucosidase–phenolic copolymers, dialyze twice against water, and concentrate about 10 times by freeze-drying.
5. Store at −20°C. Stable for more than 1 yr.

   Fixation of β-D-glucosidase–phenolic copolymers on activated bentonite:

1. In a 50-mL Erlenmeyer reaction flask, suspend 30 mg of β-D-glucosidase–phenolic copolymers in 25 mL of 0.1 M sodium acetate buffer pH 4.5 and add 1 g activated bentonite.
2. Stir the suspension for 6 h at 4°C.
3. Centrifuge at 20,000 g for 30 min at 4°C. Collect the supernatant and the precipitate.
4. Wash the precipitate with distilled water, centrifuge at 20,000 g for 30 min at 4°C. Repeat washing and centrifugation at least five times.
5. Collect the washed precipitate and freeze-dry. Store at −20°C. Stable for more than 1 yr.

15–6.2.1.5 Activity Assays

Measure the activity of the free β-D-glucosidase in the controls and supernatants and the activity of the immobilized β-D-glucosidase after resuspensions in suitable volumes of acetate buffer.

1. In screw-cap tubes, add a suitable volume (from 0.05 to 0.1 mL) of enzyme solution or suspension with 3 mL of 15 mM p-nitrophenyl β-D-glucopyranoside in 0.1 M sodium acetate buffer pH 5.4. Cap the tubes and put them in a shaker (30 rev^{-1}) set at 37°C for 1 h.
2. Add 0.1 mL of 1.2 M neutral Pb acetate to stop the enzymatic reaction.
3. Centrifuge at 10,000 g, for 20 min at 4°C, then add 0.1 mL of 1.2 M potassium oxalate to remove excess Pb from supernatant and centrifuge again for 20 min at 10,000 g at 4°C.
4. Determine the concentration of p-nitrophenol, the reaction product, at 400 nm.

15–6.2.1.6 Calculations

The activity of the free or immobilized β-D-glucosidase is calculated by Eq. [5] and expressed in Units mL^{-1}:

$$U\ mL^{-1} = (A_{400} \cdot V_r \cdot F_d)/(18.5 \cdot t \cdot V_s) \tag{5}$$

where A_{400} is the absorbance at 400 nm, V_r is the reaction volume in mL, F_d is the dilution factor by Pb acetate addition, 18.5 is the extinction molar coefficient of p-nitrophenol at 400 nm in cm^{-1} mM^{-1}, V_s is the volume of solution or suspension in mL, and t is the reaction time in hours.

   The specific activity of the free β-D-glucosidase is calculated by dividing U mL^{-1} as from Eq. [5] for the protein content expressed as mg mL^{-1}.

   The activity of the immobilized β-D-glucosidase is expressed as μmoles of p-nitrophenol produced in 1 h by 1 mg of freeze-dried copolymers.

15–6.2.2 Comments

The same experimental procedure used with glucosidase and resorcinol was applied by Garzillo et al. (1996) to phosphatase, and phosphatase–polyresorcinol complexes were prepared and characterized. Refer to Garzillo et al. (1996) for reagents, standards, activity assays and analytical procedures, and calculations.

15–6.2.3 Method: Formation of Urease–Tannate Complexes by Iron and Manganese Catalysis (Gianfreda et al., 1995b)

15–6.2.3.1 Apparatus

As described in section 15–3.2.2.1 plus:

- Magnetic stirrer and magnetic bars
- pH meter
- Oven
- Freeze-dryer
- Dialysis tubing, cutoff 10,000 Da

15–6.2.3.2 Reagents

1. Sodium phosphate monobasic (NaH_2PO_4), 0.1 M: Prepare as described in section 15–3.2.3.2.

2. Sodium phosphate bibasic (Na_2HPO_4), 0.1 M: Prepare as described in section 15–3.2.3.2.

3. Sodium phosphate buffer, 0.1 M pH 7.0–1 mM EDTA: Prepare as described in section 15–3.2.3.2.

4. Aluminum nitrate [$Al(NO_3)_3$], 0.036 M: Add 13.12 g of $Al(NO_3)_3$ to 700 mL of bidistilled water, mix, and make up to 1 L final volume with additional water. Store at room temperature for a max. of 3 mo.

5. Iron nitrate [$Fe(NO_3)_3$], 0.360 M: Add 145.45 g of $Fe(NO_3)_3$ to 700 mL of bidistilled water, mix, and make up to 1 L final volume with additional water. Store at room temperature for a max. of 3 mo.

6. Stock solution [0.180 M $Fe(NO)_3$ and 0.018 M $Al(NO_3)_3$]: Mix equal volumes of Reagent 4 and of Reagent 5.

7. Potassium hydroxide (KOH), 1 M: Add 61.72. g of KOH (90% reagent grade) to 700 mL of bidistilled water, mix, and make up to 1 L final volume with additional water. Store at room temperature up to 1 yr.

8. Tannic acid: Use reagent grade as powder (Fluka 96311).

9. Manganese oxide (MnO_2): Use a commercial MnO_2 without further purification. Analyze the sample by X-ray diffraction analysis. The material used showed peculiar characteristics of pyrolusite (β-MnO_2).

10. Iron hematite (α-Fe_2O_3) (Colombo et al., 1994): In a 1000 mL beaker, place 1000 mL of stock solution [0.180 M $Fe(NO)_3$ and 0.018 M $Al(NO_3)_3$]. Add 1 M KOH at a rate of 1 mL min^{-1} under vigorous stirring until pH 5.5 is reached. After aging for 1 mo at 97°C, collect the precipitate obtained by centrifugation at 10,000 g for 30 min. Wash with 20 mL of 0.5 M HCl to desorb anions,

dialyze (cutoff 10,000 Da membrane) in deionized water to an equilibrium concentration of ~5 dS m^{-1}, and freeze-dry.

11. Urease (amidohydrolase E.C. 3.5.1.5 from jack bean, Sigma U-0376) solution: Prepare as described in section 15–3.2.3.2.

12. Urea (substrate), 0.150 mM: Prepare as described in section 15–3.2.3.2.

13. Reagents for determination of NH$_4^+$ (hypochlorite–alkaline phenol method, Charney and Myrback, 1962): Prepare as described in section 15–3.2.3.2.

14. Protein assay kit: Use a commercial protein reagent such as Protein Assay ESL (Roche, Germany) method that does not present interference with phenolic molecules. Another possibility is the use of the Lowry assay as modified by Winters and Minchin (2005).

15–6.2.3.3 Standard Solutions and Calibration Curves

1. Prepare a calibration curve with BSA as described in No. 1 to 3 in section 15–3.2.2.3 and using the appropriate protein reagent.

2. Stock solution of NH$_4^+$: Prepare as described in section 15–3.2.3.3.

3. Plot a standard calibration curve using the different NH$_4^+$ concentrations and determine a factor (F_{NH4}) as the slope of the curve as described in section 15–3.2.3.3.

15–6.2.3.4. Preparation of Complexes

Tannic–enzyme complexes:

1. Weigh 0.085, 0.17, 0.85, and 1.7 mg of tannic acid into 12-mL reaction tubes, add 2 mL of 0.1 M sodium phosphate buffer pH 7.0–1 mM EDTA (Reagent 3) containing 0.17 mg of urease (tannic acid/urease ratio w/w of 0.5, 1, 5, and 10) and place in a thermostatically controlled room or a thermostatic bath at 10°C.

2. Also prepare a control with only urease, following the same experimental procedure.

3. Incubate the suspension at 10°C for 1 h. Shake every 10 min.

4. Centrifuge the tubes at 4°C and 10,000 g for 30 min and collect separately the supernatants and the precipitates. Wash the solid precipitates (tannic–enzyme complexes) with 2 mL of Reagent 3; centrifuge at 10,000 g for 30 min at 4°C; collect the washing. Repeat the washing procedure three to four times, collecting the washing.

5. Suspend the enzyme–tannic complex in 2 mL of 0.1 M sodium phosphate buffer pH 7.0–1 mM EDTA and use immediately.

6. When necessary, store at 10°C and pH 7.0 and measure the residual activity periodically. Discard the complexes when their activity is reduced by more than 20% compared with initial one.

Tannic–metal oxide–enzyme complexes:

1. Weigh 20 mg of manganese oxide (MnO$_2$) or iron oxide (Fe$_2$O$_3$) into a 12-mL reaction tube. Add 0.85 mg of tannic acid and 2 mL of 0.1 M sodium phosphate buffer pH 7.0–1 mM EDTA (Reagent 3) containing 0.17 mg of urease

(tannic acid/urease ratio w/w of 5) and place in a thermostatically controlled room or a thermostatic bath at 10°C.

2. Also prepare two controls: one with urease and Fe or Mn oxide, and the other one with tannic acid and Fe or Mn oxide.

Continue as in No. 3 to 8 of the tannic–enzyme complexes subsection.

15–6.2.3.5 Activity Assays

Measure the activity of urease in controls, supernatants, and washings (free enzyme) and in the suspension of the immobilized enzyme as described in section 15–3.2.3.5.

15–6.2.3.6 Calculations

Activity of the free enzyme:

The activity of free urease is calculated by Eq. [2] reported in section 15–3.2.3.6 and is expressed in Units per mL, i.e., μmol of NH_4^+ min^{-1} mL^{-1} of free enzyme solution.

Activity of the immobilized enzyme:

The activity of immobilized urease is calculated by Eq. [2] reported in section 15–3.2.3.6 where V_s is the volume of immobilized enzyme suspension, and is expressed in Units per mL, i.e., μmol of NH_4^+ min^{-1} mL^{-1} of immobilized enzyme suspension.

Specific activity of the free and immobilized enzyme:

The specific activity is calculated as described in section 15–3.2.3.6 if it is possible to measure the protein content.

15–6.2.4 Comments

The same experimental procedure reported for urease was applied to phosphatase (Rao and Gianfreda, 2000). In these cases, experimental conditions differed in terms of pH, temperature, tannic acid/enzyme ratio, and amounts of MnO_2 or Fe_2O_3 used. The reagents, standards, activity assays and analytical procedures, and calculations were those described in section 15–3.2.4.

15–6.3 General Comments to Section 15–6

Interactions of soil enzymes with humic substances can be studied following an alternative approach to the extraction of humic–enzyme complexes directly from soil. The extraction is very difficult, time-consuming, and has a low degree of reproducibility.

Several difficulties may arise with the formation and characterization of synthetic enzymatic humic-like complexes. First, as described in section 15–6.1, a loss of activity can occur and vary according to the nature of phenolic compounds, enzyme, and the sequence of component additions. The use of a phenol rather than another substance results in more than 55% of the activity being retained for β-D-glucosidase (with resorcinol) or even less than 15% with protocatechuic acid (Sarkar and Burns, 1984). A higher residual activity was observed with another enzyme, acid phosphatase, which retained more than 90% of the free enzyme activity when entrapped in polyresorcinol copolymers (Garzillo et al., 1996).

The use of an abiotic catalyst such as birnessite can modify the oxidative process with respect to the biotic catalysis (i.e., with peroxidase) and lead to the formation of phenolic–enzyme copolymers with different activity levels. Higher or lower activity levels have been measured depending on the type of phenolic substance and the enzyme involved in the oxidative process (Rao et al., 2010).

To better characterize the formed copolymers, the E_4/E_6 ratio could be calculated by measuring with a UV-Vis spectrophotometer the absorbance values at 472 nm and 664 nm of the supernatants, previously obtained through centrifugation of suspensions (as obtained in No. 4 of tannic–enzyme complexes in section 15–6.2.3.4) at 10,000 g and by dividing the two absorbance values. The E_4/E_6 ratio can give information about the properties of copolymers formed on phenol polymerization and in particular on their solubility and size. Usually, the lower the value of the ratio the higher is the condensation of polymers.

Inactivation as well as inhibition phenomena due to quinones, humates, and humic derivatives could affect enzyme molecules. The sequence of addition of each component participating in the formation of copolymers is a very important factor. The formation of copolymers in which enzyme molecules could be simultaneously involved in phenolic polymer aggregation leads to active immobilized enzymes. These conditions favor the gradual and homogeneous interaction of enzymes with the slow-growing phenolic homopolymers giving highly active complexes (Sarkar and Burns, 1984; Garzillo et al., 1996). Nonetheless, with more reactive and higher molecular weight phenolic compounds (like pyrogallol or tannic acid rather than resorcinol), less active enzymatic copolymers can be formed. A stronger involvement and consequent entrapment of the enzyme molecules usually occur, resulting either in enzyme molecules being deactivated or in the active site not being available to the substrate (Rao et al., 1999).

Conversely, the addition of enzymes after phenol polymer formation can lead to enzymatic complexes with low activity (Sarkar and Burns, 1984).

Unlike complexes prepared as described in section 15–4, the activity assay should be performed at a higher temperature (up to 37°C) than that of the copolymer formation process.

Similar to all complexes obtained with organic or organo–mineral components described in previous sections, the presence of phenols, soluble polymers, and humic-like substances in this case precludes the measurement of protein concentration in solution to determine the exact amount of enzyme immobilized in the formed copolymers. Methods such as Protein Assay ESL (Roche, Germany) or the modified Lowry method (Winters and Minchin, 2005) are capable of quantitatively detecting protein molecules in the presence of phenolic compounds and could be adopted.

REFERENCES

Ahn, M.-Y., A.R. Zimmerman, C.E. Martínez, D.D. Archibald, J.-M. Bollag, and J. Dec. 2007. Characteristics of *Trametes villosa* laccase adsorbed on aluminum hydroxide. Enzym. Microb. Technol. 41:141–148. doi:10.1016/j.enzmictec.2006.12.014

Boyd, S.A., and M.M. Mortland. 1990. Enzyme interactions with clays and clay-organo matter complexes. p. 1–28. *In* J.-M. Bollag and G. Stozky (ed.) Soil biochemistry. Vol. 6. Marcel Dekker, New York.

Bradford, M. 1976. A rapid and sensitive method for the quantification of microgram quantities of protein utilizing the principle of protein-dye binding. Anal. Biochem. 72:248–254.

Burns, R.G. 1982. Enzyme activity in soil: Location and a possible role in microbial ecology. Soil. Biol. Biochem. 14:423–427. doi:10.1016/0038-0717(82)90099-2

Burns, R.G. 1986. Interaction of enzymes with soil mineral and organic colloids. p. 429–451. In P.M. Huang and M. Schnitzer (ed.) Interactions of soil minerals with natural organics and microbes. SSSA, Madison, WI.

Carrasco, M.S., J.C. Rad, and S. Gonzales-Carcedo. 1995 Immobilization of alkaline phosphatase by sorption on Na-sepiolite. Bioresour. Technol. 51:175–181. doi:10.1016/0960-8524(94)00115-H

Chang, M.-Y., and R.-S. Juang. 2004. Stability and catalytic kinetics of acid phosphatase immobilized on composite beads of chitosan and activated clay. Process Biochem. 39:1087–1091. doi:10.1016/S0032-9592(03)00221-8

Chang, M.-Y., and R.-S. Juang. 2005. Activities, stabilities, and reaction kinetics of three free and chitosan–clay composite immobilized enzymes. Enzyme Microbial. Technol. 36:75–82. doi:10.1016/j.enzmictec.2004.06.013

Chang, M.-Y., and R.-S. Juang. 2007. Use of chitosan–clay composite as immobilization support for improved activity and stability of β-glucosidase. Biochem. Eng. J. 35:93–98. doi:10.1016/j.bej.2007.01.003

Charney, A.L., and E.P. Myrback. 1962. Determination of NH_4^+ ions with hypochlorite-alkaline phenol method. Clin. Chem. 8:130–135.

Cheng, J., S.M. Yu, and P. Zuo. 2006. Horseradish peroxidase immobilized on aluminum-pillared interlayered clay for the catalytic oxidation of phenolic wastewater. Water Res. 40:283–290. doi:10.1016/j.watres.2005.11.017

Ciurli, S, C. Marzadori, S. Benini, S. Deiana, and C. Gessa. 1996. Urease from the soil bacterium Bacillus Pasteurii: Immobilization on Ca-polygalacturonate. Soil Biol. Biochem. 28:811–817. doi:10.1016/0038-0717(96)00020-X

Colombo, C., V. Barron, and J. Torrent. 1994. Phosphate adsorption and desorption in relation to morphology and crystal properties of synthetic hematites. Geochim. Cosmochim. Acta 58:1261–1269. doi:10.1016/0016-7037(94)90380-8

Deng, S., and I. Popova. 2011. Carbohydrate hydrolases. p. 185–210. In R.P. Dick (ed.) Methods of soil enzymology. SSSA Book Ser. 9. SSSA, Madison, WI. (This volume.)

Dick, R.P., and R.G. Burns. 2011. A brief history of soil enzymology research. p. 1–34. In R.P. Dick (ed.) Methods of soil enzymology. SSSA Book Ser. 9. SSSA, Madison, WI. (This volume.)

Dick, W.A., and M.A. Tabatabai. 1999. Use of immobilized enzymes for bioremediation. p. 315–338. In D.C. Adriano, J.-M. Bollag, W.T. Frankenberger Jr., and R.C. Sims (ed.) Bioremediation of contaminated soils. Series No. 37. ASA, Madison, WI.

Durán, N., M.A. Rosa, A. D'Annibale, and L. Gianfreda. 2002. Applications of laccases and tyrosinases (phenoloxidases) immobilized on different supports: A review. Enzym. Microb. Technol. 31:907–931. doi:10.1016/S0141-0229(02)00214-4

Fornasier, F., Y. Dudal, and H. Quiquampoix. 2011. Enzyme extraction from soil. p. 371–384. In R.P. Dick (ed.) Methods of soil enzymology. SSSA Book Ser. 9. SSSA, Madison, WI. (This volume.)

Fusi, P., G.G. Ristori, L. Calamai, and G. Stozky. 1989. Adsorption and binding of protein on "clean" (homoionic) and "dirty" (coated with Fe oxyhydroxides) montmorillonite, illite and kaolinite. Soil Biol. Biochem. 21. 911–920. doi:10.1016/0038-0717(89)90080-1

Garzillo, A.M., L. Badalucco, F. De Cesare, S. Greco, and V. Buonocore. 1996. Synthesis and characterization of an acid phosphatase-polyresorcinol complex. Soil Biol. Biochem. 28:1155–1161. doi:10.1016/0038-0717(96)00113-7

Gianfreda, L., and J.-M. Bollag. 1994. Effects of soils on the behavior of immobilized enzymes. Soil Sci. Soc. Am. J. 58:1672–1681. doi:10.2136/sssaj1994.03615995005800060014x

Gianfreda, L., and J.-M. Bollag. 1996. Influence of natural and anthropogenic factors on enzyme activity in soil. p. 123–194. In G. Stotsky and J.-M. Bollag (ed.) Soil biochemistry. Vol. 9. Marcel Dekker, New York.

Gianfreda, L., A. De Cristofaro, M.A. Rao, and A. Violante. 1995a. Kinetic behavior of synthetic organo– and organo–mineral–urease complexes. Soil Sci. Soc. Am. J. 59:811–815. doi:10.2136/sssaj1995.03615995005900030025x

Gianfreda, L., M. A. Rao, F. Sannino, and F. Saccomandi. 2002. Enzymes in soil: Properties, behavior and potential applications. p. 301–328. *In* A. Violante, P.M. Huang, J.-M. Bollag and L. Gianfreda (ed.) Soil mineral–organic matter–microorganism interactions and ecosystem health. Vol. A. Elsevier, London.

Gianfreda, L., M.A. Rao, and A. Violante. 1991. Invertase (β-fructosidase): Effects of montmorillonite, Al-hydroxide and Al(OH)x-montmorillonite complex on activity and kinetic properties. Soil Biol. Biochem. 23:581–587. doi:10.1016/0038-0717(91)90116-2

Gianfreda, L., M.A. Rao, and A. Violante. 1992. Adsorption, activity and kinetic properties of urease on montmorillonite, aluminium hydroxide and Al(OH)x–montmorillonite complexes. Soil Biol. Biochem. 24:51–58. doi:10.1016/0038-0717(92)90241-O

Gianfreda, L., M.A. Rao, and A. Violante. 1993. Interactions of invertase with tannic acid, OH-Al species and/or montmorillonite. Soil Biol. Biochem. 25:671–677. doi:10.1016/0038-0717(93)90106-L

Gianfreda, L., M.A. Rao, and A. Violante. 1995b. Formation and activity of urease–tannate complexes as affected by different species of Al, Fe, and Mn. Soil Sci. Soc. Am. J. 59:805–810. doi:10.2136/sssaj1995.03615995005900030024x

Gianfreda, L., and P. Ruggiero. 2006. Enzyme activities in soil. p. 257–311. In P. Nannipieri and K. Smalla (ed.) Nucleic acids and proteins in soil. Soil biology. Vol. 8. Springer Verlag, Berlin.

Gianfreda, L., and M.R. Scarfi. 1991 Enzyme stabilization: State of the art. Molec. Cell. Biochem. 100:97–128.

Gopinath, S., and S. Sugunan. 2005. Glucoamylase immobilized on montmorillonite: Synthesis, characterization and starch hydrolysis activity in a fixed bed reactor. Catal. Commun. 6:525–530. doi:10.1016/j.catcom.2005.04.016

Gopinath, S., and S. Sugunan. 2007. Enzymes immobilized on montmorillonite K 10: Effect of adsorption and grafting on the surface properties and the enzyme activity. Appl. Clay Sci. 35:67–75. doi:10.1016/j.clay.2006.04.007

Grego, S., A. D'Annibale, M. Luna, L. Badalucco, and P. Nannipieri. 1990. Multiple forms of synthetic pronase-phenolic copolymers. Soil Biol. Biochem. 22:721–724. doi:10.1016/0038-0717(90)90021-Q

Hamzehi, E., and W. Pflug. 1981. Sorption and binding mechanisms of polysaccharide cleaving soil enzymes by clay minerals. Z. Pflanzenernaehr. Bodenkd. 144:505–513. doi:10.1002/jpln.19811440509

Harter, R.D., and G. Stotzky. 1971. Formation of clay-protein complexes. Soil Sci. Soc. Am. Proc. 35:383–398. doi:10.2136/sssaj1971.03615995003500030019x

Huang, P.M., 1990. Role of soil minerals in transformations of natural organics and xenobiotics in soil. p. 29–115. In J.-M. Bollag and G. Stotzky (ed.) Soil biochemistry. Vol. 6. Marcel Dekker, New York.

Huang, Q., M. Jiang, and X. Li. 1999. Adsorption and properties of urease immobilized on several iron, aluminum oxides and kaolinite. p. 167–174. In J. Berthelin, P.M. Huang, J.-M. Bollag and F.Andreux (ed.) Effect of mineral–organic–microorganism interactions on soil and fresh water environments. Kluwer Academic/Plenum Publ., New York.

Huang, Q., and H. Shindo. 2000. Effects of copper on the activity and kinetics of free and immobilized acid phosphatase. Soil Biol. Biochem. 32:1885–1892. doi:10.1016/S0038-0717(00)00162-0

Huang, Q., H. Shindo, and T.B. Goh, 1995. Adsorption, activities and kinetics of acid phosphatase as influenced by montmorillonite with different interlayer materials. Soil Sci. 159:271–278. doi:10.1097/00010694-199504000-00006

Kandeler, E., C. Poll, W.T. Frankenberger, Jr., and M.A. Tabatabai. 2011. Nitrogen cycle enzymes. p. 211–246. In R.P. Dick (ed.) Methods of soil enzymology. SSSA Book Ser. 9. SSSA, Madison, WI. (This volume.)

Kiss, S., M. Dragan-Bularda, and D. Radulescu. 1975. Biological significance of enzymes accumulated in soil. Adv. Agron. 27:25–87. doi:10.1016/S0065-2113(08)70007-5

Klibanov, A.M. 1983. Stabilization of enzymes against thermal inactivation. Adv. Appl. Microbiol. 29:1–28. doi:10.1016/S0065-2164(08)70352-6

Klose, S., S. Bilen, M.A. Tabatabai, and W.A. Dick. 2011. Sulfur cycle enzymes. p. 125–160. In R.P. Dick (ed.) Methods of soil enzymology. SSSA Book Ser. 9. SSSA, Madison, WI. (This volume.)

Ladd, J.N., and J.H. Butler. 1975. Humus–enzyme systems and synthetic, organic polymer-enzyme analogs, p. 143–194 In E.A. Paul and A.D. McLaren (ed.) Soil biochemistry, Vol. 4. Marcel Dekker, New York.

Lai, C.M., and M.A. Tabatabai. 1992. Kinetic parameters of immobilized urease. Soil Biol. Biochem. 24:225–228. doi:10.1016/0038-0717(92)90222-J

Martinek, K., and V. V. Mozhaev. 1985 Immobilization of enzymes: An approach to fundamental studies in biochemistry. Adv. Enzymol. 57:179–249.

Marzadori, C., C. Gessa, and S. Ciurli. 1998a. Kinetic properties and stability of potato acid phosphatase immobilized on Ca-polygalacturonate. Biol. Fertil. Soils 27:97–103 doi:10.1007/s003740050406

Marzadori, C., S. Miletti, C. Gessa, and S. Ciurli. 1998b. Immobilization of jack bean urease on hydroxyapatite: Urease immobilization in alkaline soils. Soil Biol. Biochem. 30:1485–1490. doi:10.1016/S0038-0717(98)00051-0

Mayaudon, J. 1986. The role of carbohydrates in the free enzymes in soil. p. 263–309. In C.H. Fisherman (ed.) Peat and water. Elsevier Appl. Sci. Pub., New York.

Naidja, A., P.M. Huang, and J.-M. Bollag. 1997. Activity of tyrosinase immobilized on hydroxylaluminum–montmorillonite complexes. J. Molec. Catal. A: Chemical. 115:305–316. doi:10.1016/S1381-1169(96)00335-4

Naidja, A., A. Violante, and P.M. Huang. 1995. Adsorption of tyrosinase onto montmorillonite as influenced by hydroxyaluminum coatings. Clays Clay Min. 43:647–655. doi:10.1346/CCMN.1995.0430601

Nannipieri, P. 2006. Role of stabilized enzymes in microbial ecology and enzyme extraction from soil with potential applications in soil proteomics. p. 75–94. In P. Nannipieri and K. Smalla (ed.) Nucleic acids and proteins in soil. Soil biology. Vol. 8. Springer Verlag, Berlin.

Nannipieri, P., and Gianfreda L. 1998. Kinetics of enzyme reactions in soil environments. p. 449–479. In P.M. Huang, N. Senesi and J. Buffle (ed.) Structure and surface reactions of soil particles. J. Wiley & Sons, New York.

Nannipieri, P, P. Sequi, and P. Fusi. 1996. Humus and enzyme activity. p. 293–328. In A. Piccolo (ed.) Humic substances in terrestrial ecosystems. Elsevier Science, London.

Nelson, H. 1944. A photometric adaptation of the Somoji methods. J. Biol. Chem. 153:375–380.

Pflug, W. 1980. Effect of humic acids on the activity of two peroxidases. Z. Pflanzenernaehr. Bodenkd. 143:432–440. doi:10.1002/jpln.19801430409

Pflug, W., and W. Ziechmann. 1981. Inhibition of malate dehydrogenase by humic acids. Soil Biol. Biochem. 13:293–299. doi:10.1016/0038-0717(81)90065-1

Rao M.A., S. Del Gaudio, and L. Gianfreda. 2010 Oxidative polymerization of pyrogallol by peroxidase and manganese oxides with and without acid phosphatase: Formation and properties of polymerized products. J. Agric. Food Chem., 58:5017–5025. doi:10.1021/jf100080u

Rao, M.A., and L. Gianfreda. 2000. Properties of acid phosphatase–tannic acid complexes formed in the presence of Fe and Mn. Soil Biol. Biochem. 32:1921–1926. doi:10.1016/S0038-0717(00)00167-X

Rao, M.A., F. Palmiero, L. Gianfreda, and A. Violante. 1996. Interactions of acid phosphatase with clays, organics and organo-mineral complexes. Soil Sci. 161:751–760. doi:10.1097/00010694-199611000-00004

Rao, M.A., A. Violante and L. Gianfreda. 1998. Interactions between tannic acid and acid phosphatase. Soil Biol. Biochem. 30:111–112. doi:10.1016/S0038-0717(97)00097-7

Rao, M.A., A. Violante, and L. Gianfreda. 1999. The fate of soil acid phosphatase in the presence of phenolic substances, biotic and abiotic catalysts. p. 175–179. In J. Berthelin, P.M. Huang, J.-M. Bollag, and F. Andreux (ed.) Effect of mineral–organic–microorganism interactions on soils and freshwater environments. Kluwer Academic/Plenum Publ., New York.

Rao, M.A., A. Violante, and L. Gianfreda. 2000. Interaction of acid phosphatase with clays, organic molecules and organo-mineral complexes: Kinetics and stability. Soil Biol. Biochem. 32:1007–1014. doi:10.1016/S0038-0717(00)00010-9

Rosa, A.H., A.A. Vicente, J.C. Rocha, and H.C. Trevisan. 2000. A new application of humic substances: Activation of supports for invertase immobilization. Fres. J. Anal. Chem. 368:730–733. doi:10.1007/s002160000535

Rowell, M.J., J.N. Ladd, and E.A. Paul. 1973. Enzymically active complexes of proteases and humic acid analogues. Soil Biol. Biochem. 5:699–703. doi:10.1016/0038-0717(73)90062-X

Ruggiero, P., and V.M. Radogna. 1988. Humic acids-tyrosinase interactions as a model of soil humic-enzyme complexes. Soil Biol. Biochem. 20:353–359. doi:10.1016/0038-0717(88)90016-8

Ruggiero, P., J.M. Sarkar, and J.-M. Bollag. 1989. Detoxification of 2,4-dichlorophenol by a laccase immobilized on soil or clay. Soil Sci. 147:361–370. doi:10.1097/00010694-198905000-00007

Safari Sinegani, A.A., G. Emtiazi, and H. Shariatmadari. 2005. Sorption and immobilization of cellulase on silicate clay minerals. J. Colloid Interface Sci. 290:39–44. doi:10.1016/j.jcis.2005.04.030

Sarkar, J.M., and R.G. Burns. 1984. Synthesis and properties of β-D-glucosidase-phenolic copolymers as analogues of soil humic-enzyme complexes. Soil Biol. Biochem. 16:619–625. doi:10.1016/0038-0717(84)90082-8

Sarkar, J.M., A. Leonowicz, and J.-M. Bollag. 1989. Immobilization of enzymes on clays and soils. Soil Biol. Biochem. 21:223–230. doi:10.1016/0038-0717(89)90098-9

Secundo, F., J. Miehe-Brendle, C. Chelaru, E.E. Ferrandi, and E. Dumitriu. 2008. Adsorption and activities of lipases on synthetic beidellite clays with variable composition. Micropor. Mesopor. Mater. 109:350–361. doi:10.1016/j.micromeso.2007.05.032

Skujiņš, J.J. 1976. Extracellular enzymes in soil. CRC Crit. Rev. Microbiol. 4:383–421. doi:10.3109/10408417609102304

Tabatabai, M.A., and M. Fu. 1992. Extraction of enzymes from soils. p. 197–227. In G. Stotzky and J.-M. Bollag (ed.) Soil biochemistry. Vol. 7. Marcel Dekker, New York.

Violante, A., and L. Gianfreda 2000. Role of biomolecules in the formation and reactivity towards organics and plant nutrients of variable charge minerals and organomineral complexes in soil. p. 207–270. In J.-M. Bollag and G. Stotzky (ed.) Soil biochemistry. Vol. 10. Marcel Dekker, New York.

Wallenstein, M.D., and R.G. Burns. 2011. Ecology of extracellular enzyme activities and organic matter degradation in soil: A complex community-driven process. p. 35–56. In R.P. Dick (ed.) Methods of soil enzymology. SSSA Book Ser. 9. SSSA, Madison, WI. (This volume.)

Winters, A.L., and F.R. Minchin. 2005. Modification of the Lowry assay to measure proteins and phenols in covalently bound complexes. Anal. Biochem. 346:43–46 doi:10.1016/j.ab.2005.07.041

Chapter 16

Enzyme Extraction from Soil

Flavio Fornasier,* Yves Dudal, and Herve Quiquampoix

16–1 INTRODUCTION

Enzymes, like proteins in general, have a strong affinity with soil surfaces (Quiquampoix, 2008). The flexibility of their polypeptide chains and the various physicochemical properties of the lateral chains of the individual amino acids give rise to a range of possible interactions with solid surfaces: electrostatic forces, Lifshitz–van der Waals forces, hydrophobic interactions, and increases of conformational entropy (Quiquampoix et al., 2002). The adsorption of enzymes on soil surfaces is thus a complex phenomenon that involves several subprocesses (release of the exchangeable cations of clays, dehydration of surfaces, incorporation of cations in the adsorbed layer, rearrangement of protein structure). The combination of all these different subprocesses is often responsible for a quasi-irreversibility of the enzyme adsorption by dilution of soil components in water. Only more drastic treatments can desorb enzymes from soil particles. The challenge is to preserve the fragile structure of enzymes, and their catalytic activity, and to minimize the co-desorption of other compartments of soil organic matter.

The stability of enzymes within the soil matrix, in addition to being controlled by surface reactions with minerals, is also controlled by various reactions with organic components of soils. Proteins can bind to humic substances through hydrogen, ionic, or covalent bonds. However, Simonart et al. (1967) showed that very little of the activity of the enzymes studied could be attributed to hydrogen bonding. Ionic binding does occur, but based on research that use ionic displacing extractants, the amount of activity is relatively low (Butler and Ladd, 1969; Hayano, 1977; Hayano and Katami, 1977; Ceccanti et al., 1978)

Reactions of enzymes with both inorganic and organic substances were summarized by Weetall (1975). It was suggested that these mechanisms included adsorption, microencapsulation, cross-linking, copolymerization, entrapment, ion exchange, a combination of adsorption and cross-linking, and covalent bonding.

Flavio Fornasier, C.R.A.- R.P.S. Consiglio per la Ricerca e la Sperimentazione in Agricoltura - Centro di Ricerca per lo Studio delle Relazioni tra Pianta e Suolo, 23, Via Trieste, 34170 Gorizia, Italy (flavio. fornasier@entecra.it), *corresponding author;
Yves Dudal, ENVOLURE SAS, Campus La Gaillarde, Bât 12, 2 place Pierre Viala, 34060 Montpellier cedex 2, France (yves.dudal@envolure.com);
Hervé Quiquampoix, INRA, UMR Eco&Sols, 2 place Pierre Viala, 34060 Montpellier, cedex 2, France (Francequiquampoix@inra.montpellier.fr).

doi:10.2136/sssabookser9.c16

It seems likely there is a physical protection of stabilized enzymes that occurs and keeps them safe from microbial attack and free proteases.

Today, the importance and interest in extracting enzymes and proteins from soil is even higher than in the past, due to the development of soil proteomics (Nannipieri, 2006; Ogunseitan, 2006). Unfortunately, extraction of soil enzymes has always been a highly difficult task because, due to the various interactions previously described, a low yield is usually obtained and enzymes are extracted together with other soil components, mainly humic substances. As Tabatabai and Fu (1992) pointed out some 20 years ago, no enzyme extracted from soil has been purified to the same extent as those extracted from biological tissues. The same is still true today and strongly inhibits soil enzymology and soil proteomic research.

16–1.1 Brief History of Soil Enzymes Extraction

As reported by Skujiņš (1967), the first three reports on extraction of a soil enzyme activity were done by Fermi in 1910, who extracted a proteolytically active fraction with phenol; by Subrahmanyan in 1927, who precipitated the glycine deamination active principle; and by Ukhtomskaya in 1952, who desorbed several enzymes from soil using phosphate solutions. Antoniani et al. (1954) detected a cathepsin-like activity in a precipitate obtained from a soil extract using ammonium sulfate and sodium tungstate. Martin-Smith (1963) isolated two enzymes that decomposed uric acid from a soil; that same year, Briggs and Segal (1963) obtained the first solid preparation (12 mg of total protein from 25 kg of soil) containing an enzyme (urease). Bartha and Bordeleau (1969) extracted a peroxidase; Getzin and Rosefield (1971) extracted an enzyme (carboxylesterase) that decomposed malathion; and Chalvignac and Mayaudon (1971) obtained an extract that was enzymatically active (oxygenase effect) toward L-tryptophan from a forest soil. Burns et al. (1972) extracted urease; Pancholy and Lynd (1972) extracted lipase; and Ladd (1972) extracted proteases from soil. Starting in 1973, an increasing number of papers regarding extraction of enzymes from soil have been published (see Tabatabai and Fu [1992] for a detailed list of enzymes extracted from soil up to 1992).

Various research efforts have investigated different parameters involved in the extraction process: pH, buffer composition, chelating resins, agitation, chaotropic agents, and others (Tabatabai and Fu, 1992; Nannipieri et al., 1996). Except for some cases when organic soils have been used (e.g., Nannipieri et al., 1980; Vepsalainen, 2001), the yield is usually very low as only a limited percent of the total soil enzymatic activity is extracted even with procedures lasting several hours. It is probably for this reason that extraction yield (i.e., the ratio between extracted activity and activity measured in whole soil) often was not reported.

16–1.2 Choosing the Extraction Procedure

An important decision to make before starting to extract enzymes from soils is which method to use. This should be guided by the research objective and by consideration of potential chemical interferences that might occur for various extractants relative to data interpretation. Important factors to consider for using a particular method are: enzyme fraction that is to be studied (free, nonadsorbed enzymes or stabilized enzymes), choice of enzymes, location (e.g., distribution across aggregate sizes), and interference from clay and/or humic substances.

No chemical extractant can exactly remove a specific enzyme fraction. However, depending on the chemistry and degree to which a given extractant solubilizes soil organic matter, a general inference can be made on the fraction of enzyme(s) that are being assayed in soil extract.

The amount of soil organic matter released during the extraction is important. First, stronger extractants likely release enzymes that may be physically or chemically protected, or humus–enzyme complexes that remain catalytic. Second, the release of humic substances may react with free enzymes. The methods outlined in this chapter utilize either mild extractants (e.g., water) that only remove enzymes in soil solution or that are unbound on soil colloids, or stronger extractants (e.g., pyrophosphate) that co-extract significant amounts of humic substance. The degree of humic substance solubilization can be determined qualitatively by the visual color intensity of extracts or quantitatively by measurement of the absorbance at 450 nm (Ladd, 1972).

Another factor to consider is whether an extractant will lyse cells. This may or may not be desirable depending on the goal. If one is interested in extracellular activity, cell lysis should be avoided. Alkaline extracts would be expected to lyse cells, releasing enzymes that are likely involved in only internal cellular reactions, not in extracellular processes that drive biogeochemical reactions. This might lead to overestimation of the enzyme activity associated with biogeochemical cycles that are important in soil functions.

16–1.3 Extractants

Enzymes are bound to soil with different strengths, as previously described, and consequently "stronger" extractants should yield a higher amount of enzyme. In fact, thus far several extractants have been used. In general, buffers such as acetate, phosphate, pyrophosphate, tris(hydroxymethyl)aminomethane, and borate have been the most used (Tabatabai and Fu, 1992).

Compounds that can specifically break ionic or hydrogen bonds, or metal bridges (chelating agents) have been used, but generally they did not improve the extraction yield (Nannipieri et al., 1996). In any case, significant amounts of enzyme can be brought out in solution by using only alkaline solution, which also yields high amounts of humic substances in solution (Nannipieri et al., 1996). Unfortunately, solutions with a high pH (i.e., higher than 8) also can denature enzymes. Complexes between enzymes and humic substances and/or minerals are thought to be dominant in soil. It should, however, be taken into account that the extraction yield does not depend strictly on the type of extractant, but also on its concentration (Busto and Perez-Mateos, 1995). Consequently, the extraction yield is not predictable.

16–1.4 Extraction Procedures

There is no single, well-defined enzyme extraction procedure. Each author has tried to optimize the extraction for the specific soil being extracted and investigations comparing several extractants in different soils are rare. Ladd (1972) found little difference among the following buffers for the extraction of proteolytic enzymes in one soil: Tris, Tris-borate, sodium phosphate, Tris-citrate, Tris-EDTA, and distilled water. The author concluded that much of the protease activity was not tightly bound to that soil. Fornasier and Margon (2007) compared extraction yield of pyrophosphate, (one of the most commonly used and efficient extractants) with that obtained using a tris(hydroxymethyl)aminomethane solution in which either the detergent Triton X-100 or bovine serum albumin (BSA), or both, were added. The simultaneous use of Triton and BSA led to an extraction yield about 10 times higher than with pyrophosphate for three enzymes in six soils.

A major issue that should be carefully considered when extracting enzymes from soil is the speciation of the enzyme in solution. Although humus–enzyme complexes originating from the soil are generally thought to be the dominant fraction, presence of free enzymes and artifacts, including complexes between extracted humic substances and enzymes formed during extraction, cannot be excluded a priori (Nannipieri et al., 1996) and cannot be distinguished from the complexes extracted from soil.

Another important factor regarding the extraction procedure is the time dedicated to the extraction: it can be set from a half hour (Fornasier and Margon, 2007) to 24 h (Nannipieri et al., 1974; Nannipieri et al., 1980). For short-time extractions (few hours), microbial proliferation is thought to be nonsignificant, but for longer extraction periods, a biostatic agent generally is required. However, such a plasmolytic agent should be chosen with great care. Chloroform is an efficient biostatic agent, but it also breaks microbial cells (Jenkinson and Ladd, 1981) leading to the simultaneous extraction of intra- and extracellular enzymes. Toluene is a biostatic agent commonly used in enzyme assays, especially for hydrolases (Alef and Nannipieri, 1995; Frankenberger and Johanson, 1986; Tabatabai, 1994). Toluene permeabilizes the cell membrane specifically to small molecules, such as enzyme substrates (Skujiņš, 1967), and therefore can lead to an increase in the measured enzymatic activity, such as for arylsulfatase. However, toluene is suspected to act as an effective plasmolytic agent for certain groups of microorganisms; this could cause release of intracellular enzymes (Skujiņš, 1967) leading to simultaneous extraction of both intracellular and extracellular enzymes.

The same co-extraction occurs when extraction of soil enzymes is performed on air-dried soil.

16–1.5 Selectivity of Extraction

Among the 10 locations of enzymes in soil listed by Burns (1982), only those in free solution can be extracted selectively and quantitatively. Unfortunately, free enzymes in solutions constitute a very small fraction of the total enzyme pool in soil.

Although selective extraction of extracellular enzymes can be performed using buffers, selective extraction of intracellular enzymes cannot be achieved. This is because enzymes released from cells are readily adsorbed by soil components (Quiquampoix, 2000) and subsequently can be extracted only together with extracellular enzymes. When using any cell lytic action, caution should be used as plasmolytic agents can interfere with enzymatic activity (Acosta-Martínez and Tabatabai, 2002; Frankenberger and Johanson, 1986). Margon and Fornasier (2008) increased extraction of arylsulfatase (an enzyme located both intra- and extracellularly) by lysing the cells before extraction using dichloromethane with a fast fumigation procedure (5 minutes). This procedure has the advantage of completely and easily removing the fumigant before the extraction and therefore does not interfere with enzymatic activity. In addition, microbial proliferation (if any) during extraction should be reduced to a minimum.

In this chapter, five extraction procedures are presented. Extractions with water, neutral phosphate, and neutral pyrophosphate buffers exhibit increasing extraction strength, respectively. Extraction in acetic acid also is presented, as it could be useful in organic soils having an acidic pH. Indeed, neutral or alkaline extractants would extract high amounts of humic substances, causing strong interferences in the activity measurement.

Finally, a new method, based on the desorption of enzymes using Triton X-100 and bovine serum albumin is presented (Fornasier and Margon, 2007).

Determination of enzymatic activity in the extracts is not reported because the buffers and substrates have been the same when using soil instead of extract. Therefore, the methods described in the other chapters of this book can be utilized providing the incubation time is long enough to detect an appreciable concentration of the product of the enzymatic reaction. Times usually range from 4 to 16 h.

16–2 EXTRACTION WITH WATER
16–2.1 Introduction

Ladd (1972) showed that on average extraction of protease with water resulted in the lowest amount of activity in comparison with extraction activity of Tris-borate, Tris, sodium phosphate, Tris-citrate, or Tris-EDTA. At the same time, the water extractant had the lowest absorbance at 450 nm, indicating it had the lowest level of humic substance extraction. Clearly, water extractants would act on extracellular enzymes that are likely free, in soil solution, or very loosely held on organic matter or mineral surfaces by ionic binding. This approach would be useful in characterizing what are likely recently released enzymes that have been excreted by viable microorganisms, or less likely from decomposing organic matter.

16–2.2 Principle

This method uses water as an extractant and would not remove enzymes bound to inorganic and organic surfaces. It would extract enzymes that are in soil solution or, if on a soil colloid, would be held by extremely weak binding mechanisms.

16–2.3 Assay Method (Ladd, 1972)
16–2.3.1 Apparatus

- Plastic bottles, 50 mL
- Centrifuge tubes, 50 mL
- End-over-end or reciprocating shaker
- Centrifuge, 4000 g
- Glass-fiber filters, 0.7 μm

16–2.3.2 Reagents

1. Distilled water

16–2.3.3 Procedure

1. Weigh 5 g (oven-dry basis) of soil into the plastic bottle, add 25 mL of distilled water, close the bottle, and place the bottle in the shaker. Adjust the shaker speed so that the soil suspension is well mixed and shake for 1 h.

2. Transfer the suspension into centrifuge tubes and centrifuge at 4000 g for 5 min; filter through glass-fiber filter.

3. Determine enzymatic activity by using a method described in the other chapters of this book, substituting the soil for the extract.

16–2.3.4 Comments

Centrifuging soil extracts at high speed (i.e., >25.000 g) can be more effective than filtering in removing colloidal particles. In any case, filtration is required to remove small fragments with a specific gravity less than 1 g mL^{-1}. Extraction can be performed more conveniently in centrifuge tubes. Toluene and other plasmolytics should not be used when extraction is performed using moist soils, as they can, in principle, cause the release of intracellular enzymes. In addition, as the duration of extraction is short, no microbial production of enzyme should occur.

If dry soil is used for extraction, the extract potentially contains intracellular enzymes released on desiccation. Extractions times longer than 1 h should not be necessary as water removes only free, nonadsorbed enzymes. A very small portion (less than 1%) of total soil enzymatic activity is usually extracted. Kandeler (1990) extracted less than 1/1000 of the total enzymatic activity of alkaline phosphomonoesterase in two moist soils using a saturated soil–water paste. However, this was sufficient for determining the kinetic parameters. By contrast, Ladd (1972) extracted with water, from a dried soil, an amount of protease activity comparable to that obtained with a variety of buffers but with a much lower humic content.

16–3 EXTRACTION WITH ACETATE BUFFER

16–3.1 Introduction

This method is useful for acidic soils where the use of buffers with higher pH could extract high amounts of organic matter. Vepsalainen (2001) extracted a high percentage of phosphodiesterase, i.e., 16, 23, and 30% of total soil activity by using respectively 0.1, 0.2, or 0.5 M sodium acetate solutions from a highly organic soil. Bollag et al. (1987) found that 50 mM pH 6 acetate buffer was as effective as 50 mM pH 6 phosphate buffer for extraction of peroxidase from a mineral, neutral (pH 6.8) soil.

16–3.2 Principle

Sodium acetate is a weak acid that would solubilize small amounts of organic matter. Extraction is performed at pH 5, close to the pK of acetate (4.8), where its buffering power is high and carboxylic groups are mostly not ionized. Given the ability of acetate to interact with the supramolecular structure of humic substances (Piccolo, 2001), it could be hypothesized, though not proven, that this extractant reduces the interactions between soil enzymes and humic substances. It would be expected to have a small amount of enzymes bound to humic substances, weakly held enzymes released from ionic binding, and free enzymes in soil solution.

16–3.3 Assay Method (Vepsalainen, 2001)

16–3.3.1 Apparatus

- Beaker, 1 L
- Volumetric flask, 1 L

- Beaker, 100 mL
- Volumetric flask, 100 mL
- Plastic bottles, 50 mL
- Centrifuge tubes, 50 mL
- Magnetic stirrer
- Magnetic bars
- pH meter
- End-over-end or reciprocating shaker
- Centrifuge, 4000 g
- Glass-fiber filters, 0.7 μm

16–3.3.2 Reagents

1. Sodium hydroxide solution, 1 M: Weigh 4 g of NaOH (MW 40.00) into a 100-mL beaker; add about 80 mL of distilled water and stir to dissolve. Transfer the solution to the 100-mL volumetric flask and make up to 100 mL with distilled water.
2. Sodium acetate solution, 100 mM pH 5: Dissolve 6.804 g (50 mmol) of sodium acetate trihydrate (MW 136.08) and 3.0025 g (50 mmol) of glacial acetic acid (MW 60.05) in a 1-L beaker with about 800 mL of distilled water. Adjust the pH to 5 using Reagent 1, then transfer the solution to the 1-L volumetric flask and adjust to 1 L with distilled water.

16–3.3.3 Procedure

1. Weigh 5 g (oven-dry basis) of moist soil in the plastic bottle, add 25 mL of Reagent 2, close the bottle, and place the bottle horizontally in the shaker. Adjust the shaker speed so that the soil suspension is well mixed and shake for 1 h.
2. Transfer the soil suspension into centrifuge tubes and centrifuge at 4000 g for 5 min; filter through glass-fiber filter.
3. Determine enzymatic activity by using a method described in the other chapters of this book, but adding soil extract rather than whole soil to the appropriate reaction mixture.

16–3.3.4 Comments

See 16–2.3.4.

The pH of the soil–buffer suspension should be checked to verify the value of pH during extraction. Extraction should not be performed over a long time as acetate can be utilized by microorganisms as a substrate for growth. For the use of plasmolytics and dry soil, see Extraction with Water method (16–2).

16–4 EXTRACTION WITH PHOSPHATE BUFFER
16–4.1 Introduction

Phosphate buffer is considered to be a mild extractant, therefore it probably extracts soil enzymes loosely bound to organic matter and free enzymes not associated with soil colloids (Kiss et al., 1975; Skujiņš and Burns, 1976). Usually both the extraction yield and amount of humic substances in solution are higher compared with water extractions but lower compared with pyrophosphate extractions. It has been one of the most used buffers for the extraction of a variety of enzymes (Tabatabai and Fu, 1992).

16–4.2 Principle

The phosphate extract is a salt solution and for this method the pH is 7. This can be viewed as a mild extractant that would release low amounts of organic matter but more than water (Ladd, 1972). It would be expected to capture free enzymes but also release loosely bound enzymes. Since this extractant would dissociate both anionic (phosphate) and cationic (Na) molecules, it could be expected that it would displace some enzymes based on ionic reactions.

16–4.3 Assay Method (Hayano et al., 1987)

16–4.3.1 Apparatus

As described in section 16–3.3.1.

16–4.3.2 Reagents

1. Hydrochloric acid solution, 1 M: Put about 500 mL of distilled water in a 1-L volumetric flask, then add 83.09 mL of 37% (w/w; specific gravity 1.186) concentrated HCl (MW 36.461). Swirl rapidly, then bring to 1 L with distilled water and mix thoroughly.

2. Sodium hydroxide solution, 1 M: Weigh 4 g of NaOH (MW 40.00) into a beaker; add about 80 mL of distilled water and stir to dissolve. Transfer the solution to the 100-mL volumetric flask and make up to 100 mL with distilled water.

3. Sodium phosphate solution, 100 mM pH 7.0: Dissolve 6.084 g (i.e., 39 mmol) of monobasic sodium phosphate dihydrate, $H_2NaO_4P\cdot2H_2O$ (MW 156.01), and 10.857 g (i.e., 61 mmol) of dibasic sodium phosphate dihydrate, $H_2NaO_4P\cdot2H_2O$ (MW 177.99), in about 800 mL of distilled water. This solution should exhibit a pH of 7.0. Check the pH with a pH meter; if necessary correct the pH with 1 M NaOH (Reagent 2) or 1 M HCl (Reagent 1). Transfer the solution to a 1-L volumetric flask and adjust the volume to 1 L.

16–4.3.3 Procedure

1. Weigh 5 g (oven-dry basis) of moist soil in the plastic bottle, add 50 mL of Reagent 3 buffer, close the bottle, and place the bottle horizontally in the shaker.

2. Adjust the shaker speed so that the soil suspension is well mixed and shake for 1 h.

3. Transfer the soil suspension in centrifuge tubes and centrifuge at 4000 g for 5 min; filter through glass-fiber filter.

4. Determine enzymatic activity by using a method described in the other chapters of this book, substituting the soil for the extract.

16–4.3.4 Comments

See 16–2.3.4.

16–5 EXTRACTION WITH PYROPHOSPHATE

16–5.1 Introduction

This extractant has been used to extract protease, catalase, urease, phosphatase, and β-glucosidase in several studies (Nannipieri et al., 1982a; Nannipieri et al., 1982b; Perez-Mateos et al., 1988; Busto and Perez-Mateos, 1995). A further advantage of this extractant is that it reduces microbial activity, which would limit changes to extracted enzymes during the extraction process. At the same time, Nannipieri et al. (1974) showed that pyrophosphate extracted preserved microorganisms that produce urease. This method extracts large amounts of organic matter as shown by Nannipieri et al. (1974).

16–5.2 Principle

Pyrophosphate has the ability to complex cations, which favors solubilization of humic substances (Stevenson, 1994), allowing extraction of humus–enzyme complexes (Busto and Perez-Mateos, 1995; Nannipieri et al., 1996). Therefore, the extract is richer both in enzyme activity and in humic substances compared with that obtained with phosphate buffer.

16–5.3 Assay Method (Nannipieri et al., 1974)

16–5.3.1 Apparatus

As described in section 16–3.3.1.

16–5.3.2 Reagents

1. Hydrochloric acid solution, 1 M: Put about 500 mL of distilled water in a 1-L volumetric flask, then add 83.09 mL of 37% (w/w; specific gravity 1.186) concentrated HCl (MW 36.461). Swirl rapidly, then bring to 1 L with distilled water and mix thoroughly.

2. Sodium hydroxide solution, 1 M: Weigh 4 g of NaOH (MW 40.00) into a 100-mL beaker; add about 80 mL of distilled water and stir to dissolve. Transfer the solution to the 100-mL volumetric flask and adjust to 100 mL with distilled water.

3. Sodium pyrophosphate solution, 100 mM pH 7.0: Dissolve 22.194 g of dibasic sodium pyrophosphate, $H_2Na_2O_7P_2$ (MW 221.94), in about 800 mL of distilled water. Stir the solution and bring the pH to 7 by adding NaOH (first about 10 M, then 1 M). Transfer the solution to 1-L volumetric flask and adjust the volume to 1 L.

16–5.3.3 Procedure

1. Weigh 5 g (oven-dry basis) of moist soil in the plastic bottle, add 50 mL of Reagent 3, close the bottle, and place the bottle horizontally in the shaker.

2. Adjust the shaker speed so that the soil suspension is well mixed and shake for 1 h.

3. Transfer the soil suspension into centrifuge tubes and centrifuge at 4000 g for 5 min; filter through glass-fiber filter.

4. Determine enzymatic activity by using a method described in the other chapters of this book, substituting the soil for the extract.

16–5.3.4 Comments

See 16–2.3.4.

Extraction times longer than 1 h can increase extraction yield; however times longer than 2 h are not necessary, as extraction yield increases slightly (Nannipieri et al., 1980).

16–6 EXTRACTION WITH TRITON X-100 AND BOVINE SERUM ALBUMIN

16–6.1 Introduction

A variety of reagents, ranging from mild extractants like salt solutions and buffers at neutral pH to strong organic matter solubilizing reagents (e.g., NaOH and sodium pyrophosphate) have been used to extract enzymes from soil (Tabatabai and Fu, 1992). A high extraction yield usually has been obtained only under conditions that brought humic substances into the solution (Nannipieri et al., 1996). This would suggest that a significant amount of the enzymes extracted would be extracted as enzyme–humus complexes.

To enable the separation of enzymes from humic substances, Fornasier and Margon (2007) tested nondenaturing detergent extractants. This research showed that extraction of arylsulphatase and acid and alkaline phosphomonoesterase with sodium pyrophosphate (0.14 M, pH 7.1) yielded extracts with a low enzymatic activity. Similarly, Tris–HCl (50 mM, pH 7.5) gave a very low extraction yield (<0.5%). But adding Triton X-100 or bovine serum albumin (BSA) to Tris buffer increased extraction yield by 2 to 8 times. When both Triton X-100 and BSA were added to the buffer, the extraction yield was 2 to 13% for acid phosphatase, 2 to 5% for alkaline phosphatase, and 3 to 6% for arylsulfatase, depending on the soil. In addition, these extracts were colorless or, at most, light yellow, showing that extracts had low levels of humic substances.

16–6.2 Principle

Sutton and Sposito (2005) reported that soil organic micelles can be disrupted with the aid of small organic molecules by penetrating the large hydrophobic molecular aggregates of soil organic matter. The nondenaturing detergent, Triton X-100, does not act as a solubilizer of humic substances that could, in turn, complex enzymes. Furthermore, Fornasier (2002) found that addition of BSA to the extraction buffer before cell lysis with diethyl ether increased the arylsulphatase extraction yield from soil. This treatment presumably presaturated sites on soil particles having high protein-binding affinity, and thus enabled greater extraction of arylsulphatase. In addition, the short amount of time for this method (<1 h) together with the antiseptic activity of Triton X-100 (Frankenberger and Johanson, 1986) overcomes problems of microbial proliferation during extraction.

16–6.3 Assay Method (Fornasier and Margon, 2007)

16–6.3.1 Apparatus

As described in section 16–2.3.1, plus:

    10-mL disposable plastic test tubes

16–6.3.2 Reagents

1. Hydrochloric acid solution, 1 M: Put about 500 mL of distilled water in a 1-L volumetric flask, then add 83.09 mL of 37% (w/w; specific gravity 1.186) concentrated HCl (MW 36.461). Swirl rapidly, then bring to 1 L with distilled water and mix thoroughly.

2. Hydrochloric acid solution, 0.1 M: Put 100 mL of Reagent 1 in a 1-L volumetric flask, then bring to 1 L with distilled water and mix thoroughly.

3. THAM–HCl buffer, 50 mM pH 7.5: Dissolve 6.057 g (i.e., 50 mmol) of tris(hydroxymethyl)aminomethane, $C_4H_{11}O_3$ (MW 121.14), in 1 L of distilled water. Stir the solution and bring the pH to 7.5 by adding HCl, first 1 M, then 0.1 M.

4. Calcium chloride solution, 1 M: Dissolve 14.701 g of calcium chloride dihydrate, $CaCl_2 \cdot 2H_2O$ (MW 147.01), in a 100-mL volumetric flask containing about 90 mL of distilled water. Stir until the salt is dissolved; adjust to 100 mL with distilled water.

5. Extracting solution: To a 100-mL volumetric flask containing about 50 mL of THAM–HCl 50 mM pH 7.5 add 1 g of reagent-grade Triton X-100 [4-(1,1,3,3-tetramethylbutyl)phenyl-polyethylene glycol] and 4 g bovine serum albumin. Insert the magnetic bar and stir slowly (high speed causes foaming) until protein and detergent are partially dissolved, then add THAM–HCl buffer up to the mark and stir until detergent and protein are completely dissolved. The resulting solution has a pale yellowish color.

16–6.3.3 Procedure

1. Weigh 2 g (oven-dry basis) of moist soil into disposable 10-mL test tubes, add 2 mL of extracting solution, close the tubes, and place the tubes in the end-over-end shaker.

2. Shake for 20 min, then add 4 mL of THAM–HCl buffer, close the tubes, and put the tubes in the end-over-end shaker for 3 min.

3. Add 1 mL calcium chloride 1 M, close, and shake well to mix.

4. Centrifuge tubes at 4000 g for 5 min and filter through glass-fiber filter.

5. Determine enzymatic activity just after extraction.

16–6.3.4 Comments

This soil-extractant suspension is highly foaming, so care is required to avoid loss of suspension when opening and handling the containers of the extractant. Centrifuging extracts at high speed (i.e., >25.000g for 30 minutes) is more effective in removing colloidal particles than filtering and gives a clear, almost colorless extract. Extraction yield was 10 times or more higher compared with extraction performed with pyrophosphate 0.14 M pH 7.1 for acid- and alkaline- phosphomonoesterase and arylsulphatase (Fornasier and Margon, 2007). This procedure seems to be selective for extracellular enzymes, as membrane disruption (by a 5-min fumigation) just before extraction increased extraction of arylsulfatase, an enzyme known for being both intra- and extracellular, in two soils (Margon and Fornasier, 2008).

REFERENCES

Acosta-Martínez, V., and M.A. Tabatabai. 2002. Inhibition of arylamidase activity in soils by toluene. Soil Biol. Biochem. 34:229–237. doi:10.1016/S0038-0717(01)00177-8

Alef, K., and P. Nannipieri. 1995. Enzyme activities. p. 311–373. In K. Alef and P. Nannipieri (ed.) Methods in applied soil microbiology and biochemistry. Academic Press, London.

Antoniani, C., T. Montanari, and A. Camoriano. 1954. Investigations on soil enzymology. I. Cathepsin-like activity. A preliminary note. (Article in Italian.) Ann. Fac. Agrar., Univ. Studi, Milan. 3:99–101.

Bartha, R., and L. Bordeleau. 1969. Cell-free peroxidase in soil. Soil Biol. Biochem. 1:139–143. doi:10.1016/0038-0717(69)90004-2<

Bollag, J.-M., C.-M. Chen, J.W. Sarkar, and M.J. Loll. 1987. Extraction and purification of a peroxidase from soil. Soil Biol. Biochem. 19:61–67. doi:10.1016/0038-0717(87)90126-X

Briggs, M.H., and L. Segal. 1963. Preparation and properties of a free soil enzyme. Life Sci. 2:69–72. doi:10.1016/0024-3205(63)90039-0

Burns, R.G. 1982. Enzyme activity in soil: Location and a possible role in microbial ecology. Soil Biol. Biochem. 14:423–427. doi:10.1016/0038-0717(82)90099-2

Burns, R.G., M.H. El-Sayed, and A.D. McLaren. 1972. Extraction of an urease-active organo-complex from soil. Soil Biol. Biochem. 4:107–108. doi:10.1016/0038-0717(72)90048-X

Busto, M.D., and M. Perez-Mateos. 1995. Extraction of humic-β-glucosidase fractions from soil. Biol. Fertil. Soils 20:77–82. doi:10.1007/BF00307845

Butler, J.H.A., and J.N. Ladd. 1969. The effect of methylation of humic acids on their influence on proteolytic enzyme activity. Aust. J. Soil Res. 7:263–268. doi:10.1071/SR9690263

Ceccanti, B., P. Nannipieri, S. Cervelli, and P. Sequi. 1978. Fractionation of humus–urease complexes. Soil Biol. Biochem. 10:39–45. doi:10.1016/0038-0717(78)90008-1

Chalvignac, M.A., and J. Mayaudon. 1971. Extraction and study of soil enzymes metabolising tryptophan. Plant Soil 34:25–31. doi:10.1007/BF01372757

Fermi, C. 1910. Sur les moyens de défense de l'estomac, de l'intestin, du pancréas et en géneral de la cellule et de l'albumine vivante vers les enzymes protéolitiques. Zbl. Bakt. Abt. I Orig. 56:55–86.

Fornasier, F. 2002. Indirect approaches for assessing intracellular arylsulfatase activity in soil. p. 345–351. In A. Violante, P.M. Huang, J.-M. Bollag, and L. Gianfreda (ed.) Developments in soil science. No. 28. Vol. B. Elsevier Science B.V., Amsterdam.

Fornasier, F., and A. Margon. 2007. Bovine serum albumin and Triton X-100 greatly increase phosphomonoesterases and arylsulphatase extraction yield from soil. Soil Biol. Biochem. 39:2682–2684. doi:10.1016/j.soilbio.2007.04.024

Frankenberger, W.T., and J.B. Johanson. 1986. Use of plasmolytic agents and antiseptics in soil enzyme assays. Soil Biol. Biochem. 18:209–213. doi:10.1016/0038-0717(86)90029-5

Getzin, L.W., and I. Rosefield. 1971. Partial purification and properties of a soil enzyme that degrades the insecticide malathion. Biochim. Biophys. Acta 235:442–453.

Hayano, K.M. 1977. Extraction and properties of phosphodiesterase from a forest soil. Soil Biol. Biochem. 9:221–223. doi:10.1016/0038-0717(77)90079-7

Hayano, K.M., and A. Katami. 1977. Extraction of β-glucosidase activity from pea field soil. Soil Biol. Biochem. 9:349–351. doi:10.1016/0038-0717(77)90008-6

Hayano, K., M. Takeuchi, and F. Ichishima. 1987. Characterization of a metalloproteinase component extracted from soil. Biol. Fertil. Soils 4:179–183. doi:10.1007/BF00270938

Jenkinson, D.S., and J.N. Ladd. 1981. Microbial biomass in soil: Measurement and turnover. Soil Biochem. 5:415–471.

Kandeler, E. 1990. Characterization of free and adsorbed phosphatases in soils. Biol. Fertil. Soils 9:199–202. doi:10.1007/BF00335808

Kiss, S., M. Dragan-Bularda, and D. Radulescu. 1975. Biological significance of enzymes accumulated in soil. Adv. Agron. 27:25–87. doi:10.1016/S0065-2113(08)70007-5

Ladd, J.N. 1972. Properties of proteolytic enzymes extracted from soil. Soil Biol. Biochem. 4:227–237. doi:10.1016/0038-0717(72)90015-6

Margon, A., and F. Fornasier. 2008. Determining soil enzyme location and related kinetics using rapid fumigation and high-yield extraction. Soil Biol. Biochem. 40:2178–2181. doi:10.1016/j.soilbio.2008.02.006

Martin-Smith, M. 1963. Uricolytic enzymes in soil. Nature 197:361–362. doi:10.1038/197361a0

Nannipieri, P. 2006. Role of stabilised enzymes in microbial ecology and enzyme extraction from soil with potential applications in soil proteomics. p. 75–94. *In* P. Nannipieri and K. Smalla (ed.) Nucleic acids and proteins in soil. Springer, Heidelberg.

Nannipieri, P., B. Cerccanti, S. Cervelli, and C. Conti. 1982a. Hydrolases extracted from soil: Kinetic parameters of several enzymes catalyzing the same reaction. Soil Biol. Biochem. 14:429–432. doi:10.1016/0038-0717(82)90100-6

Nannipieri, P., B. Cerccanti, S. Cervelli, and S. Matarese. 1980. Extraction of phosphatase, urease, proteases, organic carbon, and nitrogen from soil. Soil Sci. Soc. Am. J. 44:1011–1016. doi:10.2136/sssaj1980.03615995004400050028x

Nannipieri, P., B. Cerccanti, S. Cervelli, and P. Sequi. 1974. Use of 0.1 M pyrophosphate to extract urease from a podzol. Soil Biol. Biochem. 6:359–362. doi:10.1016/0038-0717(74)90044-3

Nannipieri, P., B. Cerccanti, C. Conti, and D. Bianchi. 1982b. Hydrolases extracted from soil: Their properties and activities. Soil Biol. Biochem. 14:257–263. doi:10.1016/0038-0717(82)90035-9

Nannipieri, P., P. Sequi, and P. Fusi. 1996. Humus and enzyme activity. p. 293–328 *In* A. Piccolo (ed.) Humic substances in terrestrial ecosystems. Elsevier Science B.V., Amsterdam.

Ogunseitan, O.A. 2006. Soil proteomics: Extraction and analysis of proteins from soil. p. 95–115. *In* P. Nannipieri and K. Smalla (ed.) Nucleic acids and proteins in soil. Springer, Heidelberg.

Pancholy, S.K., and J.Q. Lynd. 1972. Quantitative fluorescence analysis of soil lipase activity. Soil Biol. Biochem. 4:257–259. doi:10.1016/0038-0717(72)90018-1

Perez-Mateos, M., S. Gonzales-Carcedo, and M.D. Busto Nuñez. 1988. Extraction of catalase from soil. Soil Sci. Soc. Am. J. 52:408–411. doi:10.2136/sssaj1988.03615995005200020018x

Piccolo, A. 2001. The supramolecular structure of humic acids. Soil Sci. 166:810–832. doi:10.1097/00010694-200111000-00007

Quiquampoix, H. 2000. Mechanisms of protein adsorption on surfaces and consequences for extracellular enzyme activity in soil. p. 171–206. *In* J.-M. Bollag and G. Stotzky (ed.) Soil biochemistry. Vol. 10. Marcel Dekker, New York.

Quiquampoix, H. 2008. Enzymes and proteins, interactions with soil constituent surfaces. p. 210–216. *In* W. Chesworth ed. Encyclopedia of soil science, Springer, Dordrecht, Berlin, Heidelberg, New York.

Quiquampoix, H., S. Servagent-Noinville, and M.H. Baron. 2002. Enzyme adsorption on soil mineral surfaces and consequences for the catalytic activity. p. 285–306. *In* R.G. Burns and R.P. Dick (ed.) Enzymes in the environment: Activity, ecology, and applications. Marcel Dekker, New York.

Simonart, P., L. Batistic, and J. Mayaudon. 1967. Isolation of protein from humic acid extracted from soil. Plant Soil 27:153–161. doi:10.1007/BF01373385

Skujiņš, J.J. 1967. Enzymes in soil. Soil Biochem. 1:371–414.

Skujiņš, J.J., and R.G. Burns. 1976. Extracellular enzymes in soil. Crit. Rev. Microbiol. 4:383–421. doi:10.3109/10408417609102304

Stevenson, F.J. 1994. Humus chemistry, genesis, composition, reactions. John Wiley & Sons, New York.

Subrahmanyan, V. 1927. Biochemistry of water-logged soil: Part II. The presence of a deaminase in water-logged soils and its role in the production of ammonia. J. Agric. Sci. 17:429–448. doi:10.1017/S0021859600018748

Sutton, R., and G. Sposito. 2005. Molecular structure in soil humic substances: The current view. Environ. Sci. Technol. 39:9009–9015. doi:10.1021/es050778q

Tabatabai, M.A. 1994. Soil Enzymes. p. 775–833. *In* R.W. Weaver, J.S. Angle, and P.S. Bottomley (ed.) Methods of soil analysis. Part 2. Microbiological and biochemical properties. SSSA, Madison, WI.

Tabatabai, M.A., and M. Fu. 1992. Extraction of enzymes from soil. Soil Biochem. 7:197–227.

Ukhtomskaya, F.I. 1952. The role of enzymes in the self-cleaning of soils. (In Russian with English abstract.) Gig. Sanit. 11:46–51.

Vepsalainen, M. 2001. Poor enzyme recovery from soils. Soil Biol. Biochem. 33:1131–1135. doi:10.1016/S0038-0717(00)00240-6

Weetall, H.H. 1975. Immobilized enzymes and their application in the food and beverage industry. Process Biochem. 10:3–24.

Index